INDUS RIVER BASIN

INDUS RIVER BASIN
Water Security and Sustainability

Edited by

SADIQ I. KHAN
*University Corporation for Atmospheric Research (UCAR) at
NOAA National Water Center, Tuscaloosa, AL, United States*

THOMAS E. ADAMS III
TerraPredictions, LLC, Blacksburg, VA, United States

ELSEVIER

Elsevier
Radarweg 29, PO Box 211, 1000 AE Amsterdam, Netherlands
The Boulevard, Langford Lane, Kidlington, Oxford OX5 1GB, United Kingdom
50 Hampshire Street, 5th Floor, Cambridge, MA 02139, United States

Notices
Knowledge and best practice in this field are constantly changing. As new research and experience
broaden our understanding, changes in research methods, professional practices, or medical treatment
may become necessary.

Practitioners and researchers must always rely on their own experience and knowledge in evaluating and using any
information, methods, compounds, or experiments described herein. In using such information or methods they
should be mindful of their own safety and the safety of others, including parties for whom they have a professional
responsibility.

To the fullest extent of the law, neither the Publisher nor the authors, contributors, or editors, assume any liability
for any injury and/or damage to persons or property as a matter of products liability, negligence or otherwise, or
from any use or operation of any methods, products, instructions, or ideas contained in the material herein.

Library of Congress Cataloging-in-Publication Data
A catalog record for this book is available from the Library of Congress

British Library Cataloguing-in-Publication Data
A catalogue record for this book is available from the British Library

ISBN: 978-0-12-812782-7

For information on all Elsevier publications visit our
website at https://www.elsevier.com/books-and-journals

Working together
to grow libraries in
developing countries

www.elsevier.com • www.bookaid.org

Publisher: Candice Janco
Acquisition Editor: Louisa Hutchins
Editorial Project Manager: Emily Thomson
Production Project Manager: Prem Kumar Kaliamoorthi
Cover Designer: Greg Harris

Typeset by SPi Global, India

Dedication

To Anayah, Haris, Eshal & Ruby for infusing all the distractions and chaos in my life.

Sadiq I. Khan

To all those who came before us and especially those who make us better people.

Thomas E. Adams III

Contents

17. Integrated Irrigation and Agriculture Planning in Punjab: Toward a Multiscale, Multisector Framework

AYESHA SHAHID, AFREEN SIDDIQI,
JAMES L. WESCOAT JR.

18. Developing Groundwater Hotspots: An Emerging Challenge for Integrated Water Resources Management in the Indus Basin

MUHAMMAD BASHARAT

19. Quantitative Estimation of Resource Linkages in Water Infrastructure Planning

AKHTAR ALI, MUHAMMAD AKRAM

Contributors

Thomas E. Adams III TerraPredictions, LLC, Blacksburg, VA, United States

Muhammad Akram Water and Power Development Authority, Pakistan

Akhtar Ali RAS Knowledge Hub, Lahore, Pakistan

David R. Archer JBA Trust, Skipton, United Kingdom

Zaheer Ahmad Babar Flood Forecasting Division, Pakistan Meteorological Department, Lahore, Pakistan

Samjwal Ratna Bajracharya International Centre for Integrated Mountain Development, Kathmandu, Nepal

Muhammad Basharat International Waterlogging and Salinity Research Institute (IWASRI), Pakistan Water and Power Development Authority (WAPDA), Lahore, Pakistan

Daniele Bocchiola Department of Civil and Environmental Engineering, Politecnico di Milano, Milano, Italy

Jürgen Böhner Centre for Earth System Research and Sustainability (CEN), Institute of Geography, University of Hamburg, Hamburg, Germany

Tobias Bolch Mountain Cryosphere Research Group, Department of Geography, University of Zurich, Zurich, Switzerland

Oleksiy Boyko Department of Civil Engineering, University of Siegen, Siegen, Germany

John F. Burkhart University of Oslo, Oslo, Norway

Muhammad Jehanzeb Masud Cheema Department of Irrigation and Drainage; Precision Agriculture, Center for Advanced Studies in Agriculture and Food Security, University of Agriculture, Faisalabad, Pakistan

Zachary Flamig Center for Data Intensive Science, University of Chicago, Chicago, IL, United States

Nathan Forsythe Water Resources Research Group, School of Engineering, Newcastle University, Newcastle, United Kingdom

Hayley Fowler Water Resources Research Group, School of Engineering, Newcastle University, Newcastle, United Kingdom

David Gochis National Center for Atmospheric Research, Boulder, CO, United States

Murali Krishna Gumma International Crops Research Institute for the Semi-Arid Tropics (ICRISAT), Patancheru, India

Yang Hong School of Civil Engineering and Environmental Sciences, The University of Oklahoma, Norman, OK, United States

Vivekanand Honnungar National Center for Atmospheric Research, Boulder, CO, United States

Vijay Ratan Khadgi International Centre for Integrated Mountain Development, Kathmandu, Nepal

Sadiq I. Khan University Corporation for Atmospheric Research (UCAR) at NOAA National Water Center, Tuscaloosa, AL, United States

Asif Khan Department of Civil Engineering, University of Engineering and Technology, Jalozai Campus, Peshawar, Pakistan

Muhammad Riaz Khan Flood Forecasting Division, Pakistan Meteorological Department, Lahore, Pakistan

Krishnan Raghavan Indian Institute of Tropical Meteorology, Pune, India

Young-Joo Kwak   International Centre for Water Hazard and Risk Management (ICHARM-UNESCO), Public Works Research Institute, Tsukuba, Japan, Department of Environmental Information, Tokyo University of Information Sciences, Chiba, Japan

Lu Li   NORCE Norwegian Research Centre; Bjerknes Centre for Climate Research, Bergen, Norway

Sudan Bikash Maharjan   International Centre for Integrated Mountain Development, Kathmandu, Nepal

Michel d.S. Mesquita   NORCE Norwegian Research Centre; Bjerknes Centre for Climate Research, Bergen, Norway

Yvan J. Orsolini   Bjerknes Centre for Climate Research, Bergen; Norwegian Institute for Air Research, Kjeller, Norway

Indrani Pal   Columbia Water Center, Columbia University; NOAA Cooperative Remote Sensing Science and Technology (CREST) Center, The City University of New York, New York, NY, United States

Ashwini M. Panandiker   The Energy and Resources Institute (TERI), Goa, India

Jonggeol Park   International Centre for Water Hazard and Risk Management (ICHARM-UNESCO), Public Works Research Institute, Tsukuba, Japan, Department of Environmental Information, Tokyo University of Information Sciences, Chiba, Japan

David Pritchard   Water Resources Research Group, School of Engineering, Newcastle University, Newcastle, United Kingdom

Muhammad Uzair Qamar   Department of Irrigation and Drainage, University of Agriculture, Faisalabad, Pakistan

Asad Sarwar Qureshi   International Center for Biosaline Agriculture (ICBA), Dubai, United Arab Emirates

Rupak Rajbhandari   Tribhuvan University, Kathmandu, Nepal

Paolo Reggiani   Department of Civil Engineering, University of Siegen, Siegen, Germany

Tom H. Rientjes   Department of Water Resources, Faculty ITC, University of Twente, Enschede, The Netherlands

Ayesha Shahid   Masters of City Planning, Massachusetts Institute of Technology, Cambridge, MA, United States

Finu Shrestha   International Centre for Integrated Mountain Development, Kathmandu, Nepal

Arun Bhakta Shrestha   International Center for Integrated Mountain Development, Kathmandu, Nepal

Mandira Singh Shrestha   International Centre for Integrated Mountain Development, Kathmandu, Nepal

Afreen Siddiqi   Massachusetts Institute of Technology, Cambridge, MA, United States

Andrea Soncini   Department of Civil and Environmental Engineering, Politecnico di Milano, Milano, Italy

Shahzad Sultan   Flood Forecasting Division, Pakistan Meteorological Department, Lahore, Pakistan

Pardhasaradhi Teluguntla   U.S. Geological Survey (USGS), Western Geographic Science Center, Flagstaff, AZ, USA

Prasad S. Thenkabail   U.S. Geological Survey (USGS), Western Geographic Science Center, Flagstaff, AZ, USA

Shabeh ul Hasson   Centre for Earth System Research and Sustainability (CEN), Institute of Geography, University of Hamburg, Hamburg, Germany; Department of Space Sciences, Institute of Space Technology, Islamabad, Pakistan

Vidyunmala Veldore   Det Norske Veritas—Germanischer Lloyd, Oslo, Norway

Nisha Wagle   International Center for Integrated Mountain Development, Kathmandu, Nepal

James L. Wescoat, Jr   Massachusetts Institute of Technology, Cambridge, MA, United States

Anthony M. Whitbread   International Crops Research Institute for the Semi-Arid Tropics (ICRISAT), Patancheru, India

Foreword

The Indus River Basin (IRB) is shared by four countries, Pakistan, India, China, and Afghanistan. Within the context of sustainability, it is appropriate to summarize how they differ in their economic activity and needs in relation to their demand for water within the basin.

Pakistan, with the largest proportion of the basin area (47%) and population (61%), has the greatest dependency on the Indus River with the Jhelum and Chenab tributaries. The Indus is the very lifeblood of Pakistan, and the demand for water is growing rapidly. According to the 2017 census, Pakistan now has a population of 212 million, making it the sixth largest country by population in the world, and most of that population is within and dependent on the Indus River Basin. Population growth remains close to 2% per annum with a total added population each year of 3.8 million. The requirement for water, food, energy, and services for such an increase in demand would challenge the most developed economy but is even more exacting for a country whose primary source of income is based on agriculture.

The Indian contribution to the basin area is approximately 39%, and its population is 35% of the total basin. While the IRB makes a considerable contribution to the Indian economy in agricultural production, it represents less than 14% of the total area of the country and less than 12% of its population. Population growth is lower than in Pakistan, at around 1.3% per annum. With a more diverse national economy, India is better positioned to withstand environmental shocks to the IRB within its jurisdiction.

Afghanistan's contribution to the basin area is just 6%, and its population is 3%, mainly in the Kabul River Basin. However, the Kabul River contributes approximately a quarter of the country's freshwater, and 23% of the Afghan population live in the Kabul Basin, including the rapidly growing city of Kabul. After decades of war and political turmoil, it is attempting to develop an agricultural economy based on irrigation and provide basic power needs. It is therefore not surprising that there is strong national pressure to use the water within its own territory, but with strong opposition from downstream Pakistan, which has already developed water infrastructures in the basin.

About 8% of the basin lies in China on the fringes of the Tibetan Plateau and is sparsely populated but significant as the source of the main Indus River and the Sutlej River.

Sustainability has political, environmental, and social dimensions. Sustainability will require active cooperation between the constituent countries. The Indus Water Treaty of 1960 provides rules of cooperation and sharing of water between India and Pakistan, but it is observed in the letter of the law rather than in a spirit of cooperation. It fails to address the legitimate interests of Afghanistan and China and is concerned with abstractions from river flow and not from the shared alluvial groundwater. It also fails to address problems of water quality or provide for environmental residual flows to ensure the health of the river to the estuary. This book fleshes out the competing and conflicting demands.

Environmental concerns focus on the role of climate change and variability and consequent influences on river flow. It is imperative therefore to improve understanding of processes, trends, and variability of climate and river flow as a basis for responding to extremes of drought (e.g., from 1998 to 2002) and floods (e.g., 2010 and 2011) and planning for future added risks that may develop as a result of climate change. While the beneficiaries of improved understanding are mainly on the irrigated plains, the flow in the river and its variability are largely driven by what happens in its mountain headwaters. Understanding and analysis demand data. Although national agencies in Pakistan and India established climate and flow networks in the mountain source areas in the 1950s, these networks remain sparse, limited to lower elevations, and largely restricted within national boundaries. Some progress has been made in understanding past trends and variability, but attempts at climate and flow projections based on the limited data provide conflicting results, as can be seen from the detailed analysis provided in this book.

The focus of analysis has shifted from ground-based data to satellite remotely sensed data combined with meteorological reassessment and a variety of modeling approaches to attempt explanation of climate patterns that are sometimes at variance with global norms and with trends in Central Himalaya. Of particular interest has been the atypical downward trend in summer temperatures that have resulted in reduced glacial ablation and downward trends in river flow from high-level catchments. This pattern of behavior is now sufficiently well established to be given the name "Karakoram anomaly."

Downstream, Pakistan faces a network of interlinked challenges in which water resources play a key role. Water resources for irrigation will gradually diminish as reservoir storage in the major controlling reservoirs of Tarbela and Mangla is taken up by sediment. Reservoir sedimentation will also limit the ability to generate hydroelectric power to meet the growing demand for domestic and industrial energy. Proposed new dams are shelved for economic and political reasons. Further pressing problems are the impact of waterlogging and salinity on productive agriculture, over-abstraction of groundwater, and competition between irrigation demand and legitimate demands for domestic, industrial, and environmental use. Water resources, especially in Pakistan, are already highly stressed and will become increasingly so with projected population changes.

The contributions to the book show a fascinating variety of approaches and represent essential steps toward the definition if not the solution of problems of increasing significance to the growing population of the IRB. It should therefore be essential reading for policymakers in each of the constituent countries, for water resources and environmental managers, and for researchers in climate and hydrology, and for all readers with a concern for the prosperity and well-being of those who live in the Indus River Basin.

David Archer

Newcastle University, Newcastle upon Tyne, England
JBA Consulting Engineers and Scientists, Skipton, England

Acknowledgments

First and foremost, the editors highly value the unconditional support of each of the contributing authors who synthesized their research that appeared in a number of published scholarly manuscripts. We express gratitude to numerous collaborators from the academic as well as the public sector. The faculty and students of the School of Civil and Environmental Engineering (SCEE) at the National University of Science and Technology (NUST), the School of Civil Engineering and Environmental Science at the University of Oklahoma, and Quaid-i-Azam University are acknowledged for their contributions. We are very grateful to the team at the Pakistan Meteorological Department for providing hydrometeorological data. Development of some of the flood forecasting work for the Indus River Basin has been significantly enhanced through funding under the Pakistan-US Science and Technology Cooperation Program managed by the United States National Academy of Sciences.

The editors would be remiss without acknowledging Emily Thomson with Elsevier for her help, guidance, and patience in putting this book together. We are humbled by the opportunity to bring these chapters to the hydrologic and water resources community.

PART I

INDUS RIVER BASIN— PAST, PRESENT AND FUTURE

Introduction of Indus River Basin: Water Security and Sustainability

Sadiq I. Khan*, Thomas E. Adams III†

*University Corporation for Atmospheric Research (UCAR) at NOAA National Water Center, Tuscaloosa, AL, United States †TerraPredictions, LLC, Blacksburg, VA, United States

1 INDUS RIVER BASIN

The Indus River flows from the mountainous terrain of the Hindu Kush, Karakoram, and Himalayan junction and meanders through the productive lands in the southern plains. It is one of the largest transboundary rivers in the world with a drainage area of about 1 million km^2 and a mean annual discharge of $7900 m^3/s$. The Indus River Basin (IRB) is shared by four countries, Pakistan, India, China, and Afghanistan; however, the largest basin area is in Pakistan with 61%, followed by India with 29%, and China and Afghanistan with 8%. The Indus River is the lifeblood for >300 million people, with 61% living in Pakistan, 35% in India, and 4% on the Afghan and Chinese sides of the basin. These countries depend on the IRB as the source of their water, and therefore water security challenges are central to sustainable development of this region. The hydrologic processes in the Indus River are linked by the glaciated mountain valleys with monsoon plains and a deltaic coastline, each of which has extensive and characteristic water management regimes.

Understanding how water becomes intertwined in the dependencies, risks, uncertainties, and opportunities of agriculture, food security, as well as hydropower development is crucial for human development in the Indus Basin. This basin drives the economic, social, and political growth of its surrounding communities. This basin is exceptionally vulnerable to the climatic variability that exacerbates human insecurity (Lutz et al., 2016, 2014; Bocchiola and Diolaiuti, 2013). Moreover, the Indus is a basin of major international importance, influencing environmental, political, and socioeconomic issues throughout the region. The sustainability of water resources in the Indus Basin faces many critical water-related issues, such as the pressures of a rising population and the incessant degradation of ecosystem services, which have resulted in an increase in extreme water conditions that are worsened by inadequate planning and management. Demand for fresh water is rising, but a variety of factors make its future availability uncertain, including population growth, water contamination, groundwater depletion, climate change, and changing land uses. Other factors that will hasten the region's water insecurity are difficult to comprehend at this point, such as the exact impact of climate change on the hydrology of the Indus Basin.

One challenging aspect in studying water resources is the phenomenon of "too much water, too little water" within the context of a changing climate. For instance, the Indus River is the backbone of the agricultural sector, and since irrigation uses 96% of the diverted water resources, a shortage of water can incapacitate the economy of the region. Moreover, floods can create economic damage, such as the devastating 2010 flood that caused an estimated US$10 billion of economic damage.

It is argued that these complex interlinkages and issues of the Indus River will be magnified by a changing climate. Uncertainties surrounding climate change only compound the problem. Therefore it is imperative to understand the dynamic relationships among climate-water-agriculture paradigms of the IRB. This book, *Indus River Basin: Water Security and Sustainability*, covers a range of hydrometeorological variables that are playing a vital role in the sustainability of the Indus Basin in a politically volatile region.

A review of recent research literature shows the use of metaphorical language to indicate the importance of the IRB, for example, calling the headwaters of the Indus River the "third pole" of our planet. Some previous studies termed the IRB one of the critical water resource laboratories in the world (Mustafa, 2013; Meadows and Meadows, 1999; Gilmartin, 2015; Akhter, 2017). Because of its unique and diverse hydrologic processes that link High Asia's water towers and glaciers, monsoon plains, and a deltaic coastline, as well as water management systems, this transboundary basin offers rich research opportunities. Scientific literature has been published on the water resources of this basin, particularly regarding the hydrology of the Upper Indus Basin (UIB; Archer, 2003; Archer et al., 2010; Immerzeel and Bierkens, 2012; Immerzeel et al., 2010, 2013; Mukhopadhyay and Khan, 2014; Lutz et al., 2014; Reggiani and Rientjes, 2015; Fowler and Archer, 2006) as well as some monographs focusing particularly on water policies (Gilmartin, 2015; Yang et al., 2014; Mustafa, 2013; Adeel and Wirsing, 2016). However, to date a comprehensive treatment of the hydrology of the IRB focusing on water resources, including consideration of hydrometeorology, ecohydrology, water and food security, water-related extremes, and water management, is lacking. Contributors to this book describe the interconnections between water extremes, food security, and water management that are critical for fully grasping the complex social and political setting of this unique region in South Asia.

The relationship between changing natural environments and changes in the hydrometeorologic processes are at the heart of this book. Readers will find discussions on how this heavily populated region has witnessed significant changes in its agricultural land use and hydrologic regime, with an increase in precipitation in the upper part of the region and a corresponding decrease in the Lower Indus Basin. Research studies have been conducted on the hydrologic processes, such as the summer monsoon that comprises more than half of the IRB's annual rainfall with a long track stretching from the ocean to the convection near the Himalayan foothills. The main focus in this book is on the quantitative aspects rather than the qualitative hydrology of the IRB.

Indus River Basin: Water Security and Sustainability aims at providing details on the existing scientific knowledge of the hydrometeorological process in this transboundary basin.

The book covers a range of water security and sustainability issues important for the Indus Basin. The authors discuss hydrologic predictions and explore how new scientific knowledge can enhance human security in critical social and political economies such as irrigated agriculture and food security. The key thematic areas of this book are presented in five parts that are summarized here and detailed later in Section 3 of this chapter.

Part I is a synopsis of the past, current, and future significance of the IRB and explores the complex dynamics of hydrology and socio-hydrology in this unique environment. There are reviews on the various facets of the Indus Basin's water budget ranging from variability in streamflow, precipitation, dynamics of the cryosphere, to projections of these hydrologic variables under climate change. In *Part II* research studies are outlined on the hydrometeorological processes, such as the summer monsoon that comprises more than half of the IRB's annual rainfall. This section provides details on the water resource dynamics in the region and further explores the problem of growing water scarcity and uncertainty of water availability into the future. *Part III* discusses the challenges in the coupled human-water system and their implications for food security. The historical and institutional analysis will reveal the increased urgency by the research community to understand water management aspects for sustainability of the region. *Part IV* outlines the water extreme monitoring, with recommendations on operational flood monitoring and forecasting systems for reducing disaster risks. There is a discussion on how a changing climate will impact the seasonal availability of water resources and exacerbate water-related disasters. *Part V* focuses on the Indus Basin Irrigation System (IBIS), which is the world's largest contiguous water management system. Readers will see how the development of water management infrastructures is leading to immense pressure on the natural and built environments.

2 A HISTORICAL PERSPECTIVE OF THE IRB

The IRB's rich history includes the Indus Valley Civilization, also known as the Harappan Civilization, which dated from approximately 4000 BCE through the early- to mid-third millennium BCE. (The Harappa site was discovered in 1921 and 1922 excavations.) The Indus Valley Civilization developed one of the most extensive urban societies in the Old World (Kenoyer, 1998; Possehl, 2002; Wright, 2010). This civilization was established on the alluvial plains of the Indo-Gangetic Basin in modern-day Pakistan and northwestern India (Wright, 2010; Possehl, 2002; Marcus and Sabloff, 2008; Kenoyer, 1998). It was coexistent with the earliest urban societies of Egypt and Mesopotamia but more extensive in area, encompassing an area estimated at ~1 million km^2 (Possehl, 2002). In contrast to the Mesopotamia and Ancient Egypt, the settlement of the Indus Valley Civilization did not build large, colossal structures. There is no conclusive evidence of palaces or temples. However, the remains of the Indus Valley Civilization's cities indicate noteworthy engineering and urban planning that included elaborate drainage systems, water supply systems, water wells, and clusters of large, nonresidential buildings.

The Indus Civilization has long been considered as a river-based society, with urban communities comprising five large settlements built along the water resources; archeologists have indicated that these settlements consisted of cities and numerous smaller urban settlements characterized by distinctive architectural elements and material culture (Kenoyer, 1998; Clark, 2013). Elaborate examples of the Indus Civilization's organization are exemplified by two of its largest cities, Harappa and Mohenjo-daro, which are located adjacent to major Himalayan rivers and monsoon-fed seasonal rivers. Some of the characteristics of these societies include effective wastewater drainage and possibly even public baths and storehouses for grain. Archeologists and anthropologists revealed that the Harappan Civilization was a well-settled civilization and that agriculture was the backbone of that society (Fig. 1). The evidence has further revealed that the Indus River was the main reason for this civilization's agricultural development, economy, and overall prosperity.

Communities along the Indus River have relied upon well-developed irrigation and water management procedures for agricultural production for many centuries. Wright (2010) and Kenoyer (1998) demonstrated that early inhabitants utilized advanced architecture with dockyards, warehouses, brick platforms, and protective walls. The massive walls likely protected the Harappans from floods and may have deterred military conflicts. It is known that these communities developed techniques for harnessing the rainwater that inundated the Indus. Some researchers attribute the decline of the Indus Civilization to the hydrometeorological extremes or natural disasters such as mega floods or droughts that impacted the

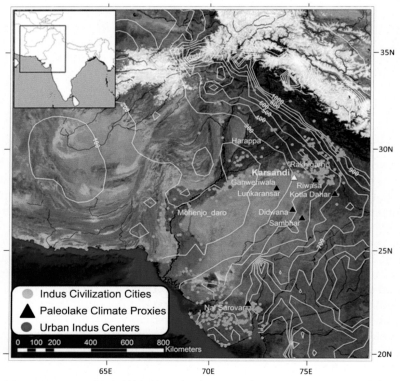

FIG. 1 Map of the Indus River Basin and distribution of Indus Civilization communities and cities. *Modified from Dixit, Y., Hodell, D.A., Giesche, A., Tandon, S.K., Gázquez, F., Saini, H.S., Skinner, L.C., Mujtaba, S.A.I., Pawar, V., Singh, R.N., Petrie, C.A., 2018. Intensified summer monsoon and the urbanization of Indus civilization in Northwest India. Sci. Rep., 8, 4225*

agriculture and economy. Consequently, the beginning and end of this civilization revolved around the water from the rivers in this unique region.

During the Buddhist, Sultanate, Mughal, and Sikh eras (in the Common Era, or CE), several communities flourished in the IRB. In chronicles of these eras, during the Sultanate era (11th–15th centuries) and Mughal era (16th–18th centuries) floods and droughts are mentioned, but the authors emphasized their impact on ferries, fords, and food supplies, rather than their frequency, duration, or magnitude (Habib, 1999; Agrawal, 1983). During the Mughal period, one of the most important developments was the construction of canals by harnessing of large rivers in northern India. Some of the examples of the famous canals include *nahr-i-faiz* or *Nahr-i-Behisht (Stream of Paradise)* also known as *Shah Nahr* (royal canal) which provided water to the capital Delhi that was built under the reign of Shah Jahan. Much later, during the colonial era, this unique river basin witnessed the development of one of the most sophisticated irrigation systems in the world. Water management structures such as canals played an important role in the groundwork of British imperial rule in the Indus Basin (Gilmartin, 2015, Naqvi, 2012). Some of the major colonial and postcolonial events in the IRB are the development of infrastructure for water management.

Since the middle of the 18th century, the Indus River has been regulated through organizations and manmade structures such as reservoirs and barrages that were constructed on the main river (Fig. 2). In Pakistan and India, many barrages, canals, and dams were built after the adoption of the famous Indus Waters Treaty of 1960 for the distribution and use of river water between the two countries. These reservoirs, canals, barrages, and dams on both sides of the border are operated for irrigation supplies and hydropower generation. A sophisticated water management system is developed that releases water from these reservoirs through a series of barrages downstream that divert available water to 26 million hectares (mha) of agricultural land in the basin (16 mha in Pakistan and 10 mha in India). As the flow in the river changes during the year, the availability of surface water varies depending on the season. Throughout the IRB, construction of new reservoirs is also in progress or planned for future construction. Fig. 2 lists some of the major events in a chronological order that are associated with the IRB. The chapters by Shahid et al. (Chapter 17) and Akhtar et al. (Chapter 19) provide details on the history of the Indus River water management system during the colonial, postcolonial, and present periods.

3 SYNOPSIS OF THIS BOOK

The book is divided into five parts to capture the full breath of water resources and related issues facing the inhabitants of the IRB and the nations in which they reside. These sections include:

1. Part I. Indus River Basin—Past, Present, and Future
2. Part II. Climate-Ecohydrology of Indus River Basin
3. Part III. Water and Food Security of Indus River Basin
4. Part IV. Water Extremes in Indus River Basin
5. Part V. Water Management in Indus River Basin

Summaries for the individual chapters follow, grouped by the major headings.

I. INDUS RIVER BASIN—PAST, PRESENT AND FUTURE

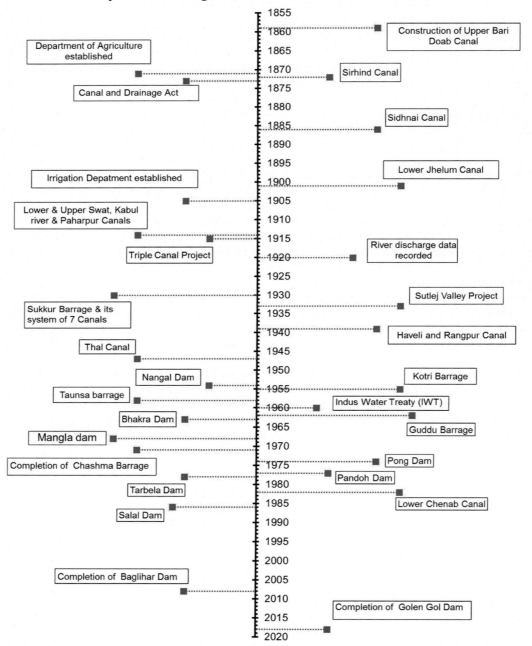

FIG. 2   Major chronological events in the Indus River Basin from 1959 to 2008. List of completed water management structures and other historically significant events.

3.1 Part I—Indus River Basin—Past, Present, and Future

- *Water resources modeling and prospective evaluation in the Indus River under present and prospective climate change*—Explores the complexity of hydrological processes in the UIB, where much of the water resources of the Pakistani Indus River are produced from seasonal snow and ice melt, and provides a feasible picture of the dynamics of future water resources, under the controls of climate and cryospheric changes. First, based on results from currently available literature and from field campaigns in the area, a data-driven depiction is given of the recent situation in terms of climate, water resources, and cryospheric dynamics, and recent changes there. Then results on recent studies aimed at modeling hydrological processes in the Indus River are provided, including precipitation, snow, and ice dynamics in stream flows, and comments about performance there. According to reference scenarios of reliable sources (e.g., from IPCC ARs), results are presented and commented on concerning predictions (i.e., projections) of climate, runoff, and cryospheric dynamics under future climatic conditions. A template is subsequently provided; this tool is based on the recent literature on glacio-hydrological modeling in high-altitude glacierized areas and subsequent hydrological projections under climate change. Additional, though more limited, attention is given to the issue of extreme floods under future climatic conditions. The template is specifically discussed and commented on based on a recent study in the UIB. The discussion section focuses in part on the need for hydrological monitoring of the area, and examples of hydrological monitoring and network design in the area are provided. Brief conclusions are given on the present and expected future dynamics of water resources in the Indus catchment, as well as the potential for accurate assessment and prediction regarding changing hydrological regimes and adaptation to them.
- *Challenges in forecasting water resources of the IRB: Lessons from the analysis and modeling of atmospheric and hydrological processes*—For millennia, the Indus Valley Civilization learned how to manage the waters of the Indus River water system. However, climate changes could have unforeseeable impacts on the growing population in the region and could hinder the emerging needs of the energy-food-water nexus for short and long-term decision making. Also, changes in monsoon variability and the timing and amount of snow and glacier melting may lead to extreme events such as droughts over the Asian plains and flooding in the neighboring Himalayas. Although global and regional climate models have advanced our knowledge about future changes in the climate system, there are still uncertainties and knowledge gaps about the precipitation systems associated with Asian summer monsoons and winter and pre-monsoon western disturbances. Hence understanding and using the summer and winter precipitation variabilities and their connection to the hydrological cycle in the Hindu Kush-Himalayan region in general and the IRB in particular as the baseline and future timescale is of utmost importance.
- *Past and expected future glacier changes in the UIB*—Meltwater from glaciers contributes significantly to the runoff of the Indus River, especially during summer months when precipitation in large areas of the basin is very low. This chapter provides a review of the current knowledge of the measured past and projected future glacial changes in the Indus Basin, which covers large parts of the Hindu Kush, Karakoram, and Western Himalaya. Glaciers have been retreating and shrinking since the end of the Little Ice Age. Glacial

changes, however, were heterogeneous, and especially in the Karakoram, many glaciers showed irregular behavior with frequent surges. In past years, the glacier mass losses have been relatively small in the Karakoram, while since about the year 2000, mass losses of glaciers in the Western Himalaya have been among the highest in all of High Mountain Asia. The average mass loss of the glaciers in the Indus Basin was approximately $0.2\,\text{m w.e. a}^{-1}$ during the past decade. Projections of future glacier changes reveal an average ice loss of about 50% by 2100. However, results of the different studies deviate, and the uncertainties are large. The glacier mass loss will at first lead to an increase in runoff, but glacier melting will probably decrease after the middle of this century.

3.2 Part II—Climate-Ecohydrology of Indus River Basin

- *Probabilistic precipitation analysis in the Central IRB*—Few studies exist on the analysis of precipitation and respective trends during past decades in the Central Indus Basin, Pakistan's principal river system. One reason for the lack of such studies is data scarcity. Modern-era reanalysis products offer the possibility to fill this knowledge gap by delivering simulated precipitation, temperature, and other atmospheric variables over poorly monitored areas. Reanalysis products are derived from atmospheric circulation reforecasts, whereby the models are adjusted by means of sequential data assimilation of past observations. Despite the fact that these products improve progressively, it is essential to post-process the data by performing systematic bias correction and to assess the predictive uncertainty of one or multiple model predictions for specific ground locations. Here we employ six reanalysis products for the Central Indus Basin region with the aim of predicting basin-average precipitation and discussing how the suggested approach can be valuable for water resource studies under conditions of climatic changes.
- *Glaciers in the IRB*—Glacier melt is essential for sustaining the water demand, food security, and livelihoods of millions of people in the mountains of the Hindu Kush, Karakoram, and Himalaya. Information on glacial areas, including elevation range, aspect slope, hypsometry, and debris cover of glaciers, plays a significant role in the assessment and analysis of water availability in river basins. An inventory of glaciers of the Hindu Kush-Himalaya (HKH) region was developed in 2011 using Landsat satellite images of 2005 ± 3 years. According to this inventory, about 3.8% of the Indus Basin within the HKH was found to be glaciated. The Indus Basin all together includes 18,495 glaciers with a total glacier area of $21,192\,\text{km}^2$. The glaciers range from an elevation of 8566 to 2409 m a.s.l., which is the lowest elevation of all glaciers in the HKH region.
- *A review of the projected changes in climate over the Indus Basin*—Inhabitants of the IRB are directly or indirectly dependent on water generated from the snow and glacier melt in the upstream part of the basin. The Indus Basin has become increasingly vulnerable due to climate change, as the basin flow is largely derived from snow and glacier melt that are sensitive to changes in climate. Current and future water availability in the basin is therefore important for future planning and management of water resources as well as for preparing effective adaptation measures to cope with changing climate scenarios. This chapter reviews the literature on future climate projections in the Indus Basin based on climate modeling of the region. A majority of the studies reviewed project an increase in

temperature, which is more prominent in the higher altitude under high emission scenarios. An increase in precipitation is projected in the UIB, while a decreasing trend is seen for the Lower Indus Basin.

- *Analysis of climate change projections for the UIB*—Key determinants for the water resources of the UIB are derived largely from cryosphere-dominated hydrological processes that includes mass and energy inputs, which are well indexed by temperature. Because the topographical relief of the UIB—or more precisely the hypsometry, that is, the distribution of surface area with elevation—is one of its defining characteristics, the ability of models to accurately represent the variation of climate inputs with respect to elevation is especially crucial. The variable skill of available climate model outputs at representing these determinants under historical conditions provides important insights for interpreting the likelihood and implications of future climate conditions simulated by these models.

3.3 Part III—Water and Food Security of Indus River Basin

- *Transboundary IRB: Potential threats to its integrity*—Complex hydrology and the geopolitical nature of the IRB underscore serious threats to the integrity of water resource governance, management, and availability. The agricultural and industrial sectors are the main economic forces in the region. Competition between these users is increasing due to intensive agriculture, population shift to cities, and industrialization. Increasing population, stagnant or reducing agricultural productivity, enhancing industrialization, increasing dependence on groundwater, high risk of climate variability, and unplanned urban growth are harsh realities of the IRB region. The situation has resulted in an alarming reduction in per capita water availability. Moreover, upstream interventions on river systems are severely affecting downstream users. Diminishing fresh water availability, groundwater overdraft, food and nutritional security, and environmental pollution are major challenges and even greater threats. Historically, the transboundary water crises of the region were considered localized with no eminent threat to the global security, but the situation now demands a more comprehensive understanding of water resource issues in a region possessing nuclear arsenals. This chapter explores the threats and vulnerability associated with the international waters of the region. It also reviews studies on recently proposed solutions to the pedagogical dimension of water-related problems encountered in the region. The chapter concludes with a discussion on the future of the IRB for efficient management, equitable distribution, and sustainable use of scarce resources.
- *IRB land use/land cover (LULC) and irrigated area mapping*—Detailed land use/land cover (LULC) mapping, showing irrigated area categories in the Ganges and Indus River Basins using near-continuous time-series 250-m resolution moderate resolution imaging spectroradiometer (MODIS) data are presented. A unique dataset was utilized, including a stack of 46 images, 23 MODIS images each of 2-bands, compiled from MODIS terra images for the years 2013 and 2014. Field-plot data were gathered from 553 precise geographic locations covering about 8000 km in the basins. Spatial information on cropland and irrigated area distribution was restricted by the district-level crop statistics published by the state or national governments in India and Pakistan. Statistics were collected by irrigation and agriculture departments, but there was discrepancy in the irrigated area

between departments. Water availability in major command areas varied frequently due to rainfall fluctuations, which leads to inadequate water supply during critical stages of crop growth. The study analyzed MODIS based 16-day normalized difference vegetation index (NDVI) time-series data acquired for 2013 and 2014 using spectral matching techniques (SMTs). The map output accuracies were evaluated based on independent ground data and compared with sub-national-level statistics. The producer's and user's accuracies of the cropland classes were between 70 and 85%. The overall accuracy and the kappa coefficient estimated for irrigated areas were both 84%.

- *Increasing water productivity in agriculture sector*—Increasing the productivity of available water resources is central to producing more food, fighting poverty, reducing competition for water, and ensuring that there is enough water for nature. For sustainable water resources management, a better understanding of the linkage between productivity changes and their impact on different users within the same basin is required. Therefore it is imperative to develop common water accounting procedures for analyzing the use, depletion, and productivity of water at the plant, field, irrigation system, and basin scales. This chapter emphasizes the need for measuring water productivity at different scales because different actions are needed to enhance water productivity at each level. It also discusses potential interventions and strategies that can be used to enhance productivity of water at different scales. The successful implementation of these interventions requires more detailed technical and economic analysis to ensure sustainability.

- *Hydrological cycle over Indus Basin at monsoon margins: Present and future*—Characterizing prevailing precipitation regimes from a comprehensive database of 11 observational datasets, this chapter first analyses the fidelity of 36 state-of-the-art global climate model experiments from the coupled model intercomparison project phase-5, and then presents changes in the surface water budgets from moderate-to-high fidelity experiments under the representative concentration pathway (RCP) 4.5 for the 21st century. Results suggest that water balance is not closed for 13 experiments while one-fourth of the analyzed experiments do not see the monsoonal precipitation regime at its extreme margins over the Indus Basin. Nevertheless, 14 moderate-to-high fidelity experiments agree on a decreasing water budget for the westerly precipitation regime, which dominates during the spring and over the UIB. In contrast, a moderate agreement suggests a consistent increase in the water budget during the monsoonal precipitation regime over an extended area. Overall water budget changes are small, suggesting a drier UIB but wetter Lower Indus Basin in the future.

3.4 Part IV—Water Extremes in Indus River Basin

- *Water resources forecasting within the IRB: A call for comprehensive modeling*—The IRB spans four countries, Pakistan, India, Afghanistan, and China. With a transboundary flow of waters, a comprehensive hydrologic modeling system is needed to capture hydrologic response from headwater regions of major tributaries that significantly impact downstream areas. Rainfall from monsoonal seasons often produces devastating flooding along the Indus River and major tributaries. Synoptic events during winter months deposit snow in the mountainous regions of the Upper IRB, which during the spring and early summer snowmelt season, coupled with glacial melt, fills reservoirs used as water resources. The lack of a comprehensive and robust transboundary hydrologic modeling system complicates flood prediction and warning on the Indus River and its major tributaries. Inadequate hydrologic modeling also presents a major impediment to water resources management, principally for irrigation of croplands and hydropower generation in Pakistan, at monthly, seasonal, annual, and climate timescales. Recent surveys of hydrometeorological monitoring, data transmission, and data archiving in Pakistan show the need for considerable infrastructure modernization and capacity building. These improvements must include greatly improved data analysis, modeling, and training of professional and technical staff at all levels.

- *Review of hydrometeorological monitoring and forecasting system for floods in the Indus Basin in Pakistan*—Climatic trends in most parts of the world show a significant increase in the amount of rainfall, its intensity, and its frequency. The major catastrophe caused by these trends comes as flooding, which is hard to predict over the mountainous region of the UIB. This chapter outlines the nature of floods in the Indus Basin and its impact. The chapter presents the mechanisms employed for collecting and processing hydrometeorological data in Pakistan that can assist in developing flood forecasts for reducing flood vulnerabilities. It introduces the hydrometeorological observation system established in the basin by various organizations and the technologies that are used for data collection. These include a combination of modern hydrometeorological stations that measure and report information automatically, as well as traditional hydrometeorological stations that record parameters at predetermined intervals using manual gauges. The chapter also describes the models that are used for flood forecasting and the dissemination process to the various users.

- *Flood monitoring system for IRB using a distributed hydrologic modeling framework*—Hydrometeorological disasters can have devastating, direct impacts on life and property as well as indirect effects on the food security, economy, and livelihood of the communities. The IRB has a high mean annual number of people at risk to floods and landslides produced by storm systems occurring during the monsoon season. Consequently, the development of operational flood monitoring and prediction systems is essential. This chapter looks at the advance flood modeling paradigm that assimilates the hydrometeorological forces, such as quantitative precipitation estimation into physically based numerical modeling for flood hazard assessment. The use of geospatial data from spaceborne sensors and geospatial modeling for hydrologic applications in data-scarce environments are demonstrated. Also discussed is the development of a distributed hydrologic modeling framework with the capability of assimilation of multi-sensor remote sensing data for operational flood forecasting systems. Improvements in IRB flood forecasting should be focused on model parameter estimation, improvements in the spatiotemporal resolution of hydrometeorological data, and development and deployment of an ensemble hydrologic forecasting system.

- *Annual flood monitoring of 2010 Indus River flood using synchronized floodwater index*—Flood detection algorithms are able to generate near-real-time flood proxy maps on regional and global scales on a daily basis using multiple spaceborne sensors. A new algorithm of annual floodwater change detection is introduced using a hybrid conditional process that is incorporated into the synchronized multiple-floodwater index (SfWI^2), a statistic-threshold-based flood detection approach coupled with in-situ hydrological data. SfWI^2 was applied to the 2010 Indus flood event, a recent extreme flood case, for more accurate flood detection at a transboundary river-basin level and effective emergency response in the early stage of a flood disaster. The resultant flood map shows good agreement between the MODIS-derived flood proxy map and high-resolution satellite images at the representative barrage stations.

3.5 Part V—Water Management in Indus River Basin

- *Water management in the Indus Basin in Pakistan: Challenges and opportunities*—Out of the total water available in Pakistan, 180 billion cubic meters (BCM) of water is withdrawn from the Indus River system, out of which about 128 BCM is diverted to the distribution system. The second source is the rainfall, with an annual quantum of about 50 BCM, and the third source is groundwater with about 50–60 BCM annual usage. However, an additional 20 BCM of groundwater can be further developed if both surface and groundwater are used and proper planning is done at both basin and local scales. In the past, underuse of groundwater resulted in waterlogging and salinity issues in many areas of the Indus Basin. In contrast, overuse is now triggering groundwater mining, saltwater intrusion, and increased surface salinity. In this chapter, several research manuscripts pertaining to canal water are reviewed and their results highlighted. Similarly, the groundwater in cities is highlighted in terms of reallocation of canal water in areas with high water tables to cities. Likewise, high water table areas are recommended for reallocation of canal water to low water table areas. Canal water reallocation is recommended from north to south and along the watercourse.
- *Re-imaging the planning of irrigation and agriculture in the IRB, Punjab, Pakistan*—Faced with rapid resource depletion, degradation, and shortages in the IRB in Punjab, Pakistan, the provincial government of Punjab has identified "integrated water resources management" as the guiding paradigm for achieving efficient, equitable, and environmentally sustainable use of natural resources in the province. However, no clear roadmap exists for how multi-sectoral, "integrated" resource management and governance can be operationalized. The larger challenge involves growing municipal, industrial, and environmental demands for water, rapidly growing need for food in the populous province, and expanding requirements for energy. This chapter focuses on the most obvious but elusive challenge of integrating irrigation and agriculture in Punjab and explores the essential nexus of water and food. It uses a combination of historical, institutional, and statistical analyses to investigate how "integrated" food and water planning can be achieved in Punjab. The historical analysis traces how the idea of "integration" in irrigated agriculture has evolved in Pakistan's colonial history and within the province of Punjab after independence. It reveals that both the Irrigation Department and the Agriculture Department have highlighted the need for vertical and horizontal integration within and between the departments throughout their existence. The institutional analysis explores how planning is currently done within and across the provincial departments of agriculture and irrigation. Finally, the chapter briefly demonstrates the statistical challenges and opportunities of integrating irrigation

and agricultural data, which are recorded in different spatial units but which can still have a measure of integration in integrated farm-level data analysis. In conclusion, this chapter uses the idea of boundary spanning to strengthen the mesoscale capacity for integrated irrigation agriculture planning in Punjab.

- *Developing groundwater hotspots: An emerging challenge for integrated water resources management in Indus Basin*—Groundwater conditions in major cities are reviewed and canal water is recommended for reallocation. Hotspot areas with high water tables are recommended for reallocation of canal water to low water table areas. In Pakistan, on average, the cropping intensity was 102.8%, 110.5%, and 121.7% during 1960, 1972, and 1980, respectively. Now it is operating at about 172% and even higher in certain areas. As a result, groundwater mining, due to higher abstraction rates as compared to the corresponding recharge, is well reported in the literature. With the dramatic increase in the intensity of groundwater exploitation in the past three decades, the policy landscape for Pakistan has changed; that is, "the main policy issues now relate to environmental sustainability and welfare." Thus it is important to avoid declining groundwater tables and deteriorating groundwater quality in fresh groundwater areas, and also to ensure equal access to this increasingly important natural resource. Moreover, urbanization is leading to immense pressure on the natural and built environment. Pakistan is an arid and water-scarce country; hence protection and sustainability of aquifers is critical and more so in a populated urban city.
- *Resource linkages for water infrastructure planning*—The IRB is a source of livelihood for about 300 million people in four neighboring countries, including Afghanistan, China, India, and Pakistan. It supplies irrigation water to over 20 mha, which greatly contributes to food security in the region. It provides over 10,000 MW of clean energy. Another 10,000-MW projects are at different stages of implementation. The IRB supports a unique ecosystem and aquatic life in cold mountains and hot and humid plains. The past 100 years witnessed a rapid growth of water infrastructure in the IRB to meet the increasing demand for water, food, and energy. Geopolitical changes and transboundary conflicts, increasing population, changes in lifestyle, and climate changes have already challenged the water security and sustainability of the basin. One of the obvious examples of water competition between irrigation and hydro-energy is the priority for irrigation in operations of the Tarbela and Mangla reservoirs in Pakistan. The high use of water for agricultural activities (>90%) and the competing demand of water for energy, household uses, and the environment require better understanding of the resource linkages and of improved water infrastructure planning in the IRB. Acknowledging resource links among water, energy, and food in planning, development, and management is critical to reducing the likelihood that decisions in one area will have negative impacts in another area. This chapter investigates the linkages between water, food, and energy and land use in the IRB context; explores the main considerations for prioritizing the water uses by the sectors, countries, and regions; and identifies gaps and factors that constrain cross-sectoral water synergies. It proposes a conducive methodology for working in competing environments and capitalizing on the cooperation.

References

Adeel, Z., Wirsing, R.G., 2016. Imagining Industan. Springer.

Agrawal, C.M., 1983. Natural Calamities and the Great Mughals. Kanchan Publications.

Akhter, M., 2017. Desiring the data state in the Indus Basin. Trans. Inst. Br. Geogr. 42, 377–389.

Archer, D., 2003. Contrasting hydrological regimes in the upper Indus Basin. J. Hydrol. 274, 198–210.

Archer, D.R., Forsythe, N., Fowler, H.J., Shah, S.M., 2010. Sustainability of water resources management in the Indus Basin under changing climatic and socio economic conditions. Hydrol. Earth Syst. Sci. 14, 1669–1680.

Bocchiola, D., Diolaiuti, G., 2013. Recent (1980–2009) evidence of climate change in the upper Karakoram, Pakistan. Theor. Appl. Climatol. 113, 611–641.

Clark, P., 2013. The Oxford Handbook of Cities in World History. OUP, Oxford.

Fowler, H., Archer, D., 2006. Conflicting signals of climatic change in the upper Indus Basin. J. Clim. 19, 4276–4293.

Gilmartin, D., 2015. Blood and Water: The Indus River Basin in Modern History. University of California Press.

Habib, I., 1999. The Agrarian System of Mughal India, 1556-1707. Oxford University Press, New Delhi.

Immerzeel, W., Bierkens, M., 2012. Asia's water balance. Nat. Geosci. 5, 841–842.

Immerzeel, W., Pellicciotti, F., Bierkens, M., 2013. Rising river flows throughout the twenty-first century in two Himalayan glacierized watersheds. Nat. Geosci. 6, 742.

Immerzeel, W.W., Van Beek, L.P., Bierkens, M.F., 2010. Climate change will affect the Asian water towers. Science 328, 1382–1385.

Kenoyer, J.M., 1998. Ancient Cities of the Indus Valley Civilization. American Institute of Pakistan Studies.

Lutz, A., Immerzeel, W., Shrestha, A., Bierkens, M., 2014. Consistent increase in high Asia's runoff due to increasing glacier melt and precipitation. Nat. Clim. Chang. 4, 587.

Lutz, A.F., Immerzeel, W., Kraaijenbrink, P., Shrestha, A.B., Bierkens, M.F., 2016. Climate change impacts on the upper Indus hydrology: sources, shifts and extremes. PLoS One 11.

Marcus, J., Sabloff, J.A., 2008. The Ancient City: New Perspectives on Urbanism in the Old and New World. School for Advanced Research Press.

Meadows, A., Meadows, P.S., 1999. The Indus River: Biodiversity, Resources, Humankind. Oxford University Press, USA.

Mukhopadhyay, B., Khan, A., 2014. A quantitative assessment of the genetic sources of the hydrologic flow regimes in upper Indus Basin and its significance in a changing climate. J. Hydrol. 509, 549–572.

Mustafa, D., 2013. Water Resource Management in a Vulnerable World: The Hydro-Hazardscapes of Climate Change. Philip Wilson Publishers.

Naqvi, S.A., 2012. Indus Waters and Social Change: The Evolution and Transition of Agrarian Society in Pakistan. Oxford University Press, Pakistan.

Possehl, G.L., 2002. The Indus Civilization: A Contemporary Perspective. Rowman Altamira.

Reggiani, P., Rientjes, T., 2015. A reflection on the long-term water balance of the upper Indus Basin. Hydrol. Res. 46, 446–462.

Wright, R.P., 2010. The Ancient Indus: Urbanism, Economy, and Society. Cambridge University Press.

Yang, Y.-C.E., Brown, C., Yu, W., Wescoat, J.J., Ringler, C., 2014. Water governance and adaptation to climate change in the Indus River Basin. J. Hydrol. 519, 2527–2537.

Further Reading

Dixit, Y., Hodell, D.A., Giesche, A., Tandon, S.K., Gázquez, F., Saini, H.S., Skinner, L.C., Mujtaba, S.A.I., Pawar, V., Singh, R.N., Petrie, C.A., 2018. Intensified summer monsoon and the urbanization of Indus civilization in Northwest India. Sci. Rep. 8, 4225.

Singh, A., Thomsen, K.J., Sinha, R., Buylaert, J.-P., Carter, A., Mark, D.F., Mason, P.J., Densmore, A.L., Murray, A.S., Jain, M., Paul, D., Gupta, S., 2017. Counter-intuitive influence of Himalayan river morphodynamics on Indus civilisation urban settlements. Nat. Commun. 8, 1617.

Ward, P.J., Jongman, B., Salamon, P., Simpson, A., Bates, P., De Groeve, T., Muis, S., de Perez, E.C., Rudari, R., Trigg, M.A., Winsemius, H.C., 2015. Nat. Clim. Chang. 5, 712–715.

Water Resources Modeling and Prospective Evaluation in the Indus River Under Present and Prospective Climate Change

Daniele Bocchiola, Andrea Soncini

Department of Civil and Environmental Engineering, Politecnico di Milano, Milano, Italy

1.1 INTRODUCTION

1.1.1 Water Resources of the Hindu Kush, Karakoram, and Himalaya

The mountain range of the Hindu Kush, Karakoram, and Himalaya (HKKH) contains a large amount of glacier ice, and it is the third pole of our planet (e.g., see Smiraglia et al., 2007; Kehrwald et al., 2008), delivering water for agriculture, drinking purposes, and power production. Some estimates indicate that more than 50% of the water flowing in the Upper Indus Basin (UIB) (Fig. 1.1), in Pakistan, is due to snow and glacier melt (Immerzeel et al., 2010). The economy in the Himalayan regions relies on agriculture and thus is highly dependent on water availability and irrigation systems (e.g., see Akhtar et al., 2008; Archer et al., 2010). The Indo-Gangetic Plain (IGP, including regions of Pakistan, India, Nepal, and Bangladesh) is challenged by increasing food production, and any perturbation in agriculture will considerably affect the food systems of the region and increase the vulnerability of the resource-poor population (e.g., see Aggarwal et al., 2004; Kahlown et al., 2007). While the southern Himalayan region is strongly influenced by the monsoon climate and by abundant seasonal precipitation, meteo-climatic conditions of Karakoram suggest a stricter dependence on snow and ice ablation as water resources, and most studies concerning present and future water resources of the Indus River focus on the high-altitude areas of Karakoram and the glaciers therein (Akhtar et al., 2008; Mayer et al., 2011; Bocchiola et al., 2011; Immerzeel et al.,

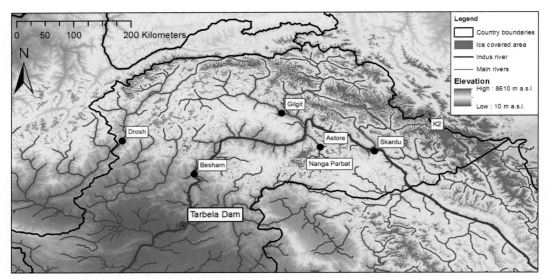

FIG. 1.1 Upper Indus Basin (UIB), showing main rivers and snow-covered areas with Tarbela Dam in the South.

2012a, b, 2013; Ragettli et al., 2013; Soncini et al., 2015a,b) nested within the UIB basin. However, most high-altitude catchments in the HKKH are not gauged or are only poorly gauged (Archer, 2003; Bocchiola et al., 2011; Hasson et al., 2017), thus posing major issues in predicting the flow in that area. The HKKH stores a significant amount of water in its extensive glacier cover at higher altitudes (about 16,300 km^2), but the lower reaches are very dry. The state of the glaciers plays an important role in future planning. Shrinking glaciers may initially provide more meltwater, but later the amount may be reduced. On the other hand, growing glaciers store precipitation and reduce summer runoff, but they can also generate local hazards. Along the HKKH range there is considerable variability in climate conditions, including varying sources and types of precipitation (e.g., see Bocchiola and Diolaiuti, 2013), influencing the behavior and evolution of cryosphere. The most recent observations of glacier fluctuations indicate that the eastern and central HKKH glaciers are subject to general retreat and have lost a significant amount of mass and area (Salerno et al., 2008; Bolch et al., 2011). Rapid decline in glacier area is reported throughout the Greater Himalaya and most of Mainland Asia (Ageta and Kadota, 1992), which is widely attributed to global warming (IPCC, 2007, 2013), but changes in the climate and the glaciers' geometry are not uniform.

1.1.2 The Karakoram Anomaly

Positive ice mass balances and advancing glaciers were reported in the Karakoram mountain range during the last decade, in spite of a worldwide decline of glaciers (Hewitt, 2005). Glaciers in the eastern part of the HKKH receive accumulation from precipitation during the

Indian summer monsoon season, whereas in the western part, snow falls occur mainly in the winter because of westerly atmospheric circulations (Bookhagen and Burbank, 2006, 2010, Kääb et al., 2012, Fowler and Archer, 2005; Winiger et al., 2005). This variability in accumulation conditions may be one reason for the large spread in glacier changes within the region (Bolch et al., 2011; Kääb et al., 2012).

Kääb et al. (2012), among others, used satellite laser altimetry to show widespread glacier wastage in the eastern, central, and southwestern parts of the HKKH during the years 2003–2008, whereas each year in the Karakoram, glaciers appear to have thinned by a few centimeters. Some studies have shown that since the 1990s, not only have balanced to slightly negative mass budgets occurred in the Karakoram range but also an expansion and thickening of the largest glaciers has taken place, mainly in the central Karakoram, accompanied by a non-negligible number of rapid glacier advances, i.e., a surge-type phenomena (see, e.g., Diolaiuti et al., 2003; Hewitt, 2005; Barrand and Murray, 2006; Belò et al., 2008; Mayer et al., 2011; Copland et al., 2011; Minora et al., 2013).

This situation of stagnant and advancing glaciers in the highest parts of central Karakoram was called the "Karakoram anomaly" by Hewitt (2005), and more recently the name "Pamir-Karakoram anomaly" was proposed by Gardelle et al. (2012). In general, glaciers in the Karakoram range seem to be less affected by the global trend of a negative glacier mass balance, with frequent observations of advancing glaciers. This behavior might be a consequence of the generally high elevations of the glaciers' bodies in this area, combined with a possible increase in orographic precipitation leading to enhanced accumulation. These observations were explained by the recent climate peculiarities: (1) a decreasing trend in maximum and minimum temperatures in some periods within the Karakoram range and (2) an increase in winter precipitation (Archer and Fowler, 2004; Bocchiola and Diolaiuti, 2013). The negative temperature trend during summer is consistent with observed advance and thickening of some Karakoram glaciers and the reducing runoff shown by some gauging station data from heavily glacierized catchments (e.g., see the Hunza River Basin; see Hewitt, 2005; Archer, 2003).

Tahir et al. (2011a) studied snow cover dynamics (using Moderate Resolution Imaging Spectroradiometer, MODIS, data during 2000–2009; see Hall et al., 2010) and the hydrological regime (using daily hydrological fluxes during 1966–2008) of the Hunza River Basin, in the northern part of Pakistan. In contrast to most regions in the world, where glaciers are melting rapidly, in the Hunza River Basin, Tahir and colleagues found a slight expansion of the cryosphere, potentially resulting from an increase in winter precipitation caused by westerly circulation, which may shield glaciers from solar radiation. The glaciological and hydrological regimes of HKKH rivers and the potential impact of climate change there have been assessed widely in the available scientific literature (e.g., see Aizen et al., 2002; Hannah et al., 2005; Kaser et al., 2010; Bocchiola et al., 2011).

Next we provide an overview of the present and future potential impact of climate change in the area by discussing (1) studies that highlight recent changes in the meteorological and hydrological cycles in the area and (2) studies that model and project forward (i.e., in the 21st century) the hydrological response of the UIB to climate change according to scenarios of reliable climate models (e.g., from IPCC report AR4/5). Further, we provide a template tool, or a guideline, for glaciohydrological modeling within high-altitude, glacierized areas and subsequent hydrological projections under climate change. We illustrate the template and discuss a recent study in the UIB in light of such an approach. Then we provide a discussion that focuses especially on the need for hydrological monitoring of the high-altitude UIB catchments, including two examples of hydrological monitoring and network design in the area. Eventually, we provide some conclusions about recent and prospective hydrology of the UIB and the necessary steps to be taken.

1.2 CLIMATE AND HYDROLOGY OF THE UPPER INDUS BASIN

The Upper Indus Basin (UIB) is a high mountainous region and contains the greatest area of perennial glacial ice outside the polar regions (22,000 km^2; e.g., see Minora et al., 2016). The area of winter snow cover is an order of magnitude greater (Hewitt, 2005). According to the Köppen-Geiger climate classification (Peel et al., 2007), the region has a typical "cold desert" climate (BWK), i.e., a dry climate with little precipitation and large daily temperature range. The mountains limit the intrusion of the monsoon whose influence weakens in the North-West. Except on the south-facing foothills, climatic controls in the UIB are therefore quite different from those in the eastern Himalayas. Over the UIB, most of the annual precipitation falls in the winter and spring and originates from the west. Monsoonal incursions bring occasional rain to trans-Himalayan areas, but even during the summer months, not all precipitation derives from monsoonal sources. Climatic variables are strongly influenced by altitude. Northern valley floors are arid with annual precipitation from 100 to 200 mm. Totals increase to 600 mm at 4400 m, and glaciological studies suggest accumulation rates of 1500–2000 mm at 5500 m (e.g., see Mayer et al., 2006; Soncini et al., 2015a). Winter precipitation (October to March) is highly spatially correlated across the UIB, north and south of the Himalayan divide (Archer and Fowler, 2004). In contrast, during the period from April to September, although there is significant correlation between neighboring northern stations, a consistent weak but occasionally significant negative correlation occurs between stations north and south of the divide. Climate records are thought to be consistent with respect to measurement practices. Standard meteorological measurement practice was established by the Indian Meteorological Department in 1891 (Fowler and Archer, 2005), and the same standards have been adopted for rainfall by the Pakistan Meteorological Department (PMD) and the Water and Power Development Authority (WAPDA).

However, studies on precipitation gradients in the mountain regions of northern Pakistan are scarce, due to the limited availability and accuracy of high-elevation data. Some studies indicate a total annual rainfall between 200 mm and 500 mm, but these amounts are generally derived from valley-based stations and are not representative of the situation encountered in the highest accumulation zones (e.g., see Archer, 2003). Recent studies show that precipitation is maximum around 5000 m a.s.l. and decreases rapidly above this elevation (Winiger et al., 2005; Immerzeel et al., 2012a, b). Recent estimates (Soncini et al., 2015a) from high-altitude accumulation pits (above 5500 m a.s.l.) indicate from 6 to 8 m of snow depth (c.3 m snow water equivalent, SWE) during 2009–2011 in the Baltoro-Biafo area. At the highest altitudes, the main contribution to the hydrological regime is from ice (and secondarily snow) melt, while the contribution from summer monsoon rainfall is quite low.

1.3 SUMMARY OF RECENT STUDIES

1.3.1 Hydrological Regime of the UIB

Analysis of relationships between seasonal climate and runoff (Archer, 2003) suggested that the UIB could be divided into three distinct hydrological regimes, namely:

(a) High-altitude catchments with large glacierized proportion (e.g., Hunza River and Shyok River) with summer runoff that is strongly dependent on concurrent energy input represented by temperature.

(b) Mainly middle-altitude catchments south of the Karakoram (e.g., Astore River and Kunhar River) that have a summer flow predominantly defined by the preceding winter's precipitation. However, the main Upper Indus catchment at Kharmong also shows the same winter precipitation control.

(c) Foothill catchments (e.g., Khan Khwar) that have a runoff regime controlled mainly by current liquid precipitation, predominantly in winter but also during the monsoon season.

Rivers in the three hydrological regimes may differ significantly in their runoff response to changes in the driving variables of temperature and precipitation as shown by linear regression analysis (Archer, 2003). For example, while the foothill catchments show significant correlation between monthly and seasonal precipitation and runoff, the high and middle altitude catchments show a negative correlation between summer precipitation and runoff, possibly the result of cloudiness and lowered temperatures associated with precipitation (Archer and Fowler, 2004).

1.3.2 Recent Evidence of Climate Change and Impact on Water Resources in the UIB

Because much of the flow abstracted from the Indus River for irrigation originates in the Himalaya, Karakoram, and Hindu Kush mountain ranges, an understanding of hydrological regimes of mountain rivers is essential for water resources management in Pakistan. Findings from the most recent results indeed indicate contrasting meteorological and hydrological patterns in the Upper Indus Basin (Archer, 2003; Archer and Fowler, 2004; Fowler and Archer, 2005; Hasson et al., 2017) in terms of expected changes in hydrological fluxes and glaciers' cover, as well as in connection with global climate indexes (Bocchiola and Diolaiuti, 2013). In Table 1.1, we provide a collection of the available literature in regard to recent evidence of the impact of climate change on the climate, the cryosphere, and water resources, and we provide a quick resume of the corresponding results.

Archer (2003), among others, investigated broad characteristics of hydrological regimes using streamflow data from 19 long-period stations in terms of annual and seasonal runoff. Regression between climatic variables and streamflow for three key basins, Hunza, Astore, and Khan Khwar, was carried out followed by regional analysis of 12 other basins. The analysis showed distinct hydrological regimes with summer volume governed by (1) melt of glaciers and permanent snow (thermal control in the current summer), (2) melt of seasonal snow (control by preceding winter and spring precipitation), or (3) winter and monsoon rainfall

TABLE 1.1   Summary of Recent Studies on Recent Climate and Hydrology in the UIB

Study	Study Area and Target Variables	Main Findings
Archer (2003)	UIB, 19 hydro stations (1960–1999, variable length). Discharge	Evidence of distinct hydrological regimes with summer volume governed by melt of glaciers and permanent snow (high altitude of Karakoram, thermal control), melt of seasonal snow (middle altitude, winter and spring precipitation), or winter and monsoon rainfall (foothill catchments, precipitation control in current season). Satisfactory levels of correlation were achieved between streamflow and measurements of temperature and precipitation at valley sites, which offers promise as a basis for assessing seasonal flow volumes
Archer and Fowler (2004)	UIB, 17 AWSs stations (1882–1999, variable length). Precipitation	Positive spatial correlation of winter precipitation, negative in summer from north and south of the Himalayan divide. During 1961–1990 significant increase in winter, summer, and yearly summer and in the annual precipitation at several stations. Significant positive correlation between the winter NAO and precipitation in the Karakoram, negative correlation between NAO and summer rainfall
Fowler and Archer (2005)	UIB, 8 AWSs stations during 1961–2000. 1 AWS during 1894–2000. Temperature	Significant increases in winter, summer and annual precipitation, and significant warming occurred in winter, however, with summer cooling. Relationships between climatic variability and large-scale climatic processes (ENSO/NAO) may be used to increase forecasting lag time for water resources
Archer et al. (2010)	UIB, based on literature	Trends in climate, most notably the decline in summer temperatures, no strong evidence of marked reductions in water resources from nival regime, glacial regime, or rainfall regime
Tahir et al. (2011a)	Hunza River, 1 hydro station (1966–2008). MODIS images of snow-covered area, SCA (2000–2009). Discharge, SCA	Analysis of remotely sensed snow cover (2000–2009) suggests slight expansion, possibly increasing river runoff at thaw
Bocchiola and Diolaiuti (2013)	Northern Pakistan, 17 AWSs stations (1980–2009). Temperature, precipitation, cloud cover	Increasing number of wet days, T_{Min} decreases in summer, T_{Max} always increasing. Cloud cover increases (1980–2009). T_{Min}, T_{Max} negatively correlated to NAO. Slight dependence of trend intensity against altitude
Minora et al. (2016)	Ice-covered area (ICA) in the CKNP (2001 and 2010). MODIS SCA (2001–2011). Three AWSs stations (1980–2009)	No significant difference in ICA 2001–2010. Increased SCA 2001–2011. Reduction of mean summer temperatures and more snowfall events (1980–2009). Potential role of debris covered and surging glaciers in the Karakoram anomaly
Hasson et al. (2017)	UIB. 18 AWSs stations (1995–2012), 6 AWSs stations (1961–2012). 10 hydro stations (1962–2011, variable length). Precipitation, temperature, discharge	During 1995–2012, warming and drying in spring, and increasing of early melt season discharge due to faster snowmelt. Cooling in the monsoon period, and hence decreasing or weakly increasing discharge. Largely consistent with the long-term trends (1961–2012); these trends indicate dominance of the nival regime, but weakening of the glacial melt regime

(precipitation control in current season). Satisfactory levels of correlation were achieved between streamflow and measurements of temperature and precipitation at valley sites, which offers promise as a basis for assessing seasonal flow volumes.

Archer and Fowler (2004) studied spatial and temporal variations (which in some cases started as early as 1882) of precipitation from 17 stations in the UIB, the teleconnections there, and potential hydrological implications. They highlighted that winter (October-March) precipitation shows a high positive correlation over the entire northern area of Pakistan, suggesting that, for forecasting of rainfall and runoff in the UIB, one may use few stations (or even only one). However, during summer (July-September) monsoon spatial coherence is low, and anticorrelation may even be found north and south of the Himalayan divide. They reported a trend in precipitation during 1961–1999 with an increase in the winter and summer and at the annual scale. They identified a positive correlation between the North Atlantic Oscillation (NAO) and winter precipitation and a negative correlation during summer.

Fowler and Archer (2005) studied temperatures (1961–2000) for seven stations in the UIB, and precipitation from seven (other) stations, and their connections with El Niño Southern Oscillation (ENSO). They used a combined index, the Multivariate ENSO Index (MEI), which uses six observed variables over the Tropical Eastern Pacific and is computed seasonally. Average seasonal temperature and precipitation for stations in the UIB are compared for maximum El Niño and La Niña years, computed for summer MEI (May-September). Fowler and Archer reported that during the years 1961–2000, there were significant increases in winter, summer, and annual precipitation and that significant warming occurred in winter, while summer showed a cooling trend. They found a very consistent contrast between enhanced winter precipitation in El Niño years and reduced early winter precipitation in La Niña years. Summer temperatures were lower in El Niño years and higher in La Niña years. Fowler and Archer suggest that any relationship between climatic variability and large-scale climatic processes may be used to increase forecasting lag time for water resource.

Bocchiola and Diolaiuti (2013) investigated (1980–2009) climate variability in the upper Karakoram. Starting from monthly data and using principal components analysis, they analyzed seasonal values of total precipitation, number of wet days, maximum (T_{Max}) and minimum (T_{Min}) air temperature, maximum precipitation in 24 h R_{Max}, and cloud cover C_C for 17 weather stations in the upper Karakoram, clustered within three climatic regions. They detected possible non-stationarity in each of these regions via (1) linear regression, (2) moving window average, and (3) the Mann-Kendall test, also in progressive form, to detect the onset date of possible trends. They evaluated linear correlation coefficients between the Northern Atlantic Oscillation (NAO) index and climate variables to assess the effectiveness of teleconnections, which recently have been claimed to affect climate in this area. Also, they compared the temperature within the investigated zone with global temperature anomalies to show evidence of enhanced warming in this area. They found mostly non-significant changes of total precipitation, except for a few stations showing an increase in the Chitral-Hindu Kush region and in northwest Karakoram in the Gilgit area, and a decrease in Western Himalaya in the Kotli region. Maximum precipitation was mostly unchanged, except for a slight increase in the Chitral and Gilgit areas and a slight decrease in the Kotli region. The number of wet days was mostly increasing in the Gilgit area and decreasing in the Chitral area, with no clear signal in the Kotli region. Minimum temperatures increased except during summer, when decreasing values were detected, especially for the Gilgit and Chitral regions. Maximum temperatures increased everywhere. Cloud cover was significantly increasing in the Gilgit area, but decreasing otherwise, especially in the Kotli region. The maximum temperature regime had a significantly positive correlation with the global thermal anomaly, while the minimum temperature regime had a non-significant negative correlation. Maximum and minimum temperatures appeared to have a mostly negative correlation with NAO. Some dependence of trend intensity for the considered variables with altitude was found, with differences in each region, suggesting that investigation of weather variables at the highest altitudes is of paramount importance in determining further climate variability in the area.

Minora et al. (2013) studied glaciers' evolution (2001−2010) within the Central Karakoram National Park (CKNP), which nested a large number of glaciers (Bocchiola et al., 2004), by means of remote sensing data (i.e., Landsat images) and found no significant area changes. Using MODIS data, they also found increasing (2001−2011) snow-covered areas at thaw (June-September). They further detected a considerable rise in supraglacial debris coverage, which could have contributed to reduce the melting of buried ice during the past decade. Based on data from a 3-year field campaign to attain an estimate of yearly ice melting in the Baltoro Glacier (at the toe of K2 peak), Minora et al. (2015) were able to provide a simplified modeling of ice melting there against debris cover and hypothesized that recently more ablation has taken place than occurred in the past three decades.

Minora et al. (2016) recently studied the Karakoram anomaly based on remote sensing assessment of glacial cover and further assessed recent local climate trends possibly affecting ice cover dynamics. They reported a slight increase in the number of wet days in Gilgit, in the Karakoram range, during 1980–2009, as well as evidence of constant or even slightly increased ice cover (the Karakoram anomaly) in the Central Karakoram National Park (13,000 km^2), where the ice-covered area (ICA) was 4605.9 km^2 in 2001 and 4606.3 km^2 in 2010, with abundant supraglacial debris cover. Further observation of a slightly increased snow cover area at fall above 1900 m a.s.l. during 2001–2011 seemingly confirmed the positive feedback of snow cover on glaciers' protection against melting. However, the results from Minora et al. (2015) and especially Soncini et al. (2015a) demonstrated that under future climate change as projected by the IPCC for 2100, a great decrease in snow cover may result in larger glaciers downwasting, thus eventually breaking down the Karakoram anomaly as we know it today.

Hasson et al. (2017) assessed hydroclimatic trends within the UIB during 1961–2012. They studied the trends in maximum, minimum, and mean temperatures, the diurnal temperature range. Using Mann-Kendall tests, they statistically assessed locally identified climatic trends for their spatial-scale significance within 10 identified hydrological subregions of the UIB. Then they qualitatively compared such trends with trends in discharges in the same subregions. They found recent (1995–2012) warming and drying in spring (most significant in March) and earlier discharge in the melt season in most of the subregions, likely due to faster snowmelt. In stark contrast, most of the subregions featured significant cooling during the monsoon period (particularly in July and September), coinciding fairly well with the main glacier melt season. Hence a decreasing or weakly increasing trend of streamflows was observed from the corresponding subregions during mid- to late-melt season (particularly in July). Such trends, being largely consistent with the long-term trends (1961–2012), most likely indicate the dominance of the snowmelt contribution, with a decrease in the glacial melt regime, which overall may modify the hydrology of the UIB in the future.

1.3.3 Projected Water Resources in the UIB During the 21st Century

Recent literature has reported the results of some sensitivity studies that project possible effects of climate changes on hydrological processes in the 21st century. Hydrological projections normally require.

(a) use of (glacio)-hydrological models capable of mimicking the local hydrological response to climate;

(b) use of climate scenarios from global circulation models (GCMs) or regional circulation models (RCMs), typically under the umbrella of the IPCC panel, to provide consistent climate change projections; and

(c) downscaling of global or regional models against local data to eliminate the biases given by scale mismatch and models' noise, especially for precipitation.

In Table 1.2 we provide a summary of the studies considered here. This includes the adopted hydrological model(s), the adopted GCMs, the IPCC reference AR and the storylines/RCPs, the projection period, the downscaling methods adopted, and the main findings. However, first, we want to comment on some of these studies.

Akhtar et al. (2008), among others, investigated hydrological conditions dependent on different climate change scenarios (PRECIS model, A2 storyline) for three glacierized watersheds in the Hindu Kush-Karakoram-Himalaya region (Hunza, 13,925 km^2, glacierized 4688 km^2; Gilgit, 12,800 km^2, glacierized 915 km^2; Astore, 3750 km^2, glacierized 612 km^2). Their results indicate an increase in temperature and precipitation toward the end of 21st century, with discharges increasing for 100% and 50% glacier cover scenarios, whereas a noticeable decrease was conjectured for 0% scenario, i.e., for depletion of ice caps. Immerzeel et al. (2009) used remotely sensed precipitation (from TRMM) and snow-covered area, or SCA, (from MODIS) together with ground temperature data and a simple snowmelt runoff model (SRM) to calibrate a hydrological model and then projected forward in time (PRECIS model, 2071–2100) the hydrological response of the strongly snow-fed Indus watershed (Pakistan, NW Himalaya, 200.677 km^2, including the Hunza and Gilgit basins). They found potential warming in all seasons, though greater at the highest altitudes, resulting in diminished snowfall. Total precipitation would increase by +20% or so against its present values. They projected snow melt peaking 1 month earlier than now, increased glacial flow due to higher temperature, and a significant increase in rainfall runoff.

Bocchiola et al. (2011) conducted a preliminarily study on the prospective (until 2059) hydrological behavior of the Shigar River (6920 km^2 at Shigar), nested within the UIB. They used a simple hydrological model, including snow and ice ablation and tuned it against historical average monthly values found in the literature to depict hydrological fluxes. Then they used future (2050–2059) locally adjusted precipitation and temperature fields from the CCSM3 model (A2 storyline) to depict future hydrological behavior of the river, under four different ice cover scenarios (unchanged, −10%, −25%, −50%). It was determined that, depending on the assumptions concerning ice cover, average yearly discharge under these scenarios will increase considerably. Namely, mean discharge will increase up to twice as much as the present values for 100% ice cover and will decrease below the present values for ice cover between 50% and 25%.

The Shigar River is paradigmatic of high-altitude catchments in the upper Karakoram and can be classified as an ungauged catchment, given that its streamflows have not been

TABLE 1.2   Summary of Recent Studies on Prospective Hydrology in UIB

Study	Study Area and Target Variables	Main Findings
Akhtar et al. (2008)	Three catchments in the UIB. Projections of streamflow Q under A2 scenario until 2100	Discharge generally increasing for 100% and 50% glacier scenarios. For the 0% glacier scenario, drastic decrease in water resources
Immerzeel et al. (2010)	Indus River at delta. Q A1B scenario until 2050	In the Indus, 40% of meltwater from glaciers. Glacier size in 2050 based on a best-guess scenario with decrease in mean water supply (−8.4%)
Bocchiola et al. (2011)	Shigar River at Shigar. Q under A2 scenario 2050–2059	Discharge $Q + 119\%$ under unchanged glaciers' cover ICA; $Q + 72\%$ with −10% ICA; $Q + 28\%$ with −25% ICA; $Q - 24\%$ with −50% ICA
Tahir et al. (2011b)	Unza River at Daynior Bridge. Q under climate scenarios until 2075	(1) A +20% increase in cryosphere area until 2075 and a 10% increase until 2050. ICA increasing under increasing precipitation scenario. Summer flow $Q_s + 7\%$ at 2050, +14% at 2075 (2) A +4 °C increase in mean temperature until 2070, +3°C increase until 2050, and +2°C by 2025 (other variables constant). $Q_s + 64\%$ at 2025, +100% at 2050 (3) An increase of +2–4°C in the mean temperature, +20% increase in snow cover area until 2075. $Q_s + 100\%$ at 2075
Immerzeel et al. (2013)	Baltoro Glacier. Q, ICA, water volume in ice IWE under RCP4.5, RCP8.5	Q increasing throughout the century, with glacier melting slightly decreasing and precipitation runoff increasing. Ice volume loss nearby −55% at 2100
Soncini et al. (2015a)	Shigar River at Shigar. Q, ICA, IWE under three RCPs scenarios, RCP2.6, RCP4.5, RCP8.5 (and three GCMs) until 2100	Annual mean discharge Q_y increases at 2050 +5%–12% for RCP2.6, between +11% and 25% for RCP4.5, +12%–23% for RCP8.5. At 2100 RCP8.5 gives similar results, RCP4.5 shows Q_y of about +3%–15%, and RCP2.6 provides decreased Q_y of about −15% for all GCMs. Future ice volume (IWE) remains constant at 2050 (−0.13% on average for all RCPs); at 2100 ice volume decreases greatly (−60%, −73%, −89% on average, RCP2.6, RCP4.5, RCP8.5, respectively)

measured since 1997 (which are available as to the authors in the form of average monthly values as reported for 1985–1997) and that little meteorological, nivological, and glaciological information is available. The study by Bocchiola et al. (2011) was indeed published under the framework of the Predictions in Ungauged Basins (PUB) initiative, launched by the International Association of Hydrological Sciences (IAHS; see also Sivapalan et al., 2003; Seibert and Beven, 2009). High-altitude glacierized catchments such as Shigar are used here to represent typical grounds of application of PUB concepts, where simple hydrological modeling based on scarce data is necessary for assessment, estimation, and prediction of the mass/water budget of glaciers under climate change conditions (e.g., see Chalise et al., 2003; Konz et al., 2007; Immerzeel et al., 2009; Bocchiola et al., 2010). Thus future research toward this aim is needed. Presentations in this chapter deal with issues around depicting the present and prospective dynamics of ice bodies in the Karakoram area and the hydrological cycle in that area.

Immerzeel et al. (2013), using a glacio-hydrological model under RCP45 and RCP85 climate scenarios, described the ice flow dynamics of the Baltoro Glacier and provided projections until the year 2100. They found that discharge may increase throughout the century, with glacier melting slightly decreasing and precipitation runoff increasing. They estimated an ice volume loss of approximately −55% by 2100. However, they did not use field data; furthermore, they made a number of overly simplified assumptions, for example, about ice ablation, ice flow, among other factors. Their results therefore must be taken only as qualitative calculations.

Soncini et al. (2015a) focused again on the Shigar River. They reported the results of a 3-year (2011 − 2013) hydrological, meteorological, and glaciological field campaign; then they used those results to set up a hydrological model depicting in-streamflows, snowmelt, and ice cover thickness. They then assessed changes in the hydrological cycle up to 2100 via climate projections provided by three state-of-the-art global climate models used in the recent IPCC AR5 under the RCP2.6, RCP4.5, and RCP8.5 emission scenarios, including detailed modeling of the glaciers' evolution, as based on field data.

1.4 A GUIDELINE AND RECENT KEY STUDY ON MONITORING AND MODELING HIGH-ALTITUDE CATCHMENTS

1.4.1 A Method to Model High-Altitude Catchments

Here we discuss a recent case study on glacio-hydrological modeling of a high-altitude catchment within the UIB river network and the subsequent assessment of present and potential future water resources for the 21st century under IPCC scenarios of climate change.

The description and comments about this paradigmatic case help illustrate the effort needed to assess and predict water resources in a typical high-altitude catchment in the UIB and generally throughout the HKKH.

In this case study, we reference a template approach recently developed and applied in a number of worldwide case studies (Soncini et al., 2015a; Viganò et al., 2015; Soncini et al., 2017; Aili et al., 2018) in order to provide a systematic guideline for scientists interested in pursuing similar efforts.

Fig. 1.2 shows a scheme of the proposed methodology. Accurate assessment of snow and especially ice cover dynamics (ice melt, mass balance, flow dynamics) is complex, given the harsh topographic and climatic conditions, and limited information is available except for a few, well-documented cases (e.g., see Soncini et al., 2015a, b, 2017). Methodologies for proper assessment of cryospheric dynamics in poorly gauged areas are therefore often required. Furthermore, high-altitude catchments are often exploited for hydropower production (e.g., see Ravazzani et al., 2016), so hydrological fluxes can be modified in the presence of flow diversion, and back calculation is needed to investigate streamflows (e.g., see Confortola et al., 2013), which is normally possible only for simple geometries. In such cases, investigation

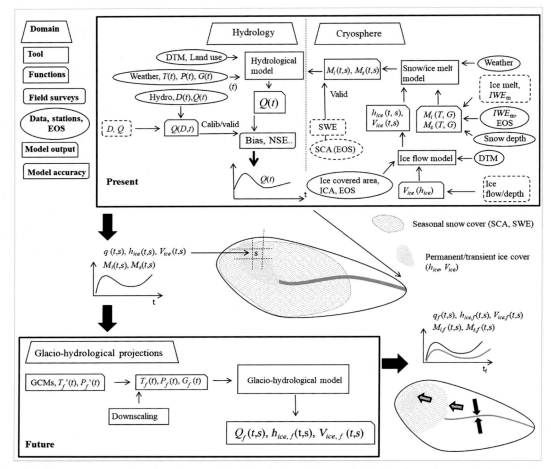

FIG. 1.2 Proposed methodology for glacio-hydrological modeling of high-altitude catchments, and projections under climate change. Explained in text. *Based on Aili, T., Soncini, A., Bianchi, A., Diolaiuti, G., Bocchiola, D., 2018. A method to study hydrology of high altitude catchments: the case study of the Mallero river, Italian Alps. Theoret. Appl. Climatol. https://doi.org/10.1007/s00704-017-2366-4 1-22.*

of natural flows, say for assessment of ecological flows (Viganò et al., 2016) is necessary but complicated. Here we comment on a template for modeling high-altitude catchments, where snow and ice dynamics are important but for which little information is available. This framework builds on the experience of the authors and their team in field work. Such experience was gathered in the last decade, by monitoring, and modeling several mountainous areas in the Italian Alps (Bocchiola et al., 2010; Groppelli et al., 2011a; Soncini et al., 2017) and worldwide (Migliavacca et al., 2015; Soncini et al., 2015a, b); in principle, these findings are portable to any catchment.

In the flow chart in Fig. 1.2, the necessary tools are reported according to seven categories: domain (e.g., hydrology, cryosphere); tools (e.g., hydrological model, snowmelt model);

functions (e.g., snowmelt M_s as a function of temperature, and radiation $M_s(T, G)$); necessary data from field surveys or other sources (ice melt from stakes, ice volume loss from topographic methods, earth observation from space EOS, etc.); data (weather, snow depth, SCA from remote sensing, etc.); model outputs (e.g., ice melt in time and space $M_i(t,s)$); and model accuracy (i.e., calculation of Bias, NSE, etc.). $T(t)$ is daily temperature, $P(t)$ is daily precipitation, $G(t)$ is solar radiation, $D(t)$ is daily flow depth at hydro station, and $Q(t)$ is daily discharge at outlet section. $M_i(t,s)$ is daily ice melt in a given place (cell) s, $M_s(t,s)$ is daily snowmelt, $q(t,s)$ is daily runoff in cell s, $h_{ice}(t,s)$ is daily ice depth, and $V_{ice}(t,s)$ is daily ice flow velocity. SCA is snow-covered area. SWE is snow water equivalent. ICA is ice-covered area, IWE_m is water equivalent of ice melt. Bias is systematic error on average, NSE is Nash-Sutcliffe Efficiency. $T'_f(t)$, $P'_f(t)$ are (future/projected) temperature and precipitation from GCMs before downscaling (biased); $T_f(t)$, $P_f(t)$ are future daily temperature and precipitation after downscaling (unbiased). Not all methods/data can be used in all cases, and adaptation to specific conditions is necessary.

A division is made between present (i.e., for assessment of present glacio-hydrological conditions), and future (i.e., for projection of future conditions using GCM scenarios). The information necessary to model each component and the interactions between components is also sketched. Data gathering for the purpose of modeling is presented according to different potential methods, which may or may not be implemented, but are still valid for the purpose. Specific implementation of the proposed method clearly requires tailoring for each case study, and depends on the characteristics of the given area and the available data and tools. The following discussion focuses on a case study of the Shigar River (Soncini et al., 2015a), and each component of the methodology is illustrated, with specific relative references.

A sketch of the procedures and tools necessary for setting up a hydrological model of the area are provided as follows:

(a) Hydrological model: This model can be either lumped, semi-distributed, or distributed, in order of complexity. Lumped models provide non-spatially, varying representation of hydrological processes, so they do not account for variations of climate in altitude, including temperature drift, and so on. Semi-distributed (typically, altitude belt-based) models allow accounting for such variability, but not for fully distributed values. However, these models are normally computationally fast and provide a good trade-off between representation of spatial variability and short computational time. Fully distributed models provide for full spatial variability (pending data availability) and are therefore very complete tools. However, their relatively high computational burden may be a liability, especially for large catchments ($>100\,km^2$). For the purpose of water resource assessment, daily simulation of streamflows is enough.

(b) Streamflow data: These data can be used for direct assessment of water resources and for calibration/validation of the hydrological models. Notice that a semi-distributed or distributed hydrological model, once calibrated, can provide flow estimates for ungauged catchments of interest, thereby increasing knowledge with respect to direct gauging.

(c) SCA pictures: In mountainous snow-fed catchments, investigation of snow-covered areas (SCA) by way of satellite data is of utmost importance. Indeed, use of SCA is widespread, as it provides a benchmark for simulation of snow cover dynamics. Satellite SCA data at medium to moderate resolutions are now available at virtually no cost and can be downloaded at least weekly (bundles) if not daily for the purpose of hydrological model calibration/validation.

(d) Snow cover data: Some information about snow cover depth and density is necessary for model setup. SCA pictures provide area coverage, but not absolute values. Also, data on snow depth may provide an indication of the winter accumulation necessary to feed the glaciers.

(e) Ice cover data: The glaciers' area cadastre is fundamental for hydrological modeling, because the latter includes ice melt and related discharge. Also, hydrological models can explicitly track ice cover over time, so validation via ice area cover is necessary.

(f) Ice melt data: Seasonal melting of ice needs to be tracked for calibration of ice melt modules within hydrological models and for ice depth.

(g) Necessary weather data: Rainfall, temperature, radiation, evaporation, wind, and air moisture need to be tracked. It is assumed that the CKNP will develop a meteorological network providing the necessary input data for the development of hydrological models. Minimal models, like the one developed and reported by the authors here, need only precipitation and temperatures. However, more sophisticated modeling requires more information.

(h) Soil moisture data: Information on soil moisture, for example, via time domain reflectometry (TDR) probes may be used to infer soil storage, which is fundamental for hydrological modeling. Exercises to assess water resources also may be conducted under climate change scenarios. This approach requires the use of a properly calibrated hydrological model, which will give input on future climate scenarios.

(i) Historical reference data base: Weather and possibly hydrometric data including at least temperature, precipitation, and discharge should be available for the past 10–30 years (especially long series can be used to assess trends). This is necessary to (1) identify recent trends for weather and water resources, (2) benchmark outputs of control runs from GCM models, and (3) project the hydrological cycle using the "what if" hypothesis of future trends mirroring past ones, which is the simplest hypothesis for climate projections.

(j) Data from GCMs (control runs, projections): General circulation GCM models suggested by IPCC can be used to provide future climate scenarios. Ideally, one should choose GCMs that provide the control runs (i.e., simulations of the past climate) closer to the climate observed locally. Under the hypothesis that reasonable coincidence may remain valid in the future, GCM projections under the IPCC storylines (A1, A1B, A2, B1, B2, etc.) or RCPS under AR5 (RCP2.6, RCP4.5, RCP6.5, and RCP8.5) can be used.

(k) Locally tailored downscaling schemes: Because GCMs do not respect small-scale climate variability, especially enhanced within mountainous area, locally tailored downscaling is required. This is especially true for precipitation, which exhibits high nonlinearity and inhomogeneity in space and time.

1.4.2 A Paradigmatic Case Study, Shigar River of Gilgit-Baltistan

1.4.2.1 *The Shigar River*

Here we report the results of recent studies within paradigmatic high-altitude ice/snow fed catchments in the UIB, and also present an application of the previously mentioned guideline. This study focuses on the Shigar River closed at Shigar (c.7000 km^2), nested within the UIB and fed by seasonal melt from two major glaciers (Baltoro and Biafo). Hydrological, meteorological, and glaciological data were gathered during 3 years of field campaigns (2011–2013) and were used to set up a hydrological model, providing a depiction of in-streamflows, snowmelt, and ice cover thickness. The model was then used to assess changes of the hydrological cycle until 2100, via climate projections provided by three state-of-the-art global climate models used in the recent IPCC AR5 under the RCP2.6, RCP4.5, and RCP8.5 emission scenarios. Under all RCPs, future flows were predicted to increase until the middle of the century, after which they will decrease, but remain mostly higher than the control run values. Snowmelt is projected to occur earlier, while the ice melt component is expected to increase, with ice thinning considerably, and even disappearing below 4000 m a.s.l. until 2100. The Shigar River flows in northern Pakistan, ranging from c.74.5°E to 76.5°E in longitude and 35.2°N to 37°N in latitude. The whole catchment is located in the Gilgit-Baltistan region, the highest altitude is reached at the K2 peak (8611 m a.s.l.), and the outlet is at the Shigar Bridge (2204 m a.s.l.). The average altitude of the catchment is 4613 m a.s.l., and more than 40% of the area lies between 4000 and 5000 m a.s.l. The catchment includes several glaciers covering an area of 2164 km^2 (28% of the basin area), the most important two being the Biafo Glacier (438 km^2) and the Baltoro Glacier (604 km^2). Looking at Askole climate data, precipitation reaches a maximum in April, and 44% of the total annual rainfall occurs in March, April, and May, with the rest equally distributed.

1.4.2.2 Data for High-Altitude Catchment Modeling

A mix of data can be used to depict the hydrological cycle of high-altitude catchments with our proposed methodology. Climate-wise, minimal hydrological modeling requires inputs of temperature and precipitation.

Our study is based on the analysis of meteorological, hydrological, and glaciological data from various in-situ stations in the Shigar River catchment, as summarized in Table 1.1. Three stations are managed by the EvK2CNR committee, two are located at Askole (3015 m a.s.l.) and Urdukas (3926 m a.s.l.), which have provided data since 2005; and one is located at Concordia (4690 m a.s.l.), which has operated since 2011. Data on the daily rainfall, air temperature, and other standard meteorological parameters are available at the three stations. The data sets show some gaps especially during winter due to sensor malfunctioning in extreme weather conditions. Monthly meteorological data for the period 1980–2009 taken at other Pakistan Meteorological Department (PMD)-managed stations located outside the catchment below 2500 m a.s.l. (Bocchiola and Diolaiuti, 2013) have also been used, most notably to assess lapse rates of temperature and precipitation.

Hydrological data are necessary in order to calibrate/validate hydrological models. We used mean monthly discharge estimates at Shigar during 1985–1997 (Archer, 2003, Table 1.3). These data come from a station managed by the Water Power Development Agency (WAPDA) of Pakistan. We further used daily discharge data available since April 2011 from a new hydrometric station (sonic ranger for water level, plus stage-discharge curve) that we installed in Shigar. A second hydrometric station (pressure transducer for water level, plus stage-discharge curve) was installed in May 2012 in Paiju (3356 m a.s.l.), near the Baltoro Glacier's snouts. Spring 2012 data from the Paiju station have been used to validate the hydrological model at this section. Also, modeling of high-altitude catchments requires information about ice and snow ablation. Seventeen ablation stakes were deployed along the main flow line of the Baltoro Glacier, from 3700 m a.s.l. to 4600 m a.s.l. in the summer of

TABLE 1.3 Case Study of Shigar River

Station	Altitude (m a.s.l.)	Longitude (°E)	Latitude (°N)	Variable	Resolution (.)	Period Used (.)
ASKOLE	3015	75.82	35.68	Temp, Precip	Daily	2005–2012
ASTORE	2168	74.87	35.36	Temp, Precip	Monthly	2005–2012
CONCORDIA	4690	76.31	35.44	Temp, Precip, Snow depth	Daily	2011–2012
SHIGAR	2221	75.73	35.42	Discharge	Daily	May 23–June 25, 2012
PAIJU	3356	76.06	35.40	Discharge	Daily	May 23–June 25, 2012
Seventeen	3693–4580	–	–	Ice ablation	Various	July 2011–June 102
Three	5600–5900	–	–	SWE, snow depth	Three samples	August 2011
CONCORDIA	4960	76.31	35.44	Ice thickness	One measure	June 2013

Data coverage.
Based on Soncini, A., Bocchiola, D., Confortola, G., Bianchi, A., Rosso, R., Mayer, C., Lambrecht, A., Palazzi E., Smiraglia C., Diolaiuti G., 2015a.
Future hydrological regimes in the upper Indus basin: a case study from a high altitude glacierized catchment, J. Hydrometeorol. 16(1), 306–326.

2011. Ice drilling was performed using a steam drill, reaching down 8–12 m. The stakes were also used to evaluate surface ice flow velocity by performing DGPS (differential GPS) surveys during two summer seasons (2011 and 2012). During the summer of 2011, three snow pits were dug in the accumulation area of the Baltoro Glacier (Table 1.1), and the data were used to validate snow accumulation by the model. Ice thickness on the Baltoro glacier was also tentatively estimated in the summer of 2013 (Soncini et al., 2015a, unpublished data), using a low frequency radar antenna (50 MHz) installed on a portable instrument (SIR 3000), to be used in a simple ice flow model. Projections of future hydrological conditions require use of data from properly chosen, locally tailored GCMs. Soncini et al. (2015a,b) used the outputs of historical and scenario (up to 2100) simulations of three global climate models participating in the Coupled Model Intercomparison Project Phase 5 (CMIP5) experiment, and contributing to the Fifth Assessment Report (AR5) of IPCC, namely the EC-Earth, ECHAM6, and CCSM4 models (Table 1.4). Climate projections are evaluated under Representative Concentration Pathway (RCP) scenarios (Moss et al., 2010; IPCC, WG-I, 2013), all including progressive decreases in aerosol (and aerosol precursors) emissions through the 21st century. They used results from RCP 2.6, RCP 4.5, and RCP 8.5, where the number indicates the radiative forcing (in $W m^{-2}$) in 2100, relative to 1850. In the framework of the SHARE-Paprika project, the EC-Earth model precipitation data for the historical period and two future scenarios were specifically analyzed for the HKKH area (Palazzi et al., 2013). Precipitation in the HKKH region simulated by EC-Earth was found to well represent the climatology of different observational and reanalysis datasets, i.e., reflecting the main seasonal precipitation patterns in the area, the wintertime western weather patterns, and the summer monsoon. EC-Earth slightly

TABLE 1.4 Case Study of Shigar River

EC-EARTH	Europe-Wide Consortium	E.U.	1.125°×1.125 degrees	62	320×160
ECHAM6	Max Planck Institute for Meteorology	GER.	1.875 × 1.875 degrees	47	192 × 96
CCSM4	National Center for Atmospheric Research	U.S.A.	1.25 × 1.25 degrees	26	288 × 144

Features of the three adopted GCMs.

Based on Soncini, A., Bocchiola, D., Confortola, G., Bianchi, A., Rosso, R., Mayer, C., Lambrecht, A., Palazzi E., Smiraglia C., Diolaiuti G., 2015a. *Future hydrological regimes in the upper Indus basin: a case study from a high altitude glacierized catchment, J. Hydrometeorol. 16(1), 306–326.*

overestimated precipitation, particularly in winter, consistently with the "wet bias" commonly seen in precipitation simulated by GCMs over high-elevated terrains. Also, this bias may be related to an underestimation of total precipitation in snow-rich areas like HKH. Typically, the coarse spatial resolution of GCMs lead to simulate poorly the effects of rapidly-changing topography, e.g., precipitation changes over short distances, and in general their spatial variability. A spatial downscaling of the outputs of climate models is required to extract local information from coarse-scale simulations, and to perform hydrological/impact studies at a basin scale (e.g., see Groppelli et al., 2011a).

1.4.2.3 Glacio-Hydrological Modeling in High-Altitude Catchments

To model ice and snow ablation, typically a degree-day approach (DD_I) can be used (Hock, 2003), either simple (i.e., using temperature only) or mixed (radiation plus temperature, Soncini et al. 2017). According to our proposed methodology in Fig. 1.2, degree day values can be estimated against data of ice melting (water equivalent of ice melt IWE_m) taken by (1) field ice melt data or (2) EOS data of ice volume reduction from satellites (Aili et al., 2018; D'Agata et a., 2018). The Shigar watershed includes large debris-covered ice that affects the melting dynamic (Scherler et al., 2011). Mihalcea et al. (2006) and Mayer et al. (2006) evaluated melt factors for both debris-covered and debris-free ice, based on field ablation data from the Baltoro Glacier. Mihalcea et al. (2008) used remote sensing and field data to develop a debris thickness map of the Baltoro. Based on data from our 17 ice ablation stakes deployed in 2011 between 3700 and 4600 m a.s.l., Soncini et al. (2015a) built a degree day factor approach for computing ice melt, for both buried and bare ice, using ice ablation data from their field experiments. Also, using data from ablation stakes and from sparse surveys of debris cover at randomly selected sites during the summer of 2011, they found a relationship between debris thickness and altitude, which they used to estimate (average) debris cover thickness within each elevation belt. Also, snowmelt can be tackled using degree day models that can be calibrate based on (1) snowmelt data from snow gauges (Bocchiola et al., 2010; Soncini et al., 2017) or (2) snow cover from satellites SCA (Bocchiola et al., 2011). Soncini et al. (2015a) used a seasonally variable degree day factor for snowmelt (DDs) estimated from MODIS snow cover images by Bocchiola et al. (2011), starting from DDs = 1.5 mm °C^{-1} d^{-1} in April (onset of snowmelt season at the lowest altitudes), and increasing monthly until DDs = 5 mm °C^{-1} d^{-1} in August, then decreasing again until DDs = 1.5 mm °C^{-1} d^{-1} in October. They used snow depth data collected at Concordia station (4690 m a.s.l.) during the 2011 winter and 2012 summer season to further validate the performance of the snow ablation model, with acceptable results. To avoid inconsistent "static" glacier cover for credible hydrological projections, one needs to use some type of ice flow model (see Fig. 1.2).

Glacio-hydrological models and flow models can be based either on altitude belts or on distributed (cell-based) modeling. Soncini et al. (2015a) used a simplified ice flow approach, by shifting a proper quantity of ice from an altitude belt to the lower one. The large valley glaciers in the Karakoram flow with maximum speeds of about 100–150 m y^{-1}, while the glacier tongues are in the order of 20–40 km long (e.g., see Mayer et al., 2006; Quincey et al., 2011). Simulations of glacier runoff with periods of 50–100 years lead to an ice movement of 5–15 km, i.e., less than half of the glacier tongue length. The velocity can be approximated by a simplified force balance, and it is proportional to ice thickness raised to $(n+1)$, with n exponent of Glen's Flow Law ($n=3$; e.g., see Wallinga and van de Wal, 1998; Cuffey and Paterson, 2010). In the simplifying hypothesis that basal shear stress τ_b be constant along the glacier, and accounting for both deformation and sliding velocity as governed by τ_b, it is possible to model depth averaged ice velocity as (Oerlemans, 2001)

$$V_{ice,i} = f_d \tau_b^n h_{ice,i} + f_s \frac{\tau_b^n}{h_{ice,i}} = K_d h_{ice,i}^{n+1} + K_s h_{ice,i}^{n-1} = K_d h_{ice,i}^4 + K_s h_{ice,i}^2, \tag{1.1}$$

with $h_{ice,i}$ (m) ice water equivalent in the belt i, and K_s (m^{-3} y^{-1}) and K_d (m^{-1} y^{-1}) parameters of basal sliding and internal deformation, which they calibrated against the observed velocity values at the stakes. Ice flow occurs then by

$$h_{ice,i}(t+\Delta t) = h_{ice,i}(t) - F_{i \to i-1}(t) + F_{i+1 \to i}(t)$$
$$F_{i \to i-1} \propto V_{ice,i} \tag{1.2}$$

where the amount of ice water equivalent $h_{ice,i}$ in the belt i at time t results from the balance of the ice passing from belt i to the lower belt $i-1$, $F_{i \to i-1}$, and that from belt $i+1$ to the lower belt i $F_{i+1 \to i}$, with ice mass passing proportional to velocity in the upper belt (and scaled by the ratio of the ice-covered area in the belt i and the one in the belt $i-1$ or $i+1$). To initialize the ice flow model, an estimated ice thickness value for each belt is necessary. The first step is the calculation of τ_b from the altitude range ΔH (difference between maximum and minimum glacier elevation), according to Baumann and Winkler (2010). Here τ_b is set to 1.5 bar, because $\Delta H > 1.6$ km. From Aster GDEM we calculated the slope in each belt α_i, and then $h_{ice,i}$ by

$$\tau_i = \rho g h_{ice,i} \sin \alpha_i = \tau_b \tag{1.3}$$

Here τ_i (Bar) is the basal shear stress in the belt i (equal to τ_b), ρ is ice density (kg m^{-3}), and g is gravity (m s^{-2}). The calculated ice thickness ranged between 50 m in the high-altitude belts and 800 m in Concordia. Processing the low-frequency radar data gathered in Concordia in the summer of 2013, we estimated c.850 m ice thickness (Table 1.1), which was similar to the preceding estimate.

Ice velocity V_{ice} can be used in our methodology to calibrate/validate the ice flow model. Soncini et al. (2015a, b) gathered data of the position of their ablation stakes' during 2011–2013, using a DGPS (Differential GPS). They used these data to estimate the surface velocity of the glacier, which could be used to calibrate the ice flow model. They approximated the depth averaged ice velocity by taking 80% of the surface velocity (Cuffey and Paterson, 2010).

The ice flow module studied by Soncini et al. (2015a) was routed once a year (on November 1) by using the ice depth resulting from the seasonal mass budget (i.e., the rate between ice ablation and snow accumulation, in mm of water equivalent). Avalanche nourishment of the glaciers was accounted for in the model according to the slope of the terrain. When the ground slope is greater than a given threshold, more snow gradually detaches (linearly increasing within 30–60 degrees), and falls in the flattest altitude belt downstream, where it can melt or transform into ice. Once a year, 10% of snow surviving at the end of the ablation season is shifted into new ice (i.e., full ice formation requires 10 years).

The hydrological model used here is the Poly-Hydro model (Bocchiola et al., 2011; Soncini et al., 2015a, 2017). The model is based on the mass conservation equation and, for each time step (here daily), evaluates the variation of the content of the soil in the ground layer. Soil water content S in two consecutive time steps $(t, t+\Delta t)$ is

$$S^{t+\Delta t} = S^t + R + M_s + M_i - ET_{eff} - Q_{g'} \tag{1.4}$$

with R the liquid rain, M_s the snowmelt, M_i the glacial ablation, ET_{eff} the effective evapotranspiration, and Q_g the groundwater discharge. Snowmelt M_s and glacial ablation M_i are estimated according to a degree day method

$$
\begin{aligned}
M_s &= D_{Ds}(T - T_t), \\
M_i &= D_{Di}(T - T_t),
\end{aligned}
\tag{1.5}
$$

with T the daily mean temperature; D_{Ds} and D_{Di} the melt factors, evaluated as reported above; and T_t the threshold temperature, $T_t = 0°C$ (Bocchiola et al., 2010). Degree day plus melt factor is a simple and parsimonious method for assessment of ablation and floods in mountain catchments, and it is used here accordingly (Singh et al., 2000; Hock, 2003; Simaityte et al., 2008). Ice melt occurs on the glacier-covered areas in each belt (see Fig. 1.3) after snow depletion is complete. The superficial flow Q_s occurs only for saturated soil

$$
\begin{aligned}
Q_s &= S^{t+\Delta t} - S_{Max} \quad \text{if} \quad S^{t+\Delta t} > S_{Max} \\
Q_s &= 0 \qquad\qquad\quad \text{if} \quad S^{t+\Delta t} \le S_{Max}
\end{aligned}
\tag{1.6}
$$

with S_{Max} the greatest potential soil storage (mm). Potential evapotranspiration is calculated using the Hargreaves equation, requiring only temperature data and monthly mean temperature excursion

$$ETP = 0.0023 S_0 \sqrt{DT_m}(T + 17.8), \tag{1.7}$$

in $mm\,d^{-1}$, where S_0 ($mm\,d^{-1}$) is the evaporating power of solar radiation (depending on the Julian date and local coordinates), and DT_m (°C) is the thermometric monthly mean excursion. Once potential evapotranspiration is known, effective evapotranspiration ET_{eff} can be calculated. ET_{eff} comprises effective evaporation from the ground E_s and effective transpiration from the vegetation T_s, both functions of ETP depending on the state of soil moisture (see Groppelli et al., 2011b). Here groundwater discharge is expressed as a function of soil hydraulic conductivity and water content (Chen et al., 2005)

$$Q_g = K \left(\frac{S}{S_{Max}}\right)^k, \tag{1.8}$$

with K the saturated permeability and k the power exponent. Eqs. (3–8) are solved using 10 equally spaced elevation belts inside the basin. The flow discharges from the belts are

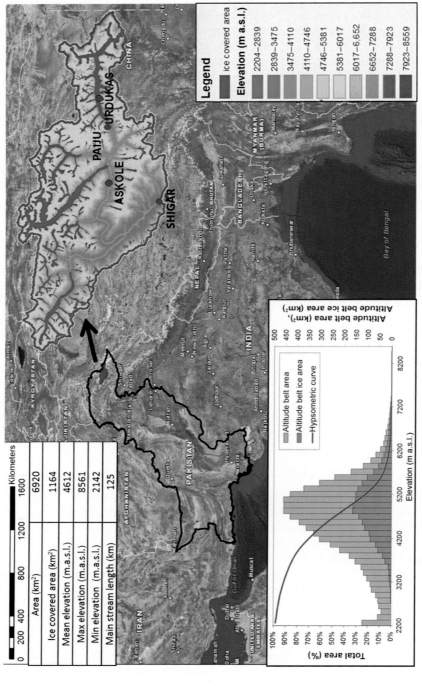

FIG. 1.3 Shigar River catchment in Gilgit Baltistan. In the lower-right subplot, the hypsometric curve, the belts area, and ice-covered area are shown.

routed to the outlet section through a semi-distributed flow routing algorithm. This algorithm is based on the conceptual model of the instantaneous unit hydrograph, IUH (Rosso, 1984). For calculation of the instream discharge, we hypothesize two (parallel) systems (groundwater, overland) of linear reservoirs (in series), each one with a given number of reservoirs (n_g and n_s). Each of the reservoirs possesses a time constant (i.e., k_g, k_s). We assume that for every belt the lag time increases proportionally to the altitude jump to the outlet section, until the greatest lag time (i.e., $T_{lag,g} = n_g k_g$ for the groundwater system and $T_{lag,s} = n_s k_s$ for the overland system). So doing, each belt possesses different lag times (and the farther the belts, the greater the lag times).

The hydrological model uses a daily series of precipitation and temperature from one representative station, here Astore, and the estimated vertical gradients to project those variables at each altitude belt. Topography is represented by a DTM model, with 30 m spatial resolution, derived from ASTER (Advanced Spaceborne Thermal Emission and Reflection Radiometer, 2006), which is used to define altitude belts and local weather variables against altitude. Glacial boundaries from Minora et al. (2013) were used.

Fig. 1.4 shows model adaptation to (monthly) average flows during 1995–1987, together with simulated flow components, i.e., rainfall and base flow, snowmelt, and ice melt. The figure clearly highlights the tremendous importance of the cryospheric flow components typical of the UIB area.

In Fig. 1.5 we show the daily flow simulation against observed discharges from a hydrometric station installed and maintained in fulfillment of the Paprika project of EVK2CNR.

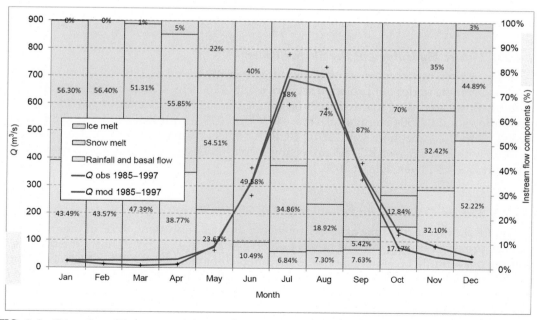

FIG. 1.4 Shigar River closed at Shigar. Monthly flows, observed and simulated during 1985–1997. Modeled flow components reported. Obviously, during summer, the ice melt may cover up to 87% of total flows.

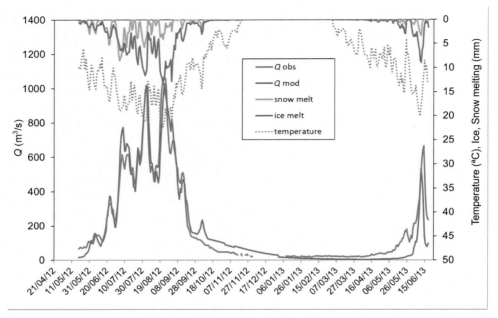

FIG. 1.5   Shigar River closed at Paiju. Daily flows, observed and simulated during May 2012 to May 2013, with measurements from a gauging station installed in fulfillment of the Paprika project. Modeled flow components reported.

1.4.2.4 Future Hydrology of the Shigar River

In Fig. 1.6 we report mean annual discharge variation (versus 1980–2012) in response to modified climate under our projected scenarios until 2100. Fig. 1.7 reports projected modified monthly flows according to three RCPs on average. Note that all scenarios provide increased streamflows until 2050 (averages of +8%, +17%, and +19% with RCP2.6, 4.5, and 8.5, respectively), with this exception: Until 2010, RCP2.6 depicts on average mainly decreased flows (−14%), with RCP4.5, and RCP8.5 projecting increased flows (+7% and +25%, respectively), but with RCP4.5 lower than at 2050.

1.4.2.5 Future Cryospheric Behavior in the Shigar River

Fig. 1.8 shows the dynamic of snow cover (snow water equivalent, SWE, at the end of summer) as projected during the 21st century, averaged per RCP. Fig. 1.9 shows the average ice volume per altitude belts in response to modified climate under the three RCPs on average until 2100. The greatest loss of cryospheric cover will happen during the second half of the century, and the warmer the scenario, the greater the loss. Comparing Figs. 1.6–1.9, one can see that with a high degree of ice melting and increased temperatures during the 21st century, streamflows will increase initially. However, when ice becomes less and less available at the altitudes with increased temperatures (and the warmer the temperature, the higher

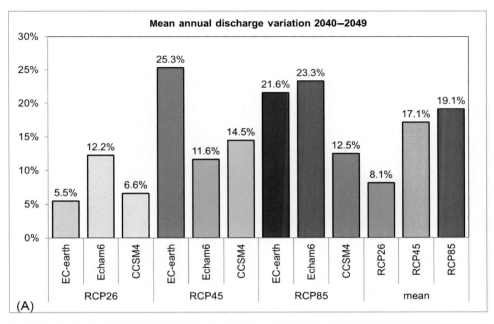

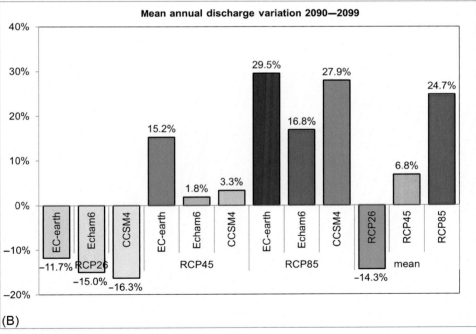

FIG. 1.6 Shigar River closed at Shigar. Mean annual discharge variation in response to modified climate under nine projected scenarios until 2100. (A) 2040–2049. (B) 2090–2099.

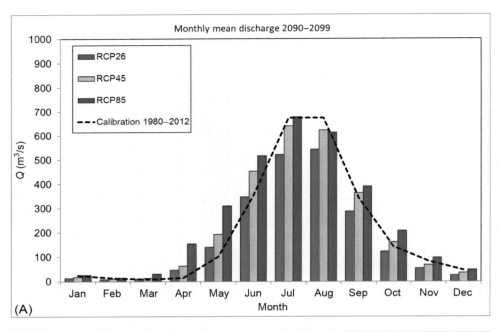

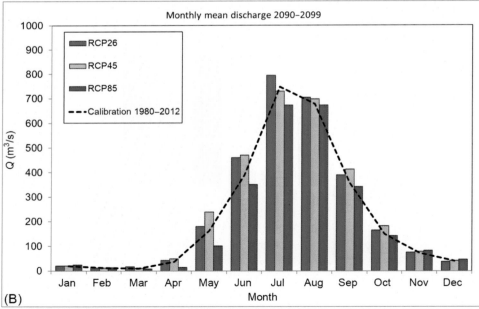

FIG. 1.7 Shigar River closed at Shigar. Mean monthly discharges in response to modified climate under the three RCPs on average until 2100. (A) 2040–2049. (B) 2090–2099.

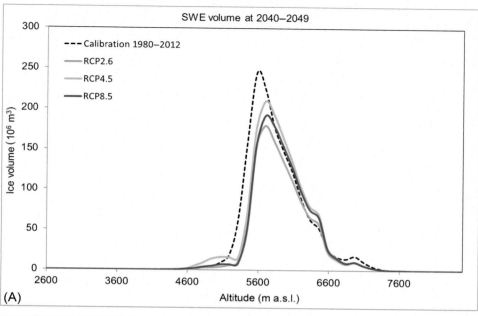

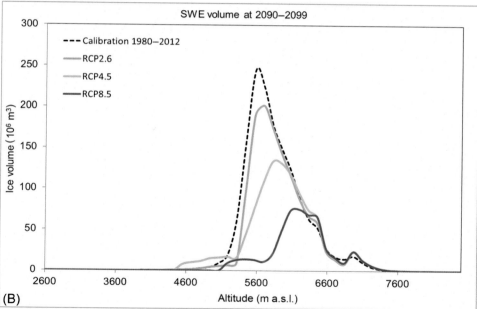

FIG. 1.8   Shigar River closed at Shigar. Average snow water equivalent (SWE) as per altitude belts in response to modified climate under the three RCPs on average until 2100. (A) 2040–2049. (B) 2090–2099.

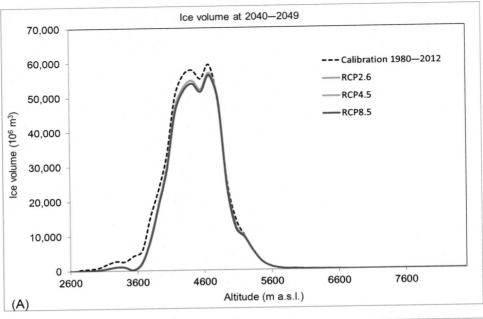

(A)

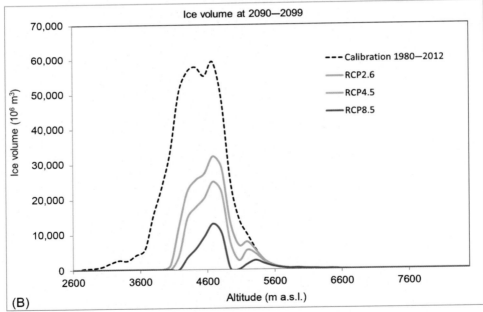

(B)

FIG. 1.9   Shigar River closed at Shigar. Average ice volume as per altitude belts in response to modified climate under the three RCPs on average until 2100. (A) 2040–2049. (B) 2090–2099.

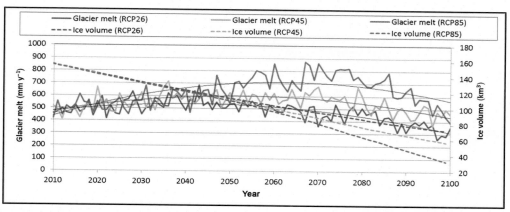

FIG. 1.10 Shigar River closed at Shigar. Total yearly ice volume in response to modified climate under the three RCPs on average until 2100.

the RCP), water resources will decrease rapidly. Under RCP8.5, at the end of century, water will remain largely available, at the cost of melting out c.90% of the available ice cover. Fig. 1.10 shows the total yearly ice volume in response to modified climate under the three RCPs on average until 2100. It is clear that after about 2080, ice melt will start decreasing even under RCP8.5 due to the large depletion of ice even at the highest altitudes (>5500 m a.s.l.; Fig. 1.9). In Fig. 1.11 we report the case of the iconic Baltoro Glacier in order to illustrate visually the expected effect of modified climate on cryospheric cover; the ice depth in response to modified climate under RCP8.5 of the EC-Earth model is shown at 2015 (present), at half century, and at end of century. As shown, even the deepest glacial area (i.e., Concordia), featuring c. 800 m of ice depth, will downwaste greatly, reaching no more than about 200 m at the end of century, significantly decreasing in the ice-covered area and disappearing at about 4000 m a.s.l.

1.4.2.6 Potential for Increased Floods Within the UIB: The Case of Shigar River

Based on projections of the future impact of climate change on the hydrology of the UIB, it is possible to analyze the sensitivity of extreme (i.e., maximum annual) floods to the modification of the climate. The authors are not aware of any similar study in the area, and some very preliminary results are reported here. In brief, an extreme floods estimation requires the assessment of the R-years flood quantile, i.e., the value of flood discharge (normally taken as the *instantaneous* peak flow within a given day) occurring on average every R years, Q_R. Such value is normally estimated via interpolation from observed (or simulated) peak flow data with a properly suited statistical distribution. A powerful tool for stream flood estimation is the GEV distribution, featuring R-years return period flood as

$$Q_R = \varepsilon + \frac{\alpha}{k}\left(1 - \exp\left(-ky_R\right)\right), \tag{1.9}$$

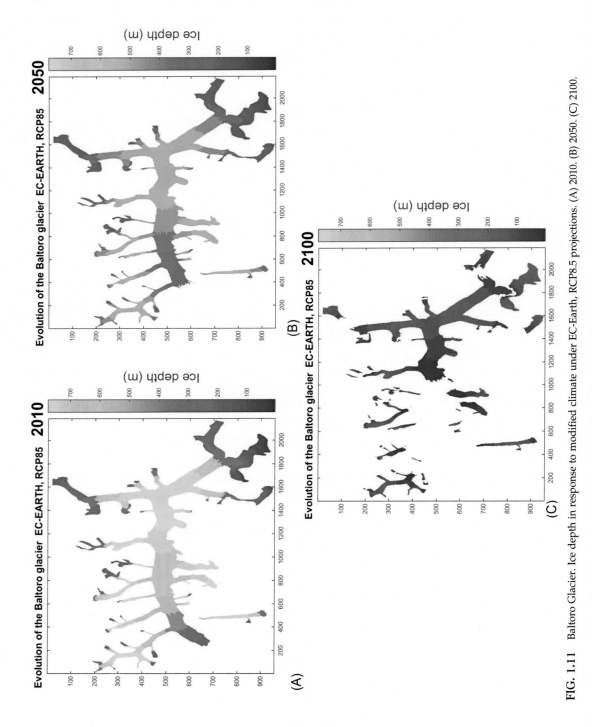

FIG. 1.11　Baltoro Glacier. Ice depth in response to modified climate under EC-Earth, RCP8.5 projections. (A) 2010. (B) 2050. (C) 2100.

with y_R Gumbel variable, $y_R = -\ln(-\ln((R-1)/R))$. The parameters of the distribution, ε location, α scale, and k shape are mostly estimated with the *L-moments* approach (e.g., see Kottegoda and Rosso, 1997).

Here we report an estimation of the present GEV parameters and R-years floods, together with corresponding estimates under climate change scenarios. Using the yearly maximum value of simulated daily discharges (1980–2012), we assessed extreme floods, according to the extreme values theory, in the Shigar River under the present climate and under prospective climate changes up to 2099.

In Fig. 1.12 we report the present assessment of R-years peak flow discharges (based on daily values) and provide samples from the nine scenarios during two future periods, namely 2024–2056 and 2068–2100.

From our results, it is clear that in the 21st century, there will be numerous chances for increased extreme flooding, caused by a combination of increased precipitation during the monsoon season under some scenarios and accelerated downwasting of glaciers due to increased temperatures under all scenarios. Nevertheless, these results are preliminary and further, more specific investigation is necessary to fully assess future floods in the area. However, we provide here some ideas to evoke creative thinking related to future adaptation to floods in the UIB.

1.5 DISCUSSION

1.5.1 Hydrological Trends Within the UIB

The Shigar River analyzed here is largely paradigmatic of the hydrology of UIB behavior, under the recently highlighted Karakoram anomaly. Fig. 1.9 shows no projected feasible changes in snow, ice cover, or ice depth until 2050, which substantiates the hypothesis of stable behavior underpinning the Karakoram anomaly. However, over a longer time scale, that is, until 2100, our projections indicate a substantial decrease of snow and ice cover, leading to a large loss of snow/ice mass in a period of about 50 years, with a consequential decrease of in-streamflows.

Immerzeel et al. (2010), among others, investigated the effects of snow and ice melt contributions to the discharge of the Indus River and concluded that streamflows can be predicted using a snowmelt runoff model (SRM) driven by remotely sensed precipitation and snow cover data calibrated based on daily discharge data. They investigated the future (2071–2100) effects of retreating glaciers on the discharge of the Upper Indus River. They found a clear warming trend in all seasons in the Upper Indus Basin, increasing with elevation, and a reduction in snow fall; whereas total precipitation was found to increase by c.20%. They also found that snowmelt peaks shifted up 1 month, that glacial flow increased due to temperature rise, and that rainfall runoff increased significantly. Immerzeel et al. (2013) used a glacio-hydrological model and also described the ice flow dynamics of the Baltoro Glacier, providing projections up to 2100, under RCP45 and RCP85 climate scenarios. They found that discharge may increase throughout the century, with glacier melting slightly decreasing and precipitation runoff increasing. They estimated an ice volume loss of nearly −55% by 2100. However, they did not use field data from the Baltoro Glacier; furthermore, they made a

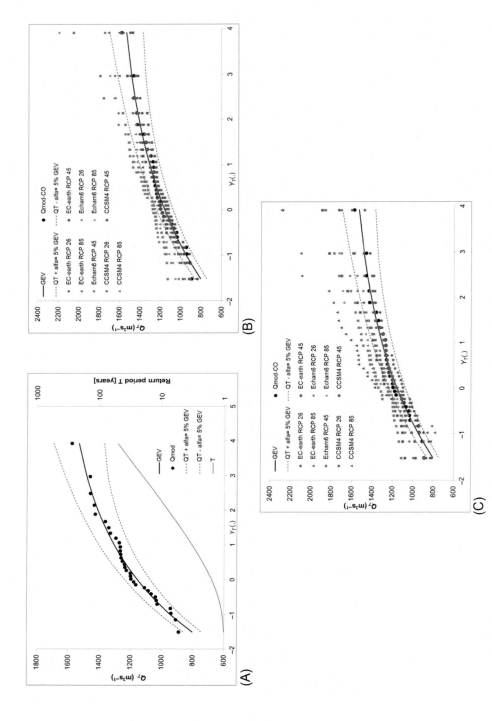

FIG. 1.12 Shigar River closed at Shigar. (A) GEV estimation based on daily simulated values (1980–2012). Confidence limits ($\alpha=95\%$) reported (see, e.g., Bocchiola and Rosso, 2009). (B) Sample values (plotting position) of Q_R values from nine future flow scenarios (2024–2056). (C) Sample values (plotting position) of Q_R values from nine future flow scenarios (2068–2100).

number of overly simplified assumptions, for example, about ice ablation and ice flow, among other factors. In contrast with our present results, their results may only be indicative in nature; therefore, their results must be taken only as qualitative calculations.

According to our projections, at mid-century the monthly hydrological cycle of the UIB (see Fig. 1.6) may exhibit different conditions, also depending on which GCM model is adopted to feed the hydrological model (not shown here; see, e.g., Soncini et al., 2015a, b). Under RCP2.6, discharge at 2100 will be lower than in the control period due to shrinking of ice at low altitudes combined with a persistent increase in temperature. The discharge variability is also significantly influenced by the precipitation regime, and for this reason, the results are dependent in great part on the projected GCM precipitation. It may be useful therefore to provide ensemble simulations to investigate the future discharge variability range under different precipitation regimes.

Analysis of snow cover dynamics also provides interesting signals (see Figs. 1.9 and 1.10). For example, they show large decreases in SWE at thaw in the second half of the century, as well as the effect of decreased albedo on ice melting.

In the Shigar River, 80% of the ice mass dwells between 3550 m a.s.l. and 5600 m a.s.l., and this is the elevation range that will be most affected by a rise of the snowline and by ice melting at the end of century, and therefore of special interest. Presently, the equilibrium line altitude for the area is placed at approximately 5200 m a.s.l. (Mayer et al., 2006), whereas in the future, it may shift, especially under RCP45 and RCP85 scenarios, reaching 5500 m a.s.l. and 5800 m a.s.l., respectively (Soncini et al., 2015a, b). In the RCP26 scenario (also called peak and decline scenario), the model shows a significant ice volume loss at the end of the century (−60%). This could be due to the glaciers' long lag time in response to a variation in the nourishment dynamics. In fact, a temperature increase causes a decrease in the solid precipitation fraction, in particular in the area between 5000 m a.s.l. and 6000 m a.s.l., where most of the glacierized area lies. The model projections are affected by uncertainties in reproducing the precipitation vertical gradient, as reported. It is clear that an underestimation of high-altitude precipitation (in particular between 5000 m a.s.l. and 6000 m a.s.l.) may lead to an underestimation of the long-term glacier mass balance (Table 1.5).

Ragettli et al. (2013), among others, assessed the uncertainties of hydrological projections of future (until 2050) runoff from the Hunza Basin, in particular uncertainties due to model parameters, to climate model, and to the natural interannual variability. They found that parametric uncertainty may have a greater effect than other sources of uncertainty (e.g., GCMs' uncertainty and natural interannual variability). They further highlighted that the lack of meteorological data at high altitudes requires extrapolation, thus introducing additional uncertainty. Accordingly, modeling of the most important components of the hydrological cycle in the target areas by calibration against observed values whenever possible is necessary as indicated in our template model (see Fig. 1.2).

1.5.2 The Need for Hydrological Monitoring in the UIB

Based on the results proposed here and to validate or question the expected trends from the hydrological studies, the first factor to seemingly emerge is the need for hydrological monitoring of the area going forward. It is necessary therefore that the hydrology of the high-altitude catchments in the area be monitored consistently, in order to

TABLE 1.5   Case Study of Shigar River

Parameter	Description	Value	Method
k_g, k_s (d)	Reservoir time constant, ground/overland	**30, 3**	Basin morphology
n_g, n_s	Reservoirs, ground/overland	3/3	Literature
K (mm d^{-1})	Saturated conductivity	2	Calibration
k (.)	Groundwater flow exponent	1	Calibration
f_v (%)	Vegetation cover, average value	30.9	Soil cover
θ_w, θ_l (.)	Water content, wilting/field capacity	0.15, 0.35	Literature
S_{Max} (mm)	Maximum soil storage, average	86.2	Soil cover
DD_S (mm °C^{-1} d^{-1})	Degree day factor for snow, average	2.33	Remote sensing
DD_I (mm °C^{-1} d^{-1})	Degree day factor for ice, average	5.65	Ablation stakes
K_d (m^{-1} y^{-1})	Ice flow internal deformation coefficient	3.1E-10	Ice stakes/Literature
K_s (m^{-3} y^{-1})	Ice flow basal sliding coefficient	5.0E-06	Ice stakes/Literature

Variable	Description	Mean Value ($m^3 s^{-1}$)	Bias (%)	RMSE (%)	NSE (.)
$Q_{mean,C}$ (m^3 s^{-1})	Observed discharge, Shigar 1985–1997	203	–	–	–
$Q_{mean,M}$ (m^3 s^{-1})	Model discharge, Shigar 1985–1997	202	−0.79	3.19	0.99
Q_{2012C} (m^3 s^{-1})	Observed discharge, Shigar May–November 2012	295	–	–	–
Q_{2012M} (m^3 s^{-1})	Model discharge, Shigar May–November 2012	305	3.42	4.58	0.94
$Q_{2012C,P}$ (m^3 s^{-1})	Observed discharge, Paiju May–June 2012	29	–	–	–
$Q_{2012M,P}$ (m^3 s^{-1})	Model discharge, Paiju, May–June 2012	25	−13.4	1.39	0.76

Hydrological model parameters. In bold values calibrated against observed discharges. Also, flow statistics, and indicators for validation are reported. Bias is percentage mean error. RMSE is percentage mean square error. NSE is Nash Sutcliffe Efficiency.
Based on Soncini, A., Bocchiola, D., Confortola, G., Bianchi, A., Rosso, R., Mayer, C., Lambrecht, A., Palazzi E., Smiraglia C., Diolaiuti G., 2015a. Future hydrological regimes in the upper Indus basin: a case study from a high altitude glacierized catchment, J. Hydrometeorol. 16(1), 306–326.

(a) assess present water resources for planning of hydropower exploitation, civil use, and agricultural purposes;
(b) build accurate modeling tools that can also be used for ungauged catchments/areas;
(c) build hydrological projections based on the most accurate hydrological information; and.
(d) verify in time the accuracy of the projected scenarios, continuously update hydrological models, and direct management efforts going forward.

There is of course a hydrological network already working and displaying a somewhat long historical data series. However, a denser network is required to monitor distributed water resources and to properly understand regional and altitudinal patterns of water resources.

Hydrological networks for the study area need to be designed accurately in order to fulfill monitoring of the largest possible share of (and ideally all) the water resources, and subsequently to build and maintain them for the acquisition of long-term data. As an example of high-altitude hydrological monitoring, the related issues, and a potential and necessary network to be installed, we next discuss a recent experience with the design of a hydrometric network for the Central Karakoram National Park CKNP (Mari et al., 2012).

1.5.3 A Proposed Hydrometric Network for the CKNP

The CKNP, an extensive ($\sim$12,000 km^2) protected area in the Karakoram mountain range in northern Pakistan, was established in 2009 and represents the main glaciated region of the Karakoram (Minora et al., 2016). Roughly 40% of this area is covered by ice. The SEED project of EVK2CNR (Mari et al., 2012) focused on producing an inventory of glaciers in the CKNP and on water management strategies for the park. Under the umbrella of the SEED project, suggestions for assessment of water resources in the CKNP were given, including the preliminary design of a hydrometric network, which we discuss here (Table 1.6).

TABLE 1.6 A Proposed Hydrometric Network in the CKNP (Mari et al., 2012)

S. No.	Name of River	Village	Length (km)	Area (km^2)	Average Discharge (m^3 s^{-1})
1	Hisper + Hoper	Sumaiyar	98	1778	Unknown
2	–	Shimshal	76	1101	Unknown
3	Hushey	Kande	73	1040	Unknown
4	–	Shingshal	59	690	Unknown
5	Talley	Doghani	44	394	Unknown
6	Bagarot	Kothi	46	431	Unknown
7	–	Hurban	42	361	Unknown
8	Astak	Astak	36	271	Unknown
9	Tormik	Dasu	32	221	Unknown
10	Baltoro	Paiju	84	1331	$\approx$30 (measur. + model)
11	Basha	Arandu	74	1049	Unknown
12	Hisper	Hisper	71	962	Unknown
13	Biafo	Biafo	66	845	Unknown
14	Shigar	–	201	6923	200 (measur. + model)

Relevant traits for deployment of a hydro network can be determined, and conclusions can be drawn. Some primary issues must be verified in order for a hydrographical station to be installed in a network, especially in this very high-altitude region. Such issues can be schematized as follows, depending on the type of device installed (i.e., sonic gauge or pressure transducer).

(1) Instrumental functioning (intrinsic to the device due to cold climate and/or for suspension load, bed load). This is especially the case for pressure gauges, but it also applies to sonic gauges, if, for example, cables and other pieces are near floodplains. The stations need to be positioned so that repairs are possible and the chances of malfunctioning due to environmental conditions and hydraulic stressing (e.g., from high turbulence and solid load) are decreased. Moreover, the stations will require continuous monitoring, likely by CKNP staff, with regard to possible malfunctioning, data downloading, and so forth.

(2) Easy accessibility to the sites of interest, including for transporting the materials necessary for installation. Perhaps the sites can intersect with roads near the river bed and, in particular, with narrow sections and bridges. Even if the sites can be accessed only by trails, the sites can be situated so that they are as accessible as possible via the trails. The necessary tools be carried by porters during an expedition.

(3) Possible positioning in a thalweg (i.e., at channel bottom, for pressure gauges). The pressure devices should be in the thalweg line, or lowest bed part, to avoid a null reading of the sensor in the presence of water. Positioning of the sensor in a thalweg will, however, expose the sensor to current and solid load. So the device should be protected by use of hoses, either plastic or metal, and shielded from the intrusion of sand and gravel.

(4) (4) Installation complicated by hanging, i.e., from bridges (sonic gauge). Mounting a sonic gauge may be complicated due to the need of a hanging frame. Also, the presence of strong winds may hamper measurement because of vibrations or movement of the device. These factors need be taken into account when planning the installation.

(5) Possible flow measurements with wading rods or tracer solutions for bigger flow sections. Wading techniques are suitable for measurement of streamflows in small mountain torrents and even more so with acceptable flow depth and velocity. The calibration of the stage-discharge equation is easier under these conditions. Use of tracers is also suitable, but some knowledge of flow depth in the section is still necessary. For higher flows, as we found in the Shigar River, a cross-section depth profile can be taken (as we did in our study), but flow velocity may not be established easily. Here, we could use a float and stopwatch. In such cases, more trials must be conducted, and surface velocity interpreted using hydraulic laws for flow velocity based on logarithmic profiles.

(6) Relative stability of morphologic conditions of the river bed, evolving over time, and modifying the stage-discharge relationship. Flow measurements and section surveys need to be made at least yearly at the onset of the thaw season to account for variations during seasonal high flows. Ideally, for the most accessible sections, surveys should be done after each noticeable flood event. Also, because immersed devices (pressure gauges) may be damaged during high flows, maintenance is necessary in such situations.

(7) Choice of a design discharge (i.e., for hydrostation dimensioning) for a given frequency of occurrence. Ideally, a design flow discharge will be estimated whenever it is necessary to install a hydrostation in order to (1) know the greatest flow conveyed in the section when it is naturally defined and (2) define the width of the section when artificial confinement is necessary (e.g., by side walls). This can be done through critical flow design, pending the choice of a reference return period. Also, optimal design of the network may be dealt with further by considering technical requirements, budget constraints, and so on.

Concerning network building, i.e., (optimal) positioning of hydrometric stations, no unique solution is available. As a rule of thumb, there are three targets:

(i) Install a number of stations that allow monitoring all (or most) of the out-flowing streams from the target area (in our current case, the CKNP).
(ii) Monitor catchments of increasing size (i.e., basin order according, e.g., to the Horton-Strahler approach).
(iii) Monitor especially interesting sections (e.g., the glaciers in our current study). However, the final choice is a subjective one.

In Table 1.6 and Fig. 1.13, we offer a preliminary sketch for deployment of the network. The CKNP is not a hydrologically closed area (i.e., the catchments there do not join within the park). Because we imagine that all the hydro stations should dwell within the park's

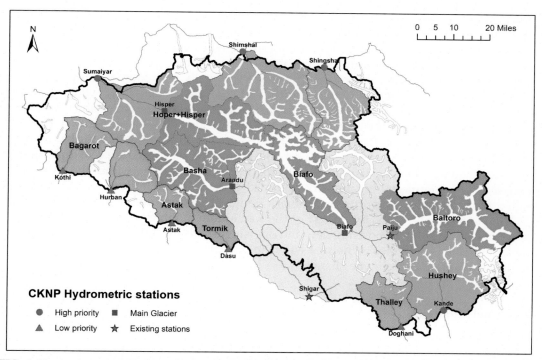

FIG. 1.13 Location of hydrostations proposed within the CKNP. *Based on Mari, F., Gallo, M., Bocci, A., Buraschi, E., Vuillermoz, E., Milanesi, D., Decè, L., Melis, M.T., Ferrari, E., Anfodillo, T., Poretti, G., Calligaris, C., Smiraglia, C., Diolaiuti, G. Bocchiola, D., Lami, A., Cristofanelli, P., Palazzi, E., Rossi, L., 2012. Integrated Park Management Plan (IPMP) for the Central Karakorum National Park CKNP, Edited by EVK2CNR.*

boundary, or at least reasonably close, we propose to install a number of stations, which allows for measurement of water from the catchments out-flowing from the park at their outlet sections. This enables assessment of water resources coming from the entire park. Also, we suggest monitoring with a special emphasis on smaller catchments at the outlets of selected significant glaciers. Because most of the water resources in the CKNP come from glaciers, and ice bodies are very sensitive to climate warming, long-term monitoring of such catchments is of the utmost importance. Shigar station, near the southern border of the park, and Paiju station, at the front of the Baltoro Glacier, are also included in Table 1.6 and Fig. 1.13. In addition, we indicate a highest and lowest priority, based on catchment size and glacial presence, that is, on the expected amount of delivered water (the more water there is, the higher the priority).

1.6 CONCLUSIONS

In this chapter, we explored the complexity of hydrological processes in the Upper Indus Basin (UIB), with a particular focus on hydrological modeling of high-altitude areas, which requires deep knowledge of cryospheric processes. We reviewed some recent findings related to ongoing hydrological variations in response to climate change, and we subsequently proposed a methodology for data-based investigation, modeling, and projection of the hydrology of such high-altitude areas. We demonstrated the application of such a method within a case study catchment, paradigmatic of Karakoram rivers, where much activity was carried out recently. We demonstrated how measurable variations of the cryospheric processes and hydrological fluxes are expected during the 21st century, initially with more water resources and then with less toward the end of the century. The "Karakoram anomaly," which appears to already exist, will persist throughout the century. However, with the increasing temperatures that are expected during the century, the Karakoram anomaly will eventually cease.

We also provided a brief assessment of future floods, while noting that further investigation is necessary, as well as the need to set up a flood forecast procedure. We can argue that such expected hydrological variability calls for accurate monitoring within the UIB, and we suggest that a river measuring network needs to be designed and built, especially to monitor high-altitude areas. We report a preliminary design for a hydrometric network within the CKNP, paradigmatic of the hydrology of the Karakoram mountain range and the UIB.

We believe that the proposed review and our suggestions for future experimental measuring and modeling activities in the area will help pave the way for future research, and eventually for adaptation to climate change.

Acknowledgments

The results proposed in this chapter are largely derived from activity in fulfillment of the SHARE-Paprika and SEED projects, funded by the EVK2CNR Association of Italy, which aimed, among other things, to evaluate the impact of climate change on the hydrology of the Upper Indus River. We hereby acknowledge the EVK2CNR Association, particularly Dr. Elisa Vuillermoz, and the Pakistani Meteorological Department (PMD) for providing weather data from their stations in northern Pakistan. Prof. Renzo Rosso; Eng. Ester Nana; Eng. Gabriele Confortola; Prof. Alberto Bianchi; Dr. Chiara Compostella, PhD; Dr. Umberto Minora, PhD; Dr. Boris Mosconi, PhD; Mr. Luigi Bonetti;

Dr. Furrukh Bashir; Dr. Habibullah Brohi; Dr. Adnan Shafiq Rana; and Dr. Faizal Dukhi are kindly acknowledged for participating in the Paprika project. Dr. Antonella Senese, PhD; Prof. Guglielmina Diolaiuti; and Prof. Claudio Smiraglia are kindly acknowledged for cooperation in the framework of the reported projects. We further acknowledge the World Climate Research Programme's Working Group on Coupled Modeling, which is responsible for CMIP, and we thank the climate modeling groups (listed in Table 1.2 of this chapter) for producing and making available their model output. CMIP, the US Department of Energy's Program for Climate Model Diagnosis and Intercomparison, provided coordinating support and led the development of software infrastructure in partnership with the Global Organization for Earth System Science Portals.

References

Ageta, Y., Kadota, T., 1992. Predictions of changes of glacier mass balance in the Nepal Himalaya and Tibetan plateau: a case study of air temperature increase for three glaciers. Ann. Glaciol. 16, 89–94.

Aggarwal, P.K., Joshi, P.K., Ingram, J.S.I., Gupta, R.K., 2004. Adapting food systems of the Indo-Gangetic plains to global environmental change: key information needs to improve policy formulation. Environ. Sci. Pol. 7, 487–498.

Aili, T., Soncini, A., Bianchi, A., Diolaiuti, G., Bocchiola, D., 2018. A method to study hydrology of high altitude catchments: the case study of the Mallero river. Italian Alps. Theoret. Appl. Climatol. https://doi.org/10.1007/s00704-017-2366-4 1-22.

Aizen, V.B., Aizen, E.M., Nikitin, S.A., 2002. Glacier regime on the northern slope of the Himalaya (Xixibangma glaciers). Quat. Int. 97–98, 27–39.

Akhtar, M., Ahmad, N., Booij, M.J., 2008. The impact of climate change on the water resources of Hindukush–Karakoram–Himalaya region under different glacier coverage scenario. J. Hydrol. 355, 148–163.

Archer, D.R., 2003. Contrasting hydrological regimes in the upper Indus Basin. J. Hydrol. 274, 198–210.

Archer, D.R., Forsythe, N., Fowler, H.J., Shah, S.M., 2010. Sustainability of water resources management in the Indus Basin under changing climatic and socio economic conditions. Hydrol. Earth Syst. Sci. 14, 1669–1680.

Archer, D.R., Fowler, H.J., 2004. Spatial and temporal variations in precipitation in the upper Indus basin, global teleconnections and hydrological implications. Hydrol. Earth Syst. Sci. 8, 47–61.

ASTER, 2006. Adv. Spaceborn Thermal Emission and Reflection Radiometer. AST14DEM-Rel. http://asterweb.jpl.nasa.gov/content/03_data/01_Data_Products/release_DEM_relative.htm

Barrand, N.E., Murray, T., 2006. Multivariate controls on the incidence of glacier surging in the Karakoram Himalaya. Arct. Antarct. Alp. Res. 38 (4), 489–498.

Baumann, S., Winkler, S., 2010. Parameterization of glacier inventory data from Jotunheimen/Norway in comparison to the European Alps and the Southern Alps of New Zealand. Erdkunde 64 (2), 155–177.

Belò, M., Mayer, C., Lambrect, A., Smiraglia, C., Tamburini, A., 2008. The recent evolution of Liligo Glacier, Karakoram, Pakistan, and its present quiescent phase. Ann. Glaciol. 48, 171–176.

Bocchiola, D., Diolaiuti, G., 2013. Recent (1980-2009) evidence of climate change in the upper Karakoram, Pakistan. Theor. Appl. Climatol. 113 (3–4), 611–641.

Bocchiola, D., Diolaiuti, G., Soncini, A., Mihalcea, C., D'Agata, C., Mayer, C., Lambrecht, A., Rosso, R., Smiraglia, C., 2011. Prediction of future hydrological regimes in poorly gauged high altitude basins: the case study of the upper Indus, Pakistan. Hydrol. Earth Syst. Sci. 15, 2059–2075.

Bocchiola, D., De Michele, C., Pecora, S. and Rosso, R., 2004. Sul tempo di risposta dei bacini idrografici italiani [Response time of Italian catchments]. L'ACQUA 1, 45–55. (In Italian with abstract in English).

Bocchiola, D., Mihalcea, C., Diolaiuti, G., Mosconi, B., Smiraglia, C., Rosso, R., 2010. Flow prediction in high altitude ungauged catchments: a case study in the Italian Alps (Pantano Basin, Adamello group). Adv. Water Resour. 33 (10), 1224–1234.

Bocchiola, D., Rosso, R., 2009. Use of a derived distribution approach for extreme floods design: a case study in Italy. Adv. Water Resour. 32 (8), 1284–1296.

Bolch, T., Pieczonka, T., Benn, D.I., 2011. Multi-decadal mass loss of glaciers in the Everest area (Nepal Himalaya) derived from stereo imagery. Cryosphere 5, 349–358. Available on line at: http://www.the-cryosphere.net/5/349/2011/tc-5-349-2011.pdf.

Bookhagen, B., Burbank, D.W., 2006. Topography, relief, and TRMM-derived rainfall variations along the Himalaya. Geophys. Res. Lett. 33. L08405: https://dx.doi.org/10.1029/2006GL026037.

Bookhagen, B. and Burbank, D.W., 2010. Towards a complete Himalayan hydrologic budget: the spatiotemporal distribution of snow melt and rainfall and their impact on river discharge. J. Geophys. Res., https://doi.org/10.1029/2009jf001426.

Chalise, S.R., Kansakar, S.R., Rees, G., Croker, K., Zaidman, M., 2003. Management of water resources and low flow estimation for the Himalayan basins of Nepal. J. Hydrol. 282, 25–35.

Chen, J.M., Chen, X., Ju, W., Geng, X., 2005. Distributed hydrological model for mapping evapotranspiration using remote sensing inputs. J. Hydrol. 305, 15–39.

Confortola, G., Soncini, A., Bocchiola, D., 2013. Climate change will affect hydrological regimes in the Alps: a case study in Italy. J. Alp. Res. 101(3): https://dx.doi.org/10.4000/rga.2176.

Copland, L., Sylvestre, T., Bishop, M.P., Shroder, J.F., Seong, Y.B., Owen, L.A., Bush, A., Kamp, U., 2011. Expanded and recently increased glacier surging in the Karakoram. Arct. Antarct. Alp. Res. 43 (4), 503–516.

Cuffey, K.M., Paterson, W.S.B., 2010. The Physics of Glaciers, fourth ed. Academic Press. ISBN 978-0123694614p. 704.

Diolaiuti, G., Pecci, M., Smiraglia, C., 2003. Liligo glacier, Karakoram, Pakistan: a reconstruction of the recent history of a surge-type glacier. Ann. Glaciol. 36 (1), 168–172.

Fowler, H.J., Archer, D.R., 2005. Hydro-climatological variability in the UpperIndus Basin and implications for water resources, proceedings: symposium S6 held during the seventh IAHS scientific assembly at Foz do Iguaçu, Brazil, April 2005. IAHS Publ. 295, 131–138.

Gardelle, J., Berthier, E., Arnaud, Y., 2012. Slight mass gain of Karakoram glaciers in the early twenty-first century. Nat. Geosci. Lett. https://dx.doi.org/10.1038/NGEO1450.

Groppelli, B., Bocchiola, D., Rosso, R., 2011a. Spatial downscaling of precipitation from GCMs for climate change projections using random cascades: a case study in Italy. Water Resour. Res. 47W03519: https://dx.doi.org/10.1029/2010WR009437.

Groppelli, B., Soncini, A., Bocchiola, D., Rosso, R., 2011b. Evaluation of future hydrological cycle under climate change scenarios in a mesoscale alpine watershed of Italy. Nat. Haz. Earth Syst. Sci. 11, 1769–1785.

Hall, D.K., Riggs, G.A., Foster, J.L., Kumar, S.V., 2010. Development and evaluation of a cloud-gap-filled MODIS daily snow-cover product. Remote Sens. Environ. 114 (3), 496–503.

Hannah, D.M., Kansakar, S.L., Gerrard, A.J., Rees, G., 2005. Flow regimes of Himalayan rivers of Nepal: nature and spatial patterns. J. Hydrol. 308, 18–32.

Hasson, S., Böhner, J., Lucarini, V., 2017. Prevailing climatic trends and runoff response from Hindukush–Karakoram–Himalaya, upper Indus Basin. Earth Syst. Dynam. 8, 337–355.

Hewitt, K., 2005. The Karakoram anomaly? Glacier expansion and the "elevation effect", Karakoram Himalaya. Mt. Res. Dev. 25 (4), 332–340.

Hock, R., 2003. Temperature index melt modelling in mountain areas. J. Hydrol. 282, 104–115.

Immerzeel, W.W., Droogers, P., de Jong, S.M., Bierkens, M.F.P., 2009. Large-scale monitoring of snow cover and runoff simulation in Himalayan river basins using remote sensing. Remote Sens. Environ. 113, 40–49.

Immerzeel, W.W., Pellicciotti, F., Bierkens, M.F.P., 2013. Rising river flows throughout the twenty-first century in two Himalayan glacierized watersheds. Nat. Geosci. 6 (8), 1–4: https://dx.doi.org/10.1038/ngeo1896.

Immerzeel, W.W., Pellicciotti, F., Shrestha, A.B., 2012a. Glaciers as a proxy to quantify the spatial distribution of precipitation in the Hunza basin. Mt. Res. Dev. 32 (1), 30–38.

Immerzeel, W.W., van Beek, L.P.H., Bierkens, M.F.P., 2010. Climate change will affect the Asian water towers. Science 328, 1382–1385.

Immerzeel, W.W., Van Beek, L.P.H., Konz, M., Shrestha, A.B., Bierkens, M.F.P., 2012b. Hydrological response to climate change in a glacierized catchment in the Himalayas. Clim. Chang. 110 (3–4), 721–736.

IPCC, Intergovernmental Panel for Climate Change, 2007. Climate Change 2007: The Scientific Basis. Cambridge University Press, Cambridge.

IPCC, Intergovernmental Panel for Climate Change, 2013. Working Group I Contribution to the IPCC Fifth Assessment Report Climate Change 2013: The Physical Science Basis Summary for Policymakers.

Kääb, A., Berthier, E., Nuth, C., Gardelle, J., Arnaud, Y., 2012. Contrasting patterns of early twenty-first-century glacier mass change in the Himalayas. Nat. Lett. 488, 495–498. https://dx.doi.org/10.1038/nature1132.

Kahlown, M.A., Raoof, A., Zubair, M., Kemper, W.D., 2007. Water use efficiency and economic feasibility of growing rice and wheat with sprinkler irrigation in the Indus Basin of Pakistan. Agric. Water Manag. 8 (7), 292–298.

Kaser, G., Großhauser, M., Marzeion, B., 2010. Contribution potential of glaciers to water availibility in different climate regimes. Proc. Natl. Acad. Sci. U. S. A. 107, 20223–20227.

Kehrwald, N.M., Thompson, L.G., Tandong, Y., Mosley-Thompson, E., Schotterer, U., Alfimov, V., Beer, J., Eikenberg, J., Davis, M.E., 2008. Mass loss on Himalayan glacier endangers water resources. Geophys. Res. Lett. 35. L22503: https://dx.doi.org/10.1029/2008GL035556.

Konz, M., Uhlenbrook, S., Braun, L., Shrestha, A., Demuth, S., 2007. Implementation of a process-based catchment model in a poorly gauged, highly glacierized Himalayan headwater. Hydrol. Earth Syst. Sci. 11 (4), 1323–1339.

Kottegoda, N., Rosso, R., 1997. Statistics Probability and Reliability for Civil and Environmental Engineers. Mc Graw-Hill.

Mari, F., Gallo, M., Bocci, A., Buraschi, E., Vuillermoz, E., Milanesi, D., Decè, L., Melis, M.T., Ferrari, E., Anfodillo, T., Poretti, G., Calligaris, C., Smiraglia, C., Diolaiuti, G., Bocchiola, D., Lami, A., Cristofanelli, P., Palazzi, E., Rossi, L., 2012. Integrated Park Management Plan (IPMP) for the Central Karakorum National Park CKNP. Edited by EVK2CNR.

Mayer, C., Fowler, A.C., Lambrecht, A., Scharrer, K., 2011. A surge of north Gasherbrum glacier, Karakoram, China. J. Glaciol. 57 (205), 904–916.

Mayer, C., Lambrecht, A., Belo, M., Smiraglia, C., Diolaiuti, G., 2006. Glaciological characteristics of the ablation zone of Baltoro glacier, Karakoram. Ann. Glaciol. 43, 123–131.

Migliavacca, F., Confortola, G., Soncini, A., Diolaiuti, G.A., Smiraglia, C., Barcaza, G., Bocchiola, D., 2015. Hydrology and potential climate changes in the Rio Maipo (Chile). Geogr. Fis. Din. Quat. 38 (2), 155–168.

Mihalcea, C., Mayer, C., Diolaiuti, G., D'agata, C., Smiraglia, C., Lambrecht, A., Vuillermoz, E., Tartari, G., 2008. Spatial distribution of debris thickness and melting from remote-sensing and meteorological data, at debris-covered Baltoro glacier, Karakoram, Pakistan. Ann. Glaciol. 48, 49–57.

Mihalcea, C., Mayer, C., Diolaiuti, G., Lambrecht, A., Smiraglia, C., Tartari, G., 2006. Ice ablation and meteorological conditions on the debris covered area of Baltoro glacier (Karakoram, Pakistan). Ann. Glaciol. 43, 292–300.

Minora, U., Bocchiola, D., D'Agata, C., Maragno, D., Mayer, C., Lambrecht, A., Mosconi, B., Vuillermoz, E., Senese, A., Compostella, C., Smiraglia, C., Diolaiuti, G., 2013. 2001–2010 Glacier Changes in the Central Karakoram National Park: A Contribution to Evaluate the Magnitude and Rate of the Karakoram Anomaly. The Cryosphere Discussion, http://www.the-cryosphere-discuss.net/7/2891/2013/tcd-7-2891-2013.html.

Minora, U., Bocchiola, D., D'Agata, C., Maragno, D., Mayer, C., Lambrecht, A., Vuillermoz, E., Senese, A., Compostella, C., Smiraglia, C., Diolaiuti, G., 2016. Glacier area stability in the Central Karakoram National Park (Pakistan) in 2001-2010: The "Karakoram anomaly" in the spotlight. Prog. Phys. Geogr. 40, 629–660.

Minora, U., Senese, A., Bocchiola, D., Soncini, A., D'agata, C., Ambrosini, R., Mayer, C., Lambrecht, A., Vuillermoz, E., Smiraglia, C., Diolaiuti, G., 2015. A simple model to evaluate ice melt over the ablation area of glaciers in the Central Karakoram National Park, Pakistan. Ann. Glaciol. 56 (70), 202–216.

Moss, R.H., Edmonds,, J.A., Hibbard K.A., Manning, M.R., Rose, S.K., van Vuuren, D.P., Carter, T.R., Emori, S., Kainuma, M., Kram, T., Meehl, G.A., Mitchell, J.F.B., Nakicenovic, N., Riahi, K., Smith, S.J., Stouffer, R.J., Thomson, A.M., Weyant, J.P., Wilbanks, T.J., 2010. The next generation of scenarios for climate change research and assessment. Nature 463, 747–756.

Oerlemans, J., 2001. Glaciers and Climate Change. A. A. Balkema Publishers, Brookfield, VT, p. 148.

Palazzi, E., von Hardenberg, J., Provenzale, A., 2013. Precipitation in the Hindu-KushKarakoram Himalaya: observations and future scenarios. J. Geophys. Res. Atmos. 118, 85–100.

Peel, M.C., Finlayson, B.L., McMahon, T.A., 2007. Updated world map of the Köppen-Geiger climate classification. Hydrol. Earth Syst. Sci. 11, 1633–1644.

Quincey, D.J., Glasser, N.F., Braun, M., Bishop, M.P., Hewitt, K., Luckman, A., 2011. Karakoram glacier surge dynamics. Geophys. Res. Lett. 38: https://dx.doi.org/10.1029/2011GL049004.

Ragettli, S., Pellicciotti, F., Bordoy, R., Immerzeel, W.W., 2013. Sources of uncertainty in modeling the glaciohydrological response of a Karakoram watershed to climate change. Water Resour. Res. 49 (9), 6048–6066.

Ravazzani, G., Dalla Valle, F., Gaudard, L., Mendlik, T., Gobiet, A., Mancini, M., 2016. Assessing climate impacts on hydropower production: the case of the Toce river basin. Climate 4 (2), 16.

Rosso, R., 1984. Nash model relation to Horton order ratios. Wat. Resour. Res. 20 (7), 914–920.

Salerno, F., Buraschi, E., Bruccoleri, G., Tartari, G., Smiraglia, C., 2008. Glacier surface-area changes in Sagarmatha national park, Nepal, in the second half of the 20th century, by comparison of historical maps. J. Glaciol. 54 (187), 738–752.

Scherler, D., Bookhagen, B., Strecker, M.R., 2011. Spatially variable response of Himalayan glaciers to climate change affected by debris cover. Nat. Geosci. 4 (3), 156–159.

Seibert, J., Beven, K.J., 2009. Gauging the ungauged basin: how many discharge measurements are needed? Hydrol. Earth Syst. Sci. 13 (6), 883–892.

Simaityte, J., Bocchiola, D., Augutis, J., Rosso, R., 2008. Use of a snowmelt model for weekly flood forecast for a major reservoir in Lithuania. Ann. Glaciol. 49, 33–37.

Singh, P., Kumar, N., Arora, M., 2000. Degree-day factors for snow and ice for Dokriani glacier, Garhwal Himalayas. J. Hydrol. 235, 1–11.

Sivapalan, M., 13, c.-a., 2003. IAHS decade on predictions in ungauged basins (PUB), 2003-2012: shaping an exciting future for the hydrological sciences. Hydrol. Sci. J. 48 (6), 857–880.

Smiraglia, C., Mayer, C., Mihalcea, C., Diolaiuti, G., Belo, M., Vassena, G., 2007. Ongoing variations of Himalayan and Karakoram glaciers as witnesses of global changes: recent studies of selected glaciers. Dev. Earth Surf. Processes 10, 235–248.

Soncini, A., Bocchiola, D., Azzoni, R.S., Diolaiuti, G.A., 2017. Methodology for monitoring and modeling of high altitude Alpine catchments. Prog. Phys. Geogr. 41 (4), 393–420.

Soncini, A., Bocchiola, D., Confortola, G., Bianchi, A., Rosso, R., Mayer, C., Lambrecht, A., Palazzi, E., Smiraglia, C., Diolaiuti, G., 2015a. Future hydrological regimes in the upper Indus basin: a case study from a high altitude glacierized catchment. J. Hydrometeorol. 16 (1), 306–326.

Soncini, A., Bocchiola, D., Confortola, G., Nana, E., Bianchi, A., Rosso, R., Diolaiuti, G., Smiraglia, C., von Hardenberg, J., Palazzi, E., Provenzale, A., Vuillermoz, E., 2015b. Hydrology of the upper Indus basin under potential climate change scenarios. In: Lollino, G., Lollino, G., Manconi, A., Clague, J., Shan, W., Chiarle, M. (Eds.), Engineering Geology for Society and Territory. In: Climate Change Ands Engineering GeologyVol. 1. Proceedings: IAEG, Turin, pp. 43–49.

Tahir, A.A., Chevallier, P., Arnaud, Y., Ahmad, B., 2011a. Snow cover dynamics and hydrological regime of the Hunza River basin, Karakoram range, northern Pakistan. Hydrol. Earth Syst. Sci. 15, 2275–2290.

Tahir, A.A., Chevallier, P., Arnaud, Y., Neppel, L., 2011b. Modeling snowmelt-runoff under climate scenarios in the Hunza River basin. J. Hydrol. 409, 104–117.

Viganò, G., Confortola, G., Fornaroli, R., Canobbio, S., Mezzanotte, V., Bocchiola, D., 2015. Future climate change may affect habitat in Alpine streams: a case study in Italy. ASCE J. Hydrol. Eng. 21 (2), 2016. https://dx.doi.org/10.1061/(ASCE)HE.1943-5584.0001293.

Wallinga, J., van de Wal, R.S.W., 1998. Sensitivity of Rhonegletscher, Switzerland, to climate change: experiments with a one-dimensional flowline model. J. Glaciol. 44 (147), 383–393.

Winiger, M., Gumpert, M., Yamout, H., 2005. Karakoram–Hindukush–western Himalaya: assessing high-altitude water resources. Hydrol. Process. 19, 2329–2338.

Further Reading

Bocchiola, D., 2010. Regional estimation of snow water equivalent using kriging: a preliminary study within the Italian Alps. Geogr. Fis. Din. Quat. 33, 3–14. http://www.glaciologia.it/gfdq/?p=1742.

Bocchiola, D., De Michele, C., Rosso, R., 2003. Review of recent advances in index flood estimation. Hydrol. Earth Syst. Sci. 7 (3), 283–296.

Groppelli, B., Bocchiola, D., Rosso, R., 2010. Precipitation downscaling using random cascades: a case study in Italy. Adv. Geosci. 8, 1–6.

Ming, J., Cachier, H., Xiao, C., Qin, D., Kang, S., Hou, S., Xu, J., 2007. Black carbon record based on a shallow Himalayan ice core and its climatic implications. Atmos. Chem. Phys. Discuss. 7, 14413–14432.

Simpson, J.J., Stitt, J.R., Sienko, M., 1998. Improved estimates of the areal extent of snow cover from AVHRR data. J. Hydrol. 204, 1–23.

Young, G.J., Hewitt, K., 1990. Hydrology research in the upper Indus basin, Karakoram Himalaya, Pakistan. Hydrol. Mountain. Areas 190, 139–152.

Challenges in Forecasting Water Resources of the Indus River Basin: Lessons From the Analysis and Modeling of Atmospheric and Hydrological Processes

Michel d.S. Mesquita,†, Yvan J. Orsolini†,‡, Indrani Pal§,¶,
Vidyunmala Veldore‖, Lu Li*,†, Krishnan Raghavan#,
Ashwini M. Panandiker**, Vivekanand Honnungar††,
David Gochis‡‡, John F. Burkhart§§*

*NORCE Norwegian Research Centre, Bergen, Norway †Bjerknes Centre for Climate Research, Bergen, Norway ‡Norwegian Institute for Air Research, Kjeller, Norway §Columbia Water Center, Columbia University, New York, NY, United States ¶NOAA Cooperative Remote Sensing Science and Technology (CREST) Center, The City University of New York, New York, NY, United States ‖Det Norske Veritas—Germanischer Lloyd, Oslo, Norway #Indian Institute of Tropical Meteorology, Pune, India **The Energy and Resources Institute (TERI), Goa, India ††CH2M HILL, Bengaluru, India ‡‡National Center for Atmospheric Research, Boulder, CO, United States §§University of Oslo, Oslo, Norway

2.1 INTRODUCTION

From ancient civilizations to modern cities, the Indus River, one of the longest Asian rivers. Originating in the Hindu-Kush-Himalayan (HKH) region (Committee on Himalayan Glaciers et al., 2012), the river has been an essential asset for the livelihoods of millions of people in South Asia. The ancient city of Dholavira exemplifies the importance of proper management of human-hydrological systems in the region. The city's complex and advanced engineering system made it possible for this Indus Valley Civilization to thrive in the past (Pande and Ertsen, 2014), and today the livelihood of the people and the agricultural and ecosystems in this region remain highly dependent on snow and glacier meltwater and the Indian summer monsoon. However, the changing climate and more extreme weather events may pose future challenges to the Indus River Basin and to those who depend on its waters.

The Indus is a transboundary river that flows for about 2880 km through four countries, China, India, Afghanistan, and Pakistan, before joining the Arabian Sea. According to the Central Water Commission in India (CWC, www.cwc.nic.in), the total catchment area in India is 321,289 km^2, which is approximately 9.8% of the country's total geographical area. The major tributaries flowing in India are the Shyok, the Nubra, the Satluj, the Beas, the Ravi, the Chenab, and the Jhelum. Being a transboundary river, the Indus River's data are classified, which may limit the calibration and verification of hydrological models.

For improved prediction, better understanding of the monsoon variability and the hydrological cycle in the HKH region is greatly needed (Mesquita et al., 2016). Half of the water in the Indus River comes from glaciers, and there is only sparse data for model verification, which presents challenges for the development of a low-carbon future pathway, such as the use of hydropower resources, and for building infrastructures (Laghari, 2013; Nepal and Shrestha, 2015). Also, agriculture is one of the major sources of income in the region. This sector could suffer from changing seasonal rainfall patterns and snow and glacier melt timing, with a predicted high river flow before the growing season and dry streams during the summer (Laghari, 2013). These changes in water resources also affect the industry and energy sectors, as well as society in general. So a better understanding of how climate changes will impact water resources is essential for making better decisions in the water-dependent sectors, and numerical models are key tools that can be used to bridge the knowledge gap.

Global and regional climate models have advanced our knowledge for the near-term and future changes in the climate system, but there is still some difficulty in representing the Asian monsoon, the tropical Pacific conditions, and the monsoon-El Niño Southern Oscillation teleconnection; and there are challenges to capturing the diurnal to interannual timescales for reliable prediction analysis (Turner and Annamalai, 2012). A proper hydrological modeling and understanding of the impacts of extreme events is still difficult due to the uncertainty in how to represent processes over the Indus River Basin (Hasson et al., 2013; Priya et al., 2016). Accurate prediction is also difficult due to the complexity of the climate system in South Asia. For instance, the western disturbances in winter and the Indian summer monsoon may not be well captured in models, and understanding them is an ongoing focus in active research (Pal et al., 2014).

The eastern and western areas of the HKH differ in climate, especially in seasonality and types of precipitation brought by singular atmospheric mechanisms, such as the aforementioned mid-latitude westerlies called western disturbances, which dominate in winter in the western part, and the south Asian monsoon activities, which dominate in summer in the eastern part (Thayyen and Gergan, 2010). The monsoon declines in strength from east to west along the Himalayas, while the westerlies weaken from west to east. There is also evidence of glacier retreat in the eastern and central parts of the Himalayas, while glaciers in the western part of the Himalayas appear to be more stable, and even may be advancing (Committee on Himalayan Glaciers et al., 2012). Glacial variation depends on many regional-scale factors, including changes in atmospheric circulation patterns and precipitation. Since the HKH region is geographically vast, both climatologically and hydrologically, the dynamic complexity may be changing in different ways in the east and west. Understanding the variability of moisture transport mechanisms for the HKH region is underway (Pal et al., 2013, 2014).

Furthermore, climate change is expected to have an impact on hydrological systems in the HKH because of changes in precipitation, temperature, and evapotranspiration. The most important sources and driving forces of climate change are at the regional scale, though. Considering that the political and technical measures to adapt to climate change also take place at a regional level, regional-scale studies on climate change are urgently needed. To do so, downscaling of General Circulation Models to a regional scale coupled with hydrological models is needed; however, recent studies have shown that there is still great uncertainty in downscaled output (Li et al., 2017a; Singh et al., 2017).

Therefore in this chapter, we discuss lessons learned from the review of a range of publications regarding atmospheric models, hydrology, and uncertainty about the future of water resources in the Indus River region. A major part of our discussion focuses on the issues regarding uncertainty. We also discuss results from NORINDIA, an Indo-Norwegian project, which studied the future of water resources in northern India (for details, see Appendix A and Mesquita et al., 2016).

2.2 LESSONS FROM ATMOSPHERIC MODELS

2.2.1 Skillful Prediction

The annual precipitation over the HKH region is largely dominated by the Indian summer monsoon (ISM). Nevertheless, in the western Himalayas, an important region for the Indus River Basin hydrology, a significant (15%–30%) fraction of the annual precipitation occurs during winter and spring, associated with the eastward-propagating synoptic systems along the subtropical jet stream, known as *western disturbances* (Pal et al., 2014; Tiwari et al., 2016). Both the ISM and the occurrence of western disturbances display strong interannual variability, partly modulated by atmospheric teleconnections.

The ISM also exhibits strong subseasonal variability in the form of break and active phases of anomalously low or high precipitation, and their frequency of occurrence and amplitude strongly impact the total monsoonal rainfall over the Indian subcontinent. The variability and the predictability of the ISM on the subseasonal-to-seasonal (s2s) scale are thus influenced by boundary conditions such as SST, snow, or soil moisture and by internal dynamics, most notably the intraseasonal oscillations (ISOs). Therefore a skillful prediction depends on adequate initialization and boundary conditions, as well as a realistic representation of air-sea interactions and land-atmosphere feedbacks.

2.2.2 Prediction Skill and Model Biases

While multi-model assessments of latest generation dynamical prediction systems (Rajeevan et al., 2012) reveal a moderate progress, the s2s prediction of the ISM rainfall remains a challenge due to persistent model biases in the representation of the mean monsoon climate and its variability, inaccuracies in initialization and boundary conditions, and unreliable coupling processes. The actual prediction skill for total land rainfall (0.45 for the multi-model mean in the ENSEMBLE comparison in Rajeevan et al. (2012)) still lags behind the potential predictability (Wang et al., 2015). The ISM onset is better predicted than the total rainfall. In a recent study using the ECMWF seasonal forecast system (Senan et al., 2016), the onset skill over the period 1981–2010 was 0.77 for forecasts initiated in April, whereas using different onset indices and prediction systems (Alessandri et al., 2015), researchers reported skill scores of 0.52 and 0.65 over the period 1989–2005 for forecasts initiated in May. A complicating factor is that the onset can be influenced by the initial northward migration of ISOs, but the atmospheric initialization contributes to the skill by improving the ISO forecast in the early monsoon period (Alessandri et al., 2015).

Model biases remain that could deteriorate skill. In the NCEP/CFS model, basin-wide, low-level circulation biases with weak anticyclones over the Indian and Central Pacific Oceans and related cold SST biases over the Arabian Sea and the equatorial Pacific could contribute to the dry bias over Central India and to the weakness of the meridional tropospheric temperature gradient (TTG) (Shukla and Huang, 2016). In the UK Met Office operational model, deficient air-sea interactions and negative bias in low-frequency surface wind variance over the Indian Ocean were identified as main causes for the unusual monsoon rainfall and propagating ISOs (Jayakumar et al., 2016; Johnson et al., 2017).

2.2.3 Spring Snow Over Eurasia

Another factor long thought to influence the ISM onset is the spring snow over Eurasia or the Himalaya-Tibetan Plateau (HTP) region. The snow-atmosphere coupling includes radiative, thermodynamical, and hydrological feedbacks. A recent observational study suggested that, through the delayed hydrological feedback, the Eurasian snowpack affects rainfall in the early monsoon period and occurs through an influence on the upper-level westerly jet and the penetration of mid-latitude troughs into the monsoon region (Halder and Dirmeyer, 2017). In free runs with the NCEP/CFS model, a high bias in Eurasian snow in winter and spring has also been suggested to contribute to anomalous surface cooling, weakening of the upper-level monsoon circulation and vertical wind shear, and the weakness of northward propagating ISOs (Saha et al., 2013). In the ECMWF seasonal forecast system, the HTP springtime snowpack influences the TTG reversal that marks the ISM onset (Senan et al., 2016).

Composite differences based on a normalized HTP snow index reveal that in high-snow years, the monsoon onset is delayed by about a week and that persisting dry and warm surface conditions prevail over India in the early monsoon period. Half of this delay could be attributed to the initialization of snow over the HTP, highlighting the importance of improving the realism of land reanalyzes over that region (Senan et al., 2016). A remaining question is the origin of the spring snowpack anomalies and of the moisture transport to the HTP region and, more specifically, the role of western disturbances in the winter-to-spring snow accumulation over the western Himalayas (Tiwari et al., 2016). Recent studies indicate a moderate increasing trend in the amplitude variations of wintertime western disturbance activity over the western Himalayas during the past few decades, which is conducive for enhanced snow-fall over the Karakoram Himalayas (Cannon et al., 2015; Madhura et al., 2015).

2.2.4 The Effect of Resolution

The effect of increasing forecast model resolution has produced mitigated results. In the NCEP/CFS model, high resolution reduced the basic state bias in seasonal rainfall, but had no clear improvement on the monthly skill in predicting ISOs; and other factors were estimated more important, such as convective parametrization and the representation of diurnal variability (Sahai et al., 2015). On the other hand, the high-resolution monthly forecasting system at ECMWF had higher interannual skill in the early monsoon precipitation than its lower-resolution seasonal counterpart (Vitart and Molteni, 2009).

2.2.5 Multidecadal Timescales

Turning now to multidecadal timescales, long-term rainfall observations reveal significant declining trends in the seasonal summer monsoon rainfall post-1950s and increasing aridity over several areas of India (Bollasina et al., 2011; Chung and Ramanathan, 2006; Guhathakurta and Rajeevan, 2008; Krishnan et al., 2013, 2015; Niranjan Kumar et al., 2013; Rajendran et al., 2012; Roxy et al., 2015; Singh et al., 2014). The decreasing trend of the monsoon rainfall is corroborated by a significant weakening of the large-scale summer monsoon circulation (Abish et al., 2013; Fan et al., 2010; Krishnan et al., 2013; Rao et al., 2004; Sathiyamoorthy, 2005), a significant increase in the duration and frequency of "monsoon-breaks" (dry spells) over India since the 1970s (Ramesh Kumar et al., 2009; Turner and Hannachi, 2010), and a declining frequency of the Bay of Bengal monsoon depressions, a primary rain-producing synoptic-scale monsoon disturbance (Krishnamurti et al., 2013; Rajeevan et al., 2000). Concomitant with the declining monsoonal rains, the Indian region additionally experienced a substantial rise in the frequency of heavy precipitation events (intensity $>10\,\mathrm{cm\,day}^{-1}$) in the post-1950s (Goswami et al., 2006; Rajeevan et al., 2008). There is emerging consensus that rapid increases in anthropogenic aerosols over the Northern Hemisphere and the Asian continent during the past few decades have significantly altered atmosphere radiative fluxes, the energy balance, and the monsoon hydrological cycle, leading to less seasonal monsoon precipitation over South Asia (Bollasina et al., 2011; Chung and Ramanathan, 2006; Ganguly et al., 2012; Krishnan et al., 2015; Ramanathan et al., 2005; Salzmann et al., 2014; Sanap et al., 2015). Additionally, the effects of land use and land cover changes and rapid warming of the equatorial Indian Ocean, the SST appears to have also contributed to the recent decline in monsoon rainfall (Krishnan et al., 2015; Roxy et al., 2015; Swapna et al., 2013).

Finally, the failure of modern dynamical prediction systems might be due to the inability to capture the weakening of the ENSO-ISM connection (Rajeevan et al., 2012) and to emerging sources of predictability during the recent era (Wang et al., 2015) that are linked to the varying flavors of ENSO events and to the springtime deepening of the Asian low. The role that increased aerosols play in the predictability of monsoonal rains on s2s scales is an important and open scientific issue, and is relevant for understanding the behavior of the ISM in the coming decades.

2.3 LESSONS FROM UNCERTAINTY

The previous section discussed some of the major challenges in modeling atmospheric processes that are relevant for South Asia. It also presented some results from the NORINDIA Project (Krishnan et al., 2015; Senan et al., 2016). We now discuss a major challenge highlighted in many studies, that of uncertainty in future projections.

Projections from hydrological models for the Indus Basin still face a major challenge related to the quality of input data. There is much uncertainty in observations, reanalysis, and CMIP (Coupled Model Intercomparison Project) projections, as well as the downscaling of these factors. In fact, more studies are needed to disentangle the uncertainty in the data used as input to hydrological models. Also, the use of more independent models is needed, as model genealogy can potentially bias the assessments and violate statistical inferences (Benestad et al., 2017; Draper, 1995; Knutti et al., 2017). In this section, we discuss the complex picture of uncertainty and the consensus related to climate extremes.

2.3.1 The Complexity of Uncertainty

There is still considerable uncertainty in observations and in model output. The former is often due to the still sparse network and the lack of access to some data sets. So it is not uncommon for different observation data sets to disagree. For instance, Turner and Annamalai (2012) note that there is a decreasing trend in the summer monsoon rainfall (All India Rainfall, AIR) for the present climate since 1950 in spite of disagreements between three of the observation data sets used in their study. Also, model output should be taken with caution. Turner and Annamalai also analyzed CMIP3 output and found considerable disagreement and uncertainty, with some models showing trends while others showed no trends. The authors point to the need for further model improvements and understanding.

Also, in a study of five selected CMIP5 models, Sharmila et al. (2015) show that, although four models project enhanced precipitation of about 10%–25% over the Indian subcontinent for June, July, August, and September for the RCP8.5 scenario (2051–2099 relative to 1951–1999), the MPI-ESM-LR shows a considerable decrease in rainfall over the northwestern Indian subcontinent. Hasson (2016), through the analysis of 30 CMIP5 models, finds that the models show a modest skill in the monsoon timing and that one-third of the models do not capture the monsoon signal over the Indus Basin. In addition, Lutz et al. (2016) conclude that the future of the Upper Indus Basin, in terms of water availability, is highly uncertain due to too much uncertainty in the projections.

Furthermore, space-time trends are not spatially uniform, and there are disagreements between studies for the Indian subcontinent (Ghosh et al., 2011). This is in spite of physical reasons for more consistent results due to the dynamics in the region: land-sea temperature contrasts under increasing greenhouse gas forcing and the Indo-Pacific Warm Pool that provides increased moisture to the monsoon regions (Turner and Annamalai, 2012).

Ghosh et al. (2011), in a study of the present climate, show that contrasting trends between the mean (a decrease) and extremes (an increase) for the ISM create a challenge for stakeholders needing to make decisions concerning water management. Through use of the VIC model, their results highlight a deficit in water yield for the Indus River from 1976 to 2000 compared with 1951–1975 (13.70% change in median).

2.3.2 Uncertainty in Downscaled Output

When CMIP models show considerable uncertainty, downscaling them may not fully overcome the uncertainties at regional and sub-regional levels, which could in turn lead to more uncertain hydrological projections at river basins. For instance, the Coordinated Regional Climate Downscaling Experiment (CORDEX), an effort to downscale CMIP simulations, could potentially be used to run hydrological models for the Indus Basin. However, there is a significant amount of room for improving the representation of the south Asian monsoon precipitation in CORDEX RCMs (Singh et al., 2017) before they can be effectively utilized for deriving value-added hydrological products.

The previously mentioned work by Singh et al. (2017) also highlights that a downscaling procedure could potentially increase the uncertainty and reduce the quality of the original GCM output. Although this does not necessarily apply to all downscaling methods and models, it emphasizes the need for careful examination of uncertainty associated with a downscaled output. In light of this, the following three sections briefly discuss a case study that uses an alternative approach to assess this uncertainty.

2.3.2.1 Assessing Uncertainty: A Case Study

To assess the uncertainty in the downscaling of future model outputs, we conducted an independent experiment using: (1) the WRF model (see Appendix B), a model not considered in Singh et al. (2017), and (2) a Bayesian approach to quantify the uncertainty and to bias-correct the WRF model output (see Appendix C). Table 2.1 shows that there is a difference of about 5°C between the model output and the ERA-Interim, of similar magnitude to the biases found in the multi-model study conducted by Niu et al. (2015) and slightly higher than that in the study conducted by Su et al. (2016). Using this bias to correct the future projections helps us appreciate the higher magnitude of the projected temperatures after the bias correction. For instance, the annual temperature mean for the period 2079–2100 is 6.87°C for the RCP4.5 scenario. After the correction, the estimate is 11.53°C.

Fig. 2.1 shows the future change estimates for the surface temperature in the Indus, Beas, and Brahmaputra basins, in agreement with Table 2.1. All the basins show an increase in surface temperature, with the Beas and Brahmaputra basins having a significant increase in all of the decades (i.e., they are different than zero). The Indus Basin shows an increase for all of the decades; however, the periods from 2039 to 2060 and 2059 to 2080 still show zero as a plausible result.

Note that these results are based on only one CMIP5 model and its downscaling using the WRF model. No ensemble members were run. So the Bayesian analysis gives us information about the uncertainty associated with downscaling the NorESM model output with the WRF model. We present this analysis to illustrate the role of uncertainty; the size of the bars for the

TABLE 2.1   Yearly Mean Surface Temperature Projections for the Indus Basin in degrees Celsius

Scenario	Variable	Decade	Mean	Std Dev.	Low	High
RCP4.5	μ	2039–2060	5.42	0.94	3.56	7.29
		2059–2080	4.90	0.83	3.27	6.52
		2079–2100	6.87	0.90	5.10	8.63
	μ'	2039–2060	10.08	0.98	8.16	12.03
		2059–2080	9.55	0.87	7.85	11.26
		2079–2100	11.53	0.94	9.69	13.37
RCP8.5	μ	2039–2060	5.12	0.91	3.33	6.92
		2059–2080	5.30	0.84	3.65	6.95
		2079–2100	8.77	0.93	6.94	10.60
	μ'	2039–2060	9.78	0.95	7.92	11.66
		2059–2080	9.96	0.89	8.21	11.71
		2079–2100	13.42	0.97	11.52	15.34
ERAi	μ	1979–2000	8.18	0.72	6.77	9.59
Hist		1979–2000	3.52	0.93	1.69	5.35

Results based on output from the ERA-Interim data set and the dynamical downscaling of the Norwegian Earth System Model using the Weather Research and Forecasting (WRF) model (see Appendix B). The means μ and μ' are estimated from a Bayesian model (see Appendix C). μ' is a corrected mean, where the bias between the ERA-Interim and the historical run (Hist.) is accounted for. Low and High represent the 95% Bayesian credible interval of the mean. The Markov chain error is <0.01 in all cases.

Indus Basin span about 5°C. To reduce this uncertainty and improve the estimates, one would need more samples, especially from different model combinations. We did this in the NORINDIA Project by considering 16 CMIP5 models (Viste and Sorteberg, 2015).

2.3.2.2 *Investigating Seasonal Changes in the Case Study*

To demonstrate the possibility of zero (i.e., no difference) or negative values in the difference between future surface temperature and the present climate in Fig. 2.1, we show in Figs. 2.2 and 2.3 the seasonal changes for the three previously mentioned basins. The size of the bars indicates an uncertainty of about 2.5°C in each estimate. The figure also highlights that most of the results are different from zero, indicating a statistically significant future change in surface temperature in the pre-monsoon, post-monsoon, and winter seasons.

The monsoon season shows a significant negative change in temperature for the Indus Basin (RCP4.5 scenario shown in Fig. 2.1). However, Fig. 2.2 shows that the cooling is not significant at a later time in the future for the RCP8.5 scenario. This cooling in the monsoon season would explain the zero and negative results, which are shown in Fig. 2.1. The cooling could be caused by the larger range of the surface temperature distribution in the future scenarios (with a bimodal distribution) compared with the shorter range in the estimate based on the historical run (unimodal) considered here (not shown). It could also be a feature of the model, since some tend to underestimate the observed temperature in the region

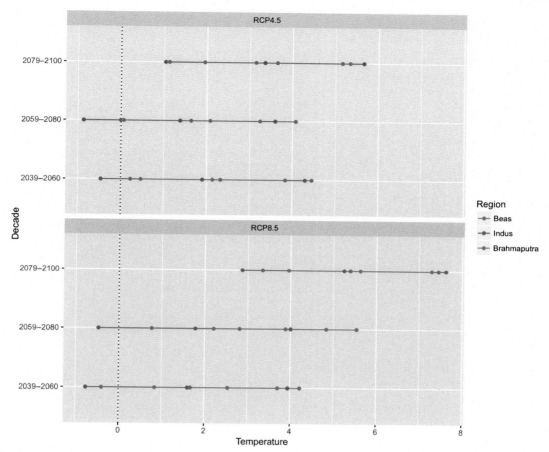

FIG. 2.1 Future change in annual surface temperature for the Beas, Indus, and Brahmaputra basins for two climate change scenarios (RCP4.5 and RCP8.5). Calculations are based on the output from the Bayesian model explained in Appendix C. Both the future and historical model output were bias-corrected. The middle dot represents the mean, and the distance between the outer dots represents the 95% Bayesian credible interval. Temperature changes are given in degrees Celsius.

(Ahmadalipour et al., 2015) or to underestimate the CMIP5 ensemble (Kapnick et al., 2014). The analysis of 16 CMIP5 models in NORINDIA shows a consistent increase in surface temperature for the Indus Basin across all months; the RCP8.5 shows an increase on the order of about 5.5°C for the period between 2071 and 2100 compared with 1971–2000 (see Viste and Sorteberg, 2015, and Fig. 8).

2.3.2.3 Is Cooling a Plausible Result?

It is interesting to note that Basha et al., (2017; Fig. 7) reports cooling over the Indian subcontinent, but only for the RCP2.6 scenario. Also, although surprising, in a study of

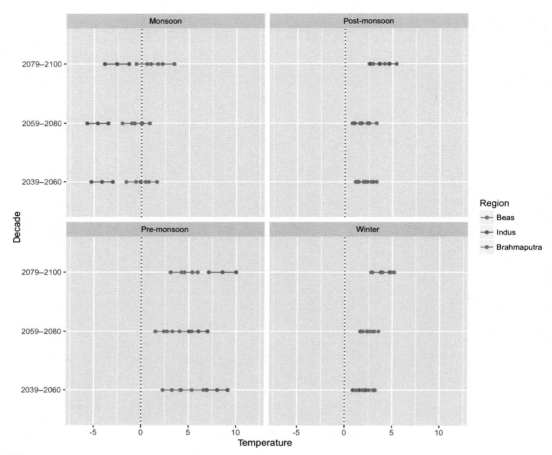

FIG. 2.2 Future change in seasonal surface temperature for the Beas, Indus, and Brahmaputra basins for the RCP4.5 scenario. Calculations are based on the output from the Bayesian model explained in Appendix C. Both the future and historical model output were bias-corrected. The middle dot represents the mean, and the distance between the outer dots represents the 95% Bayesian credible interval. Temperature changes are given in degrees Celsius.

the current climatic conditions in the Indus region, Fowler and Archer (2006) find a decrease in the summer mean temperature in the Upper Indus Basin (UIB) from 1961 to 2000. Negative trends are also described in other studies (see Fowler and Archer, 2006, and the references at the end of the chapter). The cooling could be attributed to increasing summer rainfall and increasing cloud cover in the region (Cook et al., 2003; Fowler and Archer, 2006). Zafar et al. (2015) also highlight that the Karakorum temperature is out of phase with the hemispheric temperature trends over the past five centuries.

This brief case study emphasizes the need for a thorough examination of uncertainty in downscaled output. Alternative methods can be useful for measuring the uncertainty from different perspectives. The next section presents the issue of heterogeneity in projections for rainfall, which are dependent on the lateral boundary condition or the regional (downscaling) model used.

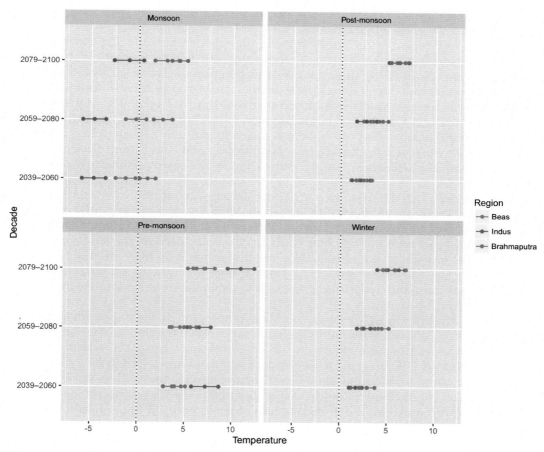

FIG. 2.3 Future change in seasonal surface temperature for the Beas, Indus, and Brahmaputra basins for the RCP8.5 scenario. Calculations are based on the output from the Bayesian model explained in Appendix C. Both the future and historical model output were bias-corrected. The middle dot represents the mean, and the distance between the outer dots represents the 95% Bayesian credible interval. Temperature changes are given in degrees Celsius.

2.3.3 Heterogeneity in the Estimates

The heterogeneity found in the projections of future changes in rainfall in the Himalayan region may be due to the choice of lateral boundary conditions or the regional model used (Choudhary and Dimri, 2017). Also, Hasson (2016) notes, "A summer cooling phenomenon is unique for the UIB with the present-day climate models being unable to represent it even at a qualitative scale." This phenomenon will affect the signal of the estimates of surface temperature changes. Due to this heterogeneity, it is important to consider altitude, regional differences, and seasons to fully understand the uncertainty in the region (Kapnick et al., 2014).

In addition, Nepal and Shrestha (2015, pp. 207–208) note, "The temperature trends reported in the Indus Basin are not homogeneous, with different studies showing different results …. Taken together, the studies suggest that there has been an overall gradual increase in temperature in the Indus Basin, but with some differences in the reported seasonal trends." So there is still divergence in the results.

According to Fowler and Archer (2006), who analyzed temperature data for seven instrumental records for seasonal and annual trends from 1961 to 2000 in the KHK mountains of the Upper Indus Basin (UIB), there are conflicting signals of climate change in the UIB region. In the Hunza and Shyok Rivers, for example, it was assessed that a decline of approximately 20% in summer runoff was caused by the observed drop of 1°C in mean summer temperature since 1961, with even larger drops in the spring months. The downward trend in temperature that was observed in summer months and the corresponding runoff were consistent with observed thickening and expansion of the Karakoram glaciers and were in contrast with the widespread decay and retreat in the eastern Himalayas (Fowler and Archer, 2006).

Immerzeel et al. (2009) identified a substantial negative trend for winter snow cover in the UIB. Their study points to an accelerated glacial melting due to regional warming, which is in turn disturbing the hydrology in the Upper Indus Basin. Another study carried out by Akhtar et al. (2008) used different glacier cover scenarios to study the impact of climate change on the water resources of the HKH. Their results from the HBV-Met model confirmed that the quality of observed data in HKH region is poor and needs more research to evaluate the uncertainties in downscaling approaches.

As part of the NORINDIA Project, Viste and Sorteberg (2015) studied the present-day and future projections of snowfall in the Himalayas. They emphasize the significant uncertainties in estimating snowfall from observations and reanalysis, which in fact can vary by factors of 2–4. In spite of the uncertainties, the model analyses of the RCP8.5 scenario also show a reduction of 30%–50% of the annual snowfall in the Indus Basin. This reduction is attributed to increasing temperatures and constant or increasing precipitation for most of the Indus region, based on a CMIP5 multi-model mean assessment.

Su et al. (2016) showed an increase in summer temperature in the Indus River Basin, in spite of the increase in precipitation. The study was based on the statistical downscaling of a multi-model CMIP5 ensemble (21 models). Under the RCP2.6, 4.5, and 8.5 scenarios, in the mid-21st century, the summer temperature is projected to increase by 1.24°C (0.72–1.87°C), 1.90°C (1.05–2.68°C), and 2.58°C (1.52–3.52°C), respectively. For the late-21st century, the increase is on the order of 1.14°C (0.72–1.63°C), 2.38°C (1.42–3.26°C), and 4.82°C (3.05–6.25°C) for the previously mentioned scenarios, respectively. Note, however, that the downscaled model output overestimates the observations on the order of about 1–2°C. Also, some of the models used in their study share the same characteristics/core (e.g., three versions of the GFDL model, two versions of the GISS model, three versions of the MIROC model, and two versions of the MRI model). Other studies have also made use of models that share a similar core (Viste and Sorteberg, 2015). However, care must be taken when analyzing the output and making inferences, since the multi-model sample is not truly independent in this case, which could potentially affect the estimates, violate the statistical assumptions, and create a biased multi-model mean (Benestad et al., 2017; Draper, 1995; Knutti et al., 2017).

2.3.4 Projections of Climate Extremes

In spite of uncertainty in estimating mean values, a clearer picture seems to emerge for extremes. For example, Sharmila et al. (2015) show the possibility of enhancement in heavy rainfall events (above 40 mm/day) over the Indian summer monsoon domain and a reduction of low rain-rate events (below 10 mm/day), and a decrease in the number of wet days.

Also, future active/break cycles could be more intense and more spatially extended, with the possibility of both short and long active/break spells. Sharmila and colleagues also project more severe drought and flood conditions over the Indian subcontinent in a future climate.

Note that Shashikanth et al. (2017) find no change to a slight increase in the spatial mean of extremes with dominance of spatial heterogeneity for the period 2081–2100. The results are based on a downscaling procedure of five GCMs. The authors argue that the increase in extremes in the future may not be visible in the projections because of a failure of their model in a non-stationary climate.

On the other hand, under the RCP4.5 scenario, Mohan and Rajeevan (2017) suggest a statistically significant increase in the hydroclimatic intensity index over the Indian region from 2010 to 2100. Seven models out of 10 suggest a consistent increase in this index. These changes are due mostly to an increase in precipitation and not to the dry spell length (both variables in the index), which showed little changes in the future climate.

Also, the work by Roxy et al. (2017) is noteworthy; it shows that for the climate from 1950 to 2015 (considered here as "current climate") there was a threefold increase in widespread extreme rain events over Central India. The authors warn that CMIP5 projections show further warming in the Arabian Sea. This could potentially increase the number of extreme events in the future, which would put lives and agriculture at great risk.

2.4 LESSONS FROM HYDROLOGICAL PROJECTIONS

So far we have discussed challenges in modeling atmospheric processes relevant for South Asia, and we have provided a general view of the complex uncertainty in estimating future changes in climate in the Indus region. The following section deals with some of the lessons taken from hydrological modeling.

2.4.1 Cascade of Uncertainty

There is a cascade of uncertainty in the use of hydrological models: uncertainty in the global model output estimates, uncertainty in downscaling of these outputs, and uncertainty in using these outputs in hydrological models. Even with bias correction, hydrological models may still fail to produce accurate forecasts due to uncertainty in the input data. This cascade of uncertainty could hinder accurate prediction (Benestad et al., 2017; Lutz et al., 2016).

An alternative approach to using downscaled model output with hydrological models is the coupled atmospheric-hydrological model system, i.e., the WRF-Hydro approach (Gochis et al., 2015). The WRF-Hydro model combines WRF with components of a terrestrial hydrological model, so that each can feed back into the other. This model was tested in the NORINDIA Project, and it has shown reasonable skill in capturing the spatial and temporal structure of high-resolution precipitation in North India (for details, see Li et al., 2017a). Also, the resulting streamflow hydrographs exhibit a good correspondence with observation at monthly timescales, although the model tends to generally underestimate streamflow amounts (Li et al., 2017a).

2.4.2 Overview of a Few Hydrological Studies

Priya et al. (2016) find an increase in the frequency and intensity of extreme precipitation events during the summer monsoon season over the western Himalayas and the Upper Indus Basin since 1950, likely due to an enhanced moisture supply produced by southerly winds blowing from the Arabian Sea into the Indus Basin. Also, Priya (2017, Chapter 7), in a comprehensive study of future hydrological changes in the Indus River Basin, used the LMDZ model at 35 km resolution over South Asia. Then for the hydrological assessment of the RCP4.5 scenario, the VIC model was used. She shows a possible reduction in seasonal mean precipitation for the first half of the 21st century and a slight positive trend toward to the second half of the century, but she recognizes the difficulty in estimating the discharge on local scales due to uncertainty. Her study points to the need for an ensemble of hydrological runs.

Lutz et al. (2016) considered the RCP4.5 and 8.5 scenarios (downscaled statistically) to force the SPHY cryospheric-hydrological model for the projections. It is estimated that changes in water availability in the Upper Indus Basin will range from −15% to 60% by the end of the 21st century (with respect to 1971–2000). They find an increase in intensity and frequency of extreme discharge for the Upper Indus Basin. They also point out that the Upper Indus Basin is likely to warm more than other parts of the world (+2.1°C to +8.0°C between 1971 and 2000 and 2071 and 2100, respectively, in contrast with the global average of +1.8°C to +4.4°C for the same period). Precipitation projections are highly uncertain. In spite of this, the mean projection shows a precipitation decrease (February-May) and increase (October-January), with the highest increase in October. The authors also find a shift in precipitation (as rain) from 58% during 1971–2000 to 66% during 2071–2100 for RCP4.5 and 75% for RCP8.5, due to changes in snowfall. However, the snowmelt contribution to total runoff increases or stays the same, which they explain is due to increased evapotranspiration and water availability in the soil and reduction in sublimation caused by a decrease in snow cover.

For a near-future scenario, Hasson (2016) finds that the mean UIB discharge could be reduced by about 8% (in a range of 6%–10%) relative to the present-day mean discharge. Also, he finds that the glacier melt contribution to the total flow could be reduced by 24% (in a range of 21%–25%). Snowmelt outflow is projected to increase by 7% (in the range of 2%–14%). From the changes in other discharge components, a 29% decrease in the rainfall outflow (range of 0%–43%) is suggested, whereas the groundwater contribution could increase by about 2% (range of 0%–3%).

2.4.2.1 Results from NORINDIA

The study by Mesquita et al. (2016), which summarizes the output from publications written related to the NORINDIA project, estimates a 50% glacier ice loss in the Beas River Basin. They also project an approximate 10% increase in precipitation in summer rainfall from 2076 to 2096 compared with the recent climate in the Indo-Pacific region for the RCP8.5 scenario. For the same scenario, there is also a reduction in annual snowfall of 30%–50% in the Indus Basin.

For the RCP4.5 scenario, the project also finds a projected increase of about 3°C in surface temperature from 2079 to 2100. From hydrological modeling, it is found that the runoff (including rainfall, ice, and snow melt) from glacier-covered areas accounts for around 19% of the total runoff measured at the Thalout station in the Beas River Basin (Li et al., 2017a,b). Climate change scenarios from the study show that precipitation may increase about 10%–33% by the middle of the century and 18%–40% by the end of 2100. At middle of the century it is predicted that glacier area loss in the Beas River Basin is about 73%–81%. Also, by the end of 2100, the glacier area loss in this basin is predicted to be about 94%–99%, which will result in a loss of discharge in the monsoon period. The precipitation increasing and glacier retreating make a complex future of total discharge in the Beas river basin, which vary with seasons with a general increase in winter and pre-monsoon periods and a large uncertainty in the monsoon period.

2.4.2.2 The Role of Glacier Melt

It is important to mention that the current version of the WRF-Hydro model does not have a glacier component, whereas glacier melting plays a significant role in the runoff in glacier-fed river basins. For instance, in the Beas River Basin, the contribution of snow and ice melting to the total runoff varies from 27.5% to about 40%, according to previous studies (Kumar et al., 2007; Li et al., 2013, 2017a; Mesquita et al., 2016). Kumar et al. (2007) used a water balance approach to estimate the average contribution of snowmelt and glacier melt runoff in the annual flow of the Beas River at Pandoh Dam, which was 35% during 1990–2004. Li et al. (2016) estimated the glacier contribution (i.e., total runoff from the glacier area, including all melted snow and glacier ice and rain water) to the runoff in the Beas River Basin up to Bhuntar, which was 27.5% during 1997–2005, using an integrated approach that coupled the HBV model and a glacier retreat model.

In addition, in a study by Li et al. (2017b), it was found that rainwater and snow and glacier melting from a glacier-covered area comprised 19% of the total runoff and that the glacier retreated 5% during 1990–2004. Kaab et al. (2015) used ICESat satellite altimetry data and indicated that 5% of the runoff was from glacier retreat during 2003–2008. There are uncertainties and challenges related to estimating the glacier melting over such high mountainous drainage basins. The lack of reliable snowfall measurements over high mountainous regions is one of the reasons for the poor understanding and great uncertainty about high-altitude precipitation over this area of North India (Viste and Sorteberg, 2015). Also, significant disagreement between precipitation in dynamical RCM simulations (WRF) and other data sources (i.e., TRMM 3B42 V7, APHRODITE, and gauge data) was found over high altitudes in the Beas River Basin (Li et al., 2017a); as a result, high-altitude precipitation may be much greater than previously thought.

2.5 CONCLUSION

In this chapter, we discussed the main challenges in estimating future climate changes in the Indus region. There are still "significant uncertainties" in the estimates (Rajbhandari et al., 2015), as highlighted in most studies considered so far. These studies also show the need to assess regional differences and seasonal changes across the region (Kapnick et al., 2014). This need is often due to the heterogeneous topography of the Himalayas, inadequacies of the observational data sets, and uncertainties around the climate models (Kapnick et al., 2014; Nengker et al., 2017), all of which may be considered as confounding factors.

Some existing measures could be used to help reduce the uncertainty or to increase our understanding of the uncertainty in hydrological modeling. Benestad et al. (2017) discusses uncertainty in depth. Here are a few recommendations:

- Use a Bayesian hierarchical approach to keep track of (and account for) the different sources of uncertainty.
- Make use of ensemble modeling and design experiments to create a representative sample.
- Compare the results of using GCMs with downscaling GCM output.
- Try both statistical and dynamical downscaling and the compare results.

In this chapter, we briefly discussed extremes. A clearer picture of extremes seems to emerge from CMIP5. The results in Sharmila et al. (2015) show, for example, that we could expect the following in the future (based on the RCP8.5 scenario for the period 2051–2099):

- enhanced heavy rainfall events (above 40 mm/day) over the Indian summer monsoon domain;
- reduced low rain-rate events (below 10 mm/day);
- decreased number of wet days;
- increased intensity of and more spatially extended active/break cycles, with the possibility of both short and long active/break spells; and
- increased severe droughts and flood conditions over the Indian subcontinent.

There has already been a threefold increase in extremes in Central India due to the increasing variability of the low-level monsoon westerlies over the Arabian Sea (Roxy et al., 2017). The warming of the Arabian Sea in the future, as shown in CMIP5, foreshadows what can be expected, that is, a further increase of extreme events. Thus new studies and projects are needed to further elucidate this picture and to devise efficient planning measures (Bhardwaj et al., 2017).

Just like the past civilizations of the Indus region, we must use ingenious methods for water management today and in the future. Some of these options are explicitly mentioned in Laghari et al. (2012). The authors highlight that climate change may result in increased water availability in the short term, but decreased water in the long term. They provide a list of measures that can be used to tackle future challenges arising from climate change: (1) water supply management options, such as reservoir management due to the strong seasonal changes in water availability; investments in water quality conservation and treatment of wastewater; use of alternative resources, such as desalination, land use planning, soil conservation, and flood management, (2) water-demand management options, which include the modernization of the current infrastructure, and (3) other demand management options, such as the use of economic instruments. It is hoped that these and other measures can help the world tackle future changes in climate and help the Indus valley population thrive in the future.

APPENDIX A. THE NORINDIA PROJECT IN A NUTSHELL

NORINDIA was a 3-year bilateral project between Norway and India, which ended in October 2015 (for details, see Mesquita et al., 2016). It was funded by the Norwegian Research Council through a proposal call to understand the hydrological impacts of climate change in India. NORINDIA has provided a fresh hydrological assessment of water resources in India through the use of IPCC AR5 scenarios, global and regional climate modeling, and the use of the new WRF-Hydro modeling system. The project had several partner institutions, both in Norway (the University of Bergen, the University of Oslo, the Norwegian Institute for Air Research, and Statkraft) and in India (the Indian Institute of Tropical Meteorology, the Centre for Mathematical Modeling and Computer Simulation, and The Energy and Resources Institute). We also collaborated with the Research Applications Laboratory of the National Center for Atmospheric Research in the United States due to their expertise in hydrological modeling and the WRF-Hydro modeling system (Gochis et al., 2015).

The project has produced 10 research papers, 5 popular science articles, 83 presentations, and 2 interviews. Another objective of the project was to work closely with an energy-sector stakeholder to discuss future water resources in India. For that, we conducted a case study of the Beas River system and the assessment of a high-resolution reanalysis product, which are discussed next. (For a detailed analysis and modeling of the Beas Basin and different observations/reanalysis products, see the NORINDIA's publications; Li et al., 2017a; Viste and Sorteberg, 2015.)

Below, we briefly consider the role of observations under two analyses conducted in the NORINDIA Project: (1) the case study of a sub-basin of the Indus River system to illustrate the need for individual basin-wide assessments and (2) the assessment of a state-of-the-art, high-resolution, reanalysis output.

A Case Study of the Beas River System

The Beas River (460 km long) is a major sub-basin of the Indus River system. It was selected as a case study in the NORINDIA project to determine future changes in the region and the potential for hydropower generation. Its total drainage area is 20,303 km^2, and it flows through the states of Himachal Pradesh and Punjab in India. Its estimated hydropower potential is 4495 MW and is linked with Sutluj through Bhakra Beas System (Jain et al., 2007).

The Bhakra Beas Management Board (BBMB) has been monitoring the key weather parameters and the flow of the Beas River at Pandoh and 14 other stations since 1965. About 45% of the Beas River Basin area at Pandoh dam is covered by snow during winter and about 15% remains covered by perpetual snow and glaciers (Kumar et al., 2007). Hence simulating the contribution of snow and glacier melt runoff in the annual flow of the Beas River is not straightforward.

The observed data indicate high summer flow consisting of monsoonal runoff and snow-melt discharge at Pandoh. The annual mean flow during the months of April to June is 15%; in July to September it is 67%; in October to December it is 10%, while in January to March it is observed to be 8% (Jain et al., 2007). The results of trend analyses of the discharge data of four rivers in the northwestern Himalayas, namely Beas, Chenab, Ravi, and Satluj, were examined by Bhutiyani et al. (2008). In the case of Satluj River, studies indicate an episodic variation in discharge in all three seasons on a longer timescale of about 82 years (1922–2004). A statistically significant decrease in the average annual and monsoon discharge and insignificant increase in winter and spring discharge, despite increasing temperatures during all the three seasons, can also be seen.

Decreasing discharge during winter and monsoon seasons in the post-1990 period, despite rising temperatures and average monsoon precipitation, strongly indicates a decreasing contribution of glaciers to the discharge and their gradual disappearance. On a shorter timescale of the past four decades of the 20th century, barring the Beas River, which shows a significantly decreasing trend, the other three rivers have shown a statistically insignificant change (at 95% confidence level) in their average annual discharge. Annual peak flood discharges show significant increasing trends in the Satluj and Chenab basins, a significant decreasing trend in the Beas River, and an insignificant trend in the Ravi River. Notwithstanding these variations, the studies indicate an increase in the number of "high-magnitude flood" events in the rivers in the northwestern Himalayas in the past three decades (Bhutiyani et al., 2008; Elalem and Pal, 2016).

B Assessment of the CFSR Reanalysis Product

One of the challenges faced in the NORINDIA project was dealing with the discrepancy between observation data sets and reanalysis model output (Viste and Sorteberg, 2015). Reanalysis information is often used as input in regional climate models and hydrological models, so their assessment is essential. An example used in the NORINDIA project was the comparison between the Climate Forecast and Reanalysis System (CFSR) with local observed data from the Indian Meteorological Department (IMD) and with the one that is monitored at Pandoh Dam by BBMB. CFSR is a coupled atmosphere-ocean-land-sea ice system, which assimilates satellite radiances instead of the retrieved temperature and humidity values. It has a spatial resolution of ~38 km, and it assimilates hydrological variables from daily gauge analysis (Saha et al., 2010).

The IMD and CFSR data sets show reasonable agreement for temperature and evapotranspiration. In the case of temperature, the mean absolute error was found to be 4.3°C and for evapotranspiration it was 9.7 mm/day. However, Fig. A.1 shows that the CFSR output underestimates precipitation considerably when compared with the IMD recorded data. Since precipitation is crucial for runoff estimation, seasonal variation was further examined (Fig. A.2). The mean absolute error for winter, spring, and the monsoon season were 77 mm/day, 71 mm/day, and 553 mm/day, clearly demonstrating that CFSR did not predict rainfall during the monsoon season very well. To further examine the CFSR reliability, the seasonal rainfall was compared with the locally monitored data at the Pandoh site (Fig. A.3). A mean absolute error of 98.04, 152.2, and 896.4 mm/day was observed for winter, spring, and the monsoon season, respectively, again indicating that CFSR did not predict precipitation during the monsoon season very well. These results highlight the need for strict quality control when using reanalysis information as input to regional climate and hydrological models, or the need to combine these data sets, as was done in NORINDIA (Viste and Sorteberg, 2015).

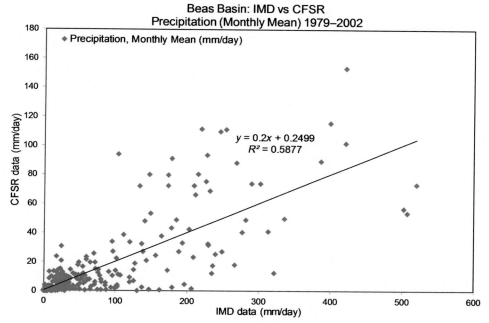

FIG. A.1   Comparison between the precipitation from IMD and CFSR.

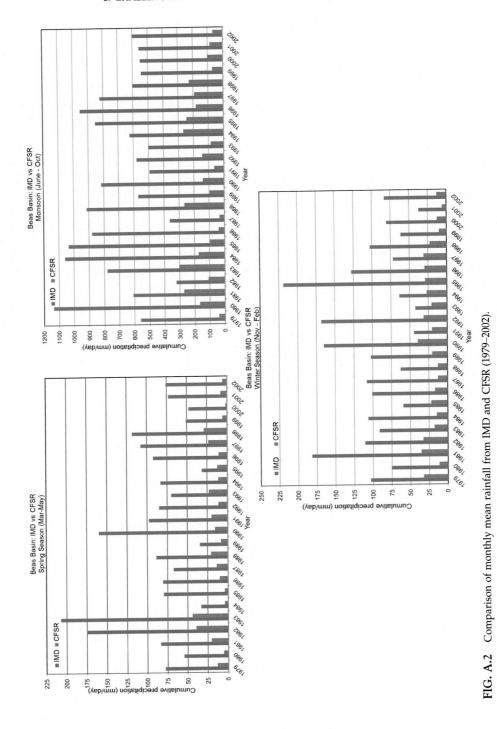

FIG. A.2 Comparison of monthly mean rainfall from IMD and CFSR (1979–2002).

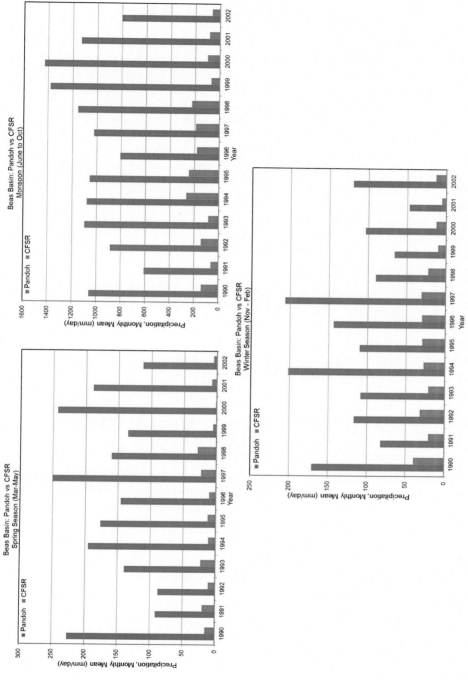

FIG. A.3 Comparison of monthly mean rainfall from Pandoh site (BBMB data) and CFSR (1990–2002).

I. INDUS RIVER BASIN—PAST, PRESENT AND FUTURE

APPENDIX B. DYNAMICAL DOWNSCALING OF NORESM AND ERA-INTERIM DATA SETS

The dynamical downscaling was made using the Weather Research and Forecasting model (Wang et al., 2016). We chose the Norwegian Earth System model as lateral boundary condition data for the historical and future scenarios because it can represent the monsoon more realistically (Sharmila et al., 2015). Although dynamical downscaling seems not to reproduce features of the monsoon for some regional climate models (Singh et al., 2017), here we adopted a different approach: we used a model, which was not used in Singh et al. (2017), and we used the same CORDEX domain size.

We ran WRF 3.3.1 for a resolution of 0.44 by 0.44 degrees centered at 43.5 degrees of latitude and 19.88 degrees of longitude. We also used 213 grid points in the west-east direction and 150 in the south-north direction. Our parameterization scheme parameters are given as follows:

- Microphysics (mp_physics) = 3
- Longwave radiation (ra_lw_physics): 3
- Shortwave radiation (ra_sw_physics): 3
- Boundary layer (bl_pbl_physics): 2
- Convective scheme (cu_physics): 1

Six hourly outputs from WRF were retained for the analysis. For the Bayesian analysis explained in Appendix B, we calculated the average surface temperature per season and per year for the following basins:

- Beas: from 76.56 to 77.53 E and 31.27 to 32.24 N
- Bhramaputra: from 88.13 to 96.95 E and 24.73 to 30.05 N
- Indus: from 72.47 to 79.65 E and 29.13 to 36.98 N

APPENDIX C. BAYESIAN METHOD FOR ESTIMATING THE MEAN SURFACE TEMPERATURE

The variable of interest we want to estimate is the yearly surface temperature $\mu^{s,d}$, during season s and decade d, as presented in Table 2.1. We now define the variables of the Bayesian model. Let.

- $X_e^{s,\,d}$ be the output from ERA-Interim
- $X_h^{s,\,d}$ be the output from the WRF-NorESM historical downscaling
- $Y_{R45}^{s,d}$ be the output from WRF-NorESM RCP4.5 scenario
- $Y_{R85}^{s,d}$ be the output from WRF-NorESM RCP8.5 scenario
- $e^{s,d}$ be the bias, defined as $X_h^{s,d} - X_e^{s,d}$

The Bayesian model is defined as follows, where τ represents the precision, that is, the inverse of the variance:

- $X_e^{s,d} \sim N\left(\mu_e^{s,d}, 1/\tau_e^{s,d}\right)$

- $X_h^{s,d} \sim N\left(\mu_h^{s,d}, 1/\tau_h^{s,d}\right)$

- $Y_{R45}^{s,d} \sim N\left(\mu_{R45}^{s,d}, 1/\tau_{R45}^{s,d}\right)$

- $Y_{R85}^{s,d} \sim N\left(\mu_{R85}^{s,d}, 1/\tau_{R85}^{s,d}\right)$

- $e^{s,d} = \mu_h^{s,d} - \mu_e^{s,d}$

The model priors are mostly flat and are given as

- $\mu_e^{s,d} \sim N(0, 1.0E - 10)$

- $\tau_e^{s,d} \sim \text{Gamma}(0.001, 0.001)$

- $\mu_h^{s,d} \sim N(0, 1.0E - 10)$

- $\tau_h^{s,d} \sim \text{Gamma}(0.001, 0.001)$

- $\mu_{R45}^{s,d} \sim N(0, 1.0E - 10)$

- $\tau_{R45}^{s,d} \sim \text{Gamma}(0.001, 0.001)$

- $\mu_{R85}^{s,d} \sim N(0, 1.0E - 10)$

- $\tau_{R85}^{s,d} \sim \text{Gamma}(0.001, 0.001)$

The bias-corrected scenario estimates were calculated by subtracting the estimated mean by the estimated bias, such that $\mu' = \mu - e$ for each scenario.

The parameters were estimated using a Markov Chain Monte Carlo procedure. Three chains of 50,000 samples each were sampled. From these, the last 10,000 samples of each chain were retained for the analysis.

Acknowledgments

This chapter would not have been possible without the kind invitation of Dr. Sadiq Khan and Dr. Thomas Adams. Furthermore, we take this opportunity to acknowledge and thank the Central Water Commission (CWC) and the Bhakra Beas Management Board (BBMB) of the Ministry of Water Resources of India for their support and for providing the hydrometeorological data for the Beas River Basin. We also thank the Indian Meteorological Department (IMD) for providing historical meteorological data. The NORINDIA project was funded through the Norwegian Research Council, grant number 216576, and through Statkraft. We thank both institutions for the funding provided.

References

Abish, B., Joseph, P.V., Johannessen, O.M., 2013. Weakening trend of the tropical easterly jet stream of the boreal summer monsoon season 1950–2009. J. Clim. 26, 9408–9414. https://dx.doi.org/10.1175/JCLI-D-13-00440.1.

Ahmadalipour, A., Rana, A., Moradkhani, H., Sharma, A., 2015. Multi-criteria evaluation of CMIP5 GCMs for climate change impact analysis. Theor. Appl. Climatol. 128, 71–87. https://dx.doi.org/10.1007/s00704-015-1695-4.

Akhtar, M., Ahmad, N., Booij, M.J., 2008. The impact of climate change on the water resources of Hindukush–Karakorum–Himalaya region under different glacier coverage scenarios. J. Hydrol. 355, 148–163. https://dx.doi.org/10.1016/j.jhydrol.2008.03.015.

Alessandri, A., Borrelli, A., Cherchi, A., Materia, S., Navarra, A., Lee, J.-Y., Wang, B., 2015. Prediction of Indian summer monsoon onset using dynamical subseasonal forecasts: effects of realistic initialization of the atmosphere. Mon. Weather Rev. 143, 778–793. https://dx.doi.org/10.1175/MWR-D-14-00187.1.

Basha, G., Kishore, P., Ratnam, M.V., Jayaraman, A., Kouchak, A.A., Ouarda, T.B.M.J., Velicogna, I., 2017. Historical and projected surface temperature over India during the 20th and 21st century. Sci. Rep. 7. https://dx.doi.org/10.1038/s41598-017-02130-3.

Benestad, R., Sillmann, J., Thorarinsdottir, T.L., Guttorp, P., Mesquita, M.D.S., Tye, M.R., Uotila, P., Maule, C.F., Thejll, P., Drews, M., Parding, K.M., 2017. New vigour involving statisticians to overcome ensemble fatigue. Nat. Clim. Chang. 7, 697–703. https://dx.doi.org/10.1038/nclimate3393.

Bhardwaj, S., Mesquita, M., Barreto, N., 2017. Preparing for the future: climate products and models for India. Eos. https://dx.doi.org/10.1029/2017EO085761.

Bhutiyani, M.R., Kale, V.S., Pawar, N.J., 2008. Changing streamflow patterns in the rivers of northwestern Himalaya: implications of global warming in the 20th century. Curr. Sci. 95, 618–626.

Bollasina, M.A., Ming, Y., Ramaswamy, V., 2011. Anthropogenic aerosols and the weakening of the south Asian summer monsoon. Science 334, 502–505. https://dx.doi.org/10.1126/science.1204994.

Cannon, F., Carvalho, L.M.V., Jones, C., Bookhagen, B., 2015. Multi-annual variations in winter westerly disturbance activity affecting the Himalaya. Clim. Dyn. 44, 441–455. https://dx.doi.org/10.1007/s00382-014-2248-8.

Choudhary, A., Dimri, A.P., 2017. Assessment of CORDEX-South Asia experiments for monsoonal precipitation over Himalayan region for future climate. Clim. Dyn. 10.014007–22. https://dx.doi.org/10.1007/s00382-017-3789-4.

Chung, C.E., Ramanathan, V., 2006. Weakening of north Indian SST gradients and the monsoon rainfall in India and the Sahel. J. Clim. 19, 2036–2045. https://dx.doi.org/10.1175/JCLI3820.1.

Committee on Himalayan Glaciers, H.C.C.A.I.F.W.S., Climate, B.O.A.S.A., Studies, D.O.E.A.L., National Research Council, 2012. Himalayan Glaciers. National Academies Press.

Cook, E.R., Krusic, P.J., Jones, P.D., 2003. Dendroclimatic signals in long tree-ring chronologies from the Himalayas of Nepal. Int. J. Climatol. 23, 707–732. https://dx.doi.org/10.1002/joc.911.

Draper, D., 1995. Assessment and propagation of model uncertainty. J. R. Stat. Soc. Ser. B. https://dx.doi.org/10.2307/2346087.

Elalem, S., Pal, I., 2016. Mapping the vulnerability hotspots over Hindu-Kush Himalaya region to flooding disasters. Weather Clim. Extrem. 8, 1–13. https://dx.doi.org/10.1016/j.wace.2014.12.001.

Fan, F., Mann, M.E., Lee, S., Evans, J.L., 2010. Observed and modeled changes in the south Asian summer monsoon over the historical period*. J. Clim. 23, 5193–5205. https://dx.doi.org/10.1175/2010JCLI3374.1.

Fowler, H.J., Archer, D.R., 2006. Conflicting signals of climatic change in the upper Indus Basin. J. Clim. 19, 4276–4293. https://dx.doi.org/10.1175/JCLI3860.1.

Ganguly, D., Rasch, P.J., Wang, H., Yoon, J.H., 2012. Fast and slow responses of the south Asian monsoon system to anthropogenic aerosols. Geophys. Res. Lett. 39224. https://dx.doi.org/10.1029/2012GL053043.

Ghosh, S., Das, D., Kao, S.-C., Ganguly, A.R., 2011. Lack of uniform trends but increasing spatial variability in observed Indian rainfall extremes. Nat. Clim. Chang. 2, 86–91. https://dx.doi.org/10.1038/nclimate1327.

Gochis, D.J., Yu, W., Yates, D.N., 2015. The WRF-Hydro Model Technical Description and user's Guide. (version 3.0. NCAR Technical Document).

Goswami, B.N., Venugopal, V., Sengupta, D., Madhusoodanan, M.S., Xavier, P.K., 2006. Increasing trend of extreme rain events over India in a warming environment. Science 314, 1442–1445. https://dx.doi.org/10.1126/science.1132027.

Guhathakurta, P., Rajeevan, M., 2008. Trends in the rainfall pattern over India. Int. J. Climatol. 28, 1453–1469. https://dx.doi.org/10.1002/joc.1640.

Halder, S., Dirmeyer, P.A., 2017. Relation of Eurasian snow cover and Indian summer monsoon rainfall: importance of the delayed hydrological effect. J. Clim. 30, 1273–1289. https://dx.doi.org/10.1175/JCLI-D-16-0033.1.

Hasson, S., Lucarini, V., Khan, M.R., Petitta, M., Bolch, T., Gioli, G., 2013. Early 21st century climatology of snow cover for the western river basins of the Indus River system. Hydrol. Earth Syst. Sci. Discuss. 10, 13145–13190. https://dx.doi.org/10.5194/hessd-10-13145-2013.

Hasson, S.U., 2016. Future water availability from Hindukush-Karakoram-Himalaya upper Indus Basin under conflicting climate change scenarios. Climate 4 (3), 40. https://dx.doi.org/10.3390/cli4030040.

Immerzeel, W.W., Droogers, P., de Jong, S.M., Bierkens, M.F.P., 2009. Large-scale monitoring of snow cover and run-off simulation in Himalayan river basins using remote sensing. Remote Sens. Environ. 113, 40–49. https://dx.doi.org/10.1016/j.rse.2008.08.010.

Jain, S.K., Agarwal, P.K., Singh, V.P., 2007. Hydrology and Water Resources of India. Springer Science & Business Media.

Jayakumar, A., Turner, A.G., Johnson, S.J., Rajagopal, E.N., Mohandas, S., Mitra, A.K., 2016. Boreal summer sub-seasonal variability of the south Asian monsoon in the Met Office GloSea5 initialized coupled model. Clim. Dyn. 49, 2035–2059. https://dx.doi.org/10.1007/s00382-016-3423-x.

Johnson, S.J., Turner, A., Woolnough, S., Martin, G., MacLachlan, C., 2017. An assessment of Indian monsoon seasonal forecasts and mechanisms underlying monsoon interannual variability in the Met Office GloSea5-GC2 system. Clim. Dyn. 48, 1447–1465. https://dx.doi.org/10.1007/s00382-016-3151-2.

Kaab, A., Treichler, D., Nuth, C., Berthier, E., 2015. Brief communication: contending estimates of 2003-2008 glacier mass balance over the Pamir-Karakoram-Himalaya. Cryosphere 9, 557–564. https://dx.doi.org/10.5194/tc-9-557-2015.

Kapnick, S.B., Delworth, T.L., Ashfaq, M., Malyshev, S., Milly, P.C.D., 2014. Snowfall less sensitive to warming in Karakoram than in Himalayas due to a unique seasonal cycle. Nat. Geosci. 7, 834–840. https://dx.doi.org/10.1038/NGEO2269.

Knutti, R., Sedlacek, J., Sanderson, B.M., Lorenz, R., Fischer, E.M., Eyring, V., 2017. A climate model projection weighting scheme accounting for performance and interdependence. Geophys. Res. Lett. 44, 1909–1918. https://dx.doi.org/10.1002/2016GL072012.

Krishnamurti, T.N., Martin, A., Krishnamurti, R., Simon, A., Thomas, A., Kumar, V., 2013. Impacts of enhanced CCN on the organization of convection and recent reduced counts of monsoon depressions. Clim. Dyn. 41, 117–134. https://dx.doi.org/10.1007/s00382-012-1638-z.

Krishnan, R., Sabin, T.P., Ayantika, D.C., Kitoh, A., Sugi, M., Murakami, H., Turner, A.G., Slingo, J.M., Rajendran, K., 2013. Will the south Asian monsoon overturning circulation stabilize any further? Clim. Dyn. 40, 187–211. https://dx.doi.org/10.1007/s00382-012-1317-0.

Krishnan, R., Sabin, T.P., Vellore, R., Mujumdar, M., Sanjay, J., Goswami, B.N., Hourdin, F., Dufresne, J.L., Terray, P., 2015. Deciphering the desiccation trend of the south Asian monsoon hydroclimate in a warming world. Clim. Dyn. 47, 1–21. https://dx.doi.org/10.1007/s00382-015-2886-5.

Kumar, V., Singh, P., Singh, V., 2007. Snow and glacier melt contribution in the Beas River at Pandoh Dam, Himachal Pradesh, India. Hydrol. Sci. J. 52, 376–388.

Laghari, A.N., Vanham, D., Rauch, W., 2012. The Indus basin in the framework of current and future water resources management. Hydrol. Earth Syst. Sci. 16, 1063–1083. https://dx.doi.org/10.5194/hess-16-1063-2012.

Laghari, J., 2013. Climate change: melting glaciers bring energy uncertainty. Nature 502, 617–618. https://dx.doi.org/10.1038/502617a.

Li, H., Xu, C.-Y., Beldring, S., Tallaksen, L.M., Jain, S.K., 2016. Water resources under climate change in Himalayan basins Hong. Water Resour. Manag. 30, 843–859. https://dx.doi.org/10.1007/s11269-015-1194-5.

Li, L., Engelhardt, M., Xu, C.-Y., Jain, S.K., Singh, V.P., 2013. Comparison of satellite-based and reanalysed precipitation as input to glacio-hydrological modeling for Beas river basin, Northern India. Proceedings of H, I.-I.-I.A In: Presented at the Cold and Mountain Region Hydrological Systems Under Climate Change Towards Improved Projections, Gothenburg, pp. 45–52.

Li, L., Gochis, D.J., Sobolowski, S., Mesquita, M.D.S., 2017a. Evaluating the present annual water budget of a Himalayan headwater river basin using a high-resolution atmosphere-hydrology model. J. Geophys. Res. Atmos. 122, 4786–4807. https://dx.doi.org/10.1002/2016JD026279.

Li, L., Hou, Y., Xu, C.-Y., Chen, H., Jain, S.K., 2017b. Projection of future glacier and runoff change in Himalayan headwater Beas basin by using a coupled glacier and hydrological model. Hydrol. Earth Syst. Sci. Discuss., 1–36. https://dx.doi.org/10.5194/hess-2017-525.

Lutz, A.F., Immerzeel, W.W., Kraaijenbrink, P.D.A., Shrestha, A.B., Bierkens, M.F.P., 2016. Climate change impacts on the upper Indus hydrology: sources, shifts and extremes. PLoS One 11, e0165630. https://dx.doi.org/10.1371/journal.pone.0165630.

Madhura, R.K., Krishnan, R., Revadekar, J.V., Mujumdar, M., Goswami, B.N., 2015. Changes in western disturbances over the western Himalayas in a warming environment. Clim. Dyn. 44, 1157–1168. https://dx.doi.org/10.1007/s00382-014-2166-9.

Mesquita, M., Veldore, V., Li, L., Krishnan, R., Orsolini, Y., Senan, R., Ramarao, M., Viste, E., 2016. Forecasting India's water future—Eos. Eos 97. https://dx.doi.org/10.1029/2016eo049099.

Mohan, T.S., Rajeevan, M., 2017. Past and future trends of hydroclimatic intensity over the Indian monsoon region. J. Geophys. Res. Atmos. 122, 896–909. https://dx.doi.org/10.1002/2016JD025301.

Nengker, T., Choudhary, A., Dimri, A.P., 2017. Assessment of the performance of CORDEX-SA experiments in simulating seasonal mean temperature over the Himalayan region for the present climate: Part I. Clim. Dyn. 13, 1–31. https://dx.doi.org/10.1007/s00382-017-3597-x.

Nepal, S., Shrestha, A.B., 2015. Impact of climate change on the hydrological regime of the Indus, Ganges and Brahmaputra river basins: a review of the literature. Int. J. Water Resour. Dev. 31, 201–218. https://dx.doi.org/10.1080/07900627.2015.1030494.

Niranjan Kumar, K., Rajeevan, M., Pai, D.S., Srivastava, A.K., Preethi, B., 2013. On the observed variability of monsoon droughts over India. Weather Clim. Extremes 1, 42–50. https://dx.doi.org/10.1016/j.wace.2013.07.006.

Niu, X., Wang, S., Tang, J., Lee, D.-K., Gutowski, W., Dairaku, K., McGregor, J., Katzfey, J., Gao, X., Wu, J., Hong, S., Wang, Y., Sasaki, H., 2015. Projection of Indian summer monsoon climate in 2041-2060 by multiregional and global climate models. J. Geophys. Res. Atmos. 120, 1776–1793. https://dx.doi.org/10.1002/2014JD022620.

Pal, I., Lall, U., Robertson, A.W., Cane, M.A., Bansal, R., 2013. Diagnostics of Western Himalayan Satluj River flow: warm season (MAM/JJAS) inflow into Bhakra dam in India. J. Hydrol. 478, 132–147. https://dx.doi.org/10.1016/j.jhydrol.2012.11.053.

Pal, I., Robertson, A.W., Lall, U., Cane, M.A., 2014. Modeling winter rainfall in Northwest India using a hidden Markov model: understanding occurrence of different states and their dynamical connections. Clim. Dyn. 44, 1003–1015. https://dx.doi.org/10.1007/s00382-014-2178-5.

Pande, S., Ertsen, M., 2014. Endogenous change: on cooperation and water availability in two ancient societies. Hydrol. Earth Syst. Sci. 18, 1745–1760. https://dx.doi.org/10.5194/hess-18-1745-2014.

Priya, P., 2017. Modelling Studies on Hydro-Meteorological Response of Indus River Basin to Heavy Monsoon Rain Events under Changing Climate. IITM, Pune.

Priya, P., Krishnan, R., Mujumdar, M., Houze, R.A., 2016. Changing monsoon and midlatitude circulation interactions over the Western Himalayas and possible links to occurrences of extreme precipitation. Clim. Dyn. 26, 9408–9414. https://dx.doi.org/10.1007/s00382-016-3458-z.

Rajbhandari, R., Shrestha, A.B., Kulkarni, A., Patwardhan, S.K., Bajracharya, S.R., 2015. Projected changes in climate over the Indus river basin using a high resolution regional climate model (PRECIS). Clim. Dyn. 44, 339–357. https://dx.doi.org/10.1007/s00382-014-2183-8.

Rajeevan, M., Bhate, J., Jaswal, A.K., 2008. Analysis of variability and trends of extreme rainfall events over India using 104 years of gridded daily rainfall data. Geophys. Res. Lett. 35, 245. https://dx.doi.org/10.1029/2008GL035143.

Rajeevan, M., De, U.S., Prasad, R.K., 2000. Decadal variability of sea surface temperature, cloudiness and monsoon depressions in the north Indian Ocean. Curr. Sci. 79, 283–285.

Rajeevan, M., Unnikrishnan, C.K., Preethi, B., 2012. Evaluation of the ENSEMBLES multi-model seasonal forecasts of Indian summer monsoon variability. Clim. Dyn. 38, 2257–2274. https://dx.doi.org/10.1007/s00382-011-1061-x.

Rajendran, K., Kitoh, A., Srinivasan, J., Mizuta, R., Krishnan, R., 2012. Monsoon circulation interaction with Western Ghats orography under changing climate. Theor. Appl. Climatol. 110, 555–571. https://dx.doi.org/10.1007/s00704-012-0690-2.

Ramanathan, V., Chung, C., Kim, D., Bettge, T., Buja, L., Kiehl, J.T., Washington, W.M., Fu, Q., Sikka, D.R., Wild, M., 2005. Atmospheric brown clouds: impacts on south Asian climate and hydrological cycle. Proc. Natl. Acad. Sci. U. S. A. 102, 5326–5333. https://dx.doi.org/10.1073/pnas.0500656102.

Ramesh Kumar, M.R., Krishnan, R., Sankar, S., Unnikrishnan, A.S., Pai, D.S., 2009. Increasing trend of "break-monsoon" conditions over India—role of ocean–atmosphere processes in the Indian Ocean. IEEE Geosci. Remote Sens. Lett. 6, 332–336. https://dx.doi.org/10.1109/LGRS.2009.2013366.

Rao, B.R.S., Rao, D.V.B., Rao, V.B., 2004. Decreasing trend in the strength of tropical easterly jet during the Asian summer monsoon season and the number of tropical cyclonic systems over Bay of Bengal. Geophys. Res. Lett. 31, 4393. https://dx.doi.org/10.1029/2004GL019817.

Roxy, M.K., Ghosh, S., Pathak, A., Athulya, R., Mujumdar, M., Murtugudde, R., Terray, P., Rajeevan, M., 2017. A threefold rise in widespread extreme rain events over central India. Nat. Commun. 8, 708. https://dx.doi.org/10.1038/s41467-017-00744-9.

Roxy, M.K., Ritika, K., Terray, P., Murtugudde, R., Ashok, K., Goswami, B.N., 2015. Drying of Indian subcontinent by rapid Indian Ocean warming and a weakening land-sea thermal gradient. Nat. Commun. 6, 7423. https://dx.doi.org/10.1038/ncomms8423.

Saha, S., Moorthi, S., Pan, H.-L., Wu, X., Wang, J., Nadiga, S., Tripp, P., Kistler, R., Woollen, J., Behringer, D., Liu, H., Stokes, D., Grumbine, R., Gayno, G., Wang, J., Hou, Y.-T., Chuang, H.-Y., Juang, H.-M.H., Sela, J., Iredell, M., Treadon, R., Kleist, D., Van Delst, P., Keyser, D., Derber, J., Ek, M., Meng, J., Wei, H., Yang, R., Lord, S., Van den Dool, H., Kumar, A., Wang, W., Long, C., Chelliah, M., Xue, Y., Huang, B., Schemm, J.-K., Ebisuzaki, W., Lin, R., Xie, P., Chen, M., Zhou, S., Higgins, W., Zou, C.-Z., Liu, Q., Chen, Y., Han, Y., Cucurull, L., Reynolds, R.W., Rutledge, G., Goldberg, M., 2010. The Ncep climate forecast system reanalysis. Bull. Am. Meteorol. Soc. 91, 1015–1057. https://dx.doi.org/10.1175/2010BAMS3001.1.

Saha, S.K., Pokhrel, S., Chaudhari, H.S., 2013. Influence of Eurasian snow on Indian summer monsoon in NCEP CFSv2 freerun. Clim. Dyn. 41, 1801–1815. https://dx.doi.org/10.1007/s00382-012-1617-4.

Sahai, A.K., Abhilash, S., Chattopadhyay, R., Borah, N., Joseph, S., Sharmila, S., Rajeevan, M., 2015. High-resolution operational monsoon forecasts: an objective assessment. Clim. Dyn. 44, 3129–3140. https://dx.doi.org/10.1007/s00382-014-2210-9.

Salzmann, M., Weser, H., Cherian, R., 2014. Robust response of Asian summer monsoon to anthropogenic aerosols in CMIP5 models. J. Geophys. Res. Atmos. 119, 11,321–11,337. https://dx.doi.org/10.1002/2014JD021783.

Sanap, S.D., Pandithurai, G., Manoj, M.G., 2015. On the response of Indian summer monsoon to aerosol forcing in CMIP5 model simulations. Clim. Dyn. 45, 2949–2961. https://dx.doi.org/10.1007/s00382-015-2516-2.

Sathiyamoorthy, V., 2005. Large scale reduction in the size of the tropical easterly jet. Geophys. Res. Lett. 32. https://dx.doi.org/10.1029/2005GL022956 n/a–n/a.

Senan, R., Orsolini, Y.J., Weisheimer, A., Vitart, F., Balsamo, G., Stockdale, T.N., Dutra, E., Doblas-Reyes, F.J., Basang, D., 2016. Impact of springtime Himalayan-Tibetan plateau snowpack on the onset of the Indian summer monsoon in coupled seasonal forecasts. Clim. Dyn. 47, 2709–2725. https://dx.doi.org/10.1007/s00382-016-2993-y.

Sharmila, S., Joseph, S., Sahai, A.K., Abhilash, S., Chattopadhyay, R., 2015. Future projection of Indian summer monsoon variability under climate change scenario: an assessment from CMIP5 climate models. Glob. Planet. Chang. 124, 62–78. https://dx.doi.org/10.1016/j.gloplacha.2014.11.004.

Shashikanth, K., Gosh, S., Vittal, H., Karmakar, S., 2017. Future projections of Indian summer monsoon rainfall extremes over India with statistical downscaling and its consistency with observed characteristics. Clim. Dyn. https://dx.doi.org/10.1007/s00382-017-3604-2.

Shukla, R.P., Huang, B., 2016. Mean state and interannual variability of the Indian summer monsoon simulation by NCEP CFSv2. Clim. Dyn. 46, 3845–3864. https://dx.doi.org/10.1007/s00382-015-2808-6.

Singh, D., Tsiang, M., Rajaratnam, B., Diffenbaugh, N.S., 2014. Observed changes in extreme wet and dry spells during the south Asian summer monsoon season. Nat. Clim. Chang. 4, 456–461. https://dx.doi.org/10.1038/nclimate2208.

Singh, S., Ghosh, S., Sahana, A.S., Vittal, H., Karmakar, S., 2017. Do dynamic regional models add value to the global model projections of Indian monsoon? Clim. Dyn. 48, 1375–1397. https://dx.doi.org/10.1007/s00382-016-3147-y.

Su, B., Huang, J., Gemmer, M., Jian, D., Tao, H., Jiang, T., Zhao, C., 2016. Statistical downscaling of CMIP5 multi-model ensemble for projected changes of climate in the Indus River Basin. Atmos. Res. 178–179, 138–149. https://dx.doi.org/10.1016/j.atmosres.2016.03.023.

Swapna, P., Krishnan, R., Wallace, J.M., 2013. Indian Ocean and monsoon coupled interactions in a warming environment. Clim. Dyn. 42, 2439–2454. https://dx.doi.org/10.1007/s00382-013-1787-8.

Thayyen, R.J., Gergan, J.T., 2010. Role of glaciers in watershed hydrology: a preliminary study of a "Himalayan catchment". Cryosphere 4, 115–128. https://dx.doi.org/10.5194/tc-4-115-2010.

Tiwari, S., Kar, S.C., Bhatla, R., 2016. Atmospheric moisture budget during winter seasons in the western Himalayan region. Clim. Dyn. 48, 1277–1295. https://dx.doi.org/10.1007/s00382-016-3141-4.

Turner, A.G., Annamalai, H., 2012. Climate change and the south Asian summer monsoon. Nat. Clim. Chang. 2, 587–595. https://dx.doi.org/10.1038/nclimate1495.

Turner, A.G., Hannachi, A., 2010. Is there regime behavior in monsoon convection in the late 20th century? Geophys. Res. Lett. 37, 1–5. https://dx.doi.org/10.1029/2010GL044159.

Viste, E., Sorteberg, A., 2015. Snowfall in the Himalayas: an uncertain future from a little-known past. Cryosphere 9, 1147–1167. https://dx.doi.org/10.5194/tc-9-1147-2015.

Vitart, F., Molteni, F., 2009. Dynamical extended-range prediction of early monsoon rainfall over India. Mon. Weather Rev. 137, 1480–1492. https://dx.doi.org/10.1175/2008MWR2761.1.

Wang, B., Xiang, B., Li, J., Webster, P.J., Rajeevan, M.N., Liu, J., Ha, K.-J., 2015. Rethinking Indian monsoon rainfall prediction in the context of recent global warming. Nat. Commun. 6. https://dx.doi.org/10.1038/ncomms8154.

Wang, W., Bruyère, C., Duda, M., Dudhia, J., Gill, D., Kavulich, M., Keene, K., Lin, H.-C., Michalakes, J., Rizvi, S., Zhang, X., Berner, J., Ha, S., Fossell, K., Beezley, J.D., Coen, J.L., Mandel, J., Chuang, H.-Y., McKee, N., Slovacek, T., Wolff, J., 2016. ARW Weather Research & Forecasting User's Manual. National Center for Atmospheric Research, Boulder.

Zafar, M.U., Ahmed, M., Rao, M.P., Buckley, B.M., Khan, N., Wahab, M., Palmer, J., 2015. Karakorum temperature out of phase with hemispheric trends for the past five centuries. Clim. Dyn. 46, 1943–1952. https://dx.doi.org/10.1007/s00382-015-2685-z.

Past and Future Glacier Changes in the Indus River Basin

Tobias Bolch

Mountain Cryosphere Research Group, Department of Geography, University of Zurich, Zurich, Switzerland

3.1 INTRODUCTION

Glaciers are an important element in the Upper Indus Basin, particularly because the melt from these glaciers provides about 40% of the total runoff in the Indus River (Lutz et al., 2014), which drains into arid lowlands. Thus changing glaciers directly impact the streamflow of the Indus River and the availability of water. In addition, glacial changes have been recognized as an important indicator of climate change, with changes in precipitation, temperature, and other climatic elements affecting the glacier area, length, and mass balance (GCOS, 2004). However, only changes in glacier mass can be directly linked to climatic forcing and runoff; changes in glacier area and length are only an indirect signal as the terminus response is delayed by flow dynamics. Moreover, area and length are affected by non-climatic factors such as debris cover and glacier surges (rapid advances). Both debris-covered and surging glaciers are abundant in the Upper Indus, especially in the Karakoram mountain range (Bhambri et al., 2017; Bolch et al., 2012; Scherler et al., 2011). Therefore interpretation of average length and area changes requires caution. Future predictions are also challenged as the influence of debris cover on glacier mass balance is, despite recent progress (e.g., Benn et al., 2012; Carenzo et al., 2016), still only partly understood and information about the characteristics of the debris cover is scarce. The mechanisms and impacts on the glacier mass balance of glacier surges are also not yet reproduced by models projecting future changes in glaciers.

The Indus drains from the Hindu Kush, Karakoram, and Himalayan mountain ranges (Fig. 3.1). The Indus Basin covers an area of about 1.12 Mio km², while glaciers cover roughly 2.3% of the basin (Table 3.1). More than 60% of the glacier area is located in the Karakoram and slightly less than 30% in the Himalaya. About 10% of coverage is located within the Hindu Kush with comparatively little ice found on the Tibetan Plateau (Table 3.2).

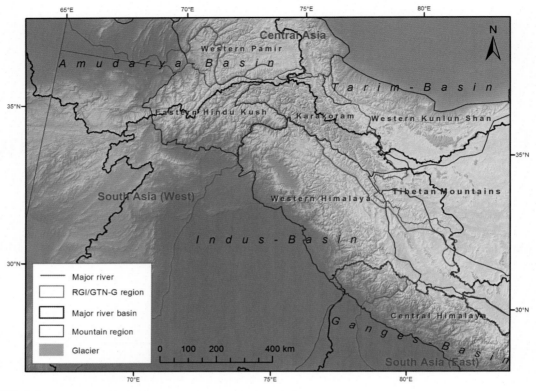

FIG. 3.1 Overview of Indus Basin, mountain regions, and glaciers.

TABLE 3.1 Glacier Coverage in the Indus Basin According to Different Recent Glacier Inventories (RGI6.0, Pfeffer et al., 2014; RGI Consortium, 2017; ICIMOD, Bajracharya and Shrestha, 2011; GAMDAM, Nuimura et al., 2015)

Region/Dataset	Area (Mio km²)	Glacier Cover (km²)			Difference (Abs. [km²], Rel. [%])	
		RGI6.0	ICIMOD	GAMDAM	RGI vs. ICIMOD	RGI vs. GAMDAM
South Asia West		33,568	Not fully covered	30,841		−2727 (−8.1)
Indus Basin	112	26,553	21,162	25,190	−5391 (−20.3)	−1363 (−5.1)

The Indus Basin covers large parts of the Hindu Kush, Karakoram, and Western Himalaya mountain subdivisions as defined by Bolch et al. (2012). Moreover, the basin covers large parts of the RGI (Randolph Glacier Inventory)/GTN-G (Global Terrestrial Network for Glaciers) glacier region "South Asia-West" (Pfeffer et al., 2014; GTN-G, 2017, Fig. 3.1); about 80% of the glacier coverage of this RGI region is located in the Indus Basin. Thus results on glacier changes reported for this region are very similar to the changes in the Indus Basin.

TABLE 3.2 Glacier Coverage (Source RGI) and Average of Published Information About Glacier Mass Changes in Different Mountain Regions That Are Part of the Indus Basin

Region	Glacier Coverage (km²)		Mass Changes (m w.e. a⁻¹)	
	Entire Region	Within Indus Basin#	Before 2000	After 2000
Hindu Kush	2948	1849	No studies	−0.30 (4)
Karakoram	21,355	16,266	−0.10 (3)	−0.06 (12)
Western Himalaya	7923	7923	−0.24 (7)	−0.50 (11)

The number of the considered published studies are shown in parentheses.
The following studies were considered: Azam et al. (2016), Bolch et al. (2017), Brun et al. (2017), Gardelle et al. (2013), Gardner et al. (2013), Kääb et al. (2015), Kaul (1986), Koul and Ganjoo (2010), Raina (2009), Rankl and Braun (2016), Vincent et al. (2013), Wei et al. (2015), WGMS (2008), Yao et al. (2012), and Zhou et al. (2017, 2018).
515 km² of glacier area within the Indus Basin are located on the Tibetan Plateau and the Central Himalaya (Figs. 3.1 and 3.4).
Modified from Bolch et al. (2019).

Knowledge about past glacier changes in the Himalaya has improved significantly over the past few years, especially due to the liberal availability of satellite imagery, but also due to the advancement of remote sensing methodologies and increased efforts by both the national and the international community to investigate glaciers. Therefore we now have a relatively sufficient overview of general trends. The availability of important baseline data such as complete glacier inventories, mass balance data, and improved climate models have also allowed significant progress in predicting future trends.

However, despite this progress detailed knowledge about past and future glacier changes is still limited and uncertain, in part because important baseline information such as glacier area differs significantly within a given region (Table 3.1). Glacier outlines of three inventories based on satellite images dating from 1999 and later are available: (1) the GAMDAM (Glacier Area Mapping for Discharge from the Asian Mountain glacier inventory (Nuimura et al., 2015)), which is based on Landsat TM/ETM + images of the period 1999–2003, (2) the glacier inventory generated by scientists at the International Centre for Integrated Mountain Development (ICIMOD; Bajracharya and Shrestha, 2011), which is based on Landsat ETM + images of the period 2004–2009, and (3) the Randolph Glacier Inventory (Pfeffer et al., 2014). The current version of the RGI is 6.0 and consists of the Indus Basin of the Glaciers_cci inventory for the Karakoram (Mölg et al., 2018) and the GAMDAM inventory elsewhere in the Indus Basin (RGI Consortium, 2017). The Glaciers_cci inventory is also based on Landsat TM/ETM + data of the period 1999–2003. The glacier areas of the Indus Basin in these three available inventories differ by about 20%. The major reasons for the discrepancy are the different treatments of the headwalls above the bergschrund, missed glaciers, and real glacier changes due to the varied acquisition dates of the satellite imagery (Fig. 3.2). However, the different acquisition dates can be only a minor reason for the differences as the glaciers in the Karakoram have been stable on average, and reported area losses in the other regions were about 0.5% a⁻¹ during the past decades (see Section 3.2.1). The differences in the upper glacier area are especially significant in the Karakoram (Fig. 3.2). Improvements in the delineations of the upper glaciers' areas are ongoing for the GAMDAM inventory (Sakai, 2018), and that is why the glacier area of the currently available (December 2015) version of the GAMDAM inventory (Table 3.1) is larger than that reported for the first version (Nuimura et al., 2015).

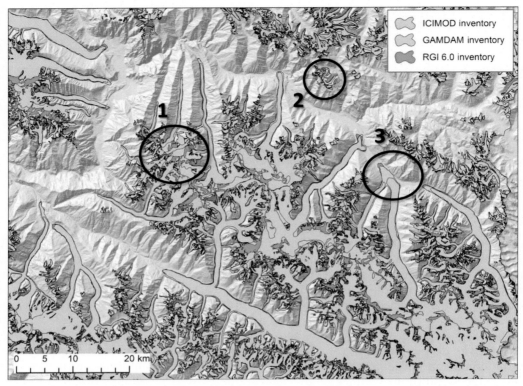

FIG. 3.2 Comparison of different major glacier inventories available for the Indus River Basin showing an example of Hunza Basin/Karakoram. Three major reasons for the differences in glacier area are (1) different methodologies of mapping the accumulation areas, (2) missing glaciers, and (3) changes in the glacier snouts due to the different years of satellite data used.

3.2 PAST GLACIER CHANGES

3.2.1 Area and Length Changes

Glaciers in the Indus Basin have been on average retreating since the mid-19th century. Existing snout fluctuation measurements indicate heterogeneous behavior with advances, retreats, and periods of stagnancy, especially between about 1920 and 1940 (Mayewski and Jeschke, 1979). Apart from the Karakoram, the vast majority of the glaciers also have been retreating and shrinking since the mid-20th century. Reported shrinkage rates in the Western Himalaya vary between 0.20 and 0.75%a^{-1} (Bolch et al., 2012; Kulkarni and Karyakarte, 2014). On average, larger glaciers are shrinking less than smaller ones (e.g., Bhambri et al., 2011), although the smaller glaciers of Ladakh show lower rates of retreat than other Himalayan glaciers (Schmidt and Nüsser, 2012). Glaciers in the Hindu Kush mountains have also experienced significant length reductions since 1973 (Haritashya et al., 2009; Sarikaya et al., 2012). Investigations of glacier area changes since the 1970s revealed on average no significant changes in the Karakoram (Bhambri et al., 2013; Minora et al., 2016). Individual glaciers, however, showed large non-synchronous variations with rapid advances ("surges," Fig. 3.3). Indeed, glacier surges are common in the Karakoram (Bhambri et al., 2017; Copland et al., 2011). This "Karakoram anomaly" was first investigated in detail by Hewitt (2005). However, the first reports on glacier surges date back to the beginning of the past century (Hayden, 1907; Mason, 1930). Mason (1935) and Tewari (1971 in Mayewski and Jeschke, 1979) reported on the periodicity of the surges and discovered an approximate 50-year surge cycle. A more detailed study reported average surge cycles of 25–40 years (Copland et al., 2011). The occurrence of surges can be related to specific topographic conditions and in particular basal processes related to subglacial hydrology and changes in the glacier bed (Sevestre and Benn, 2015), but the exact reasons for glacier surges still are not well understood. Glaciers in the Zanskar Range and glaciers at Nanga Parbat,

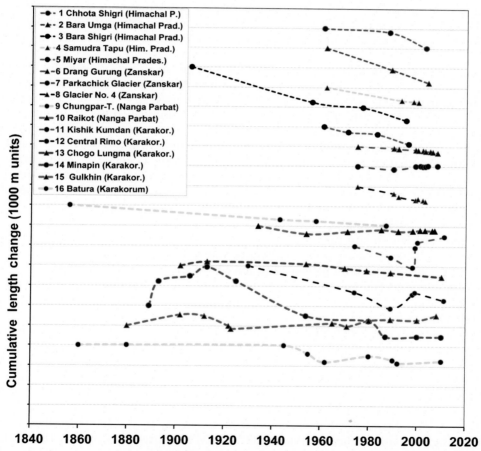

FIG. 3.3 Cumulative length changes of selected glaciers in the Indus Basin. Data sources: Bhambri et al. (2013), Bhambri and Bolch (2009), Bolch et al. (2012), Kamp et al. (2011), Schmidt and Nüsser (2009), and WGMS (2008).

located in the western part of Western Himalaya relatively close to the Karakoram show on average slight retreat but also some advances (Fig. 3.3, Kamp et al., 2011; Schmidt and Nüsser, 2009).

3.2.2 Mass Changes

On average, the mass budget of glaciers in the Indus Basin was negative since the 1970s. A recent study reported a mass loss of -0.16 ± 0.08 m w.a. a^{-1} for the period 2000–16, calculated based on ASTER stereo data (Brun et al., 2017). This number fits relatively well with the average mass loss of -0.21 w.a. a^{-1}, calculated based on existing studies that provide information on mass changes in similar regions (Table 3.2). However, mass changes are highly heterogeneous. Mass loss rates in the Western Himalaya region are the highest in the Indus Basin (see Table 3.1) while glaciers in the Karakoram exhibit on average only slight mass losses for the similar period (see Table 3.1; Bolch et al., 2017; Kääb et al., 2015; Rankl and Braun, 2016). Gardelle et al. (2013) found positive (although insignificant) mass budgets for Central Karakoram. In the adjacent Hindu Kush, mass changes were negative (Table 3.2).

Fewer studies address the longer-term mass changes in sub-regions of the Indus Basin based on remote sensing data. Those that do, confirm that the Karakoram anomaly with heterogeneous mass changes with only slight mass changes overall can be dated back to at least the 1970s (Bolch et al., 2017; Zhou et al., 2017). The only published in situ estimate of mass changes in the Karakoram exists for the Siachen Glacier and is based on the hydrological method (estimated based on measured outflow of the glaciers and estimated precipitation). The value of -0.51 m w.a. a^{-1} for 1986–91 by Bhutiyani (1999) was later corrected to -0.23 m w.e. a^{-1} (Zaman and Liu, 2015). If all the potential error sources are taken into account, the glacier may even have been in balance (Zaman and Liu, 2015). A geodetic estimate of this glacier revealed a mass balance of -0.03 ± 0.21 m w.e. a^{-1} for the period 1999–2007 (Agarwal et al., 2017).

To date very few in situ mass balance measurements exist. All these measurements are concentrated in the Western Himalaya region and are relatively short-term measurements (Bolch et al., 2012). The only glacier with long-term in situ glaciologic mass balance data is the Chhota Shigri Glacier, located in Lahaul-Spiti in Western Himalaya where measurements started in 2002 (Azam et al., 2012). No in situ glaciological measurements are available to date from glaciers in the Karakoram, but a few measurements were recently initiated. The in situ measurements of Chhota Shigri were extended back by in-situ GPS measurements and modeling (Azam et al., 2014; Vincent et al., 2013). The results showed significantly different mass losses before and after about 2000. While the glacier was on average in balance between 1986 and 2000, there was a significant mass loss (-0.57 ± 0.36 m w.a. a^{-1}) afterward. As the glacier showed similar mass loss rates as the entire Lahaul-Spiti after 2000, it was postulated that the entire Lahaul-Spiti experienced this period of little mass loss (Vincent et al., 2013). This general tendency have been confirmed recently by geodetic mass balance calculations using 1971 corona measurements, 2000 SRTM measurements, and ~2010 ASTER and Cartosat-1 data (Mukherjee et al., 2018). Zhou et al. (2018) also found only slightly negative

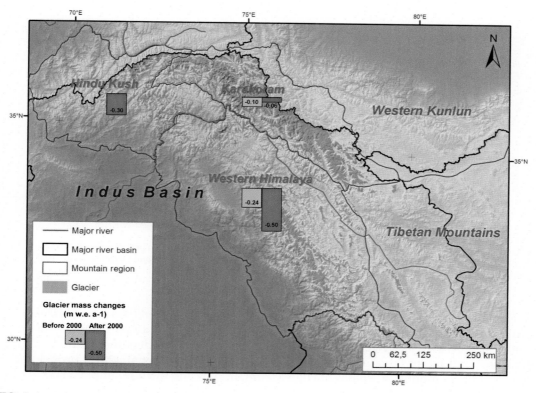

FIG. 3.4 Average glacier mass changes for the periods ~1970–2000 and ~2000–15 for Hindu Kush, Karakoram, and Western Himalaya regions.

mass loss for the period 1975–2000 in Lahaul-Spiti. Considering existing short-term in situ mass balance measurements and the geodetic results, the mass loss for the entire Western Himalaya was on average ~0.24 m w.e. a^{-1} before 2000 and then doubled afterward (Table 3.2, Fig. 3.4).

3.3 FUTURE TRENDS

A few studies provide modeling results of future glacier changes in the Indus Basin as part of a global or large regional study or as part of hydrological modeling. These studies agree that glaciers in the RGI region of Southwest Asia will significantly lose mass until the end of this century under typical climate scenarios. Modeling results project volume reductions between −18.6% and −30.3% for Representative Concentration Pathway (RCP) 4.5% and −19.1% and −35.9% for RCP 8.5 until 2050 (Fig. 3.5). These studies project an ice volume loss

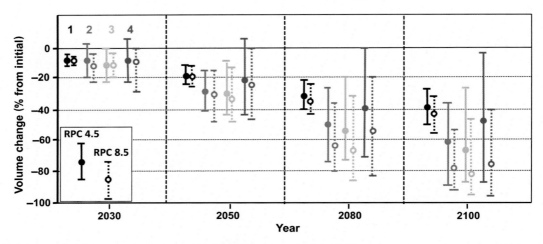

FIG. 3.5 Selected projections of future glacier volume changes in the Indus Basin based on the RPC 4.5 and RPC 8.5. Modified from Bolch et al. (2019). Data sources: 1: Marzeion et al. (2012), 2: Giesen and Oerlemans (2013), 3: Huss and Hock (2015), and 4: Radić et al. (2014).

from 40% to more than 60% by 2100. The smallest reduction is predicted by Marzeion et al. (2012) and the highest by Huss and Hock (2015). The latter study parametrizes important physical processes like glacier flow and calving and also considers the distributed glacier thickness and might therefore be more realistic. Kraaijenbrink et al. (2017) provide results for all different mountain regions of High Mountain Asia and include simple schemes of glacier flow and the impact of debris cover. This study predicts a more moderate glacier volume loss in the Karakoram (between 15% and 22% by 2050 for RPC 4.5 and 8.5 and between 30% and 50% by 2100), the highest mass loss for the western Himalayan region (between 50% and 60% by 2050 and between 70% and 90% by 2100), and a slightly lower loss for the Hindu Kush (between 48% and 56% by 2050 and between 70% and 85% by 2100).

The overall runoff of the Upper Indus River is projected to increase until at least 2050, mainly due to increased glacier melt (Lutz et al., 2014). However, further projected glacier mass and area losses will lead to glacier melt and water decreases after a tipping point is reached. The timing of the tipping point varies and depends on the characteristics of climate change, especially their effects on the glacier volume and elevation distribution. Overall glacier runoff in the Indus Basin is projected to decline after 2050, but the timing of the peak water is later (∼2070) for the large glacierized basins (Huss and Hock, 2018) that are characteristic of the Karakoram.

3.4 DISCUSSION AND CONCLUSIONS

Past glacier changes in the Indus Basin are contrasting. The mass loss rates in the Western Himalaya region have not only been the highest of the regions in the Indus Basin (see Table 3.1) but also are among the highest in all of High Mountain Asia (HMA) since the year 2000. Only the glaciers in the outer ranges of the Tien Shan and in the eastern Himalayan and southeastern Tibetan mountains (Hengduan Shan, eastern Nyainqentanglha) exhibited similar high mass loss rates between 0.40 and 0.60 m w.e. a^{-1} (Brun et al., 2017; Farinotti et al., 2015; Gardner et al., 2013; Kääb et al., 2015; Neckel et al., 2014). In contrast, glaciers in the Karakoram were shown to be among those with the lowest mass loss rates of HMA. Similar low mass losses or even slight mass gains have been found in the neighboring Western Kunlun and Eastern Pamir (Brun et al., 2017; Holzer et al., 2015; Kääb et al., 2015; Lin et al., 2017; Neckel et al., 2014). Glacier mass loss was found to be moderate in the Hindu Kush. Since glacier cover and glacier volume in the Karakoram are much higher than in the other two sub-regions, the average mass loss rate of the entire Indus Basin was moderate since about the 1970s (about −0.20 m w.e. a^{-1}).

An interesting phenomenon is that glacier mass loss at least doubled in Lahaul-Spiti after about 2000 in comparison with about 1970–2000 (Mukherjee et al., 2018), while glacier mass loss rates remained similar or even less in the Karakoram in the recent period (Bolch et al., 2017). The anomalous behavior of glaciers in the Karakoram is also manifested in the fact that the region has one of the highest concentrations of surge-type glaciers on Earth (Sevestre and Benn, 2015). Surge-type glaciers are also relatively common in the neighboring Pamir (Kotlyakov et al., 2008; Osipova and Khromova, 2010) and Western Kunlun (Yasuda and Furuya, 2013). A few glacier surges were also reported for the Tien Shan (Mukherjee et al., 2017), but to date none have been reported for the Hindu Kush and Western Himalaya. Although surge-type glaciers have a characteristic cycle between the active phase with a strong thickening and typically advance at the snout and the quiescence phase with strong downwasting of the terminus, the exact timing of a surge is hard to predict. As the surge-type glaciers are on average relatively large (Barrand and Murray, 2006), the impact on the entire mass changes are not negligible but have not been considered in studies about future predictions. However, not considering the surge-type glaciers separately may not have a significant effect on the overall modeling results of long-term projections, as several studies found no significant differences in mass changes between surge-type and non-surge-type glaciers (Bolch et al., 2017; Gardelle et al., 2013).

In addition, most studies do not find significant differences in mass changes of debris-covered and debris-free glaciers (Gardelle et al., 2013; Kääb et al., 2012). However, debris-covered tongues typically reach lower elevations (Mölg et al., 2018), and the highest surface elevation changes are not close to the termini as with clean-ice glaciers but are a bit up-glacier (Bolch et al., 2011; Ragettli et al., 2016). Thus not considering the effect of debris-cover would overestimate the ice melt, but thus far the effect of debris cover has been integrated in only one published study addressing large regions (Kraaijenbrink et al., 2017). Besides differences in glacier outlines, glacier volume estimates also vary strongly for the Indus Basin (Frey et al., 2014). Future predictions of glacier mass changes therefore are subject to high uncertainty, which is indicated by the large error bars shown in Fig. 3.5. All projections, however, indicate that the glaciers will significantly lose mass until the end of this century, which will probably also lead to a decrease in runoff of the Indus River in the long run (Huss and Hock, 2018).

Acknowledgments

The author acknowledges funding from the ESA projects Glaciers_cci (Project no. 4000109873/14/I-NB) and Dragon 4 (No. 4000121469/17/I-NB), and the Swiss National Science Foundation (No. 200021E_177652/1).

References

Agarwal, V., Bolch, T., Syed, T.H., Pieczonka, T., Strozzi, T., Nagaich, R., 2017. Area and mass changes of Siachen Glacier (East Karakoram). J. Glaciol. 63 (237), 148–163. https://dx.doi.org/10.1017/jog.2016.127.

Azam, M.F., Wagnon, P., Ramanathan, A., Vincent, C., Sharma, P., Arnaud, Y., Linda, A., Pottakkal, J.G., Chevallier, P., Singh, V.B., Berthier, E., 2012. From balance to imbalance: a shift in the dynamic behaviour of Chhota Shigri Glacier (Western Himalaya, India). J. Glaciol. 58 (208), 315–324. https://dx.doi.org/10.3189/2012JoG11J123.

Azam, M.F., Wagnon, P., Vincent, C., Ramanathan, A., Linda, A., Singh, V.B., 2014. Reconstruction of the annual mass balance of Chhota Shigri glacier, Western Himalaya, India, since 1969. Ann. Glaciol. 55 (66), 69–80. https://dx.doi.org/10.3189/2014AoG66A104.

Azam, M.F., Ramanathan, A., Wagnon, P., Vincent, C., Linda, A., Berthier, E., Sharma, P., Mandal, A., Angchuk, T., Singh, V.B., Pottakkal, J.G., 2016. Meteorological conditions, seasonal and annual mass balances of Chhota Shigri Glacier, western Himalaya, India. Ann. Glaciol. 57 (71), 328–338. https://dx.doi.org/10.3189/2016AoG71A570.

Bajracharya, S.R., Shrestha, B.R. (Eds.), 2011. The Status of Glaciers in the Hindu Kush-Himalayan Region. International Centre for Integrated Mountain Development (ICIMOD), Kathmandu, Nepal. 127 pp.

Barrand, N.E., Murray, T., 2006. Multivariate controls on the incidence of glacier surging in the Karakoram Himalaya. Arct. Antarct. Alp. Res. 38 (4), 489–498. https://dx.doi.org/10.1657/1523-0430(2006)38.

Benn, D.I., Bolch, T., Hands, K., Gulley, J., Luckman, A., Nicholson, L.I., Quincey, D., Thompson, S., Toumi, R., Wiseman, S., 2012. Response of debris-covered glaciers in the Mount Everest region to recent warming, and implications for outburst flood hazards. Earth Sci. Rev. 114 (1–2), 156–174. https://dx.doi.org/10.1016/j.earscirev.2012.03.008.

Bhambri, R., Bolch, T., 2009. Glacier mapping: a review with special reference to the Indian Himalayas. Prog. Phys. Geogr. 33 (5), 672–704. https://dx.doi.org/10.1177/0309133309348112.

Bhambri, R., Bolch, T., Chaujar, R.K., Kulshreshtha, S.C., 2011. Glacier changes in the Garhwal Himalayas, India 1968-2006 based on remote sensing. J. Glaciol. 57 (203), 543–556. https://dx.doi.org/10.3189/002214311796905604.

Bhambri, R., Bolch, T., Kawishwar, P., Dobhal, D.P., Srivastava, D., Pratap, B., 2013. Heterogeneity in glacier response in the Shyok valley, northeast Karakoram. Cryosphere 7, 1384–1398. https://dx.doi.org/10.5194/tc-7-1385-2013.

Bhambri, R., Hewitt, K., Kawishwar, P., Pratap, B., 2017. Surge-type and surge-modified glaciers in the Karakoram. Sci. Rep. 7 (1), 15391. https://dx.doi.org/10.1038/s41598-017-15473-8.

Bhutiyani, M.R., 1999. Mass-balance studies on Siachen Glacier in the Nubra valley, Karakoram. Himalaya, India. J. Glaciol. 45 (149), 112–118. https://dx.doi.org/10.3198/1999JoG45-149-112-118.

Bolch, T., Pieczonka, T., Benn, D.I., 2011. Multi-decadal mass loss of glaciers in the Everest area (Nepal, Himalaya) derived from stereo imagery. Cryosphere 5, 349–358. https://dx.doi.org/10.5194/tc-5-349-2011.

Bolch, T., Kulkarni, A., Kääb, A., Huggel, C., Paul, F., Cogley, J.G., Frey, H., Kargel, J.S., Fujita, K., Scheel, M., Bajracharya, S., Stoffel, M., 2012. The state and fate of Himalayan glaciers. Science 336 (6079), 310–314. https://dx.doi.org/10.1126/science.1215828.

Bolch, T., Pieczonka, T., Mukherjee, K., Shea, J., 2017. Brief communication: glaciers in the Hunza catchment (Karakoram) have been nearly in balance since the 1970s. Cryosphere 11 (1), 531–539. https://dx.doi.org/10.5194/tc-11-1-2017.

Bolch, T., Shea, J.M., Liu, S., Azam, F.M., Gao, Y., Gruber, S., et al., 2019. Status and change of the HKH cryosphere. In: Wester, P., Mishra, A., Mukherji, A., Shrestha, A.B. (Eds.), The Hindu Kush Himalaya Assessment—Mountains, Climate Change, Sustainability and People. Springer, in press.

Brun, F., Berthier, E., Wagnon, P., Kääb, A., Treichler, D., 2017. A spatially resolved estimate of High Mountain Asia glacier mass balances from 2000 to 2016. Nat. Geosci. 10, 668–673. https://dx.doi.org/10.1038/NGEO2999.

Carenzo, M., Pellicciotti, F., Mabillard, J., Reid, T., Brock, B.W., 2016. An enhanced temperature index model for debris-covered glaciers accounting for thickness effect. Adv. Water Resour. 94, 457–469. https://dx.doi.org/10.1016/j.advwatres.2016.05.001.

Copland, L., Sylvestre, T., Bishop, M.P., Shroder, J.F., Seong, Y.B., Owen, L.A., Bush, A., Kamp, U., 2011. Expanded and recently increased glacier surging in the Karakoram. Arct. Antarct. Alp. Res. 43 (4), 503–516. https://dx.doi.org/10.1657/1938-4246-43.4.503.

Farinotti, D., Longuevergne, L., Moholdt, G., Duethmann, D., Mölg, T., Bolch, T., Vorogushyn, S., Guntner, A., 2015. Substantial glacier mass loss in the Tien Shan over the past 50 years. Nat. Geosci. 8 (9), 716–722. https://dx.doi.org/10.1038/ngeo2513.

Frey, H., Machguth, H., Huss, M., Huggel, C., Bajracharya, S., Bolch, T., Kulkarni, A., Linsbauer, A., Salzmann, N., Stoffel, M., 2014. Estimating the volume of glaciers in the Himalayan-Hindu Kush-Karakoram region using different methods. Cryosphere 8 (6), 2313–2333. https://dx.doi.org/10.5194/tc-8-2313-2014.

Gardelle, J., Berthier, E., Arnaud, Y., Kääb, A., 2013. Region-wide glacier mass balances over the Pamir-Karakoram-Himalaya during 1999-2011. Cryosphere 7, 1263–1286. https://dx.doi.org/10.5194/tc-7-1263-2013.

Gardner, A.S., Moholdt, G., Cogley, J.G., Wouters, B., Arendt, A.A., Wahr, J., Berthier, E., Hock, R., Pfeffer, W.T., Kaser, G., Ligtenberg, S.R.M., Bolch, T., Sharp, M.J., Hagen, J.O., van den Broeke, M.R., Paul, F., 2013. A reconciled estimate of glacier contributions to sea level rise: 2003 to 2009. Science 340, 852–857. https://dx.doi.org/10.1126/science.1234532.

GCOS, 2004. Implementation Plan for the Global Observing System for Climate in Support of the UNFCCC. GCOS-92. WMO/TD No.1219. No.1219. GCOS-92. WMO/TD.

Giesen, R.H., Oerlemans, J., 2013. Climate-model induced differences in the 21st century global and regional glacier contributions to sea-level rise. Clim. Dyn. 41, 3283–3300. https://dx.doi.org/10.1007/s00382-013-1743-7.

GTN-G, 2017. GTN-G Glacier Regions. Global Terrestrial Network for Glaciers. https://dx.doi.org/10.5904/gtng-glacreg-2017-07.

Haritashya, U.K., Bishop, M.P., Shroder, J.F., Bush, A.B.G., Bulley, H.N.N., 2009. Space-based assessment of glacier fluctuations in the Wakhan Pamir, Afghanistan. Clim. Chang. 94, 5–18. https://dx.doi.org/10.1007/s10584-009-9555-9.

Hayden, H.H., 1907. Notes on certain glaciers in Northwest Kashmir. Rec. Geol. Surv. India 35, 127–137.

Hewitt, K., 2005. The Karakoram Anomaly? Glacier expansion and the "Elevation Effect" Karakoram Himalaya. Mount. Res. Dev. 25 (4), 332–340.

Holzer, N., Vijay, S., Yao, T., Xu, B., Buchroithner, M., Bolch, T., 2015. Four decades of glacier variations at Muztagh Ata (eastern Pamir): a multi-sensor study including Hexagon KH-9 and Pléiades data. Cryosphere 9 (6), 2071–2088. https://dx.doi.org/10.5194/tc-9-2071-2015.

Huss, M., Hock, R., 2015. A new model for global glacier change and sea-level rise. Front. Earth Sci. 3, 54. https://dx.doi.org/10.3389/feart.2015.00054.

Huss, M., Hock, R., 2018. Global-scale hydrological response to future glacier mass loss. Nat. Clim. Chang. https://dx.doi.org/10.1038/s41558-017-0049-x.

Kääb, A., Berthier, E., Nuth, C., Gardelle, J., Arnaud, Y., 2012. Contrasting patterns of early twenty-first-century glacier mass change in the Himalayas. Nature 488 (7412), 495–498. https://dx.doi.org/10.1038/nature11324.

Kääb, A., Teichler, D., Nuth, C., Berthier, E., 2015. Brief communication: contending estimates of 2003–2008 glacier mass balance over the Pamir–Karakoram–Himalaya. Cryosphere 557–564. https://dx.doi.org/10.5194/tc-9-557-2015.

Kamp, U., Byrne, M., Bolch, T., 2011. Glacier fluctuations between 1975 and 2008 in the Greater Himalaya Range of Zanskar, southern Ladakh. J. Mt. Sci. 8 (3), 374–389. https://dx.doi.org/10.1007/s11629-011-2007-9.

Kaul, M.N., 1986. Mass balance of Lidder glaciers. Trans. Inst. Indian Geograph. 8, 95.

Kotlyakov, V.M., Osipova, G.B., Tsvetkov, D.G., 2008. Monitoring surging glaciers of the Pamirs, central Asia, from space. Ann. Glaciol. 48, 125–134.

Koul, M., Ganjoo, R., 2010. Impact of inter- and intra-annual variation in weather parameters on mass balance and equilibrium line altitude of Naradu Glacier (Himachal Pradesh), NW Himalaya, India. Clim. Chang. 99 (1), 119–139. https://dx.doi.org/10.1007/s10584-009-9660-9.

Kraaijenbrink, P.D.A., Bierkens, M.F.P., Lutz, A.F., Immerzeel, W.W., 2017. Impact of a global temperature rise of 1.5 degrees celsius on Asia's glaciers. Nature. https://dx.doi.org/10.1038/nature23878.

Kulkarni, A.V., Karyakarte, Y., 2014. Observed changes in Himalayan glaciers. Curr. Sci. 106 (2), 237–244.

Lin, H., Li, G., Cuo, L., Hooper, A., Ye, Q., 2017. A decreasing glacier mass balance gradient from the edge of the Upper Tarim Basin to the Karakoram during 2000–2014. Sci. Rep. 7 (1), 6712. https://dx.doi.org/10.1038/s41598-017-07133-8.

Lutz, A.F., Immerzeel, W.W., Shrestha, A.B., Bierkens, M.F.P., 2014. Consistent increase in High Asia's runoff due to increasing glacier melt and precipitation. Nat. Clim. Chang. 4 (7), 587–592.

Marzeion, B., Jarosch, A.H., Hofer, M., 2012. Past and future sea-level change from the surface mass balance of glaciers. Cryosphere 6 (6), 1295–1322. https://dx.doi.org/10.5194/tc-6-1295-2012.

Mason, K., 1930. The glaciers of the Karakoram and neighbourhood. Rec. Geol. Surv. India 63 (2), 214–278.

Mason, K., 1935. The study of threatening glaciers. Geogr. J. 85 (1), 24–35.

Mayewski, P.A., Jeschke, P.A., 1979. Himalayan and trans-Himalayan glacier fluctuations since AD 1812. Arct. Alp. Res. 11 (3), 267–287.

Minora, U., Bocchiola, D., D'Agata, C., Maragno, D., Mayer, C., Lambrecht, A., Vuillermoz, E., Senese, A., Compostella, C., Smiraglia, C., Diolaiuti, G.A., 2016. Glacier area stability in the Central Karakoram National Park (Pakistan) in 2001–2010: The "Karakoram Anomaly" in the spotlight. Prog. Phys. Geogr. https://dx.doi.org/10.1177/0309133316643926.

Mölg, N., Bolch, T., Rastner, P., Strozzi, T., Paul, F., 2018. A consistent glacier inventory for the Karakoram and Pamir derived from Landsat data: distribution of debris cover and mapping challenges. Earth Syst. Sci. Data 10, 1807–1827. https://doi.org/10.5194/essd-10-1807-2018.

Mukherjee, K., Bolch, T., Goerlich, F., Kutuzov, S., Osmonov, A., Pieczonka, T., Shesterova, I., 2017. Surge-type glaciers in the Tien Shan (Central Asia). Arct. Antarct. Alp. Res. 49 (1), 147–171. https://dx.doi.org/10.1657/AAAR0016-021.

Mukherjee, K., Bhattacharya, A., Pieczonka, T., Ghosh, S., Bolch, T., 2018. Moderate glacier mass loss followed by strongly accelerated loss in western Himalaya with an increase in temperature and precipitation. Clim. Chang. https://dx.doi.org/10.1007/s10584-018-2185-3.

Neckel, N., Kropáček, J., Bolch, T., Hochschild, V., 2014. Glacier mass changes on the Tibetan Plateau 2003–2009 derived from ICES at laser altimetry measurements. Environ. Res. Lett. 9 (1), 14009. https://dx.doi.org/10.1088/1748-9326/9/1/014009.

Nuimura, T., Sakai, A., Taniguchi, K., Nagai, H., Lamsal, D., Tsutaki, S., Kozawa, A., Hoshina, Y., Takenaka, S., Omiya, S., Tsunematsu, K., Tshering, P., Fujita, K., 2015. The GAMDAM glacier inventory: a quality-controlled inventory of Asian glaciers. Cryosphere 9 (3), 849–864. https://dx.doi.org/10.5194/tc-9-849-2015.

Osipova, G.B., Khromova, T.E., 2010. Elektronnaja baza dannykh "Pulsiruyuchne ledniki Pamira". Ice Snow 4, 15–24.

Pfeffer, W.T., Arendt, A.A., Bliss, A., Bolch, T., Cogley, J.G., Gardner, A.S., Hagen, J.-O., Hock, R., Kaser, G., Kienholz, C., Miles, E.S., Moholdt, G., Mölg, N., Paul, F., Radić, V., Rastner, P., Raup, B.H., Rich, J., Sharp, M.J. a.t.R.C., 2014. The Randolph Glacier Inventory: a globally complete inventory of glaciers. J. Glaciol. 60 (221), 537–552. https://dx.doi.org/10.3189/2014JoG13J176.

Radić, V., Bliss, A., Beedlow, A.C., Hock, R., Miles, E., Cogley, J.G., 2014. Regional and global projections of twenty-first century glacier mass changes in response to climate scenarios from global climate models. Clim. Dyn. 42 (1–2), 37–58. https://dx.doi.org/10.1007/s00382-013-1719-7.

Ragettli, S., Bolch, T., Pellicciotti, F., 2016. Heterogeneous glacier thinning patterns over the last 40 years in Langtang Himal, Nepal. Cryosphere 10 (5), 2075–2097. https://dx.doi.org/10.5194/tc-10-2075-2016.

Raina, V.K., 2009. Himalayan Glaciers: A State-of-Art Review of Glacial Studies, Glacial Retreat and Climate Change. New Delhi, 56 pp.

Rankl, M., Braun, M., 2016. Glacier elevation and mass changes over the central Karakoram region estimated from TanDEM-X and SRTM/X-SAR digital elevation models. Ann. Glaciol. 57 (71), 273–281. https://dx.doi.org/10.3189/2016AoG71A024.

RGI Consortium, 2017. Randolph Glacier Inventory (RGI)—A Dataset of Global Glacier Outlines: Version 6.0. Technical ReportGlobal Land Ice Measurements from Space, Boulder, CO.

Sakai, A., 2018. Brief communication: updated GAMDAM glacier inventory over the high mountain Asia. Cryosphere Discuss. in review. https://doi.org/10.5194/tc-2018-139.

Sarikaya, M.A., Bishop, M.P., Shroder, J.F., Olsenholler, J.A., 2012. Space-based observations of Eastern Hindu Kush glaciers between 1976 and 2007, Afghanistan and Pakistan. Remote Sens. Lett. 3 (1), 77–84. https://dx.doi.org/10.1080/01431161.2010.536181.

Scherler, D., Bookhagen, B., Strecker, M.R., 2011. Spatially variable response of Himalayan glaciers to climate change affected by debris cover. Nat. Geosci. 4, 156–159. https://dx.doi.org/10.1038/NGEO1068.

Schmidt, S., Nüsser, M., 2009. Fluctuations of Raikot Glacier during the past 70 years: a case study from the Nanga Parbat massif, northern Pakistan. J. Glaciol. 55 (194), 949–959.

Schmidt, S., Nüsser, M., 2012. Changes of high altitude glaciers from 1969 to 2010 in the trans-Himalayan Kang Yatze Massif, Ladakh, Northwest India. Arct. Antarct. Alp. Res. 44 (1), 107–121. https://dx.doi.org/10.1657/1938-4246-44.1.

Sevestre, H., Benn, D.I., 2015. Climatic and geometric controls on the global distribution of surge-type glaciers: implications for a unifying model of surging. J. Glaciol. 61 (228), 646–662. https://dx.doi.org/10.3189/2015JoG14J136.

Vincent, C., Ramanathan, A., Wagnon, P., Dobhal, D.P., Linda, A., Berthier, E., Sharma, P., Arnaud, Y., 2013. Balanced conditions or slight mass gain of glaciers in the Lahaul and Spiti region (northern India, Himalaya) during the nineties preceded recent mass loss. Cryosphere 7, 569–582. https://dx.doi.org/10.5194/tc-7-569-2013.

Wei, J., Shiyin, L., Wanqin, G., Junli, X., Weijia, B., Donghui, S., 2015. Changes in glacier volume in the North Bank of the Bangong Co Basin from 1968 to 2007 based on historical topographic maps, SRTM, and ASTER stereo images. Arct. Antarct. Alp. Res. 47 (2), 301–311. https://dx.doi.org/10.1657/AAAR00C-13-129.

WGMS, 2008. Fluctuation of Glaciers 2000–2005. Vol. IX. ICSU (FAGS) - IUGG (IACS) - UNEP-UNESCO-WMO, Zürich.

Yao, T., Thompson, L.G., Yang, W., Yu, W., Gao, Y., Guol, X., Yang, X., Duan, K., Zhao, H., Xu, B., Pu, J., Lu, A., Xian, Y., Kattel, D.B., Joswiak, D., 2012. Different glacier status with atmospheric circulations in Tibetan Plateau and surroundings. Nat. Clim. Chang. 2, 663–667. https://dx.doi.org/10.1038/NCLIMATE1580.

Yasuda, T., Furuya, M., 2013. Short-term glacier velocity changes at West Kunlun Shan, Northwest Tibet, detected by synthetic aperture radar data. Remote Sens. Environ. 128, 87–106. https://dx.doi.org/10.1016/j.rse.2012.09.021.

Zaman, Q., Liu, J., 2015. Mass balance of Siachen Glacier, Nubravalley, Karakoram Himalaya: facts or flaws? J. Glaciol. 61 (229), 1012–1014. https://dx.doi.org/10.3189/2015JoG15J120.

Zhou, Y., Li, Z., Li, J.I.A., 2017. Slight glacier mass loss in the Karakoram region during the 1970s to 2000 revealed by KH-9 images and SRTM DEM. J. Glaciol. 63 (238), 331–342. https://dx.doi.org/10.1017/jog.2016.142.

Zhou, Y., Li, Z., Li, J., Zhao, R., Ding, X., 2018. Glacier mass balance in the Qinghai–Tibet Plateau and its surroundings from the mid-1970s to 2000 based on Hexagon KH-9 and SRTM DEMs. Remote Sens. Environ. 210, 96–112. https://dx.doi.org/10.1016/j.rse.2018.03.020.

PART II

CLIMATE-ECO-HYDROLOGY OF INDUS RIVER BASIN

Probabilistic Precipitation Analysis in the Central Indus River Basin

Paolo Reggiani*, Oleksiy Boyko*, Tom H. Rientjes[†], Asif Khan[‡]

[*]Department of Civil Engineering, University of Siegen, Siegen, Germany [†]Department of Water Resources, Faculty ITC, University of Twente, Enschede, The Netherlands
[‡]Department of Civil Engineering, University of Engineering and Technology, Jalozai Campus, Peshawar, Pakistan

4.1 INTRODUCTION

Pakistan's national borders encompass a surface area of 880,000 km^2 and a population of about 199 million people (2018). In general, the climate is arid, and mean annual precipitation ranges from less than 100 mm in parts of the Lower Indus Plain to over 750 mm in the Upper Indus Plain (UNESCO, 2012). The Himalayan foothills are characterized by heavy precipitation up to 2000 mm and beyond due to orographic effects (Singh and Kumar, 1997; Archer and Fowler, 2003), while the Upper Indus Basin (UIB) is once again dryer, with a mean annual precipitation estimated between 600 and 700 mm (Reggiani and Rientjes, 2015). The Indian monsoon advecting moist air from the Bay of Bengal and the Arabian Sea, and the western disturbances, mainly extratropical storms that originate in the Mediterranean and over the Black Sea, are the chief sources of rainfall, two-thirds of which usually falls between July and September. On the Punjab plains, most of the rain falls during the monsoons in early July.

Originating on the Tibetan plateau at about 5000 m.a.s.l., the river Indus flows initially in the westerly direction across the Upper Indus Valley, carved in-between the Karakoram range to the north and the Himalayan range to the south, and then to the mountainous region across a narrow gorge toward the Punjab Plains. The river Indus and its tributaries are the country's most important source of fresh water. With a total drainage area of 1.1 million km^2, the Indus River basin covers approximately 65% of the territory of Pakistan (FAO, 2011), and extends into neighboring China, India, and Afghanistan. The Indus River has two main tributaries, the Kabul on the right bank and the Panjnad on the left. The Panjnad is the merged flow of the Jhelum and Chenab rivers, known as the western rivers together with the river Indus, and the Ravi, Beas, and Sutlej, known as the eastern rivers. This division came into effect at the time of settlement of a water dispute between India and Pakistan in 1960 (Gulhati, 1973). The closure of the UIB at the Himalayan foothills is marked by the Tarbela reservoir, which plays a central role for Pakistan's hydropower production and is the head node of the Indus irrigation scheme, one of the largest of its kind.

Due to rapid population growth during recent decades, Pakistan has become one of the most water-stressed countries in the world, a situation that eventually is going to approach the status of outright water scarcity (UNESCO, 2012). As of 2009, Pakistan had an internal reusable water resource of 323 m^3/inhabitant and year, which is far below 1700 m^3/inhabitant (FAO, 2011), a threshold under which there are indications of water stress. The status of the total available water resources is equally serious with less than 1700 m^3/inhabitant available, slowly approaching a situation of chronic water scarcity once the resource availability has fallen under the 1000 m^3/inhabitant threshold. As of 2010, Pakistan also belongs to the 10 primary groundwater abstracting countries of the world (UNDP, 2016; UNESCO, 2012) and has severely depleted important aquifers due to unsustainable water mining for irrigation during past decades.

According to assessments by the World Bank (Brisoce and Qamar, 2005), there is no feasible intervention at present that would enable Pakistan to mobilize appreciably more water than it now uses. Pakistan's natural dependence on a single major river system means it lacks the robustness in terms of water supply redundancy that most countries enjoy by virtue of drawing water from multiple river basins. To the contrary, Pakistan is also exposed to high risks of floods. As recently as 2010, the country has been hit by an extreme monsoonal rainfall event with extensive flooding in the Punjab and Middle Indus Valley, leading to a considerable destruction of infrastructure and crop production with two thousand lives and more than a million homes lost. Another similarly extreme, albeit less harmful, flood event occurred in 2011 in the Sindh province, affecting also eastern Balochistan and the southern Punjab with a considerable number of casualties and loss of crop production and property.

In view of long-term highly strained water resources and simultaneous high flood risk in Pakistan, it is of primary importance to correctly assess the spatial distribution of precipitation and related uncertainty for a series of reasons. First, past and present precipitation is a crucial input variable into hydrological decision support tools, for instance, rainfall-runoff and flow forecasting models as well as water resources allocation models. For example, incorrect estimation of areal precipitation may lead to erroneous water resources assessments, with potential malinvestments for infrastructure. Second, precipitation fields cannot be measured or estimated deterministically, as their space-time structure features a strongly random behavior. As a result, the uncertainty associated with precipitation estimation or prediction must be accounted for in any decision-making processes for integrated water resources management. The quantification of uncertainty associated with precipitation is also relevant in the case of poorly gauged basins due to lack of sufficiently dense and continuous ground observations for reliable precipitation estimation. Third, hydrometeorological variables such as precipitation, but also temperature and evaporation, may be subject to future change and thus require reliable trend quantification by means of climate change indicators (Sankarasubramanian et al., 2001; Ray and Brown, 2015). Without probabilistic assessment methods that combine information from multiple predictors (there are currently outputs from more than 20 climate change models available at the CMIP5 portal: https://esgf-node.llnl.gov/search/cmip5/, which provide the atmospheric modeling basis for the latest IPCC report (IPCC, 2014)) such indicators can only be evaluated deterministically, with the risk of misrepresenting possible climate change effects attributable to the uncertainty of projections.

Very few studies on precipitation analysis of relevance for the Central Indus Basin can be found in the literature, despite its high water-related vulnerability. Examples include the interpolated daily precipitation products CRU TS4.01 (Harris et al., 2013), the GPCC gridded precipitation (Schneider et al., 2014) and APHRODITE (Yatagai et al., 2012; Ali et al., 2012) dataset for the South-East Asian region. In a somewhat different geographical context, Reggiani et al. (2016) presented a study on high-altitude monthly precipitation and temperature estimation in the Shigar River Basin, Karakoram, Upper Indus, in which six reanalysis products conditioned on a single nearby ground observing station were used. The methodology applied in the analysis is based on a Bayesian processor of uncertainty (Coccia and Todini, 2011), which combines, and thus maximizes, the compound information on monthly precipitation from numerical precipitation reanalyses, whereby ground observations are used for bias removal and processor calibration. The adopted Bayesian approach is referred to as the model conditional processor (MCP) (Todini, 2008) and as such is one among several proven quantitative assessment approaches for uncertainty such as Bayesian Model Averaging (Raftery, 1993; Raftery et al., 2005) or Quantile Regression (QR) analysis (Koenker, 2005). Without entering the details on the advantages and disadvantages of different methodological choices, we devote this chapter to a specific application of the MCP to the Central Indus Basin. The aim of the analysis is the use of six reanalysis products to assess monthly areal precipitation for selected subbasin areas in the Punjab region. In this way, we demonstrate the strength of the chosen approach and open the way for further applications of the methodology on uncertainty quantification of hydrometeorological variables such as precipitation, temperature, and evaporation, and specific climate change indicators. The extended goal is to present a method for rational decision making to support integrated water resources planning and infrastructure investments.

The present chapter is structured as follows: In Section 4.2 we introduce data and methods, in Section 4.3, we apply the Bayesian processor and present the results, Section 4.4 is devoted to the discussion of the outcome, while Section 5.5 contains the summary and conclusions. The details of the MCP approach are summarized in the Appendix.

4.2 DATA AND METHODS

4.2.1 Study Area

In this study we focus exclusively on the Central Indus Basin, including the subbasins of the Ravi, the Jhelam, the Upper and Lower Chenab Rivers, and the Middle Indus Valley Basin, as depicted in Fig. 4.1. The total study area covers a surface of 256,462 km^2 and is mostly located in the Pakistani part of the Indus Basin, with the exception of the Northern Ravi and the Chenab basins reaching into the Indian states of Jammu-Kashmir and Himachal

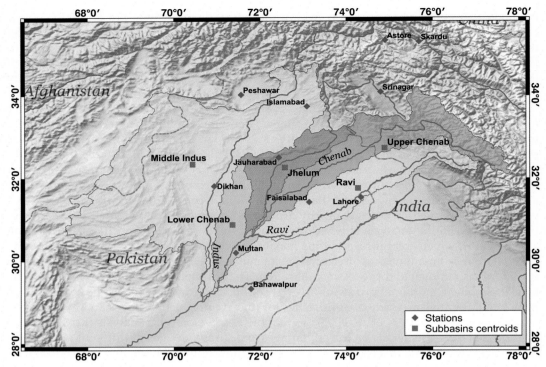

FIG. 4.1   Central Indus study area with subbasin basins and observing stations for precipitation.

Pradesh. The western fringe of the Middle Indus subbasin lies partly in Afghan territory. The study region is largely characterized by a near-flat topography in the Pakistani Punjab, while the northeastern part of Ravi and Chenab reaches into the southern Himalayas. Precipitation varies considerably in the north-south direction with a maximum mean annual precipitation of 1142 mm recorded in Islamabad, 697 mm in Srinagar, 600 mm in Lahore, and falling to a low of 188 and 143 mm in Multan and Bahawalpur, respectively. The high precipitation recorded in Islamabad is due to the proximity of the city to the Himalayan foothills, where orographic effects become dominant. Skardu, which is located in the Upper Indus Valley, also shows precipitation levels of 200 mm, typical for the arid climate found in the low valleys of the UIB. Table 4.1 summarizes the gauging station network used for this study and the mean annual precipitation recorded at the respective sites. The station positions are also indicated on the map in Fig. 4.1.

From an analysis of a regional digital terrain model, we extract five subbasins (Fig. 4.1) making up the study region. Their IDs, centroid position, and individual areal extent are summarized in Table 4.2. The area is subject to the influence of the Indian monsoon, which leads to heavy rainfall from June to September, and light rain due to western disturbances between mid-November and mid-February. The region experiences mild spring weather from mid-February to mid-April, which gives way to very hot and dry weather lasting until the onset of the monsoon in the month of June. Temperatures can get close to or exceed 40°C during

TABLE 4.1 Position and Elevation of Recording Stations for Precipitation, Available Record Length, and Mean Annual Precipitation

Station	Position	Elevation (m)	Period	P (mm)	Gaps (%)
Astore	35°20′ N, 74°54′ E	2394	1954–2010	427	0
Bahawalpur	29°20′ N, 71°47′ E	110	1931–2010	143	0
D.I. Khan	31°49′ N, 70°56′ E	171.2	1931–2010	269	0
Faisalabad	31°26′ N, 73°08′ E	185.5	1951–2010	346	0
Islamabad	33°36′ N, 73°05′ E	508	1976–2010	1142	14
Jauharabad	32°30′ N, 72°26′ E	187	2007–10	389	87
Lahore	31°33′ N, 74°20′ E	214	1931–2010	628	0
Multan	30°12′ N, 71°26′ E	121.95	1950–2010	188	0
Peshawar	34°02′ N, 71°56′ E	372	1950–2010	404	3
Skardu	35°18′ N, 75°41′ E	2210	1952–2010	202	<1
Srinagar (IN)	34°05′ N, 74°47′ E	1587	1961–2010	697	0

TABLE 4.2 Central Indus Subbasins Used for the Precipitation Analysis Including Centroid Position and Surface Area

Subbasin	Centroid	Area (km²)
Jhelam	32°15′ N, 72°34′ E	22,046
Middle Indus	32°20′ N, 70°25′ E	131,079
Upper Chenab	32°44′ N, 74°53′ E	46,427
Lower Chenab	30°53′ N, 71°21′ E	17,817
Ravi	31°46′ N, 74°16′ E	39,093

May and June and fall toward or below the freezing point during December and January, depending on location. In the Himalayas, snow is encountered above the snowfall line during winter months.

4.2.2 In Situ Observations

The first step in our analysis is the elaboration of the records of monthly cumulative precipitation provided by the Pakistan Meteorological Department. The series with original length indicated in Table 4.1 were trimmed to the 1979–2010 period selected for the present analysis. Most series were complete with only a few months missing. The respective percentages of missing values for the 1979–2010 period are indicated in Table 4.1. The incomplete records are Jauharabad (87% missing) with only 3 years of data available, followed by Islamabad (14% missing) and Peshawar (3% missing).

To reconstruct smaller dispersed gaps in the records of Islamabad and Lahore we used the statistical procedure described in Pegram (1997) that is based on multiple linear regression, singular value decomposition, and pseudo-expectation-maximization. In this context we used only the record of Lahore to fill the 3-year gap for Islamabad, as the precipitation patterns between the two locations were statistically most similar. Using additional stations with very different precipitation statistics would have distorted the results. Filling the gaps for Peshawar posed no problems at all using all available stations as support.

To fill the 87% gap in Jauharabad, we resorted to Kriging, using all remaining stations as a basis. Kriging is a geostatistical method in which the mutual spatial correlation structure of the records is represented by an empirical semivariogram, which is approximated in terms of a parametric model. One needs to note that the semivariogram changes each time step and thus needs to be fitted in the present case on a monthly basis. We chose from four different semivariogram models, which were selected on the basis of optimal weighted least squares fitting (Cressie, 1985). The optimal parameters were found by means of the conjugate gradient method by minimizing the least squares error used as cost function. Fig. 4.2 shows four semivariograms and respective parametric models (Gaussian, exponential, spherical, and rational quadratic models) for four arbitrary months of the study period that were used for representing the change of variance as a function of the horizontal distance between stations. Once all series were completed, we performed Block Kriging from the 11 stations in Table 4.1 toward the centroids of the subbasins (Table 4.2), obtaining complete estimates of observed monthly areal precipitation in the study basins. These will be used as a reference for comparison with postprocessed precipitation from atmospheric reanalysis products. Finally, we separated the records for each year into a 5-month subrecord representing the monsoon period (July–November) and a 7-month record representing the intermonsoon period or dry season (December–June).

4.2.3 Atmospheric Reanalyses

As stated in Section 4.1, we resort to an ensemble of six atmospheric reanalyses to be used as predictors for monthly precipitation. We have selected the most contemporary reanalysis products developed by source organizations that apply different models and data assimilation procedures to obtain a six-member ensemble of independent physically based estimates of monthly precipitation: (1) ERA-Interim (Dee et al., 2014), (2) ERA20C (Stickler et al., 2010), (3) Japanese 55-year reanalysis (Kobayashi et al., 2015), (4) NCEP-NCAR reanalysis R1 (Kistler et al., 2001), (5) NCEP-CFSR (Suranjana et al., 2010), and (6) NASA MERRA (Rienecker et al., 2011). The products and their most important characteristics are summarized in Table 4.3. For reasons of record length consistency among reanalysis projects, only data from 1979 onward were used.

Next, the precipitation values for individual model cells are spatially averaged over the five subbasins (Table 4.2, Fig. 4.1). The reanalysis outputs have different spatial resolutions and therefore involve a varying number of grid cells. The averaging procedure requires the calculation of weighted mean of the respective variable X using the subbasin mask to identify the relevant model cells. All reanalysis cells that overlap with the area enclosed by the mask

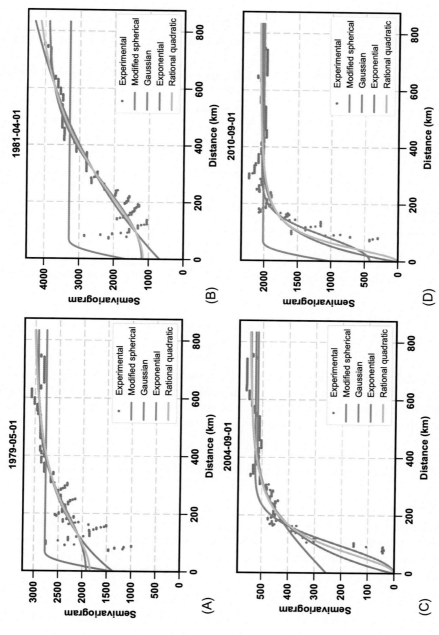

FIG. 4.2 Empirical and fitted semivariograms using four fitting models for different time periods in subplots (A)–(D).

TABLE 4.3 Overview of the Six Atmospheric Reanalyses and Gridded Observed Precipitation Indicating the Product Name, Source Agency, Model Grid Projection, and Spatial Output Resolution for the Analyzed Fields

Data Set	Origin	Grid Type	Spatial Res.
ERAI	ECMWF	N128 Gaussian	0.75 degrees × 0.75 degrees (~83 km)
NCEP/NCAR R1	NCEP/NCAR	1.875 degrees × 1.875 degrees	1.875 degrees × 1.875 degrees (~209 km)
MERRA	NASA	1/2 degrees × 2/3 degrees	1/2 degrees × 2/3 degrees (~55 × 73 km)
NCEP CFSR	NCEP	T382 Gaussian	0.313 degrees × 0.313 degrees (~38 km)
55Yr Japanese Reanalysis	Jpn Met. Agency	TL390L60	1.25 degrees × 1.25 degrees (~135 km)
ERA 20C Reanalysis	ECMWF	N80 Gaussian	1.125 degrees × 1.125 degrees (~125 km)

are used to perform the averaging. The average $\overline{X}$ is given by the area-weighted mean precipitation is

$$\overline{X}^j = \frac{\sum_{i=1}^{n} X_i^j \cdot A_i}{A_{tot}} \tag{4.1}$$

where the index i indicates cell i, j indicates the monthly time step, A_i is the basin area portion covered by the ith reanalysis pixel, A_{tot} is the subbasin area, and n is the total number of cells found within the subbasin mask. The resulting series of basin-average data for each reanalysis series are supposed to hold approximately at the basin centroid.

4.3 APPLICATION OF THE MODEL CONDITIONAL PROCESSOR

4.3.1 Predictive Uncertainty

Bayesian inference and forecasting approaches have become increasingly popular in assessing the uncertainty of numerical predictions in flow and weather forecasting, especially after "predictive uncertainty" (PU) was defined (Krzysztofowicz, 1999) as a conditional probability density,

$$PU = f(y|\hat{y}_{1,t_o}, \hat{y}_{2,t_o}, \hat{y}_{3,t_o}, \ldots, \hat{y}_{n,t_o}) \tag{4.2}$$

where f is a conditional probability density function, y is the predictand describing an uncertain process such as precipitation at a given time t and $\hat{y}_{i,t_o}, i = 1, \ldots, n$ are n random predictor variables at time t for a forecast starting at time t_o. In a forecasting context the chief interest lies in the assessment of PU for prediction times $t > t_o$. A reforecast on the other hand, as addressed in this analysis, reproduces the past and thus $t_o = t$ at each time.

The model-conditional processor (MCP) (Todini, 2008) is a probabilistic tool designed to estimate conditional posterior distribution in Eq. (4.2). The prior distributions of both the predictand and the predictors are assumed equal to their climatological distributions and are mapped to the Gaussian space by means of the Normal Quantile Transform (NQT) (Wilks, 1995). In the normal space, the interdependency structure of predictand and predictors is assumed to be multivariate normal. Coccia and Todini (2011) proposed an extended version of the MCP that allows for the inclusion of multiple forecasting models with heteroscedastic dependency structures. If the assumption of normality of the multivariate distribution is not acceptable due to significant heteroscedasticity of the residuals in the normal space, the joint distribution of predictand and predictors is separated into multiple truncated normal distributions (TNDs), which are applicable to subsegments of the normal variable value range. An application of the MCP to monthly precipitation in the Karakoram has been proposed in Reggiani et al. (2016). Given the prior distributions of predictand and predictors and their joint multivariate normal distribution, Bayes' theorem can be applied to derive normal conditional densities of the predictand given multiple predictors. Finally, the conditional density is back-transformed from the normal into the original space by inverse NQT. A more detailed mathematical recapitulation of the MCP is provided in the Appendix.

4.3.2 Normalization of Variables

To process the precipitation observations and atmospheric reanalysis data, we apply the MCP in a series of consecutive steps. First, we transform the series of monthly basin-average observed precipitation, which were separated into monsoon and dry season into the Gaussian space using the NQT. For a realization $\mathbf{y}$ of the random precipitation process $\mathbf{Y}$ consisting of m data points with respective empirical distribution Γ, the NQT is defined as the transformation

$$\boldsymbol{\eta} = Q^{-1}(\Gamma(\mathbf{y})) \quad \mathbf{y} = y_k; \ \boldsymbol{\eta} = \eta_k; \ k = 1, \ldots, m \tag{4.3}$$

where Q^{-1} is the inverse of the standard normal distribution and the transform of $\mathbf{y}$ is denoted with $\boldsymbol{\eta}$ and is $N(0, 1)$. The same is done for the subbasin-average precipitation series of the six atmospheric reanalysis outputs, all to be used as indicators. We note that we use monthly precipitation, leading to very few values of zero precipitation in the series. For this reason we model monthly precipitation as a continuous random process with sporadic zero occurrences, in contrast to precipitation at higher temporal resolution, which needs to be modeled as a binary-continuous process (Todorovic and Yevjevich, 1969) by separating storm from interstorm periods. This assumption simplifies the following analysis considerably.

4.3.3 Precipitation: Normal Distributions

We refer to the notation introduced in the Appendix., which indicates realizations at time t of the standard normal random process η and $\hat{\eta}_i$ as η, respectively, $\hat{\eta}_i, i = 1, \ldots, n$, whereby we omit an index referring to the time of the realization for notational simplicity. Fig. 4.3 shows two examples of the NQT-transformed empirical $\eta - \hat{\eta}_i$ relationships as scatterplots for two

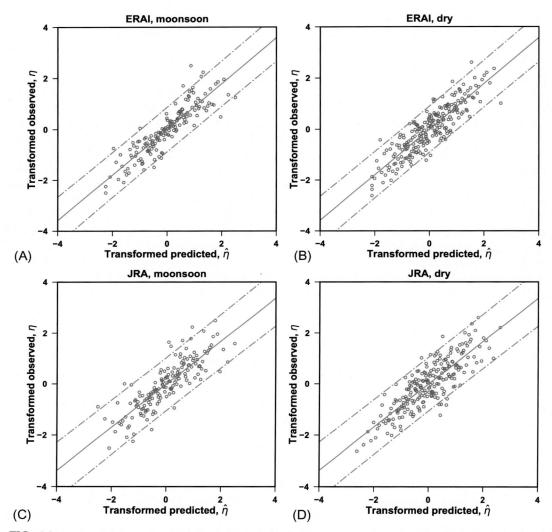

FIG. 4.3 Jhelam subbasin: scatterplots and regression line showing the linear dependency between predictand and predictor for two reanalysis products and two analysis periods; ERAI monsoon (A), ERAI dry period (B), JRA monsoon (C), and JRA dry period (D).

reanalysis reforecasts (ERAI and JRA) for the monsoon and the dry season, respectively, in the Jhelam subbasins.

One needs to envisage these scatterplots as the projection of the multivariate-normal dependency $(\eta, \hat{\eta}_i), i = 1, \ldots, n$ onto the respective $\eta - \hat{\eta}_i$ plane as a bivariate-normal process $(\boldsymbol{\eta}, \hat{\boldsymbol{\eta}}_i)$. We note that in all plots there are no data points for the Gaussian variate η below a threshold value (corresponding to zero precipitation in the Gaussian space) because precipitation is always nonnegative. The red solid line indicates the 50% quantile or conditional median obeying to the linear relationship $\eta|\hat{\eta}_i(\hat{\eta}_i) = \rho_{\eta\hat{\eta}^n} \cdot \hat{\eta}_i + \mu_\eta$ with $\mu_\eta = 0$, while conditional mean and variance are equal to

$$E(\eta|\hat{\eta}_i) = \mu_{\eta|\hat{\eta}_i} = \rho_{\eta\hat{\eta}^n} \cdot \hat{\eta}_i \tag{4.4}$$

$$Var(\eta|\hat{\eta}_i) = \sigma^2_{\eta|\hat{\eta}_i} = 1 - \rho^2_{\eta\hat{\eta}_i} \tag{4.5}$$

where $\mu_{\eta|\hat{\eta}_i}$ is the conditional mean and $\rho_{\eta\hat{\eta}_i}$ is the Pearson correlation. We note that thanks to the transformation of original data into standard normal variates $N(0, 1)$ the Pearson correlation is equal to the covariance and the conditional median coincides with conditional mean and modal values. The red dash-dotted lines indicate the 95% credible interval and are at a parallel distance of two standard deviations from the mean. The parametric conditional multivariate normal density function for a single predictor $\hat{\eta}_i$ is given by the expression:

$$\phi(\eta|\hat{\eta}_i) = \frac{\exp\left[-\frac{1}{2}(\eta - \mu_{\eta|\hat{\eta}_i})^2 / (1 - \rho^2_{\eta\hat{\eta}_i})\right]}{\sqrt{2\pi \cdot (1 - \rho^2_{\eta\hat{\eta}_i})}} \tag{4.6}$$

We note that the conditional variance $\sigma^2_{\eta|\hat{\eta}_i} = (1 - \rho^2_{\eta\hat{\eta}_i})$ is at the denominator and thus with $\rho_{\eta\hat{\eta}_i} \to 1$ and $\sigma^2_{\eta|\hat{\eta}_i} \to 0$ the conditional density curve becomes steeper with probability mass narrowly concentrated. This is tantamount to minimizing the PU by using an increasingly "optimal" model. As described in more detail in the Appendix., the conditional density can be extended to include multiple predictors as conditioning variables, leading to the parametric multinormal density function of η, conditional on n potential predictors

$$\phi(\eta|\hat{\eta}_1, \dots, \hat{\eta}_n) = \frac{\exp\left[-\frac{1}{2}(\eta - \mu_{\eta|\hat{\eta}^n})^2 / \sigma^2_{\eta|\hat{\eta}^n}\right]}{\sqrt{2\pi} \cdot \sigma_{\eta|\hat{\eta}^n}} \tag{4.7}$$

where conditional mean and variance are expressed in terms of the $(n + 1) \times (n + 1)$-dimensional variance-covariance matrix $C_{\eta,\hat{\eta}^n}$ with submatrices $C_{\eta\hat{\eta}^n}$ and $C_{\hat{\eta}^n\hat{\eta}^n}$:

$$\mu_{\eta|\hat{\eta}^n} = C_{\eta\hat{\eta}^n} \cdot C^{-1}_{\hat{\eta}^n\hat{\eta}^n} \cdot \hat{\eta}^n \tag{4.8}$$

$$\sigma^2_{\eta|\hat{\eta}^n} = 1 - C_{\eta\hat{\eta}^n} \cdot C^{-1}_{\hat{\eta}^n\hat{\eta}^n} \cdot C^T_{\eta\hat{\eta}^n} \tag{4.9}$$

In Eq. (4.7) the variance-covariance matrix combines n different models with observations. The covariances are evaluated over a sufficiently long calibration period, for which predictions as well as observations are available, and specify the added value of the different forecast models by weighting the predictor values at a given time $\hat{\eta}^n$ accordingly in the conditional variance $\sigma^2_{\eta|\hat{\eta}^n}$. Model forecasts that correlate poorly with the predictand are weighted lower with respect to those with higher correlation. Eq. (4.7) is the conditional of the multivariate normal density $\phi(\eta, \hat{\eta}_1, \dots, \hat{\eta}_n)$ defined in an $(n + 1)$-dimensional data space. Assuming a homoscedastic dependence structure $(\eta, \hat{\eta}_i)$ between variables, Eqs. (5.7)–(5.9) hold over the entire value range of $\hat{\eta}_i$, as is clearly visible in Fig. 4.3. However, in a situation of heteroscedastic dependency it may become necessary to derive separate dependence structures for two or multiple subintervals of the $\hat{\eta}_i$ value range. For the purpose of the present case study, we proceed assuming homoscedastic $(\eta, \hat{\eta}_i)$ dependencies between observations and models.

4.3.4 Back-Transformation Into the Original Space

Once the series of conditional means and the 95% confidence intervals of the conditional densities have been obtained in the Gaussian space, the normal variates need to be back-transformed into the real space. This is achieved by inverting the NQT (4.3):

$$\mathbf{y}|\hat{\mathbf{y}} = \Gamma^{-1}(Q(\boldsymbol{\eta}|\hat{\eta})) \quad \mathbf{y} = y_k, \boldsymbol{\eta} = \eta_k; \quad k = 1, \ldots, m \tag{4.10}$$

For this purpose, the empirical cumulative distribution function Γ, which is monotonically ascending, is interpolated by means of piecewise linear segments, while the distribution tails below the 15% and above the 85% quantile are modeled with a Pareto distribution. Approximating the tails with Pareto distributions ensures a smooth back-transformation of the extreme high and low range of the standard normal precipitation variate for extreme events. For example, in the very high precipitation range, inaccurate modeling of the distribution tails can cause large deviations of back-transformed precipitation from actual values given small changes in the Gaussian variate, leading inevitably to a distorted estimation of the conditional mean and the credibility intervals in the original space.

4.4 APPLICATION AND RESULTS

4.4.1 Calibration

First, we calibrate the processor for the Central Indus region by applying it to the precipitation series for the centroids of the five subbasins. For calibration, we select the 26-year period from 1979 to 2005. The results of the postprocessed reanalysis data for the calibration period and the five subbasins in Table 4.2 is shown in the subplots of Fig. 4.4. The figure shows the MCP-processed and combined monthly reanalysis data for 1979–2005. The processor performs bias-removal and returns estimates of the observed precipitation at the subbasin centroids in terms of the expected value of the back-transformed conditional density (4.9), that is, the conditional mean. The navy blue line in the figures represents the observed areal average monthly values at the centroid, while the red line is the conditional mean estimating the monthly precipitation given n reanalysis precipitation values as reforecast predictors. The dark and light gray-shaded areas represent the 50% and 95% confidence intervals, respectively. In other words, given multiple atmospheric model predictions of the mean monthly basin-average precipitation, the processor returns at each time step a probability density function indicating the probability mass distribution and uncertainty bandwidth for actual precipitation. The calibrated processor can be utilized in forecast mode to estimate monthly mean areal precipitation given multiple forecasts.

4.4.2 Performance Indicators

The performance of the processor can be examined by means of a series of indicators evaluated in the normal space, which quantify the skill of every single prediction and the Bayesian combination of all predictions. A first quantitative indicator of performance is the intercomparison of the correlations of observed and predicted monthly means. A more in-depth analysis of processor performance can be obtained by looking at the correlation and variance of the residuals. Tables 4.4 and 4.5 show an overview of the correlations

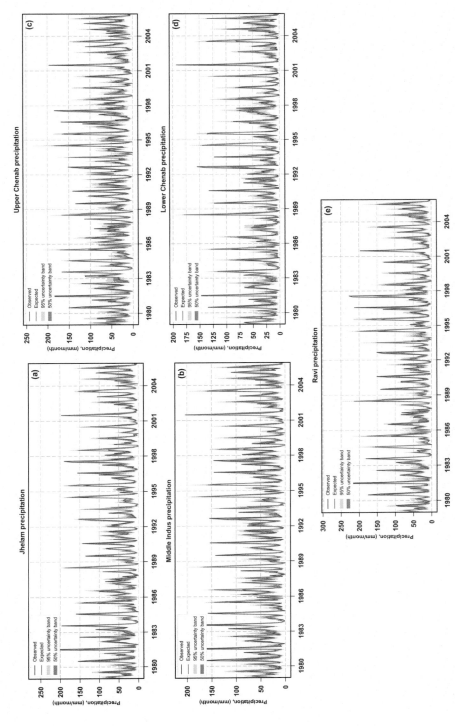

FIG. 4.4 Processor calibration 1979–2005: conditional mean, confidence intervals, and precipitation observed for the five subbasins of the study area in subplots (A)–(E).

TABLE 4.4 Jhelam, 1979–2005, Monsoon Season: Summary of Performance Indicators After Uncertainty Processing for Individual Models and the Combination of All Models

Indicator	Def.	CFRS	ERA20C	ERAI	JRA	MERRA	R1	All
Correlation	$\rho_{\eta\hat{m}_i}$	0.8	0.77	0.89	0.84	0.87	0.81	0.93
Var. res.	$1 - \rho^2_{\eta\hat{m}_i}$	0.34	0.38	0.19	0.28	0.23	0.33	0.14
Expl. var.	$\rho^2_{\eta\hat{m}_i}$	0.49	0.17	0.75	0.66	0.68	0.62	0.65
Frac. var. unexpl.	$\dfrac{(1 - \rho^2_{\eta\hat{m}_i})}{\sigma^2_{\hat{\eta}_i}}$	0.41	0.69	0.2	0.3	0.25	0.35	0.17
Frac. var. expl.	$\dfrac{\rho^2_{\eta\hat{m}_i}}{\sigma^2_{\hat{\eta}_i}}$	0.59	0.31	0.8	0.7	0.75	0.65	0.83
Signal/noise	$\dfrac{\rho^2_{\eta\hat{m}_i}}{(1 - \rho^2_{\eta\hat{m}_i})}$	1.43	0.44	3.97	2.36	2.95	1.88	4.81
Bias		−31.19	−10.13	34.27	40.95	−25.82	−23.5	−1.65
RMSE		48.39	47.96	55.67	71.32	41.52	50.63	23.39

TABLE 4.5 Jhelam, 1979–2005, Dry Season: Summary of Performance Indicators After Uncertainty Processing for Individual Models and the Combination of All Models

Indicator	Def.	CFRS	ERA20C	ERAI	JRA	MERRA	R1	All
Correlation	$\rho_{\eta\hat{m}_i}$	0.62	0.67	0.89	0.85	0.74	0.62	0.91
Var. res.	$1 - \rho^2_{\eta\hat{m}_i}$	0.58	0.53	0.2	0.26	0.44	0.59	0.16
Expl. var.	$\rho^2_{\eta\hat{m}_i}$	0.31	0.14	0.73	0.7	0.51	0.36	0.6
Frac. var. unexpl.	$\dfrac{(1 - \rho^2_{\eta\hat{m}_i})}{\sigma^2_{\hat{\eta}_i}}$	0.65	0.79	0.22	0.27	0.46	0.62	0.21
Frac. var. expl.	$\dfrac{\rho^2_{\eta\hat{m}_i}}{\sigma^2_{\hat{\eta}_i}}$	0.35	0.21	0.78	0.73	0.54	0.38	0.79
Signal/noise	$\dfrac{\rho^2_{\eta\hat{m}_i}}{(1 - \rho^2_{\eta\hat{m}_i})}$	0.53	0.26	3.63	2.66	1.18	0.61	3.74
Bias		−8.03	−16.38	7.0	7.88	−19.76	−27.94	−0.74
RMSE		23.72	24.9	16.57	20.53	27.49	35.31	10.29

(first row), the variance of the residuals (also called variance unexplained), and the variance explained (second and third rows) for the postprocessing of the Jhelum subbasin data, monsoon, and dry season, respectively. The first type of variance is an overall indicator of the Gaussian scatter around the linear regression model indicated in Fig. 4.3, while the latter is the variance, which can be explained solely by the regression model. In addition, we calculate the fractions of variance explained and variance unexplained with respect to total variance (fourth and fifth rows) to show the percentages of each. As can be seen in the tables, all six models have a correlation with observations between 0.62 and 0.89 over both seasons. The combination of all models (first row, last column) increases the correlations to a value >0.9 for both seasons, proving that there is net added informative value in performing a Bayesian combination of multiple predictors. However, the fraction of variance unexplained varies between 0.86 (86%) down to 0.21 (21%) of total variance, indicating that precipitation prediction is clearly achieved much better by some models (ERAI, JRA, and MERRA) than the remaining ones (ERA20, CFSR, and NCAR R1) and is highest (i.e., 0.17) when combining all models for the monsoon period (fourth row, last column).

Next, we report the "signal-to-noise" ratio (sixth row), a decision-theoretic measure of informativeness of output (Krzysztofowicz, 1999) for individual models and the combination of all models. It shows values that are smallest for ERA20C, R1, and CFSR and highest for JRA55, ERAI, and MERRA. The combination of all models brings the ratio to a value of 4.77 for the monsoon season and 3.69 for the dry season, indicating that the coprocessing of all models leads to an improvement of several times with respect to the worst performing predictor when processed as a single model. In the hypothetical case of a totally uninformative model, which would be completely uncorrelated with observations, that is, $\rho_{\eta\hat{m}_i} = 0$, the total variance becomes unexplained, and the signal-to-noise ratio $\rho_{\eta\hat{m}_i}^2 / (1 - \rho_{\eta\hat{m}_i}^2)$ drops to zero. To the contrary, if the model is "perfect," $\rho_{\eta\hat{m}_i} = 1$ and the signal-to-noise ratio $\rightarrow \infty$. The bias and RMSE, both calculated on the back-transformed variables, are reported in rows 7 and 8. We note that before processing the reanalysis data are significantly biased, while bias has been removed after processing and combining all reanalyses. The last row reports the RMSE between observations and conditional means.

Finally, in Table 4.6 we show a summary of the observations-model correlations for individual models (columns 2–7) and subbasins before processing and after Bayesian combination of all models (column 8). We note that the correlations before processing are in the range 0.52–0.86 and increase to a value of 0.9 after processing, when correlating the conditional mean with observations.

4.4.3 Validation

Finally, we validate the calibrated processor on the 2006–10 period by using the reanalysis output as a predictor and estimating the actual precipitation through the conditional mean value. As we are working in the past, we can compare directly the processor output with the observations. The results of the processor run for the validation period are shown in Fig. 4.5.

TABLE 4.6 Correlations Between Observations Versus Raw Predictions ($corr(y, \hat{y}_i)$) and Observations (Columns 2–7) Versus Posterior Conditional Mean ($corr(y, E[y|\hat{y}^n])$) (Last Column), 1979–2005

Subbasin	CFRS	ERA20C	ERAI	JRA	MERRA	NCAR	All Combined
Jhelam	0.66	0.73	0.87	0.83	0.79	0.62	0.91
Middle Indus	0.52	0.71	0.86	0.79	0.82	0.78	0.9
Upper Chenab	0.75	0.79	0.83	0.84	0.83	0.7	0.89
Lower Chenab	0.52	0.69	0.86	0.78	0.79	0.71	0.9
Ravi	0.81	0.77	0.84	0.84	0.85	0.72	0.89

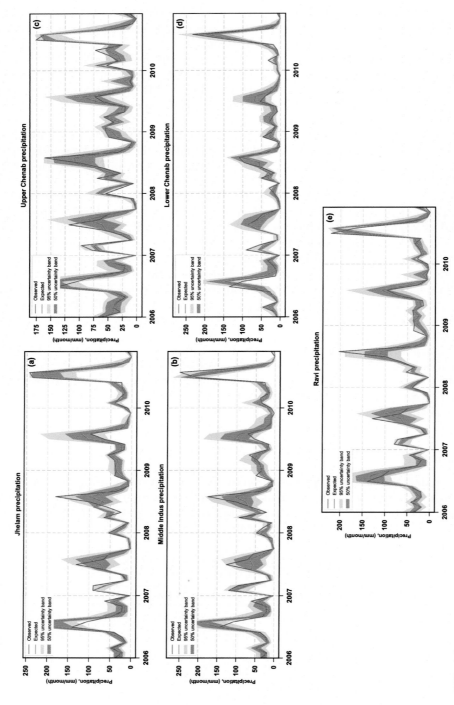

FIG. 4.5   Processor validation 2006–10: conditional mean, confidence intervals, and precipitation observed for the five subbasins of the study area in subplots (A)–(E). The extreme precipitation event causing the 2010 Indus floods has been adequately captured by the processor and lies on the upper edge of the 95% confidence interval.

We note that the observations lie mostly within the 95% confidence interval and also that the precipitation event of July 2010, which caused the large Indus floods, is captured as lying on the upper fringe of the 95% confidence interval. Even though this event was not included in the calibration period, the processor managed to identify it as probable on the basis of the six reanalysis predictions.

4.5 CONCLUSIONS

In this study Bayesian forecasting was applied to quantitatively assess the uncertainty of monthly precipitation for five subbasins in the Central Indus region using a six-member ensemble of independent reanalysis outputs. A probabilistic processor was calibrated on the 1979–2005 year period and applied in validation mode for 2006–10. In the validation period, the reanalysis was used as a pseudo-forecast to show the capabilities of the method in predicting the basin-average precipitation and respective uncertainty expressed in terms of 50% and 95% confidence intervals. As the example case, the extreme July–August precipitation event causing the 2010 Indus flood in Pakistan has been reforecasted. From the results, it has been shown that by using multiple predictor sources and combining them by means of a Bayesian processor allows maximizing the compound information expressed in terms of a signal-to-noise ratio for predicted precipitation.

This methodology had been applied earlier in the UIB, Pakistan (Reggiani et al., 2016) to quantify precipitation uncertainty for a poorly gauged area in the Karakoram and has been extended in the present article to the Central Indus region. The strength of the methodology lies in the possibility to use the processor to perform educated estimation of areal precipitation, including related uncertainty by making use of multiple predictors. Such probabilistic estimates of precipitation are of considerable importance for future water allocation analysis and the planning of infrastructure and related investments, where a narrow determination of uncertainty can create a new perspective on risk perception or a more narrowly defined estimate of investment needs.

While we used reanalysis products as predictors in the current example, these can be extended to include satellite or radar precipitation estimates, or involve different predictors such as relative air humidity. The present methodology can also be employed to perform an analysis of precipitation, temperature, or evaporation forecasts from climate projections, which can be used in the same way as reanalysis products in the present example. Climate model performance on historical data can be evaluated over a historical period and the calibrated probabilistic processor adapted to evaluate climate indicators such as precipitation and temperature change indexes for the mid-21st century and beyond.

ACKNOWLEDGMENTS

Special thanks go to the Pakistan Meteorological Department, which provided the precipitation data. Without their support, this research would not have been possible. The authors also acknowledge the organizations and the teams responsible for the creation and publication of the JRA55, NCEP-NCAR (R1), NCEP-CFSR, ERA20C, ERA-Interim, and MERRA reanalysis archives. The JRA55 data were obtained from the Japan Meteorological Agency (http://jra.kishou.go.jp), NCEP-NCAR (R1) data were obtained from NOAA/OAR/ESRL PSD (http://www.esrl.noaa.gov/psd), NCEP-CFSR data from NOAA (http://cfs.ncep.noaa.gov/cfsr/), ERA 20th Century and ERA-Interim reanalysis data from ECMWF (http://www.ecmwf.int/en/research/climate-reanalysis), and Modern-Era Retrospective Reanalysis (MERRA) data from NASA (http://gmao.gsfc.nasa.gov/research/merra/).

APPENDIX THE MODEL CONDITIONAL PROCESSOR

The MCP was first presented (Todini, 2008) in the context of flood forecasting as a Bayesian uncertainty processor that uses the output of a single forecasting model for water levels or flows to estimate the PU on respective values to be observed,

$$PU = f(\mathbf{y}|\hat{\mathbf{y}}) \qquad (A.1)$$

where f is a conditional probability density function, $\mathbf{y}$ is the random time series vector of observations that is to be predicted, and $\hat{y}$ the random model output vector acting as a predictor. The empirical probability distributions of predictand and predictor can be mapped into the normal space by applying the nonparametric NQT (Wilks, 1995). The transformed standard normal variables, $N(0, 1)$, are indicated with η and $\hat{\eta}$, and can be modeled by respective parametric expressions of normal distributions. In the normal space the joint distribution of the two variables, $\Phi(\eta, \hat{\eta})$ is bivariate normal for which the conditional density $\phi(\eta|\hat{\eta})$ can be evaluated analytically. The conditional density can later be back-transformed from the normal to the original space, giving the PU (Eq. A.1).

The single-model case can be extended by analogy to include multiple models as predictors (Coccia and Todini, 2011). Based on the properties of the multivariate normal distribution (Mardia et al., 1980), the MCP allows evaluating the density of the predictand conditional on the forecasts by n models via multiple regression in the normal space.

The derivation of the predictive density is performed by first converting observations $\mathbf{y}$ and the forecasts by the n prediction models, $\hat{y}^n = \hat{y}_1, \hat{y}_2, \ldots, \hat{y}_n$, into the Gaussian space by NQT. The transformed variables are denoted with the Greek letter η. If m is the number of data in the observed series, $\mathbf{y}$ and its transform η are vectors of length m, while the predictions $\hat{y}^n$ and their respective transforms $\hat{\eta}^n$ are organized in $m \times n$ matrices. In the normal space predictand and predictor are related through a joint probability distribution with vector of means $\mu_{\eta, \hat{\eta}^n}$ and variance-covariance matrix $C_{\eta, \hat{\eta}^n}$. Because the transformed variables are standard normal, the vector of means of the ensemble of variables (observations and predictors) is equal to the null vector,

$$\boldsymbol{\mu}_{\eta, \hat{\eta}^n} = \begin{bmatrix} 0 \\ \vdots \\ 0 \end{bmatrix} \qquad (A.2)$$

while the variance-covariance matrix is structured as follows:

$$C_{\eta,\hat{\eta}^n} = \begin{bmatrix} C_{\eta\eta} & C_{\eta\hat{\eta}^n} \\ C_{\eta\hat{\eta}^n}^T & C_{\hat{\eta}^n\hat{\eta}^n} \end{bmatrix} \tag{A.3}$$

Moreover, as the standard normal variance is equal to 1, the covariances coincide with the Pearson product moment correlations between the variables, and can be interpreted as slopes of the linear regressions between variables. Consequently $C_{\eta\eta} = 1$, $C_{\eta\hat{\eta}^n} = [\rho_{\eta\hat{\eta}_1}, \dots, \rho_{\eta\hat{\eta}_n}]$ is a $1 \times n$ vector of correlations, while

$$C_{\hat{\eta}^n\hat{\eta}^n} = \begin{bmatrix} 1 & \rho_{\hat{\eta}_1\hat{\eta}_2} & \cdots & \rho_{\hat{\eta}_1\hat{\eta}_n} \\ \rho_{\hat{\eta}_2\hat{\eta}_1} & \ddots & \ddots & \vdots \\ \vdots & \ddots & \ddots & \rho_{\hat{\eta}_{n-1}\hat{\eta}_n} \\ \rho_{\hat{\eta}_n\hat{\eta}_1} & \cdots & \rho_{\hat{\eta}_n\hat{\eta}_{n-1}} & 1 \end{bmatrix} \tag{A.4}$$

is a $n \times n$ matrix of correlations. The joint observations-forecast probability density is the multivariate normal density,

$$\phi(\eta,\hat{\eta}^n) = \frac{\exp\left(-\frac{1}{2} \cdot [(\eta,\hat{\eta}^n) - \mu_{\eta,\hat{\eta}^n}]^T \cdot C_{\eta,\hat{\eta}^n}^{-1} \cdot [(\eta,\hat{\eta}^n) - \mu_{\eta,\hat{\eta}^n}]\right)}{\sqrt{(2\pi)^{(n+1)} \cdot |C_{\eta,\hat{\eta}^n}|}} \tag{A.5}$$

from which we obtain an analytical expression of the conditional density by exploiting the properties of the multivariate normal distributions (Mardia et al., 1980) and dividing by the marginal density,

$$\phi(\eta|\hat{\eta}^n) = \frac{\phi(\eta,\hat{\eta}^n)}{\phi(\hat{\eta}^n)} = \frac{\phi(\eta,\hat{\eta}^n)}{\int_{-\infty}^{\infty} \phi(\eta,\hat{\eta}^n)\,\phi(\eta)\,d\eta} = \frac{\exp\left[-\frac{1}{2}(\eta - \mu_{\eta|\hat{\eta}^n})^2 / \sigma_{\eta|\hat{\eta}^n}^2\right]}{\sqrt{2\pi} \cdot \sigma_{\eta|\hat{\eta}^n}} \tag{A.6}$$

where mean and variance of the conditional density are as follows:

$$\mu_{\eta|\hat{\eta}^n} = C_{\eta\hat{\eta}^n} \cdot C_{\hat{\eta}^n\hat{\eta}^n}^{-1} \cdot \hat{\eta}^{n\,T} \tag{A.7}$$

$$\sigma_{\eta|\hat{\eta}^n}^2 = 1 - C_{\eta\hat{\eta}^n} \cdot C_{\hat{\eta}^n\hat{\eta}^n}^{-1} \cdot C_{\eta\hat{\eta}^n}^T \tag{A.8}$$

We observe that Eqs. (A.5)–(A.7) are vectorial with each vector position referring to a given observing and prediction time t. The predictive density at time t can be obtained by evaluating the respective expressions for realizations of the random observations η and model forecasts $\hat{\eta}^n$ processes at time t. The realizations at a given time are denoted with η and $\hat{\eta}^n$. After the predictive density in the normal space has been obtained, it is back-transformed into the original space by applying the inverse NQT.

At this stage we note that the variance given by Eq. (A.8) is a scalar value is constant over the entire value range of the random variables. This is a consequence of the implicit assumption that the dependency $(\eta,\hat{\eta}^n)$ is homoscedastic. However, for many random variables, such as flow levels, discharges, or precipitation, such an assumption is not appropriate. Very low or high flows or water levels in a river can show higher variances than their mid-range.

Similar characteristics are observed for precipitation. In such cases it is inaccurate to apply a single linear regression model assuming homoscedastic behavior. A proven solution is to apply TND, which are fitted to the variables over smaller subdomains of the whole value range, in which homoscedasticity can be assumed. This approach has been shown (Coccia and Todini, 2011) to yield satisfactory results in typical situations of heteroscedastic variables with a subdivision of the random variable into two or at most three subdomains.

References

Ali, G., Rasul, G., Mahmood, T., Zaman, Q., Cheema, S.B., 2012. Validation of APHRODITE precipitation data for humid and sub humid regions of Pakistan. Pak. J. Meteorol. 9 (17), 57–69.

Archer, D.R., Fowler, H.J., 2003. Spatial and temporal variations in precipitation in the Upper Indus Basin, global teleconnections and hydrological implications. J. Hydrol. 274 (1), 198–210.

Brisoce, J., Qamar, U., 2005. Pakistan's water economy running dry. The World Bank, Washington, DC. 44375.

Coccia, G., Todini, E., 2011. Recent developments in predictive uncertainty assessment based on the model conditional processor approach. Hydrol. Earth Syst. Sci. 15, 3253–3274.

Cressie, N., 1985. Fitting variogram models by weighted least squares. Math. Geol. 17 (5), 563–586.

Dee, D.P., Balmaseda, M., Balsamo, G., Engelen, R., Simmons, A.J., Thepaut, J.N., 2014. Toward a consistent reanalysis of the climate system. Bull. Am. Meteorol. Soc. 95 (8), 1235–1248.

FAO, 2011. Irrigation in Southern and Eastern Asia in figures. FAO, Rome. 44375.

Gulhati, N.D., 1973. The Indus Waters Treaty: An Exercise in International Mediation. Allied Publishers, Mumbai, p. 472.

Harris, I., Jones, P.D., Osborn, T.J., Lister, D.H., 2013. Updated high-resolution grids of monthly climatic observations—the CRU TS3.10 dataset. Int. J. Climatol. 34, 623–642.

IPCC, 2014. Climate Change 2014: synthesis report. Contribution of working groups I, II and III to the fifth assessment report of the intergovernmental panel on climate change. IPCC, Geneva, Switzerland.

Kistler, R., Kalnay, E., Collins, W., Saha, S., White, G., Woollen, J., Chelliah, M., Ebisuzaki, W., Kanamitsu, M., Kousky, V., van den Dool, H., Jenne, R., Fiorino, M., 2001. The NCEP-NCAR 50-year reanalysis: monthly means CD-ROM and documentation. Bull. Am. Meteorol. Soc. 82, 247–267.

Kobayashi, S., Ota, Y., Harada, Y., Ebita, A., Moriya, M., Onoda, H., Onogi, K., Kamahori, H., Kobayashi, C., Endo, H., Miyaoka, K., Takahashi, K., 2015. The JRA-55 reanalysis: general specifications and basic characteristics. J. Meteorol. Soc. Jpn 93 (1), 5–48.

Koenker, R., 2005. Quantile Regression. Econometric Society Monographs, Cambridge University Press, Cambridge, MA, p. 349.

Krzysztofowicz, R., 1999. Bayesian theory of probabilistic forecasting via deterministic hydrologic model. Water Resour. Res. 35 (9), 2739–2750.

Mardia, K.V., Kent, J.T., Bibby, J.M., 1980. Multivariate Analysis (Probability and Mathematical Statistics). Academic Press, London, p. 536.

Pegram, G., 1997. Patching rainfall data using regression methods. 3. Grouping, patching and outlier detection. J. Hydrol. 198 (1), 319–334. https://dx.doi.org/10.1016/S0022-1694(96)03284-2.

Raftery, A.E., 1993. Bayesian model selection in structural equation models. In: Bollen, K.A., Long, J.S. (Eds.), Testing Structural Equation Models, Sage Publishing, pp. 163–180.

Raftery, A.E., Gneiting, T., Balabdaoui, F., Polakowski, M., 2005. Using Bayesian model averaging to calibrate forecast ensembles. Mon. Weather Rev. 133 (5), 1155–1174.

Ray, P.A., Brown, C.M., 2015. Confronting climate uncertainty in water resources planning and project design the decision tree framework. The World Bank Group, Washington, DC. 99180.

Reggiani, P., Rientjes, T.H.M., 2015. A reflection on the long-term water balance of the Upper Indus Basin. Hydrol. Res. 46 (3), 446–462. https://dx.doi.org/10.2166/nh.2014.060.

Reggiani, P., Coccia, G., Mukhopadhyay, B., 2016. Predictive uncertainty estimation on a precipitation and temperature reanalysis ensemble for Shigar Basin, Central Karakoram. Water 8 (6), 263. https://dx.doi.org/10.3390/w8060263.

Rienecker, M.M., et al., 2011. MERRA: NASA's modern-era retrospective analysis for research and applications. J. Clim. 24, 3624–3648.

Sankarasubramanian, A., Vogel, R.M., Limbrunner, J.F., 2001. Climate elasticity of streamflow in the United States. Water Resour. Res. 37 (6), 1771–1781.

Schneider, U., Becker, A., Finger, P., Meyer-Christoffer, A., Ziese, M., Rudolf, B., 2014. GPCC's new land surface precipitation climatology based on quality-controlled in situ data and its role in quantifying the global water cycle. Theor. Appl. Climatol. 115, 15–40.

Singh, P., Kumar, N., 1997. Effect of orography on precipitation in the western Himalayan region. Hydrol. Earth Syst. Sci. 199 (1), 183–206.

Stickler, A., Brönnimann, S., Valente, M.A., Bethke, J., Sterin, A., Jourdain, S., Roucaute, E., Vasquez, M.V., Reyes, D.A., Allan, R., Dee, D., 2010. ERA-CLIM: historical surface and upper-air data for future reanalyses. Bull. Am. Meteorol. Soc. 95 (9), 1419–1430.

Suranjana, S., et al., 2010. The NCEP climate forecast system reanalysis. Bull. Am. Meteorol. Soc. 91 (8), 1015–1057.

Todini, E., 2008. A model conditional processor to assess predictive uncertainty in flood forecasting. Int. J. River Basin Manag. 36, 3265–3277.

Todorovic, P., Yevjevich, V., 1969. Stochastic processes of precipitation. Colo. State Univ. Hydrol. Pap. 35, 1–61.

UNDP, 2016. Water security in Pakistan: issues and challenges. UNDP Office Pakistan, Islamabad, Pakistan. www.pk.undp.org.

UNESCO, 2012. The United Nations—World Water Development Report 4: Managing Water Under Uncertainty and Risk. UNESCO, Paris. http://publishing.unesco.org/.

Wilks, D.S., 1995. Statistical Methods in the Atmospheric Sciences: An Introduction. Academic Press, London, p. 467.

Yatagai, A., Arakawa, O., Kamiguchi, K., Kawamoto, H., Nodzu, M.I., Hamada, A., 2012. APHRODITE: constructing a long-term daily gridded precipitation dataset for Asia based on a dense network of rain gauges. Bull. Am. Meteorol. Soc. 93, 1401–1415.

5

Glaciers in the Indus Basin

Samjwal Ratna Bajracharya, Sudan Bikash Maharjan, Finu Shrestha

International Centre for Integrated Mountain Development, Kathmandu, Nepal

5.1 INTRODUCTION

The Indus River (The Father River) is a major river that flows through the Indian sub-continent. The river originates from the Tibetan plateau in the vicinity of Lake Mansarovar in the Tibet Autonomous Region, China. It follows a northwesterly course through Tibet and extends to Afghanistan, India, and Pakistan in the Hindu Kush, Karakoram, and Himalaya mountains (Fig. 5.1). The river has a total basin area of 1,116,086 km^2, out of which 555,450 km^2 lie within the Hindu Kush-Himalayan (HKH) boundary (Bajracharya and Shrestha, 2011). The river's estimated annual flow is around 207 km^3, making it the 21st largest river in the world in terms of annual flow (Shaikh et al., 2014).

The Indus River is the main water resource in Pakistan. The Kabul, Upper Indus, and Upper Panjnad rivers are the main tributaries of the Indus River (Fig. 5.1). The Kabul Basin intersects with Pakistan and Afghanistan and includes the Panjsher-Ghorband Rod, Alingar-Alishing-Nuristan Rod (AAN), Kunar Rod, and Swat sub-basins. The Rod represents the local name for the river. The Upper Indus Basin has extensions into the territory of the Tibetan Autonomous Region (TAR) of China, India, and Pakistan, and the main glaciated sub-basins are Gilgit, Hunza, Shigar, Shyok, Zanskar, Shingo, and Astor basins (Fig. 5.1). The Upper Panjnad Basin intersects with India and Pakistan and includes the sub-basins of the Jhelum, Chenab, Ravi, Beas, and Satluj basins. Part of the Satluj Basin lies within the territory of the People's Republic of China, and the information on debris cover, morphological classification, and hypsography from this territory are not described here.

Glaciers in the Hindu Kush-Himalaya are indeed melting and retreating, especially at the lower elevations (Bajracharya et al., 2014a,b). As an anomaly, glaciers in the Karakoram region of the Indus Basin are known to show stable and advancing terminus positions or surging behavior, which contrasts with the worldwide retreat of many mountain glaciers (Mason, 1935; Hewitt, 1969, 1998, 2007; Kotlyakov et al., 2008; Barrand and Murray, 2006;

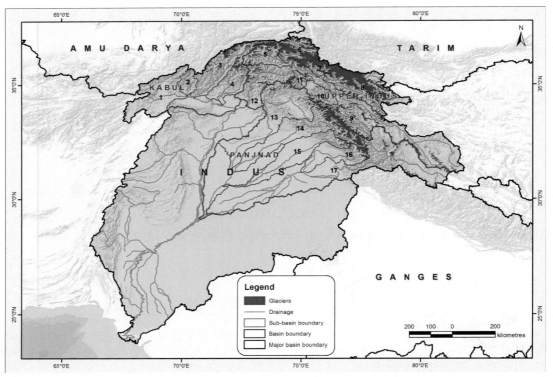

FIG. 5.1   The Indus River system, showing the sub-basins in the upper reaches. *Kabul Basin*: 1, Panjsher-Ghorband; 2, Alingar-Alishing-Nuristan; 3, Kunar; 4, Swat. Upper Indus Basin: 5, Gilgit; 6, Hunza; 7, Shigar; 8, *Shyok; 9, Zanskar; 10, Shingo; 11, Astor; 12, *Upper Indus. *Upper Panjnad Basin*: 13, Jhelum; 14, Chenab; 15, Ravi; 16, Beas; 17, *Sutlej (Satluj). (*Note*: *Transboundary drainage catchment with China).

Gardelle et al., 2012). Fowler and Archer (2006) found evidence of a cooling trend between 1961 and 2000 consistent with glacier thickening. Copland et al. (2011) found evidence of increased glacier surging in the 1990s compared with previous decades, consistent with increased precipitation. Some studies indicated climate change as a potential cause for the Karakoram anomaly, while others claimed it was due to the existence of large areas of debris-covered glaciers that have reduced sensitivity to change and thus contributed to the spatial heterogeneity (Scherler et al., 2011). Individual glacier advances have been reported in the Shyok Valley in the eastern Karakoram during the past decade (Raina and Srivastava, 2008). Additionally, other studies reported on the surging phenomenon of individual glaciers such as the Rimo, Chong Kumdan, Kichik Kumdam, and Aktash glaciers in the eastern Karakoram (Raina and Srivastava, 2008; Tangri et al., 2013). The in situ data source for the Karakoram indicates an average budget of $-0.51 \mathrm{myr}^{-1}$ w.e. for the Siachen Glacier (1986–1991) (Bhutiyani, 1999). Eighteen surge-type glaciers have been identified, with a total area of $1001.1 \pm 32.1 \mathrm{km}^2$ covering ~34%t of the glacier area of the Upper Shyok Valley, indicating that the average size of the surge-type glaciers is significantly larger (~56 km^2) than the average size of the non-surge-type glaciers ($\sim 1\,km^2$) (Bhambri et al., 2013). In the period from 1989 to 2002, surge-type glaciers showed an average area increase of $+0.5\pm 3.8\%$, while the non-surge glaciers were more or less stable ($-0.1\pm 3.8\%$). However, during the decade 2002–11, the non-surge glaciers also showed an indication of a slight area increase (Bhambri et al., 2013). To monitor these glaciers at regular intervals at Karakoram, multispectral and multi-temporal satellite data offer great potential and complement conventional field surveys, which are laborious and may be dangerous. Investigations of long-term glacier changes through remote sensing techniques will help determine and provide the clearest picture of glacier fluctuations and behavior.

In the context of global climate change, there has been an increasing concern that the water resources of this river system may be vulnerable (IPCC, 2007) and could have considerable implications for the livelihoods of the people in the downstream areas (Eriksson et al., 2009). Rising temperatures and changes in precipitation could affect the hydrological regime through changes in seasonal extremes, changes in glacier and snowmelt (Lutz et al., 2014), and changes in glacier volume (Bolch et al., 2012). Shrinking of glaciers might result in a marked reduction in water availability (Bolch et al., 2012; Immerzeel et al., 2012; Kääb et al. 2012). To understand the extent of a possible reduction in long-term glacier melt, continuous monitoring of the glacier dynamics must be carried out (Immerzeel et al., 2010; Lutz et al., 2014).

5.2 DATA AND METHODOLOGY

The Landsat scenes of glaciated basins were downloaded from the United States Geological Survey (USGS) site, and the topographic characteristics were derived by using the digital elevation model (DEM) from Shuttle Radar Topography Mission (SRTM) data. The SRTM is an international project spearheaded by the US National Geospatial-Intelligence Agency (NGA) and the US National Aeronautics and Space Administration (NASA). The current study used the SRTM DEM at a spatial resolution of 90 m with horizontal and vertical accuracy near 20 m and 16 m (linear error at 90% confidence). The SRTM data reveals the full resolution of the world's landforms as originally measured by SRTM in 2000. The Landsat images used are from the period 2004 to 2009.

To map the vast number and area of glaciers of the HKH (including the Indus Basin), ICIMOD (International Centre for Integrated Mountain Development) developed a semi-automatic mapping methodology separately to delineate glacier boundaries of clean ice and debris cover in eCognition software using Landsat satellite images from the year 2005 ± 3 (Table 5.1). The clean-ice (CI) glacier was mapped using the normalized difference snow index (NDSI); however, the threshold value used for the NDSI also captures the snow cover and other unnecessary objects (features) such as vegetation, shadow, water bodies, bare rock, and debris, which creates an error in the delineation of the glacier. Various filters were then used to reduce the error (Fig. 5.2). Similarly, a separate methodology with different filters was applied to delineate the debris-covered (DC) glaciers (Fig. 5.2). The glacier polygons were then exported, and the area was calculated in the GIS environment. The thickness and glacier volume were estimated using an empirical

TABLE 5.1 Summary of Glaciers in the Indus Basin and Sub-Basins

Basin	Sub-Basin	Basin Area (km²)	Number of Glaciers	Glacier Area (km²)	Estimated Ice Reserves (km³)	Elevation (m a.s.l.) Highest	Elevation (m a.s.l.) Lowest	Largest Glacier Area (km²)
Kabul	Panjsher-Ghor	29,823	88	14.63	0.442	5242	3857	2.5
	AAN	6217	37	5.82	0.16	5284	4162	1.5
	Kunar	25,925	1149	1574	176.8	7578	3114	189.5
	Swat	14,728	327	127.4	5.296	5580	3772	4.9
	Total	94,290	1601	1722	182.7	7578	3114	189.5
Upper Indus	Gilgit	13,540	968	938.3	71.32	7730	2703	61.8
	Hunza	13,734	1384	2754	310.6	7749	2409	345.7
	Shigar	7046	439	2374	601.9	8566	2774	631.5
	Shyok*	33,429	3357	5938	981.7	7803	3231	925.9
	Zanskar	15,856	1197	975.5	82.13	6368	3997	62.6
	Shingo	10,502	882	612.7	42.88	7027	3656	46.3
	Astor	3988	372	239.6	16.88	8032	2991	31.0
	Upper Indus*	75,117	2814	1230	66.06	7820	2760	51.9
	Total	17,3213	11,413	15,062	2174	8566	2409	925.9
Upper Panjnad	Jhelum	50,844	733	222.8	8.974	6285	3404	6.8
	Chenab	44,840	2039	2341	210.7	7103	3001	109.3
	Ravi	30,590	217	113.6	5.508	5824	3276	9.2
	Beas	19,500	384	416.6	31.78	6196	3079	29.0
	Sutlej*	105,735	2108	1315	82.89	6652	3606	49.6
	Total	286,383	5481	4409	339.9	7103	3001	109.3
Indus		1,116,085	18,495	21,193	2696	8566	2409	925.9

Note: *Trans-boundary drainage catchment with China.

formula. Other glacier attributes, such as elevation, slope, and hypsometry were generated by draping the glacier polygons over SRTM DEM. The morphology of each glacier was classified based on subjective judgment. The details of the methodology are given in Bajracharya and Shrestha (2011).

The threshold value 0.02 km² was taken as an area for the smallest glacier because with an ice mass smaller than this value, it was uncertain whether it was perennial ice or temporary snow patches. Thus all the perennial ice equal to or larger than 0.02 km² was mapped as a glacier of the region.

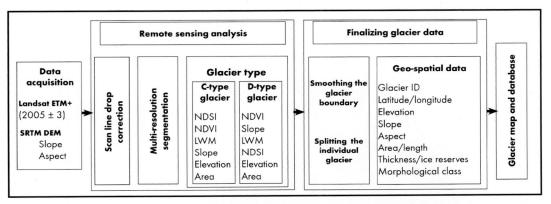

FIG. 5.2   Methodology to capture the glaciers from satellite images and to develop attribute database of captured glaciers.

A few measurements of glacial ice thickness have been undertaken for the Nepal Himalaya by using radio eco-sounding and ground penetrating radar (GPR). In HKH, the volume of the glacier ice was obtained from the area–thickness relationship based on the glacial ice thickness measurement in the Tianshan Mountains in China. The results showed that glacial thickness increased with an increase in area (LIGG et al., 1988). The relationship between ice thickness (H) in meter and glacial area (F) in square kilometer was obtained there as

$$H = -11.32 + 53.21\, F^{0.3}$$

This formula has been used to estimate the mean ice thickness in the glacier inventory of the HKH (Chaohai and Liangfu, 1986; Mool et al., 2001a,b; Kulkarni et al., 2007; Bajracharya and Shrestha, 2011, 2014a,b). The same method is also used here to estimate the ice thickness. The ice reserves are calculated by mean ice thickness multiplied by the glacial area.

5.3 GLACIER CHARACTERISTICS IN THE INDUS BASIN

Glaciers are a major landscape feature of the Indus Basin. About 3.8% of the Indus Basin area within the HKH was found to be glaciated (Bajracharya and Shrestha, 2011). Snow and glacier melt is estimated to contribute >50% of the total flow into the Indus River system (Winiger et al., 2005; Xu et al., 2007). The glaciers are distributed between a latitude of 30.45°N and 37.08°N and a longitude of 69.36°E and 81.65°E (Fig. 5.3). The distribution and characteristics of the glaciers in each sub-basin are summarized in Table 5.1.

5.3.1 Number, Area, and Estimated Ice Reserves

The Indus Basin comprises 18,495 glaciers with a total glacier area of 21,192 km² and estimated ice reserve of 2696 km³ (see Table 5.1). The elevation of the glaciers ranges from 8566 to 2409 m a.s.l. The lowest elevation of the glacier in the Indus Basin is the lowest elevation of all

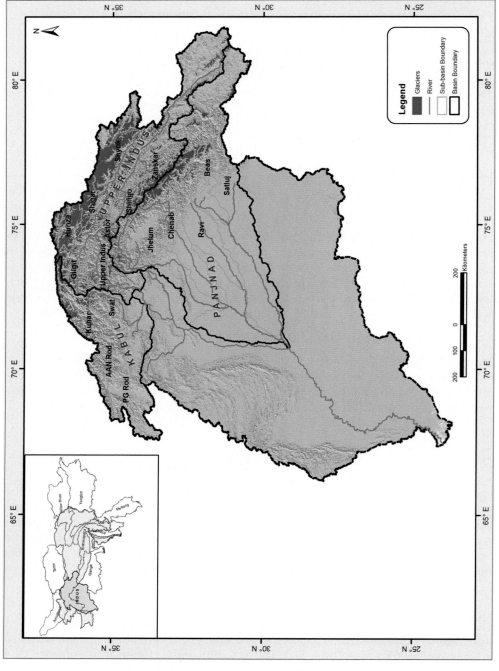

FIG. 5.3   Distribution of glaciers in the Indus Basin.

glaciers in the HKH. This basin also contains the largest glacier of the HKH, the Siachen Glacier in the Shyok sub-basin, which covers about $926 \, km^2$, whereas the average glacier size in the region is $1.1 \, km^2$.

The Upper Indus Basin is the largest Indus River basin, which contains 11,413 glaciers with a total glacier area of $15,061 \, km^2$ and $2173 \, km^3$ of estimated ice reserves. The highest number of glaciers is found in the Shyok sub-basin and the lowest in the Astor sub-basin. The glacier area is much higher in the Shyok, Hunza, and Shigar sub-basins and is at its lowest in the Astor sub-basin. Similarly, the largest ice reserve is in the Shyok sub-basin ($981 \, km^3$), and the lowest is in the Astor sub-basin ($16.88 \, km^3$).

The main glaciated sub-basins of the Upper Panjnad Basin are Jhelum, Chenab, Ravi, Beas, and Satluj. This basin comprises 5481 glaciers with an area of $4409.23 \, km^2$ and an ice reserve of $339.85 \, km^3$ (see Table 5.1). Higher concentrations of glaciers are seen at the Chenab, Ravi, and Beas sub-basins, while the glaciers are scattered in the sub-basins of the Jhelum and Satluj. The Satluj sub-basin has the highest number of glaciers (2108), with the highest glacier area coverage occurring in the Chenab sub-basin ($2341 \, km^2$). The upper catchment of the Satluj sub-basin lies in Chinese territory, with the remainder in India. The largest glacier (G077681E32169N) of the Upper Panjnad Basin lies in the Chenab sub-basin, with an area of $109.33 \, km^2$, and the elevation extends from $6268 \, m$ a.s.l. to $3904 \, m$ a.s.l. The Upper Panjnad Basin contains mostly smaller glaciers with an average glacier area of $<1 \, km^2$ in the Jhelum, Ravi, and Satluj sub-basins, and with glaciers slightly $>1 \, km^2$ in the Chenab and Beas sub-basins. The overall average glacier area of the Upper Panjnad Basin is only $0.81 \, km^2$.

The Kabul Basin is the smallest basin in the Indus system. It contains 1601 glaciers with $1721.69 \, km^2$ of glacier area and an estimated $182.67 \, km^3$ of ice reserves. The Kunar Rod sub-basin has both the highest number of glaciers (1149) and the largest glaciers (GLIMS ID—G073722E36798N) in the Kabul Basin. The GLIMS ID (identity of glacier coded by the Global Land and Ice Measurement from Space, project of the National Snow and Ice Data Center at the University of Colorado) is based on the location of the longitude and latitude of a center point in the polygon of the glacier. It consists of a 14-digit code, including G for Global, E for East, and N for North. The longitude and latitude are in degrees with three decimal places, e.g., 73.722 (longitude), 36.798 (latitude). The largest glacier is a compound basins-type glacier and has an area of $189.47 \, km^2$, extending from 6077 to $3715 \, m$ a.s.l.

5.3.2 Glacier Area Classes

The number, area, and estimated ice reserves of glaciers in the different size classes in the Indus Basin are summarized in Table 5.2. The glacier area is group into five area classes: class 1 ($\leq 0.5 \, km^2$), class 2 ($0.51–1.00 \, km^2$), class 3 ($1.01–5.00 \, km^2$), class 4 ($5.01–10.00 \, km^2$), and class 5 ($\geq 10.00 \, km^2$), respectively.

In the Indus Basin the majority of glaciers (71%) are in class 1, the smallest class, and contribute 11% of the glacier area and 2% of the ice reserves (Fig. 5.4). The glacier number distribution in classes 2 and 3 are similar (13%), but the glacier area contribution in class 3 is three times higher than in class 2. The largest class, class 5, has the smallest number of glaciers (1.29%) but contributes the second highest glacier area (48%). The mean area per glacier is only $1.15 \, km^2$ (see Table 5.3).

TABLE 5.2 Glacier Area Classes in the Indus Basin

Area Class	Size (km²)	Glacier Numbers		Glacier Area		Estimated Ice Reserve		Mean Area Per Glacier (km²)
		Total	%	km²	%	km³	%	
1	≤0.5	13,225	71.51	2290.06	10.81	52.54	1.95	0.17
2	0.51–1.00	2341	12.66	1643.65	7.76	60.17	2.23	0.70
3	1.01–5.00	2396	12.95	4986.96	23.53	285.38	10.59	2.08
4	5.01–10.00	294	1.59	2010.46	9.49	166.43	6.17	6.84
5	≥10.00	239	1.29	10,261.26	48.42	2131.52	79.06	42.93
Total		18,495	100.00	21,192.39	100.00	2696.05	100.00	1.15

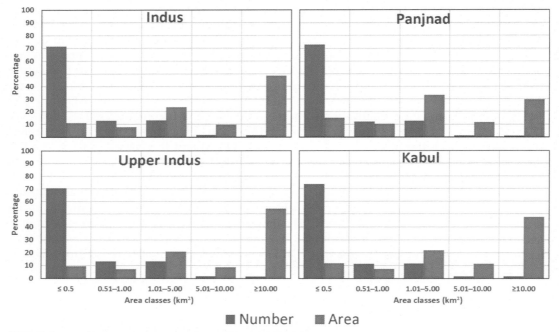

FIG. 5.4 Distribution of glacier number and area in the Indus Basin and sub-basins.

Out of 1601 glaciers in the Kabul Basin, 1178 glaciers (73%) are classified as the smallest glaciers (≤0.5 km²) and contribute 204.31 km² (12%) to the total glacier area and 5 km³ (2.57%) to the estimated ice reserves. Glaciers in class 5 contribute the lowest number of glaciers (1.62%), but their contribution of glacier area (47.52%) and estimated ice reserves (73.98%) are highest of all the classes. Glaciers in class 4 are more than six times fewer than those in class 2, but the distribution of glacier area (11.38%) and estimated ice reserves (9.12%) are higher in class 4 (Table 5.3).

TABLE 5.3 Glacier Area Classes in the Kabul Basin

Area Class	Size (km²)	Glacier Number		Glacier Area		Estimated Ice Reserve		Mean Area Per Glacier (km²)
		Total	%	km²	%	km³	%	
1	≤0.5	1178	73.58	204.31	11.87	4.69	2.57	0.17
2	0.51–1.00	183	11.43	126.34	7.34	4.63	2.54	0.69
3	1.01–5.00	186	11.62	376.82	21.89	21.54	11.79	2.03
4	5.01–10.00	28	1.75	195.99	11.38	16.67	9.12	7
5	≥10.00	26	1.62	818.23	47.52	135.13	73.98	31.47
Total		1601	100	1721.69	100	182.67	100	1.08

Out of 11,413 glaciers in the Upper Indus Basin, 8052 glaciers (71%) are of the smallest size (≤0.5 km²) and contribute 1430 km² (9.5%) of the total glacier area and 33 km³ (1.52%) of the estimated ice reserves. Glaciers from classes 2 and 3 contribute a similar number (1500 in class 2 and 1513 in class 3), but the area and estimated ice reserve contributions are higher in class 3 (21% and 8.2%) than in class 2 (7% and 1.8%). The area distribution in classes 5 and 4 contributes about 54% and 9% of glacier area and 83% and 5% of estimated ice reserves, respectively. The average glacier size of class 5 glaciers is 52.15 km², whereas the mean area per glacier of the entire sub-basin is only 1.32 km² (Table 5.4).

The Upper Panjnad Basin has 5481 glaciers, among which 3995 (73%) glaciers are of the smallest size (class 1), and their contribution in terms of glacier area (15%) and ice reserves (4.33%) are smaller, too. The large glaciers in class 5 are fewer in number (1.04%) compared to the other four classes, but their contribution is significantly higher in terms of the total glacier area (30%) and estimated ice reserves (53%). Glaciers in class 3 have the second highest number of glaciers (697 or 13%) and the highest total area (1468 km² or 33%), with an average glacier size of 2.11 km² and ice reserves of 84.54 km³ or 25% (Table 5.5).

TABLE 5.4 Glacier Area Classes in the Upper Indus Basin

Area Class	Size (km²)	Glacier Number		Glacier Area		Estimated Ice Reserve		Mean Area Per Glacier (km²)
		Total	%	km²	%	km³	%	
1	≤0.5	8052	70.55	1430.23	9.5	33.14	1.52	0.18
2	0.51–1.00	1500	13.14	1055.54	7.01	38.78	1.78	0.7
3	1.01–5.00	1513	13.26	3142.41	20.86	179.30	8.25	2.08
4	5.01–10.00	192	1.68	1298.22	8.62	106.46	4.9	6.76
5	≥10.00	156	1.37	8135.06	54.01	1815.84	83.54	52.15
Total		11,413	100	15,061.46	100	2173.52	100	1.32

TABLE 5.5 Glacier Area Classes in the Upper Panjnad Basin

Area Class	Size (km²)	Glacier Number		Glacier Area		Estimated Ice Reserve		Mean Area Per Glacier (km²)
		Total	%	km²	%	km³	%	
1	≤0.5	3995	72.89	655.52	14.87	14.71	4.33	0.16
2	0.51–1.00	658	12.01	461.77	10.47	16.76	4.93	0.7
3	1.01–5.00	697	12.72	1467.73	33.29	84.54	24.88	2.11
4	5.01–10.00	74	1.35	516.25	11.71	43.3	12.74	6.98
5	≥10.00	57	1.04	1307.97	29.66	180.55	53.13	22.95
Total		5481	100	4409.24	100	339.86	100	0.8

5.3.3 Glacier Elevation

In the Upper Indus Basin glaciers are distributed at an elevation ranging from 2409 m a.s.l. to 8566 m a.s.l. The highest elevation is found in the Shigar sub-basin and the lowest in the Hunza sub-basin. A wide range of glacier elevations is found in the Shigar sub-basin, mostly due to the scattered distribution of glaciers, while a short elevation range is found in the Zanskar sub-basin (Fig. 5.5).

The elevation of glaciers in the Kabul Basin ranges from 3114 m a.s.l. to 7578 m a.s.l. Both of these extreme elevations are found in glaciers of the Kunar Rod sub-basin. The highest elevation is characterized by clean-ice glaciers and the lowest by debris-covered glaciers. Most of the glaciers are located at 4000–7000 m a.s.l., and are more dominant at 4000–6000 m a.s.l.

In the Upper Panjnad Basin glaciers are distributed from an elevation of 3001–7103 m a.s.l. Most of the glaciers distributed at higher altitudes are small mountain glaciers (<1 km²). These mountain glaciers have a restricted elevation range, while the valley glaciers have a

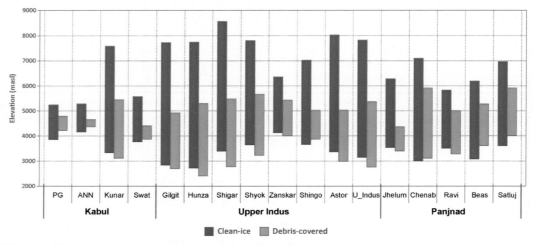

FIG. 5.5 Clean-ice and debris-covered glacier elevation in the sub-basins of the Indus Basin.

large range of elevation from crown to snout. The elevation of glaciers also varies from basin to basin depending on the local topography. Both the highest glacier (CI) elevation and the lowest glacier (DC) elevation in the Upper Panjnad Basin lie in the Chenab sub-basin (Fig. 5.5).

5.3.4 Aspects

The aspect of the glacier is the orientation of the glacier, represented in eight cardinal directions. Each of the cardinal directions is divided, with a 45-degree zone extending 22.5° on both sides: N: 337.5°–22.4°; NE: 22.4°–67.4°; E: 67.5°–112.4°; SE: 112.5°–157.4°; S: 157.5°–202.4°; SW: 202.5°–247.4°; W: 247.5°–292.4°; and NW: 292.5°–337.4°. Glaciers capping the mountain peak are divided into several glaciers depending on the water divides. The orientations of both the areas (accumulation and ablation) are the same for most of the glaciers.

Almost 20% of the glaciers in the Kabul Basin have east and west aspects, which is the highest percentage; the next highest orientation, around 15%, has southeast, south, and southwest aspects. <10% of glaciers have northeast and northwest aspects, and north aspect glaciers make up <0.5% of glaciers (Fig. 5.6).

Upper Indus: Most glaciers have an eastern aspect, while the north aspects are the least common. The second highest aspects, about 17%, are southeast, followed by south and southwest aspects with 15% and 14%, respectively. The northeast and west aspects are each 12% (Fig. 5.6), while the northwest aspects are <10%.

Upper Panjnad: The distribution of glaciers on various aspects is shown in Fig. 5.6. The majority of glaciers have south and southwest aspects of about 16% each. The east and southeast aspect contributions are 15% each. The west and northeast aspect contributions are 14% and 11%, respectively. The north and northwest aspects contribute 5% and 7% of glaciers (Fig. 5.6).

5.3.5 Slope

In many mountain ranges, the lowest reaches of big glaciers are covered by a continuous layer of sand, gravel, and rock. Because these layers are loose with no fines, the angle of repose of these layers is <20 degrees (Fig. 5.7). The glaciers covered by debris are known as debris-covered glaciers, and they are particularly common in the mountains of the HKH (Kääb et al., 2012). The slope of the clean-ice and debris-covered glaciers have different characteristics. The CI glaciers can maintain a steeper slope compared to DC glaciers.

Kabul: The mean slope of the CI glaciers in the sub-basins range from 22 to 27°, with an average of 25°, whereas in the DC glaciers, it ranges from 9 to 11°, with an average of ~9 degrees (Fig. 5.6).

The mean slopes of the glaciers in the sub-basins are divided into seven classes at an interval of 10°. The slope of the glaciers in the basin ranges from 10 to 50°. The slope classes 1, 6, and 7 with slopes <10° and above 50° are not distinct. Class 5 with slopes from 40 to 50° are distributed in all cardinal directions but don't exceed >1% of the area. In the slope class 3 (20 to 30°), glaciers are predominant, followed by class 2 (10 to 40°) and class 4 (10 to 20°) slopes (Fig. 5.6).

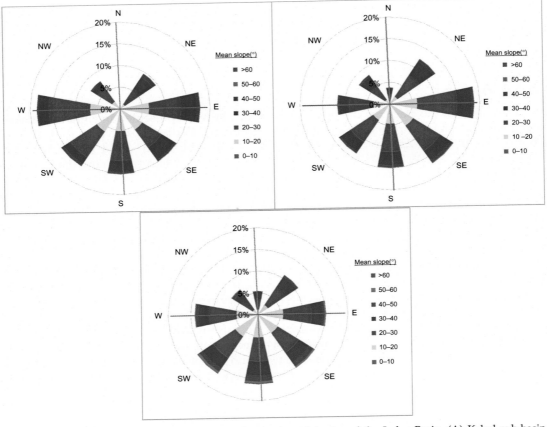

FIG. 5.6 Distribution of aspect and slope classes in the sub-basins of the Indus Basin. (A) Kabul sub-basin. (B) Upper Indus sub-basin. (C) Upper Panjnad sub-basin.

Upper Indus: The mean slope of CI glaciers ranges from 22 to 29°, with an average of 25°, comparatively higher than in other basins and consistent with a greater dominance of mountain glaciers. The mean slope of DC glaciers ranges from 9 to 12 degrees (Fig. 5.6), with an average of 10°.

The slope is generally absent for class 1 and 7 and has a very small contribution of <1% for class 5 and 6. The slope classes 2, 3, and 4 are dominant in all directions, except for northern aspects with less than or equal to 2% (Fig. 5.6).

Upper Panjnad: The average slope of clean-ice (CI) and debris-covered glaciers in the Upper Panjnad Basin is 27 and 11°, respectively. The CI glaciers with a maximum mean slope of 30° belong to the Jhelum sub-basin. Likewise, the DC glaciers with a maximum slope of 12° are located in the Jhelum and Satluj Basins. There are very few glaciers with slopes >60° in the south, southwest, and west aspects. Classes 2, 3, and 4 are predominant in all cardinal directions (Fig. 5.6).

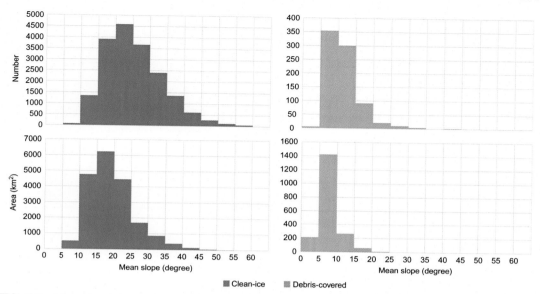

FIG. 5.7   Mean slope of clean-ice and debris-covered glaciers in the Indus Basin.

5.3.6 Morphological Glacier Type

Glaciers are classified into mountain glaciers and valley glaciers (Muller et al., 1977). The glaciers on the mountain slope that end in the middle before approaching the main river are known as mountain glaciers. When a mountain glacier approaches the main valley, then the whole glacier, including the mountain glacier, is known as a valley glacier. Mountain glaciers are classified into five types (miscellaneous, ice apron, cirque, niche, and basins) based on the form of the glaciers. The valley glaciers are classified into compound basins, compound basin, and simple basin; however, in this study only valley glaciers are represented.

Out of 18,495 glaciers in the Indus Basin, only 16,964 glaciers are classified morphologically. About 90% of the glaciers are of the mountain type and 10% are valley trough type (Table 5.6). But the area covered by the mountain type is about 40%, while the valley type covers an area of >60%. Similarly, the ice reserves contribution is around 84%. The mountain glaciers are larger in number but smaller in area and ice reserve contribution.

In the Kabul Basin about 90% of the glaciers are mountain type and about 10% are valley trough type (Fig. 5.8). Mountain basin glaciers and valley-type glaciers are abundant in the Kabul Basin compared to other basins. A total of 974 mountain basin glaciers and 168 valley trough glaciers are classified, which is equivalent to 61% and 10%, respectively, of the total number of glaciers in the basin. The mountain basin glaciers are smaller in size, with an average area of $0.44 \, km^2$; thus the area contribution is $429 \, km^2$, equivalent to 25% of the total glacier area and 9% of the total ice reserve. The average area of the valley trough glaciers is $7.24 \, km^2$ and contributes 70% of the total glacier area and 89% of the total ice reserve. The basin-type glaciers (mountain or valley) are predominant in the Kabul Basin (Fig. 5.8).

TABLE 5.6 Morphological Classification of Glaciers in the Indus Basin

Glacier Type		Glacier Number		Glacier Area		Estimated Ice Reserve		Mean Area Per Glacier (km²)
		Total	%	km²	%	km³	%	
Mountain	Miscellaneous	37	0.22	12.66	0.06	0.52	0.02	0.34
	Ice apron	4780	28.18	1032.23	5.04	54.36	2.05	0.22
	Cirque	241	1.42	48.75	0.24	1.28	0.05	0.20
	Niche	753	4.44	61.74	0.30	1.35	0.05	0.08
	Basin	9440	55.65	6819.90	33.30	358.35	13.51	0.72
Valley trough		1713	10.10	12,501.86	61.05	2235.98	84.32	7.30
Total		16,964	100.00	20,477.15	100.00	2651.83	100.00	1.21

Note: 1561 glaciers in Chinese territory are not classified.

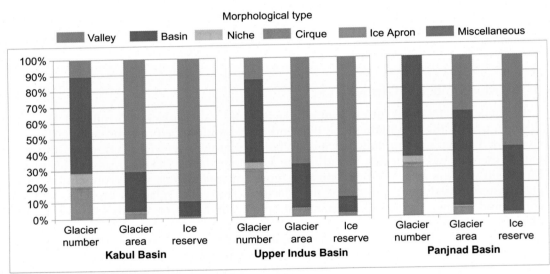

FIG. 5.8 Distribution of glacier types in the sub-basins of the Indus Basin.

Out of 11,413 glaciers in the Upper Indus Basin, 10,709 glaciers are classified morphologically, and of that number, >85% of the glaciers are mountain glaciers. Among the mountain glaciers, 52% are mountain basin, 29% are ice apron, and <4% are other types. Almost 28% of the glacier area and 10% of ice reserves are contributed by mountain basin glaciers. There are 1445 (13.5%) valley trough glaciers, but they contribute 67% of the total glacier area and 88% of the total ice reserve. The basin-type glacier, including valley and mountain glaciers, comprises 66% of the total number, 94% of glacier area, and 98% of the ice reserve, which indicates they are mature mountains, unlike the much younger Himalaya. The average area per glacier ranges from 0.10 to 6.88 km^2, whereas the average area of mountain basin-type glaciers is 0.73 km^2 and of valley troughs 6.88 km^2. The overall average glacier area of the basin is 1.39 km^2.

The valley glaciers in the region are morphologically characterized as either compound basin or simple basin. In order to extend far into the valleys, a large accumulation area is normally needed. Therefore the average area of the valley glaciers is normally greater than that of the mountain glaciers. However, the valley glaciers in the region are fewer in number (1445) compared with mountain glaciers, but they contribute the highest glacier area and ice reserves.

Most of the glaciers in the Upper Panjnad Basin are mountain glaciers. They are dominant in number (about 98%), and they contribute about 65% and 43% of the glacier area and ice reserve, respectively. Among the mountain glaciers, the basin-type glaciers are dominant in the Upper Panjnad Basin and contribute about 62%, 59%, and 41% in number, glacier area, and ice reserve, respectively. The remaining 2% of glaciers are valley type and contribute 35% and 57% of glacier area and ice reserve, respectively.

The valley glaciers in the Upper Panjnad Basin are characterized as both compound basin and simple basin. The area and ice reserves of the valley glaciers are generally large, owing to the fact that the ice thickness increases with the increase in glacier area. The longitudinal profiles of the valley glaciers from crown to toe/snout show an even or regular shape. The headwater is steeper and continues to a gentle slope in the lower reaches, and the profile makes the curve concave upward. They are mainly nourished by snow and drift snow at the headwaters and by snow and ice avalanches in the lower valley.

5.3.7 Clean-Ice and Debris-Covered Glaciers

Debris cover plays an important role, as it has an insulating effect and reduces melting rates. Approximately, 10,200 glaciers with a total area of 9.3% are debris-covered glaciers in the Indus Basin. The debris-covered (DC) glaciers are mostly valley glaciers with a thick debris cover on the glacier tongue. They have an average slope of around 12°, whereas clean-ice (CI) glaciers are much steeper, with average slopes of around 25° (Table 5.7).

About 6% of the glaciers in the Kabul Basin, contributing 9% of glacier area, are DC glaciers. The highest (20%) and lowest (2%) of DC glaciers are in the Alingar-Alishing Nuristan (AAN) Rod and Swat sub-basins of the Indus Basin respectively (Fig. 5.9). The elevations of the CI glaciers range from 3335 to 7578 m a.s.l., and the DC glaciers range from 3114 to 5447 m a.s.l.

About 13% of the glacier area in the Upper Indus Basin consists of DC glaciers. These glaciers' contribution in the sub-basins ranges from about 2% to 16%, with the highest in the Shigar sub-basin and the lowest in the Shingo sub-basin (Fig. 5.9).

Upper Panjnad: The clean-ice glaciers in the Upper Panjnad Basin cover an area of 3610 km^2. The maximum contribution of clean-ice (855 km^2) glacier is from the Satluj sub-basin. Out of 4910 glaciers, only 253 (5%) glaciers, with a total area of 410 km^2, are (partly) debris-covered. The Chenab sub-basin has the maximum number (123) and area (276 km^2) of DC glaciers.

TABLE 5.7 Number, Area, Elevation, and Mean Slope of Clean-Ice and Debris-Covered Glaciers in the Sub-Basins of the Indus Basin

S-N	Basin	Sub-Basin Name	Glacier Number			Area (km²)			Elevation (m a.s.l)				Mean Slope (Deg)	
									CI		DC			
			CI	DC	Total	CI	DC	Total	Min	Max	Min	Max	CI	DC
1	Kabul	PGR	88	10	88	12.11	2.52	14.63	3857	5242	4224	4788	24	9
2		AAN Rod	37	4	37	4.67	1.15	5.82	4162	5284	4363	4658	27	9
3		Kunar Rod	1148	85	1149	1419.95	153.93	1573.88	3335	7578	3114	5447	25	11
4		Swat	327	2	327	125.00	2.35	127.35	3772	5580	3872	4408	22	9
		Total	1600	101	1601	1561.73	159.96	1721.69	3335	7578	3114	5447	25	9
5	Upper Indus	Gilgit	968	68	968	857.33	81.00	938.33	2840	7730	2703	4925	23	11
6		Hunza	1384	64	1384	2344.04	409.56	2753.6	2723	7749	2409	5297	27	10
7		Shigar	439	45	439	1984.83	389.22	2374.05	3395	8566	2774	5481	29	12
8		Shyok	3357	152	3357	5565.68	372.03	5937.71	3646	7803	3231	5666	26	12
9		Zanskar	1197	29	1197	926.986	48.526	975.51	4133	6368	3997	5433	25	9
10		Shingo	882	29	882	588.42	24.26	612.68	3656	7027	3868	5023	22	9
11		Astor	372	12	372	208.71	30.88	239.59	3367	8032	2991	5031	22	10
12		U Indus	2814	29	2814	1186.04	43.93	1229.98	3149	7820	2760	5364	24	10
		Total	11,413	428	11,413	13,662.05	1399.4	15,061.45	2723	8566	2409	5666	25	10
13	Upper Panjnad	Jhelum	733	14	733	213.95	8.82	222.77	3536	6285	3404	4368	30	12
14		Chenab	2039	123	2039	2064.67	276.57	2341.24	3001	7103	3115	5913	28	11
15		Ravi	217	13	217	96.63	16.99	113.62	3511	5824	3276	5008	27	11
16		Beas	384	17	384	379.78	36.82	416.60	3079	6196	3612	5274	25	10
17		Satluj	2107	94	2108	1236.56	78.45	1315.02	3606	6973	4006	5918	24	11
		Total	5480	261	5481	3991.59	417.65	4409.24	3001	7103	3115	5918	27	11
	Grand total		18,493	790	18,495	19,215.37	1977.01	21,192.37	2723	8566	2409	5918	25	10

PGR, *Panjsher-Ghorband Rod*; AAN, *Alingar-Alishing-Nuristan Rod*.

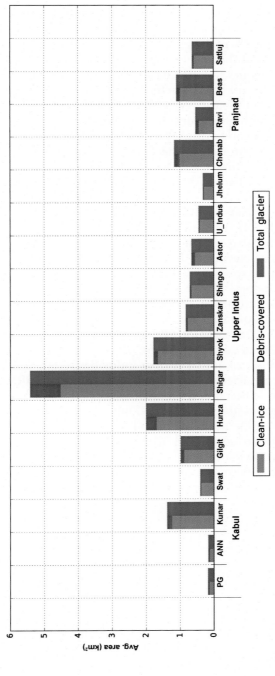

FIG. 5.9 Distribution of clean-ice and debris-covered glaciers in the sub-basins of the Indus Basin.

5.3.8 Hypsography

Glaciers are mapped across a wide range of elevations, from 2400 to 8600 m a.s.l., in the Indus Basin. The glaciers distributed at the highest and lowest elevation are few in number and area. An estimated 96% of the glacier area of the Indus Basin is found between 3700 and 6400 m a.s.l. About 59% of the glacier area is distributed at 4800 to 5800 m a.s.l. The highest concentration of glaciers in the Indus Basin is found at 5200 to 5300 m a.s.l.

Kabul: The distribution of glaciers in the Kabul Basin ranges from 3114 to 7578 m a.s.l., with both in the Kunar Rod sub-basin. A glaciated area >10 km^2 has an elevation ranging from 3800 to 6500 m a.s.l. The highest concentration of glaciers of about 20.59 km^2 lie at 4500 to 4600 m a.s.l. in the Swat sub-basin; 150.66 km^2 lie at 4900–5000 m. a.s.l. in the Kunar Basin; 2.73 km^2 lie at 4600–4700 m a.s.l. in the Panjsher-Ghorband sub-basin; and 1.70 km^2 lie at 4500–4600 m a.s.l. in the Alingar-Alishing Nuristan Rod sub-basin. The Kunar sub-basin has the highest elevation band (7500 m a.s.l.), followed by the Swat sub-basin (5500 m a.s.l.) (Fig. 5.10).

Upper Indus: The glaciers in the Upper Indus Basin range from an elevation of 8566 to 2409 m a.s.l. (see Table 5.4). The lowest and highest elevations are found in the Hunza and Shigar sub-basins. The glaciers are distributed across a wide range of elevations, but are mostly concentrated at elevations between 6400 and 3700 m a.s.l., with a glacier area of >100 km^2. The elevation zone of 4600 to 6000 m a.s.l. contains >500 km^2 within a 100 m rise in elevation and the highest glacier area of 977 km^2 at the elevations of 5300 to 5400 m a.s.l. The peak of the glacier area-distribution of 100 km^2 occurs at 4800 to 4900 m a.s.l. for Gilgit; 197 km^2 at 5100 to 5200 m a.s.l. for Hunza; 170 km^2 at 5000 to 5100 m a.s.l. for Shigar; 589 km^2 at 5700 to 5800 m a.s.l. for Shyok; 130 km^2 at 5400 to 5500 m a.s.l. for Zanskar; 64 km^2 at 5200 to 5300 m a.s.l. for Shingo; 27 km^2 at 4600 to 4700 m a.s.l. for Astor; and 75 km^2 at 4700 to 4800 m a.s.l. for the Upper Indus Basin (Fig. 5.10).

Upper Panjnad: The glacier area of 100 m bin distribution at different elevations for the Upper Panjnad Basin shows that the elevation of 4300–5800 m a.s.l. has more than 100 km^2 at each 100 m bin. The maximum glacier area is 347 km^2 at an elevation of 5200 to 5300 m a.s.l. The highest glacier areas in the individual sub-basins are 38 km^2 at 4400 to 4500 m a.s.l. in the Jhelum sub-basin; 218 km^2 at 5100 to 5200 m a.s.l. in the Chenab sub-basin; 17 km^2 at 4700 to 4800 m a.s.l. in the Ravi sub-basin; 52 km^2 at 5100 to 5200 m a.s.l. in the Beas sub-basin, and 97 km^2 at 5400 to 5500 m a.s.l. in the Satluj sub-basin (Fig. 5.10). The Chenab sub-basin has higher elevation glaciers compared with other sub-basins of the Upper Panjnad Basin.

5.4 CONCLUSIONS

Remote sensing-based mapping of glaciers smaller than 0.02 km^2 from the Landsat images of the HKH are difficult and unrealistic. Thus the glaciers of an area equal to or >0.02 km^2 are mapped from the HKH region.

A consistent semi-automated methodology based on remote sensing using the Landsat satellite images from the year 2005 ± 3 was used to map the glaciers of the Indus River Basin. A total of 18,495 glaciers were identified, with an overall area of 21,192 km^2 and estimated ice reserves of 2696 km^3. It was found that the average glacier size in the Indus Basin is

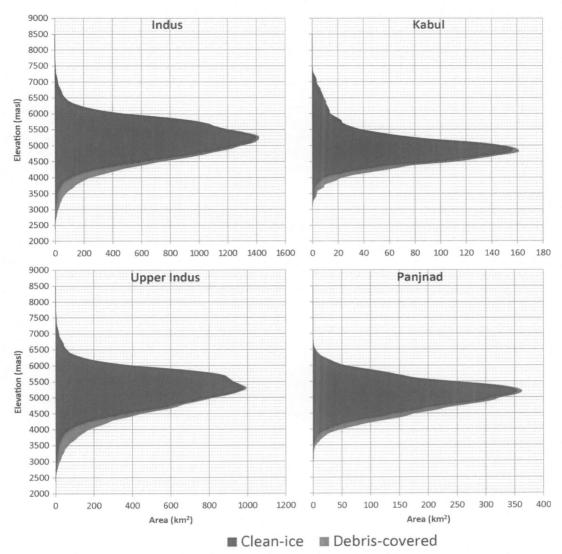

FIG. 5.10   Distribution of glacier area at different elevations in the Indus Basin and its sub-basins.

1.15 km^2. The glaciers are distributed at a wide range of elevations, from 2409 to 8566 m a.s.l. The Indus Basin contains the largest glacier in the HKH region, the Siachen glacier in the Shyok sub-basin, with a total area of 926 km^2.

Overall, 10,200 glaciers with a glaciated area of 9.3% were analyzed as debris-covered glaciers. The debris-covered (DC) glaciers were found to be mostly valley glaciers with a thick debris cover on the glacier tongue. They have an average slope of around 12°, whereas the clean-ice (CI) glaciers are much steeper, with average slopes of around 25°. There is a large variation in glacier elevation range. Almost 96% of the glacier area of the Indus Basin lies between 3700 and 6400 m a.s.l. About 59% of the glacier area is distributed at 4800 to 5800 m a.s.l. However, the highest concentration of glaciers is at 5200 to 5300 m a.s.l. The glaciers distributed at the highest and lowest elevation are very few in both number and area.

Small class 1 glaciers ($\leq$0.5 km^2) account for 71% of the glaciers, but contribute 11% of the area and 2% of the estimated ice reserves. Glaciers in class 3 are second highest in number, total area, and estimated ice reserves. Class 5 ($\geq$10 km^2) has the smallest number of glaciers but contributes nearly half of the glacier area and 79% of the estimated ice reserves.

Small glaciers ($\leq$0.5) contribute >70% in number, but the area and ice volume contribution is <11% and 2%, respectively. The large glaciers number <2% in the basin but contribute >48% and 79% of glacier area and ice reserves, respectively. However, although the small glaciers contribute less water equivalent, they are very important from the perspective of climate change assessment.

The glacier volume was estimated based on the area-volume relation. This relation shows a higher glacier volume with respect to a higher glacier area compared with other methods. However, the smaller glaciers are equally important in the contribution of glacier melt. Therefore the smaller glaciers are also included in the inventory.

The glaciers are distributed at a wide range of elevations, from 2409 to 8566 m a.s.l. The highest concentration of glaciers is from 4000 to 6000 m a.s.l. The wider range of elevations is derived mostly from to the scattered distribution of glaciers in the sub-basin.

Overall, the analysis provides baseline information about the glaciers and their characteristics in the Indus River Basin. There is still a need to better understand and conduct a comprehensive assessment of the future state of the glaciers. Water from glacier melt is of great importance to the highly populated areas downstream. The reduction in the amount and changes in the meltwater could have a major impact on water availability. Therefore a better understanding of glacier dynamics and of downstream hydrological planning and water resources management will play a critical role in the water security in the Indus River Basin.

Acknowledgments

Landsat data are courtesy of NASA and the USGS. The SRTM elevation model version is courtesy of NASA and was further processed by the Consultative Group for International Agriculture Research (CGIAR). This study was supported by the Cryosphere Monitoring Project of the Swedish International Development Cooperation Agency and the Norwegian Ministry of Foreign Affairs. The financial support from these donors is gratefully acknowledged.

References

Bajracharya, S.R., Shrestha, B. (Eds.), 2011. The Status of Glaciers in the Hindu Kush-Himalayan Region. ICIMOD, Kathmandu.

Bajracharya, S.R., Maharjan, S.B., Shrestha, F., 2014a. The status and decadal change of glaciers in Bhutan from 1980's to 2010 based on the satellite data. Ann. Glaciol. 55 (66), 159–166. https://dx.doi.org/10.3189/2014AoG66A125.

Bajracharya, S.R., Maharjan, S.B., Shrestha, F., Bajracharya, O.R., Baidya, S., 2014b. Glacier Status in Nepal and Decadal Change From 1980 to 2010 Based on Landsat Data. ICIMOD, Kathmandu.

Barrand, N.E., Murray, T., 2006. Multivariate controls on the incidence of glacier surging in the Karakoram Himalaya. Arct. Antarct. Alp. Res. 38, 489–498.

Bhambri, R., Bolch, T., Kawishwar, P., Dobhal, D.P., Srivastava, D., Pratap, B., 2013. Heterogeneity in glacier response in the upper Shyok valley, northeast Karakoram. Cryosphere 7 (5), 1385.

Bhutiyani, M.R., 1999. Mass-balance studies on Siachen glacier in the Nubra valley, Karakoram Himalaya, India. J. Glaciol. 45 (149), 112–118.

Bolch, T., Kulkarni, A., Kääb, A., Huggel, C., Paul, F., Cogley, J.G., Frey, H., Kargel, J.S., Fujita, K., Scheel, M., Bajracharya, S., Stoffel, M., 2012. The state and fate of Himalayan glaciers. Science 336 (6079), 310–314.

Chaohai, L., Liangfu, D., 1986. The newly progress of glaciers inventory in Tianshan Mountains. J. Glaciol. Geocryol. 8 (2), 168–169.

Copland, L., Sylvestre, T., Bishop, M.P., Shroder, J.F., Seong, Y.B., Owen, L.A., Bush, A., Kamp, U., 2011. Expanded and recently increased glacier surging in the Karakoram. Arct. Antarct. Alp. Res. 43, 503–516.

Eriksson, M., Jianchu, X., Shrestha, A.B., Vaidya, R.A., Nepal, S., Sandström, K., 2009. The Changing Himalayas Impact of Climate Change on Water Resources and Livelihoods in the Greater Himalaya. International Centre for Integrated Mountain Development (ICIMOD), Kathmandu.

Fowler, H.J., Archer, D.R., 2006. Conflicting signals of climatic change in the upper Indus Basin. J. Clim. 19, 4276–4293.

Gardelle, J., Berthier, E., Arnaud, Y., 2012. Slight mass gain of Karakoram glaciers in the early twenty-first century. Nat. Geosci. 5, 322–325.

Hewitt, K., 1969. Glacier surges in the Karakoram Himalaya, Central Asia. Can. J. Earth Sci. 6, 1009–1018.

Hewitt, K., 1998. Recent glacier surges in the Karakoram Himalaya, South Central Asia. EOS, Electronic Supplement.

Hewitt, K., 2007. Tributary glacier surges: an exceptional concentration at Panmah Glacier, Karakoram Himalaya. J. Glaciol. 53, 181–188.

Immerzeel, W.W., van Beek, L.P.H., Bierkens, M.F.P., 2010. Climate change will affect the Asian water towers. Science 328, 1382–1385. https://dx.doi.org/10.1126/science.1183188.

Immerzeel, W.W., Beek, L.P.H., Konz, M., Shrestha, A.B., Bierkens, M.F.P., 2012. Hydrological response to climate change in a glacierized catchment in the Himalayas. Clim. Chang. 110, 721–736.

IPCC, 2007. Climate change 2007. Impacts, adaptation and vulnerability: working group II contribution to the fourth assessment report of the IPCC intergovernmental panel on climate change. In: Parry, M.L., Canziani, O.F., Palutikof, J.P., van der Linden, P.J., Hanson, C.E. (Eds.), Assessment. Cambridge University Press, Cambridge, UK. 976 pp.

Kääb, A., Berthier, E., Nuth, C., Gardelle, J., Arnaud, Y., 2012. Contrasting patterns of early twenty-first-century glacier mass change in the Himalayas. Nature 488 (7412), 495–498.

Kotlyakov, V.M., Osipova, G.B., Tsvetkov, D.G., 2008. Monitoring surging glaciers of the Pamirs, Central Asia, from space. Ann. Glaciol. 48, 125–134.

Kulkarni, A.V., Bahuguna, I.M., Rathore, B.P., Singh, S.K., Randhawa, S.S., Sood, R.K., Dhar, S., 2007. Glacial retreat in Himalaya using Indian Remote Sensing satellite data. Curr. Sci. 92 (1), 69–74.

LIGG/WECS/NEA, 1988. Report on First Expedition to Glaciers and Glacier Lakes in the Pumqu (Arun) and Poique (Bhote-Sun Kosi) River Basins, Xizang (Tibet), China. Sino-Nepalese Investigation of Glacier Lake Outburst Floods in the Himalaya. Science Press, Beijing, China.

Lutz, A.F., Immerzeel, W.W., Shrestha, A.B., Bierkens, M.F.P., 2014. Consistent increase in High Asia's runoff due to increasing glacier melt and precipitation. Nat. Clim. Chang. 4, 587–592. https://dx.doi.org/10.1038/nclimate2237.

Mason, K., 1935. The study of threatening glaciers. Geogr. J. 85, 24–41.

Mool, P.K., Bajracharya, S.R., Joshi, S.P., 2001a. Inventory of Glaciers, Glacial Lakes, and Glacial Lake Outburst Flood Monitoring and Early Warning System in the Hindu Kush-Himalayan Region, Nepal. ICIMOD and UNEP/RRC-AP. 364 pp.

Mool, P.K., Wangda, D., Bajracharya, S.R., Joshi, S.P., Kunzang, K., Gurung, D.R., 2001b. Inventory of Glaciers, Glacial Lakes, and Glacial Lake Outburst Flood Monitoring and Early Warning System in the Hindu Kush-Himalayan Region, Bhutan. ICIMOD and UNEP/RRC-AP. 227 pp.

Muller, F., Caflish, T., Muller, G., 1977. Instruction for Compilation and Assemblage of Data for a World Glacier Inventory. Temporary Technical Secretariat for World Glacier Inventory, Swiss Federal Institute of Technology, Zurich. 74 pp.

Raina, V.K., Srivastava, D., 2008. Glacier Atlas of India. Geological Society of India, Bangalore. 316 pp.

Scherler, D., Bookhagen, B., Strecker, M.R., 2011. Spatially variable response of Himalayan glaciers to climate change affected by debris cover. Nat. Geosci. 4 (3), 156–159.

Shaikh, H., Memon, N., Bhanger, M.I., Nizamani, S.M., 2014. GC/MS based non-target screening of organic contaminants in river Indus and its tributaries in Sindh (Pakistan). Pak. J. Anal. Environ. Chem. 15 (1), 24.

II. CLIMATE-ECO-HYDROLOGY OF INDUS RIVER BASIN

Tangri, A.K., Chandra, R., Yadav, S.K.S., 2013. Signatures and evidences of surging glaciers in the Shyok valley, Karakoram Himalaya, Ladakh region, Jammu & Kashmir State, India. In: Sinha, R., Ravindra, R. (Eds.), Earth System Processes and Disaster Management, Volume 1, Society of Earth Scientists Series. Springer Berlin Heidelberg, pp. 37–50. https://dx.doi.org/10.1007/978-3-642-28845-6_4.

Winiger, M., Gumpert, M., Yamout, H., 2005. Karakoram-Hindu Kush-Western Himalaya: assessing high-altitude water resources. Hydrol. Process. 19 (12), 2329–2338.

Xu, J., Shrestha, A., Vaidya, R., Eriksson, M., Hewitt, K., 2007. The Melting Himalayas: Regional Challenges and Local Impacts of Climate Change on Mountain Ecosystems and Livelihoods. ICIMOD, Kathmandu, ISBN 978 92 9115 047 2.

Further Reading

American Geophysical Union, http://www.agu.org/pubs/eos-news/supplements/1995-2003/97106e.shtml.

NASA JPL, 2014. U.S. Releases Enhanced Shuttle Land Elevation Data. NASA Jet Propulsion Laboratory (JPL). http://www.jpl.nasa.gov/news/news.php?release=2014-321. (18 January 2018).

A Review on the Projected Changes in Climate Over the Indus Basin

Arun Bhakta Shrestha*, Nisha Wagle*, Rupak Rajbhandari†
*International Center for Integrated Mountain Development, Kathmandu, Nepal †Tribhuvan University, Kathmandu, Nepal

6.1 INTRODUCTION

The Indus River is one of the oldest documented rivers in history, and its transboundary basin comprises a total area of 1.12 million km^2 that is shared by four countries: Pakistan (47%), India (39%), China (8%), and Afghanistan (6%) (FAO, 2011). The Indus River originates near Mount Kailash in the Gangdise Range in Tibet and flows into the Arabian Sea, covering a distance of about 3000 km (Inam et al., 2008). Among the 27 major tributaries in the Indus system, the Chenab, Ravi, Sutlej, Jhelum, Beas, and Indus rivers are the most significant branches flowing westward; whereas the most significant branch flowing eastward is the Kabul River, which originates in Afghanistan and flows into Pakistan (Rauf, 2013). As shown in Fig. 6.1, the Indus is divided into two physiographic divisions: the upper basin, which occupies the glaciated and barren majestic high mountains of the Himalayas (the Karakoram, the Hindu Kush, Siwalik, Suleiman, and Kirthar ranges); and the lower basin, which is formed by the deposition of the Indus River and tributaries, including the alluvial plains of Punjab and Sindh (Khan, 2013).

The climate in the Indus Basin varies from subtropical arid and semi-arid to temperate sub-humid in Sindh and Punjab to alpine in the mountainous highlands in the north, with average temperatures ranging between 2°C and 49°C (Ali, 2013; FAO, 2011). The mean annual rainfall varies between 90 and 500 mm in the downstream and midstream segments, while it is more than 1000 mm in the upstream catchment. The mean evaporation ranges between 1650 and 2040 mm (Ali, 2013), which makes the basin semi-arid to arid. Bimodal precipitation distribution in the basin is observed due to the westerly disturbance and summer monsoon system (Hasson et al., 2016). The Indus River flow consists of runoff from seasonal rainfall and glacier melt and snowmelt. Glaciers account for almost 80% of the water in the Upper Indus Basin

145

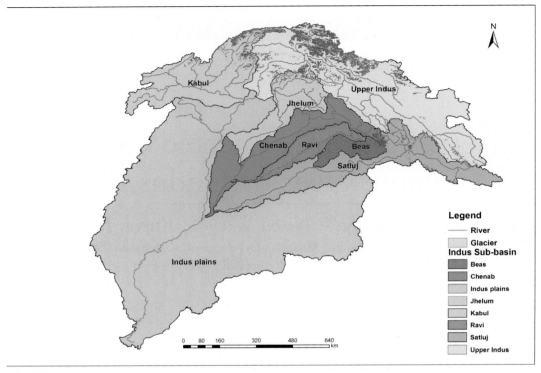

FIG. 6.1   Indus Basin and its tributaries.

(UIB), serving as natural storage as well as providing perennial supplies to the Indus River and its tributaries (FAO, 2011; Lutz et al., 2014; Khan, 2013). The climate in the Indus Basin is complex with mountainous climate types and monsoon dynamics playing important roles (Lutz et al., 2016).

The Indus Basin sustains the lives of nearly 268 million people. With its high population density and an approximate water availability of 1329 m^3/capita (Rajbhandari et al., 2015), the basin plays an important role in the socioeconomic development of both Pakistan and India, as people of these countries are directly or indirectly dependent on it for agriculture, tourism, hydroelectricity, and river transportation. The Indus Basin Irrigation System (IBIS) is the largest contiguous irrigation system in the world (Nepal and Shrestha, 2015; Qureshi, 2011). Considering the importance of the basin, alterations in the availability of its water resources due to climate change and other factors will seriously impact the food security and environment in the area (Archer et al., 2010). The rapidly changing demographics, the climatic conditions, and the increasing demand for water are further stressing the basin's water resources (ICIMOD, 2017). For example, about 13 million ha of cultivable land remain barren because of a water shortage in Pakistan (ICIMOD, 2010).

Kraaijenbrink et al. (2017) reported consistent warming of Asia's glacierized high mountain areas and that it is occurring at much higher rates than the global average, indicating the loss of glacier mass under present climatic conditions. They also noted that, even if temperatures were to stabilize, glacier loss would continue for decades. The perennial snow and ice cover of the UIB is about 20,000 km^2 (Lutz et al., 2014). The majority of the flow comes from glaciers, so the Indus basin is particularly vulnerable in terms of climate change that causes higher warming trends and loss of glacial mass. Global warming will initially increase the water flow, causing flash floods and glacial lake outburst floods, but will be followed by a reduction in the water flow. However, the future response of the glaciers in the Indus Basin to the runoff is not very clear (Hasson et al., 2013). Any changes in precipitation and temperature in the basin are important parameters and must be taken into consideration. Therefore information on possible future climatic states is critical for policymakers and regional actors dealing with both short-term and long-term management, planning, and adaptation policies regarding water resources (Lutz et al., 2014). Although several studies have been carried out to project future temperature and precipitation, a comprehensive assessment of the current state of climatic components is largely missing (Hasson et al., 2017).

This chapter reviews key research literature on the projected future climate changes in the Indus Basin based on several climate models. The aim of this chapter is to provide the present state of knowledge about future climate projections for the Indus Basin.

6.2 CLIMATE PROJECTION IN THE INDUS BASIN

A number of general circulation models (GCMs) and regional climate models (RCMs) have been used for climate projections in the Indus Basin. Due to the coarse resolution outputs from GCMs and their inability to resolve sub-grid scale features, GCMs are generally downscaled to a higher resolution or are used as boundary conditions for RCMs (Lutz et al., 2016; Wilby and Dawson, 2007), which generally show reduced bias and provide greater spatial detail (Huang et al., 2017). However, it is important first to consider whether a climate model reproduces regional climatology or whether large biases exist (Forsythe et al., 2014).

Hasson et al. (2013) analyzed the Coupled Model Intercomparison Project Phase 3 (CMIP3) models, which were also used for the Intergovernmental Panel on Climate Change (IPCC) Fourth Assessment report. Due to its limitation in simulating the summer monsoon, the CMIP3 resulted in uncertainty in simulating the hydrological cycle in the studied basin. The authors reported disagreement between the observations of most of the GCMs, resulting in a large inter-model spread. The Providing Regional Climates for Impact Studies (PRECIS) system was used by Rajbhandari et al. (2015), which was able to capture the seasonal climate scenario quite well, although there were overestimations of rainfall. Warm bias in the maximum temperature during the summer, slight cold bias during the winter and post-monsoon seasons, and cold bias in minimum temperatures during the winter season, post-monsoon season, and most of the pre-monsoon season were observed.

The Coupled Model Inter-Comparison Project Phase 5 (CMIP5) model, was used for the IPCC Assessment Report and went through extensive development, with higher vertical and horizontal resolution and improved interactions between components of atmosphere, land use, and vegetation (Taylor et al., 2012). In recent years CMIP5 models were used by several authors for climate study in the Indus Basin (e.g., Hasson et al., 2016; Gebre and

Ludwig, 2014; Su et al., 2016). Su et al. (2016) reported that the model captured features of seasonal variations in both temperature and precipitation, but the quantity value was difficult to reproduce. Despite advancements, uncertainty in simulation of precipitation associated with the South Asian summer still exists (Hasson et al., 2016; Gebre and Ludwig, 2014). The case is even worse for higher altitudes (Su et al., 2016). The absence of irrigation representation in the CMIP5 models commonly results in systematic errors in the patterns of precipitation (Hasson et al., 2016). However, Su et al. (2016) reported that the CMIP5 performed better in quantity and annual mean cycle of precipitation over the Indus River Basin (IRB) compared with the PRECIS used by Rajbhandari et al. (2015).

GCMs contain high uncertainties mainly due to their coarse spatial resolutions. Downscaling is performed in order to avoid the uncertainties associated with GCMs. Dynamical downscaling using RCMs can be useful to improve the simulation of a monsoonal regime (Hasson et al., 2016). Recently, Huang et al. (2017) used RCM, COSMO-CCLM (CCLM), which produced the annual cycle of temperature as well as spatial variations of precipitation quite well. However, quantitative biases, especially in the precipitation pattern, still existed, which might be due to the regional atmospheric circulation produced by global circulation models. In comparison with PRECIS and CMIP5 GCMs, the CCLM model performed better in reproducing the amounts and annual cycles of precipitation as well as spatial patterns (Huang et al., 2017). The overall results show that the CCLM is able to satisfactorily capture the dominant spatial characteristics in annual temperature and precipitation time series, including the warming trend and the seasonal variations (Huang et al., 2017). Ali et al. (2015) projected future climate in the UIB using the Conformal-Cubic Atmospheric Model (CCAM) and the RegCM4. The CCAM was integrated from a coarse resolution (200 km) to a much higher resolution (15 km), which resulted in more specific and accurate projections. RegCM is more flexible, user-friendly, and portable, while also making necessary updates in source code easier to conduct. Significant biases between the observed and simulated data for the base period were observed. Both CCAM and RegCM underestimated the temperature and overestimated the precipitation over the UIB. Statistical correction techniques were used in the simulated data for a future projection that showed realistic results. Although both models handled the snowmelt, they did not include a comprehensive parameterization of glaciers. Thus models particularly related to glacier melting are required.

Because RCMs can be time consuming and computationally heavy, Statistical Downscaling Model (SDSM) is more promising, as it provides a low-cost option for the assessment of localized climate (Wilby and Dawson, 2007). It also incorporates large-scale state and regional/local physiographic features. Kazmi et al. (2014) used the model to analyze the climate of Pakistan and reported the similarity between the SDSM output and the observed data. Likewise, Forsythe et al. (2014) developed a point-based stochastic Weather Generation (WG) downscaling method for three stations in the UIB, which was first WG to be used in a semi-arid climate regime with substantial monsoonal influence. However, uncertainties in climate projection persist, requiring more advanced downscaling methods to evaluate changes in variability and sequencing of events in the region (Forsythe et al., 2014).

Despite the improvement and availability of several models, uncertainty about the future climate still exists (Lutz et al., 2016; Nepal and Shrestha, 2015). In practice, limited models with specific projections are used for a specific study (Nepal and Shrestha, 2015). Multiple models are needed for quantification of uncertainty. Model selection depends on multiple criteria, which is crucial for the studies on the impact of climate change. An approach in which models covering all ranges of projections for more than one climatological variable are selected from a pool of climate models is called the envelope approach. However, the major drawback of this approach is that it does not consider the capacities of the models to simulate climate and instead only considers annual mean change. Other approaches consider either one or more variables using cluster analysis algorithms. This limitation is overcome in an advanced envelope approach in which a past-performance approach for model selection is based on the model's capability to simulate present and near-past climates (Lutz et al., 2016).

6.2.1 Projected Temperature Trends

A study jointly conducted by the Pakistan Meteorological Department (PMD) and the Global Change Impact Studies Center projected an increase in both maximum and minimum temperatures. The climate scenarios are based on SRED-IPCC 2007 using A2, A1B, and B1 scenarios. Temperature increase in the northern part of Pakistan during 2011–2050 under A2, A1B, and B1 scenarios was projected to be 0.76°C/decade, 0.63°C/decade, and 0.39°C/decade, respectively (Chaudhry et al., 2009).

Rajbhandari et al. (2015) projected climate change in the Indus Basin using the PRECIS system based on the IPCC SRES A1B scenario for three time periods—2011–40, 2041–70, and 2071–98—as compared to 1961–90 based on three simulations[1]: Q0, Q1, and Q14. Increase in both seasonal and annual maximum and minimum temperatures were projected in the Indus Basin, particularly in higher altitudes. For annual maximum temperature, the highest increase was projected to occur during 2071–98, with an increase of 4°C, 3.4°C, and 4.6°C under Q0, Q1, and Q14 simulations, respectively. Similarly, annual minimum temperature was projected to increase by 4.3°C, 3.9°C, and 5.1°C under the three respective simulations. In the case of seasonal temperature, an increasing trend in winter temperatures was projected for 2011–40 (Fig. 6.2A, D, and G), 2041–70 (Fig. 6.2B, E, and H), and 2071–98 (Fig. 6.2C, F, and I) for three simulations Q0 (Fig. 6.2A, B, and C), Q1 (Fig. 6.2D, E, and F), and Q14 (Fig. 6.2G, H, and I), as shown in Fig. 6.2.

A year-round increase in annual mean temperature of 4.8°C during 2071–2100 in the UIB basin, as compared with 1961–90, was projected by Forsythe et al. (2014) using a stochastic rainfall model (RainSim) combined with a rainfall conditioned weather generator (CRU WG). Gebre and Ludwig (2014) analyzed CMIP5 global climate models and reported increases in maximum and minimum temperatures for both time periods—2030s and 2070s—under RCP4.5 and RCP8.5 as compared to the base period 1971–2005. The increase in maximum temperature ranged between 1°C and 7°C. A similar pattern was observed for average monthly and seasonal minimum temperatures. Kazmi et al. (2014) analyzed data from 44 meteorological stations in Pakistan using a statistical downscaling method and projected temperature for the period of 1961–2099; they reported an increase in both minimum and maximum temperatures, particularly in the northern areas, as compared to the base

---

[1]Q0, Q1, and Q14 are three simulations made under the Quantifying Uncertainties in Model Projection (QUMP) experiments.

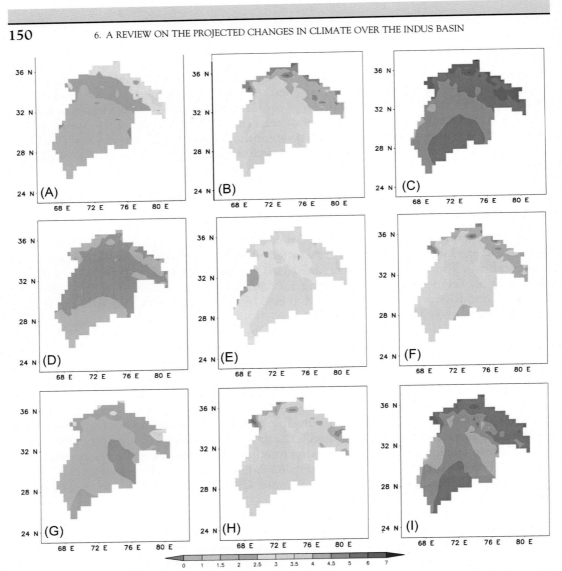

FIG. 6.2   Projected change in winter season minimum temperature in degrees Celsius in 2011–40 (A, D, G), 2041–70 (B, E, H), and 2071–98 (C, F, I) for three simulations Q0 (A, B, C), Q1 (D, E, F), and Q14 (G, H, I) (Rajbhandari et al., 2015).

period 1961–90. Higher warming in the upper altitudes has been reported by other authors as well (e.g., Chaudhry et al., 2009; Forsythe et al., 2014; Rajbhandari et al., 2015). Several other studies also report elevation-based warming in temperature, and it is suggested to be more pronounced in the future (Lutz et al., 2016) due to several factors, such as snow albedo, water vapor changes, latent heat release, temperature changes, and aerosols (Kraaijenbrink et al., 2017).

Ali et al. (2015) used CCAM data under RCP4.5 and RCP8.5 scenarios and RegCM data under RCP8.5 for climate projection in the UIB for the periods 2006–35, 2041–70, and 2071–2100 and reported warming as compared to 1976–2005. The seasonal mean temperatures of MAM (Fig. 6.3A, E, I, M, and Q), JJA (Fig. 6.3B, F, J, N, and R), SON (Fig. 6.3C, G, K, O, and S), and DJF (Fig. 6.3D, H, I, P, and T) of CRU and CCAM under RCP8.5 are shown in Fig. 6.3. CCAM projected, respectively, an increase in maximum temperature of 2.2°C, 4.2°C, and 5.8°C under RCP8.5; and of 0.5°C, 1.5°C, and 2°C under RCP4.5 during 2006–35, 2041–70, and 2071–2100. The minimum temperature was also projected to increase for all future periods. The temperature increase was projected to be higher during the spring and winter than in the summer, except in the northern part where the increase in summer temperature is higher. RegCM also projected an increase in temperature in the basin, with spring and winter increases higher than in the summer. The total increases in temperature during 2041–50 and 2071–80 were 1.8°C and 4.3°C, respectively.

Su et al. (2016) analyzed CMIP5 output under three RCP scenarios (RCP2.6, RCP4.5, and RCP8.5) for mid- (2046–65) and late 21st century (2081–2100), relative to the base period 1986–2005, and reported a consistent increment and spatial distribution in the annual mean temperature from 2046 to 2065 (Fig. 6.4A–C) and 2081–2100 (Fig. 6.4D–F) over the entire basin, as shown in Fig. 6.4. The area with the maximum increase, exceeding 5°C, is located in the UIB under RCP8.5. The annual mean temperature is projected to increase by 1.21°C, 1.93°C, and 2.71°C in the mid-21st century and by 1.10°C, 2.49°C, and 5.19°C in the late 21st century for the

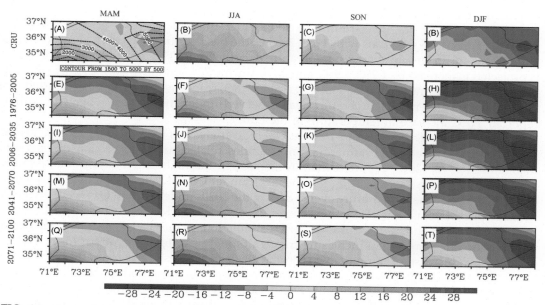

FIG. 6.3 Seasonal mean temperature of MAM (A, E, I, M, Q), JJA (B, F, J, N, R), SON (C, G, K, O, S), and DJF (D, H, L, P, T) of CRU and CCAM with emission scenario RCP8.5 (°C). *Dash line* in (A) indicates the elevation (m) (Ali et al., 2015).

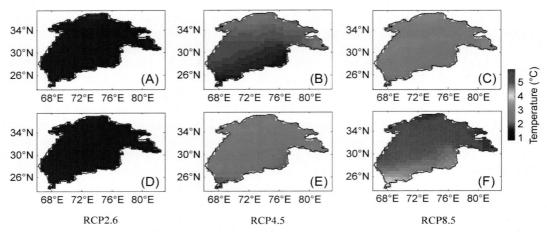

FIG. 6.4 Change of annual mean temperature from 2046 to 2065 (A–C) and 2081–2100 (D–F) under 2.6, 4.5, and 8.5 scenarios relative to 1986–2005 in the Indus River Basin. *Reprinted from Su, B., Huang, J., Gemmer, M., Jian, D., Tao, H., Jiang, T., Zhao, C., 2016. Statistical downscaling of CMIP5 multi-model ensemble for projected changes of climate in the Indus River Basin. Atmos. Res. 178, 139—149. Copyright (2016), with permission from Elsevier.*

three respective RCP scenarios. In terms of seasonal variation, the summer temperature is projected to increase. In the latter part of the 21st century, the summer temperature is projected to increase by 4.82°C under RCP 8.5.

Huang et al. (2017) used RCM, COSMO-CLM (CCLM) to project the temperature in the Indus River Basin for two future periods—2047–61 and 2081–2100—under RCP2.6, RCP4.5, and RCP8.5 scenarios with 1986–2005 as the base period. The result showed consistent increments in annual average temperature over the basin. The mean annual temperature is projected to increase by 1.2°C, 1.9°C, and 2.7°C during 2047–61 and by 0.9°C, 2.3°C, and 5°C during 2081–2100 under RCP2.6, RCP4.5, and RCP8.5, respectively. It can be concluded that the highest increment in temperature was observed under the highest RCP. The authors reported a 5°C increase in temperature in the UIB.

For the most part, the literature does not include analysis of extreme temperature projections, likely because of the numerous uncertainties related to the projection of extremes. Rajbhandari et al. (2015) reported that an increase in the highest maximum temperatures in the basin will occur in the 2050s and the 2080s under Q0, Q1, and Q14 simulations, while a decrease in extreme temperature events during the 2020s under Q1 and Q14 were projected in small pocket areas. The maximum temperature was projected to increase by 4°C to 8°C under Q0 and Q14 and by 4°C to 6°C under Q1 simulations in the 2080s. The lowest minimum temperature also showed an increasing trend, although a systematic pattern for it was not obtained. An increase of more than 4°C in the 2080s in the basin and more than 8°C in the border between the Upper and Lower Indus Basin was also projected for the 2080s. Huang et al. (2017) showed that the Probability Density Curves (PDF) model projected an increase in average temperature as well as extreme temperatures during the mid- and late 21st century under three RCPs (RCP2.6, RCP4.5, and RCP8.5). This will likely lead to an increase in heat waves.

6.2.2 Projected Precipitation Trends

Immerzeel et al. (2010) projected an increase in mean precipitation in the Indus Basin by 25% during 2046–65 under the A1B scenario as compared to 2000–07. Similarly, Forsythe et al. (2014) projected a 27% increase in seasonal mean precipitation and an 18% increase in annual mean precipitation during 2071–2100 as compared to 1961–90.

A non-uniform change in precipitation over the Indus Basin was widely reported by several authors. A study jointly conducted by the Pakistan Meteorological Department (PMD) and the Global Change Impact Studies Center reported mixed trends of increases and decreases in precipitation in the northern part of Pakistan. Precipitation changes of +4.6 mm/decade, +2.9 mm/decade, and −1.3 mm/decade were projected during 2011–50 under A2, A1B, and B1 scenarios respectively (Chaudhry et al., 2009). Hasson et al. (2013) also reported inconsistent results for precipitation projections for the 21st and 22nd centuries, which could be due to the use of GCMs that do not implement irrigation or water diversion. Rajbhandari et al. (2015) reported an increase in precipitation over the Upper Indus Basin and a decrease over the Lower Indus Basin during the winter season. As shown in Fig. 6.5, a marked increase in precipitation during the monsoon season in the near future (Fig. 6.5A, D, and G) and a decrease in precipitation over southern parts of the area in the distant future (Fig. 6.5E, F, H, and I) have been projected.

Gebre and Ludwig (2014) reported an increase in precipitation during summer and a decrease during winter under both emission scenarios for the 2030s and 2070s as compared to the base period 1971–2005. The projection of five different GCMs shows positive change in major parts of the basin, particularly in the mid part of the basin, while under two GCMs, a decrease in average seasonal precipitation ranging from −20% to 0% was projected.

Ali et al. (2015) also reported that model representation of precipitation is highly variable during the projected period. Precipitation is projected to increase under RCP4.5 and RCP8.5 scenarios by 10% to 11% during 2006–35 and 13% during 2041–70, while it is projected to increase under the two scenarios by 12% and 20% during 2071–2100, respectively. Future seasonal precipitation was overestimated by CCAM, showing more increase during autumn (Fig. 6.6K, O, and S). Similarly, RegCM also projected 14% and 23% increases in precipitation during 2041–50 and 2071–80, respectively.

Su et al. (2016) also reported distinct spatial differences in mean annual precipitation for the mid- and late 21st century compared to 1986–2005 in the Basin. An increase in mean annual precipitation was projected for 2046–65 (Fig. 6.7G–I) and 2081–2100 (Fig. 6.7J–L) under all RCPs as compared to the baseline. Moreover, a prominent increase of more than 50% was projected for late 21st century, especially in high altitudes. Seasonal precipitation, however, showed variation, as monsoon precipitation was projected to increase under all scenarios while winter and spring precipitation showed both increasing and decreasing trends.

Hasson et al. (2016) used CMIP5 and reported both an increase and a decrease in precipitation during 1961–2000 over the Indus Basin, even as historical data sets showed negligible precipitation. The authors showed that almost one-third of the models failed to capture the monsoon signal over the basin, although the majority of the models agreed in delayed and shortening of the monsoon season. The seasonality index was lower than observed for the Indus Basin during the monsoon season, followed by an underestimation of precipitation related to the shift of the westerly storm track; while an increase in the number of dry days and a

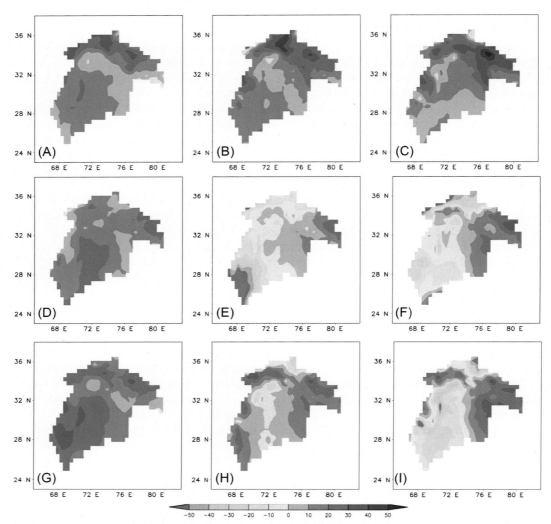

FIG. 6.5   Projected change in mean summer monsoon rainfall (%) with respect to baseline 1961–90 in the 2011–40 (A, D, G), 2041–70 (B, E, H), and 2071–98 (C, F, I) periods for the three PRECIS simulations Q0 (A, B, C), Q1 (D, E, F), and Q14 (G, H, I) (Rajbhandari et al., 2015).

decrease in precipitation in winter with occurrence of westerly disturbance over the Karakorum were also reported.

Huang et al. (2017) also reported a decreasing trend for annual precipitation over the entire basin, except in the far north and the south of the basin, for the periods 2046–66 and 2081–2100 under three emission scenarios (RCP2.6, RCP4.5, and RCP8.5) relative to the baseline period 1986–2005. During 2046–66 the mean precipitation is projected to decrease by 5.5%, 12.4%, and 10.5% under RCP2.6, RCP4.5, and RCP8.5, respectively; while during 2081–2100 the

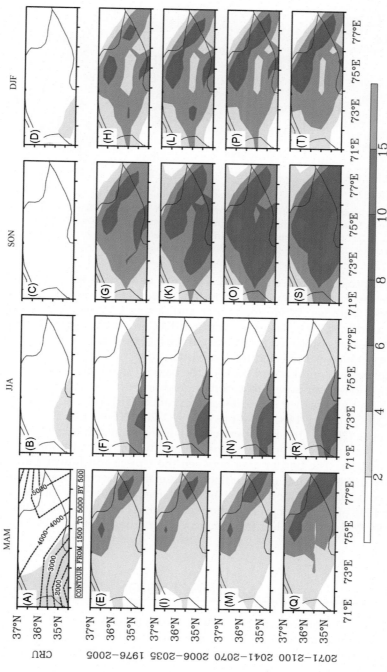

FIG. 6.6 Seasonal mean precipitation of MAM (A, E, I, M, Q), JJA (B, F, J, N, R), SON (C, G, K, O, S), and DJF (D, H, L, P, T) of CRU and CCAM with emission scenario RCP8.5 (mm). *Dash line* in (A) indicates the elevation (m) (Ali et al., 2015).

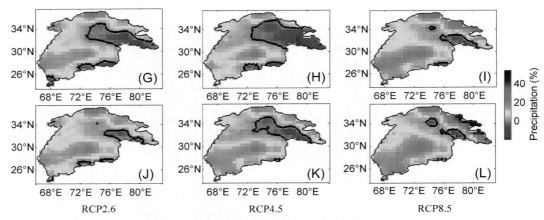

FIG. 6.7   Change of annual mean precipitation from 2046 to 2065 (G–I) and 2081–2100 (J–L) under 2.6, 4.5, and 8.5 scenarios relative to 1986–2005 in Indus River Basin. *Reprinted from Su, B., Huang, J., Gemmer, M., Jian, D., Tao, H., Jiang, T., Zhao, C., 2016. Statistical downscaling of CMIP5 multi-model ensemble for projected changes of climate in the Indus River Basin. Atmos. Res. 178, 139—149. Copyright (2016), with permission from Elsevier.*

precipitation is projected to decrease by 2.4%, 9.6%, and 17.5%, respectively. Seasonal variation shows a decrease in monsoon precipitation, particularly in the central and southern plains of lower altitudes. While for winter and spring precipitations—except for the late 21st century under RCP2.6—a decreasing trend is projected. Under the RCP4.5 scenario, the greatest increase is projected for the eastern part of the Indus Basin, and the strongest decrease in precipitation is projected for the western part, which is due to the contrast in the climate of the western and eastern parts of the basin (Lutz et al., 2016).

As in the case of temperature, projections regarding extreme precipitations are still lacking. An increase in rainy days over the northeast part of the UIB and a decrease over the southern part under Q0, Q1, and Q14 simulations in the 2020s, 2050s, and 2080s were projected by Rajbhandari et al. (2015). The study projected an increase in rainfall intensity over the border area between the upper and lower basins, where the amount of rainfall is projected to be highest in the 2080s. The border area between the Upper and Lower Indus Basin is prone to flash floods and is a source of large floods. Thus the projected changes in extreme events are likely to exacerbate the flood and flash flood hazards in this part of the basin. In contrast, the projected PDF of monsoon precipitation for the mid- and late 21st century under RCP2.6, RCP4.5, and RCP8.5 indicated less frequent extreme rainfall events in the Indus Basin (Huang et al., 2017).

6.3 CONCLUSION

Overall, the climate projections in the Indus Basin indicate significant warming in the future, while a variation in the seasonal temperature has been reported. The majority of the studies report variation in the annual and seasonal precipitation in the basin. The main findings of these studies are summarized here:

- An increase in temperature is agreed upon across most of the GCMs, while poor agreement on precipitation patterns is observed.
- Temperature in the upper basin is projected to increase by more than 5°C by the late 21st century compared to the late 20th century, which will likely affect the snow and glacier melt, leading to variations in the availability of water.
- Seasonal increases of summer temperature, especially in the upper altitudes, will likely result in a negative glacial mass balance.
- Studies generally project an increase in mean annual precipitation in the late 21st century, especially in the high altitudes, and report an increase in monsoon precipitation, which might cause more floods and flash floods.
- However, decreasing winter and spring precipitations will likely deplete the availability of water resources. An increase in the frequency of westerly disturbances could offset this and is likely to even cause a positive mass balance for some glaciers.
- The complex topography of the basin combined with the fact that the majority of the models do not incorporate the vast irrigation system—which likely impacts atmospheric circulation as well as monsoon precipitation—is a major limitation of current climate models. Including appropriate representations of irrigation water and of the orography is therefore important.
- Several downscaling techniques used to analyze the climate do not evaluate changes in variability and sequencing of events. Therefore more advanced downscaling techniques with higher spatial resolutions are required.

Acknowledgments

This work was carried out by the Indus Basin Initiative of ICIMOD, which contributes to the Sustainable Development Investment Portfolio and is supported by the Department of Foreign Affairs and Trade, Government of Australia. The study was partially supported by ICIMOD core funds (contributed by the governments of Afghanistan, Australia, Austria, Bangladesh, Bhutan, China, India, Myanmar, Nepal, Norway, Pakistan, Switzerland, and the United Kingdom).

References

Ali, A., 2013. Indus Basin Floods Mechanisms, Impacts, and Management. Asian Development Bank, Mandaluyong City, Philippines.

Ali, S., Li, D., Congbin, F., Khan, F., 2015. Twenty first century climatic and hydrological changes over upper Indus Basin of Himalayan region of Pakistan. Environ. Res. Lett. 1014007.

Archer, D.R., Forsythe, N., Fowler, H.J., Shah, S.M., 2010. Sustainability of water resources management in the Indus Basin under changing climatic and socio economic conditions. Hydrol. Earth Syst. Sci. 14, 1669–1680.

Chaudhry, Q.-Z., Mahmood, A., Rasul, G., Afzaal, M., 2009. Climate Change Indicators of Pakistan. Pakistan Meteorological Department. Technical Report No. PMD-22/2009 42.

FAO, 2011. Indus River Basin 1–14. Irrigation in Southern and Eastern Asia in Figures—AQUASTAT Survey—2011.

Forsythe, N., Fowler, H.J., Blenkinsop, S., Burton, A., Kilsby, C.G., Archer, D.R., Harpham, C., Hashmi, M.Z., 2014. Application of a stochastic weather generator to assess climate change impacts in a semi-arid climate: the upper Indus Basin. J. Hydrol. 517, 1019–1034.

Gebre, S.L., Ludwig, F., 2014. Spatial and temporal variation of impacts of climate change on the hydrometeorology of Indus River Basin using RCPs scenarios, South East Asia. J. Earth Sci. Clim. Change 5.

Hasson, S., Lucarini, V., Pascale, S., 2013. Hydrological cycle over south and southeast Asian river basins as simulated by PCMDI/CMIP3 experiments. Earth Syst. Dyn. 4, 199–217.

Hasson, S.u., Pascale, S., Lucarini, V., Böhner, J., 2016. Seasonal cycle of precipitation over major river basins in south and southeast Asia: a review of the CMIP5 climate models data for present climate and future climate projections. Atmos. Res. 180, 42–63.

Hasson, S., Böhner, J., Lucarini, V., 2017. Prevailing climatic trends and runoff response from Hindukush-Karakoram-Himalaya, upper Indus Basin. Earth Syst. Dyn. 8, 337–355.

Huang, J., Wang, Y., Fischer, T., Su, B., Li, X., Jiang, T., 2017. Simulation and projection of climatic changes in the Indus River Basin, using the regional climate model COSMO-CLM. Int. J. Climatol. 37, 2545–2562.

ICIMOD, 2010. Climate Change Impacts on the Water Resources of the Indus Basin, International Centre for Integrated Mountain Development, Kathmandu, Nepal.

ICIMOD, 2017. Upper Indus Basin Network and Indus Forum Collaboration Meeting. Workshop Report, ICIMOD, Kathmandu.

Immerzeel, W.W., van Beek, L.P.H., Bierkens, M.F.P., 2010. Climate change will affect the Asian water towers. Science 328, 1382–1385. https://dx.doi.org/10.1126/science.1183188.

Inam, A., Clift, P.D., Giosan, L., Tabrez, A.R., Tahir, M., Rabbani, M.M., Danish, M., 2008. The geographic, geological and oceanographic setting of the Indus River. In: Large Rivers: Geomorphology and Management. pp. 333–346 (Chapter 16).

Kazmi, D.H., Li, J., Rasul, G., Tong, J., Ali, G., Cheema, S.B., Liu, L., Gemmer, M., Fischer, T., 2014. Statistical downscaling and future scenario generation of temperatures for Pakistan region. Theor. Appl. Climatol. 120, 341–350.

Khan, R.N.A., 2013. Geographical Profile of Indus Basin, pp. 47–70. (Chapter 3). Retrieved from: http://shodhganga.inflibnet.ac.in/bitstream/10603/14384/7/chapter_3.pdf.

Kraaijenbrink, P.D.A., Bierkens, M.F.P., Lutz, A.F., Immerzeel, W.W., 2017. Impact of a global temperature rise of 1.5 degrees Celsius on Asia's glaciers. Nature 549, 257–260.

Lutz, A.F., Immerzeel, W.W., Shrestha, A.B., Bierkens, M.F.P., 2014. Consistent increase in high Asia's runoff due to increasing glacier melt and precipitation. Nat. Clim. Chang. 4, 587–592.

Lutz, A.F., ter Maat, H.W., Biemans, H., Shrestha, A.B., Wester, P., Immerzeel, W.W., 2016. Selecting representative climate models for climate change impact studies: an advanced envelope-based selection approach. Int. J. Climatol. 36, 3988–4005.

Nepal, S., Shrestha, A.B., 2015. Impact of climate change on the hydrological regime of the Indus, Ganges and Brahmaputra river basins: a review of the literature. Int. J. Water Resour. Dev. 31, 201–218.

Qureshi, A.S., 2011. Water Management in the Indus Basin in Pakistan: Challenges and Opportunities. Mt. Res. Dev. 31, 252–260.

Rajbhandari, R., Shrestha, A.B., Kulkarni, A., Patwardhan, S.K., Bajracharya, S.R., 2015. Projected changes in climate over the Indus river basin using a high resolution regional climate model (PRECIS). Clim. Dyn. 44, 339–357.

Rauf, I.A., 2013. Environmental Issues of Indus River Basin: An Analysis, ISSRA paper, pp. 91–112.

Su, B., Huang, J., Gemmer, M., Jian, D., Tao, H., Jiang, T., Zhao, C., 2016. Statistical downscaling of CMIP5 multimodel ensemble for projected changes of climate in the Indus River Basin. Atmos. Res. 178–179, 138–149.

Taylor, K.E., Stouffer, R.J., Meehl, G.A., 2012. An overview of CMIP5 and the experiment design. Bull. Am. Meteorol. Soc. 93, 485–498.

Wilby, R.L., Dawson, C.W., 2007. SDSM 4.2—A Decision Support Tool for the Assessment of Regional Climate Change Impacts, Version 4.2 User Manual. Lancaster Univ. Lancaster/Environment Agency Engl. Wales, pp. 1–94.

A Hydrological Perspective on Interpretation of Available Climate Projections for the Upper Indus Basin

Nathan Forsythe*, David R. Archer[†], David Pritchard*, Hayley Fowler*

*Water Resources Research Group, School of Engineering, Newcastle University, Newcastle, United Kingdom [†]JBA Trust, Skipton, United Kingdom

7.1 INTRODUCTION

Any attempt to foresee the evolution of a system as chaotic as the climate, particularly over a region with a land surface as complex as the Upper Indus Basin (UIB), requires focus on the key determinants of its present behavior in order to rigorously study potential future conditions. In terms of the water resources of the UIB, which derive largely from cryosphere-dominated hydrological processes, these key determinants are mass inputs—in the form of snow or rain—and energy inputs (including radiation), which are well indexed by temperature. Because the topographical relief or more precisely the hypsometry, that is, the distribution of surface area with elevation, of the UIB is one of its defining characteristics, the ability of models to accurately represent the variation of climate inputs with respect to elevation is especially crucial. The variable skill of available climate model outputs at representing these determinants under historical conditions provides important insights for interpreting the likelihood and implications of future climate conditions simulated by these models.

This chapter provides a summary of the importance of the UIB's water resources, a primer on how these resources are controlled—as quantified through ground-based observations from recent decades—by climate inputs, followed by a systematic assessment of the skill of present climate models in representing these inputs. The assessment then serves as a framework for interpreting the available climate projections.

Water resources context of the UIB. With the exception of Egypt, Pakistan is the only country in the world that depends on a single water system for its livelihood and economy, in Pakistan's case, the Indus water system. With its predominantly semi-arid climate, Pakistan's agricultural industry is based on irrigation that is sourced primarily from the Indus River and the Jhelum River (a tributary to the Indus River). However, the flow in the rivers is largely driven by what happens in the mountain headwaters from which a combined annual average volume of about 175,109 m^3 (billion cubic meters, or bcm) (equivalent to an average discharge of 5550 m^3 s^{-1}) for all major rivers is discharged into the Indus plains. Knowledge of the climate, hydrology, and glacial behavior of the Upper Indus Basin (UIB) is therefore essential for the management of current water resources and for future planning, particularly in light of the increasing demands of a growing population and the risks associated with climate variability and change (Archer et al., 2010).

7.2 INITIAL ANALYSIS OF GROUND-BASED DATA

This chapter begins by summarizing and analyzing findings based on ground-based climate and river-flow station data from the Indus Basin and the conclusions that may be drawn on seasonal, spatial, and temporal variability. However, until the 1990s ground-level climate stations were limited to valley stations below 2500 m, and even the remotely accessed climate stations established in the 1990s are, with one exception, below the mean elevation (4500 m) of the catchment to its lowest river-flow gauging station at Besham, Pakistan. Climate processes that drive accumulation and ablation of glaciers and river flow, on the other hand, mainly act above 3500 m. Climate station mass inputs are represented by precipitation (as snow and rain), and energy is represented by temperature. Interpretation initially involved extrapolating information from low elevations; however, increasingly satellite imagery and a variety of modeling techniques are being used, and these approaches are the main focus of this chapter. Models are used first to simulate and then to extend observations and to make predictions under a range of climatic scenarios.

7.2.1 Precipitation

The recorded total annual rainfall in the Upper Indus Basin (UIB) varies widely in amount, ranging from an annual total of ~1400 mm at stations in the southern slopes of the Himalaya to <200 mm at valley floor stations in the Karakoram. The annual total also varies in terms of seasonal distribution. The southern foothills receive ~45% of their annual total from the southwest monsoon from July to September, whereas Chitral receives a mere 5% in those months, and Kabul even less. Intervening mountain ranges block the monsoon progressively from southeast to northwest. The main rainfall season over much of the Karakoram and Hindu Kush is mainly from January to March but can extend to May, especially in the Gilgit and Hunza catchments, and is a result of westerly disturbances with moisture originating from the Mediterranean Sea. Most of this winter precipitation falls as snow and contributes either to glacier accumulation or to spring and summer snowmelt runoff. Since remote measurement of precipitation as snow is unreliable, the influence of altitude on precipitation totals is perhaps the most uncertain element of the hydrological cycle. An isolated glaciological study (Wake, 1989) suggested annual accumulation rates of 1500–2000 mm at 5500 m. A glacier mass balance and model-driven study by Immerzeel et al. (2015) estimated the catchment-wide precipitation to be 913 mm yr^{-1}, albeit with a wide uncertainty band (±323 mm).

In spite of differing seasonal totals, winter rainfall (October to March) is significantly correlated across the region both north and south of the Himalayan divide, an important attribute with respect to modeling and also to predicting current and future seasonal flows (Archer and Fowler, 2004). In contrast, spring and summer rainfall (April to September) shows positive correlation between stations north of the divide but is significant only for neighboring stations; there is a consistent weak but occasionally significant negative correlation between stations north and south of the divide.

Over the past century there were no statistical significant trends in annual or seasonal precipitation time series. However, since 1961 most stations have shown an increasing trend in annual rainfall with four of them being statistically significant ($P < .05$). Both winter (October to March) and summer (April to September) rainfall show predominantly increasing trends.

7.2.2 Temperature

Mountainous regions are said to require a much greater density of climate stations than neighboring lowlands to achieve the same reliability in areal estimates. However, in the UIB, despite the separation of stations in distance and by major topographic barriers, correlation coefficients are positive for seasonal temperatures between all stations (Fowler and Archer, 2006), but are higher during the spring and summer months (with few exceptions significant at $P < .01$) than the winter months. The high correlation in mean temperature between valley stations separated by intervening mountain barriers suggests that these correlations will be reflected over much shorter vertical distances. This spatially conservative behavior is critical for modeling and predicting river flows from energy inputs from snow and glacier melt. Nevertheless, there is uncertainty in extrapolating temperature and lapse rates from valley locations to higher elevations, especially across the boundary from snow-free to snow- and ice-covered areas with the effects of changing albedo.

Analysis of ground-based temperature trends in the UIB shows patterns of behavior that in some cases are critically different from those observed globally. The most significant difference in terms of influence on river flow is the trend of significant cooling of mean temperature during spring (March to May) and summer (June to August) since 1961, but greater cooling for minimum temperatures than for maximum temperatures. Thus there is an increase in diurnal temperature range (DTR) that is consistently observed in all seasons in contrast to a narrowing of observed DTR in most parts of the world and to GCM projections. Winter temperatures (December to February) show warming for both maximum and minimum temperatures for most stations, but such trends are less significant for influencing river flow. A remote sensing-based study by Forsythe et al. (2015) that also drew upon meteorological reanalyses found that differential changes in the prevalence of daytime and night-time clouds, which impact near-surface temperature differently through opposing shortwave and longwave radiative effects, may contribute to these asymmetrical changes in minimum and maximum temperatures and hence to changes in DTR.

7.3 HYDROLOGICAL REGIMES AND CLIMATE RUNOFF RELATIONSHIPS

Analysis of streamflow in conjunction with climate records shows that hydrological regimes and historical trends in runoff reflect the combined influences of moisture and energy inputs and how these operate over the great range of elevation. Effects of inputs may operate differently at seasonal and daily levels. Archer (2003) postulated three basic but overlapping hydrological regimes:

1. Foothill catchments have a seasonal runoff and flood regime that is controlled mainly by liquid precipitation on snow-free ground both in winter and during the summer monsoon. Hydrograph response is flashy, dependent on incident rainfall thus exhibiting pluvial regime behavior. Energy plays a role mainly in controlling evaporative loss and soil moisture status.

2. Middle altitude catchments south of the Karakoram (Jhelum, Kunhar, and Swat) (Archer and Fowler, 2008) as well as some major tributaries of the Upper Indus (e.g., Gilgit) have seasonal summer flow predominantly defined by preceding winter precipitation, thus a nival (snowpack-governed) regime. Since both winter and summer precipitation show increasing trends, streamflow volume also exhibits a predominantly increasing trend (Sharif et al., 2013). The magnitude of winter precipitation also influences timing of the runoff hydrograph with higher annual flow volume being linked with an earlier center of volume. Although winter precipitation defines seasonal runoff volume, short-term peaks including flood flows are strongly influenced by incident rainfall.

3. High altitude Karakoram catchments with large glacierized proportions (Hunza, Shigar, and Shyok, but notably the Hunza) due to seasonal temperature show a falling trend in runoff and a declining proportion of glacial contribution to the main stem of the Indus, which is attributed primarily to a falling trend in summer temperatures. Extreme but rare monsoon incursions are paradoxically accompanied by a decrease in river flow as precipitation in the form of snow is accompanied by a sharp fall in temperature and reduced ablation (Archer, 2004).

Archer's findings, derived from ground-based observations, were confirmed by remote sensing-based studies (Forsythe et al., 2012a,b) that utilized catchment-aggregate values of snow-covered areas (SCAs) and the fractional area experiencing above-freezing night-time temperatures—derived from land surface temperatures (LST)—as indexes, respectively, of mass and energy inputs to the UIB cryo-hydrological system. Forsythe's findings demonstrated that Archer's nival and glacial hydrological regimes corresponded, respectively, to the limiting mass and energy constraints on meltwater runoff generation from snow and ice.

Trends in runoff and climate are consistent with reported changes in glacial behavior. A falling trend in summer energy input is expected to change glacier mass balance in favor of increased storage and reduced runoff while a trend of increasing precipitation also influences glacier mass balance. Field and remote sensing of glaciers (Hewitt, 2005; Minora et al., 2013) shows that, in contrast to glacier retreat worldwide including those in the central Himalayas, Karakoram glaciers are generally stable and in some cases thickening. Combined evidence of climate, streamflow, and glacier behavior confirms the existence of the "Karakoram anomaly." Recent studies (Forsythe et al., 2017; Mölg et al., 2017) have linked the summer temperature declines associated with the Karakoram anomaly to variability in the behavior of mid-latitude circulation systems, in particular to interactions between the westerly (mid-latitude) jet and South Asian monsoon.

7.4 EVALUATION AND INTERPRETATION OF AVAILABLE CLIMATE PROJECTIONS FOR THE UIB

Given their importance for the economic development of Pakistan, as well as their wider implications for food security through global commodities markets, the likely trajectories of water resources in the UIB in the coming decades are a matter of great concern (Barnett et al., 2005), particularly in downstream areas, that is, below the Tarbela reservoir. To assess likely future resource availability, there is a need for climate scenarios, the primary sources of which are projections from general circulation models (GCMs). Lutz et al. (2016) summarized a range of projections for the UIB from models included in the CMIP5 ensemble—used for the IPCC 5th Assessment Report—from the high emissions scenario (RCP8.5) as follows:

- Precipitation (annual total): −10.2% to +31.0%
- Temperature (annual mean): +5.6 to +8.0°C

These changes are the difference between conditions simulated for the period 2071 to 2100 (late 21st century) and of 1961 to 1990 (reference historical climate). The exceedingly widespread potential changes in mean annual total precipitation highlight the lack of consensus among GCMs regarding evolution of precipitation in areas such as the UIB, which not only is characterized by extreme topographic gradients but also sits at the intersection of mid-latitude (westerly) and monsoonal circulation systems.

Furthermore, in light of previously identified sensitivity of the UIB tributary catchment runoff to climate inputs (Archer, 2003, 2004; Forsythe et al., 2012a,b), these projected changes, if they come to pass, will have serious hydrological impacts. The temperature changes in particular could result in eventual shifts of the hydrological regime, that is, from glacial to nival or from nival to pluvial. While such regime changes could well take more than a century to become effective, in the next few decades, variability in the timing and magnitude of runoff could begin to exceed conditions experienced in the historical record. This is particularly the case given that the reported mean changes aggregate future interannual variability and potential shifts in seasonality.

7.5 AVAILABLE METHODS FOR DOWNSCALING GCM PROJECTIONS TO FINER SPATIAL SCALES

GCMs are the root source of future climate projections as they are able to integrate the cumulative changes in the Earth system resulting from increased radiative forcing—due to increased atmospheric greenhouse gas (GHG) concentrations—acting on the atmosphere, oceans, and land surface. Since the development of the CMIP5 ensemble for the IPCC 5th Assessment Report, the spatial resolution of GCMs has remained relatively coarse (~150–300 km) due to computational (processing capacity) limitations. There are ongoing research initiatives to improve substantially on this situation, for example, the High Resolution Model Intercomparison Project (HighResMIP, Haarsma et al., 2016), which is part of the larger Coupled Model Intercomparison Project Phase 6 (CMIP6) that will be a major input to the forthcoming sixth assessment report (AR6) by the Intergovernmental Panel on Climate Change (IPCC). HighResMIP aims to deliver higher resolution (~25 km) GCMs, but the results of these initiatives are not yet (at the time of this writing) available to scientists working to assess climate change impacts on particular regions and sectors. Hydrological impact studies in particular, however, are carried out at much finer spatial scales than currently available GCMs. Thus GCM outputs must be downscaled to a sufficiently fine spatial resolution to capture key subregional climate features. Furthermore, prior to using climate model outputs as inputs for impact assessment simulations inherent biases must, as much as possible, be corrected (removed). A variety of downscaling techniques exist that have been comprehensively reviewed (e.g., Fowler et al., 2007) and can be broadly grouped into two categories:

1. Statistical downscaling, which incorporates a range of techniques, many drawing on identified relationships between predictors of local climate and local climate variables (predictands). A commonly used method is the simple delta change approach where relative changes between control and future GCM runs are applied to time series of local observations.
2. Dynamical downscaling, in which a higher spatial resolution Regional Climate Model (RCM: ~25–50 km) is run over a limited spatial domain using GCM outputs as boundary conditions (see Rummukainen, 2010). Furthermore, an RCM can then be nested within another limited area model, potentially multiple times, even achieving a resolution (>~4 km) where development of convection can be simulated directly (Chan et al., 2018) rather than approximated through a parameterized algorithm. While convection may not be a dominant mode of precipitation in the majority of the UIB, achieving a similar spatial resolution may well be necessary to accurately capture the influence of topographic gradients (orographic forcing) on precipitation occurrence and magnitude (Rasmussen et al., 2011). Nevertheless these nesting procedures remain costly in terms of computational capacity (processing power), and thus simulations of substantive length and extensive spatial domain remain, for now, quite rare.

Therefore the default option for extracting climate change projection information to utilize in hydrological climate change impact assessments remains simple (non-nested) RCMs. This is the case for the UIB where the current generation of RCMs largely comprises the outputs of modeling experiments conducted for the South Asia domain of the Coordinated Regional Downscaling Experiment (CORDEX-South Asia) overseen by the World Regional Climate

Programme (WRCP). All of the currently available CORDEX-South Asia RCM outputs are at ~50 km spatial resolution.

7.6 EVALUATION OF OUTPUTS FROM AVAILABLE RCM HISTORICAL PERIOD SIMULATIONS

Before available dynamically downscaled projections for the Upper Indus Basin can be evaluated in terms of their implications for the availability of future water resources, a number of aspects must be considered in order to provide a context in which to frame the projections. These aspects include the skill with which the regional climate model (RCM) simulations reproduce the historical climate of the target domain. Some of these aspects relate to the representation of large-scale circulation patterns by the general circulation models (GCMs) that provide the forcing/driving boundary conditions for the RCMs. Other aspects relate to the simulation of local-scale processes—precipitation initiation and surface energy balance resolution—by the atmospheric and land surface components (respectively) of the RCMs themselves. The following sections of this chapter present a framework for evaluation of these critical aspects of RCM simulations drawing on a subset of CORDEX-South Asia outputs forming a nine-member/3×3 mini-ensemble of three individual RCMs driven by three GCMs. The insights from this evaluation provide the context or comparison of projections by the nine-member ensemble for the UIB's precipitation and temperature distributions for the late 21st century under the "business-as-usual" (high) emissions scenario (RCP8.5).

7.6.1 RCM Skill in Representation of Large-Scale Climate

To assess the skill in individual RCM-GCM pairings at representing the historical climate over a given target domain, it is first necessary to choose (or construct) a credible baseline data set against which the RCM outputs can be evaluated. It is common practice for climate modeling centers to assess RCM performance against gridded observational data sets compiled from geostatistical aggregation and interpolation of available ground-station observational records. In regions with an adequate spatial density of observing sites, this may be a reasonable approach. Unfortunately, these data sets rarely (if ever) utilize physically based approaches for interpolation and thus overly data-scarce areas with purported gridded observations—interpolated from a handful of sites hundreds of kilometers away—can yield physically implausible values. This has particularly been the case for studies of the UIB and of the wider High Mountain Asia (HMA) over the past decade or more.

Global meteorological reanalyses, however, offer a substantially stronger option. We opted to regrid—by subdivision and reaggregation—ERA-Interim, JRA-55, and NASA MERRA2 to CORDEX (~50 km/0.5 degrees) resolution, and then we derived the ensemble mean, minimum and maximum values for each grid cell for each selected variable. This does not solve spatial resolution-driven issues in critical processes (orography, convection), but is the best region-wide option at present.

Given the critical role of winter (solid) precipitation over the UIB in constituting seasonal (preablation) snowpack in nival regime tributary catchments and in providing annual accumulation in glacial regime zones, it is logical to begin ensemble performance evaluation by examining regional patterns of bias in January precipitation (Fig. 7.1). In all ensemble members, there is an area of substantial dry bias over western Pakistan and Afghanistan, although there is substantial variation within the ensemble on the spatial extent and magnitude of excess dryness in this subregion. There is similar ensemble homogeneity in dry bias over Southeast Asia/Southwest China with contrasting wet bias over the Indian Ocean south of 20 N latitude. Focusing on differences in performance within the ensemble, six of nine members show a mild wet bias over the Tibet Plateau. Over the Himalayan mountain arc, three members show strong wet bias, three moderate wet bias, and three dry bias. From these results in particular, it becomes evident that there is greater similarity among the triptychs of individual RCMs forced by different GCMs than among those of individual GCMs used to force three separate RCMs. From this it can be inferred that for the South Asia domain, precipitation bias is more strongly shaped by differences in analytical schemes or parameterizations of precipitation initiation or generation in the RCMs than by moisture fluxes simulated by the GCMs. Although not shown here, patterns of agreement and disagreement within the ensemble for July (monsoonal) precipitation bias support the same interpretation.

Summer night-time temperatures play a critical role in governing glacial meltwater generation. This effectively determines which areas within the sub-region retain a permanent ice pack and which become nival, that is, with seasonal snowpack fully ablated by late summer. Thus the monthly mean of July daily minimum 2 m air temperatures (Tmin) is an important element for evaluation of a regional bias pattern in the ensemble. Arguably, the first feature of July Tmin bias in the ensemble that becomes visually apparent is the nearly complete lack of warm bias (although the spatial extent of null/neutral bias varies widely among members). A second readily apparent (nearly homogenous aspect) is the large magnitude of cold bias over the highest elevation areas in the South Asia domain. This bias could stem from a range of model deficiencies, including, for example, over-simplification—or pure lack—of the simulation of glacier processes or from flaws in the surface energy balance algorithms/parameterizations leading to excess frozen soil.

The strong cold bias in the control (historical) condition is particularly relevant for the UIB in terms of the accuracy of projected changes in regional glaciological and hydrological controls. Strong bias in July Tmin could indicate excess annual minimum snow and ice cover with a corresponding bias in albedo and implications for the surface energy balance (via shortwave radiation absorption). Commonly applied bias correction schemes utilize change factor approaches that calculate the incremental change between projected future and simulated historical conditions. This approach is reasonable when there is continuity or linearity of potential change. However, in the presence of state changes and physical thresholds (e.g., freeze and thaw), a change factor approach may not be valid. For example, when simulated land surfaces are erroneously projected to change from snow- or ice-covered to bare or thawed conditions, the associated temperature increase could differ substantially from warming that would be calculated in the absence of a state change.

Regarding influence on the precipitation phase in winter, cold bias (not shown) in the monthly mean of January daily maximum temperature (Tmax) is also pervasive in the mini-ensemble over high elevation areas of the South Asia domain. Given the severely cold temperatures in the UIB during winter under (recent) historical conditions, this may not imply substantial bias in the fractional prevalence of precipitation by phase, that is, snow versus

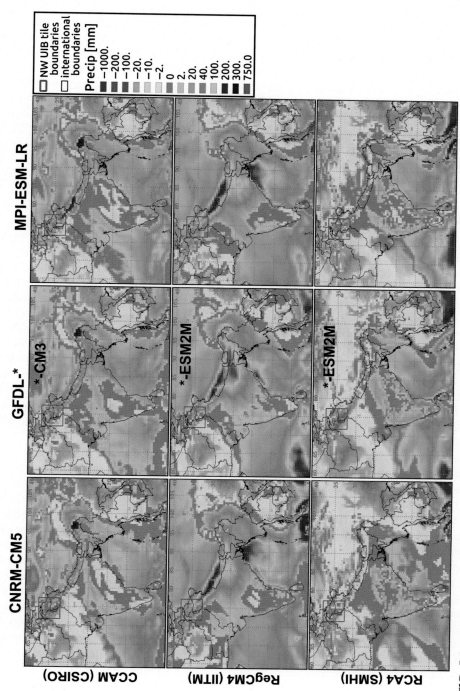

FIG. 7.1 Regional bias in January precipitation. In the grid of tiles, columns are driven by boundary conditions from a common GCM. Rows are dynamically downscaled by a common RCM. This yields a mini-ensemble allowing comparison of GCM and RCM influence.

rain, as the likelihood of errors would increase in both marginal elevations (near the freezing isotherm) and especially in transitional seasons. Thus robust assessment of implications of temperature bias on cryosphere-influenced hydrological processes—either snowpack accumulation or rain-on-snow events—would require detailed evaluation on a model-by-model (individual ensemble member) basis at a monthly or (ideally) daily time-step.

Thus it is clear from the 3-by-3 mini-ensemble grids for both January precipitation (see Fig. 7.1) and July Tmin (Fig. 7.2) that regional bias patterns are overwhelmingly driven by the choice of RCM. This suggests that for the historical period (control conditions) local process representation in the RCMs—land surface schemes, parameterizations of cloud formation, and precipitation initiation—have greater influence (on simulated climate) than large-scale atmospheric circulation.

7.6.2 RCM Skill in Representation of the UIB-Karakoram Climate

Shifting from the broad regional patterns of RCM and GCM bias in key climate variables over South Asia to the UIB as the specific target (sub)region, different metrics and aspects of climate model skill become relevant as focus shifts to accurate representation of the key determinants of runoff variability. As noted at the beginning of this chapter, the determinants for the UIB are mass and energy inputs, respectively, quantified by precipitation and temperature. At the regional level, that is, South Asia domain, there are substantial apparent similarities between mini-ensemble members. Differentiating between results from an individual RCM driven by various GCMs thus requires focusing on fine (subregional) details. For example, differences at the scale of the UIB are much more apparent. This is likely due to the positioning of the UIB subdomain at the intersection of westerly and monsoonal circulation systems. Additionally, relatively small differences (in continental-scale terms) between GCMs in tracking westerly weather systems could yield substantial differences in both cumulative precipitation and in surface conditions, for example, snow-influenced surface albedo, that have an impact on near-surface air temperature.

Both for purposes of simplicity and to assess subregional variations influenced by a substantial range of elevation differences, the target subdomain for this section of analyses is defined as a square tile from within the broader CORDEX-South Asia domain. This tile—with (grid cell center) corner coordinates of 32.5°N–72.0°E and 37.5°N–77.0°E—overlays the highest yielding UIB tributaries in terms of annual runoff. Fig. 7.3 provides a comparison between the real-world (900 m spatial resolution, from GTOPO30) topography and topography aggregated to the spatial resolution of CORDEX-South Asia along with an intercomparison of the hypsometry from these two spatial discretizations plus those at the resolution of various potential reference data sets from meteorological reanalyses. These intercomparisons highlight how simulation of UIB surface processes and atmosphere-surface exchanges at the spatial resolution of the current CORDEX-South Asia oversimplifies the surface and also leads to over-representation (by area) in the climate model of elevation ranges between 3500 m and 5000 m with corresponding under-representation of surfaces above 5000 m. Given the ranges where 0°C isotherms fluctuate in the UIB, these structural biases could aggravate errors in simulated prevalence of snow and ice.

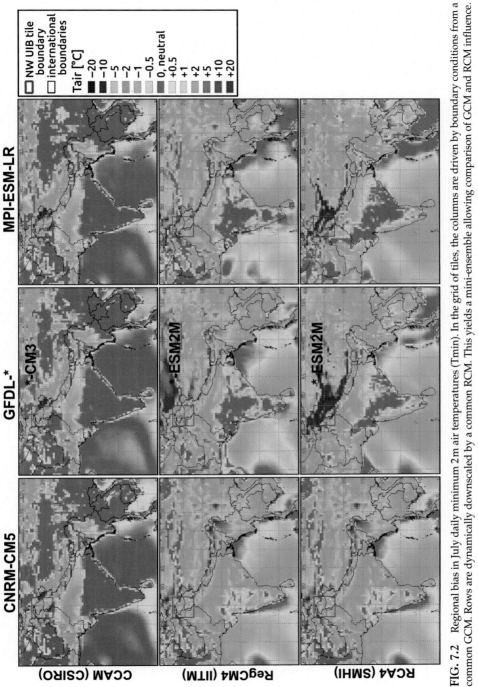

FIG. 7.2 Regional bias in July daily minimum 2 m air temperatures (Tmin). In the grid of tiles, the columns are driven by boundary conditions from a common GCM. Rows are dynamically downscaled by a common RCM. This yields a mini-ensemble allowing comparison of GCM and RCM influence.

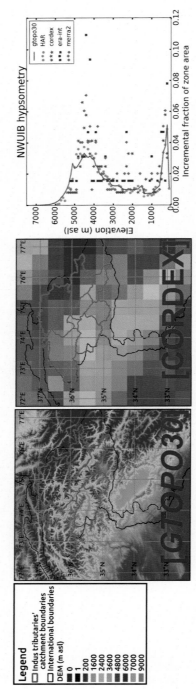

FIG. 7.3 Intercomparison of topography and hypsometry for the UIB using spatial resolutions for various available climate data sets, including the CORDEX-South Asia RCM ensemble and various meteorological reanalyses. The underlying elevation data is from the GTOPO30 data set.

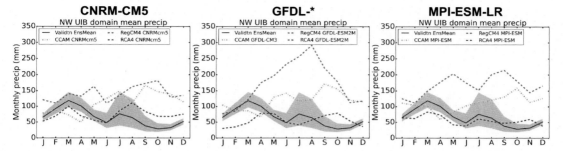

FIG. 7.4   Spatially aggregated monthly mean precipitation over 5×5 latitude-and-longitude-degree square tiles covering the UIB and adjacent areas. Individual panels show various RCMs driven by a common GCM. The RCMs are differentiated by line color, that is, *red* for RegCM4, *green* for CCAM, and *blue* for RCA4. The reference climatology derived from three meteorological reanalyses are shown with a *solid black line* (mean estimate) and *gray* bounds (ranged from minimum to maximum estimates).

Given the accumulating character of mass outflows in the form of runoff discharging through its tributaries to the Indus River's main channel, spatial aggregate mean precipitation (accumulation) is the logical metric for evaluating CORDEX-South Asia ensemble members' skill at representing present precipitation behavior. Owing to its location at the intersection of circulation systems, detailed consideration of the UIB subdomain repeatedly reveals stark differences in the results from an individual RCM forced by various GCMs (Fig. 7.4). This is especially the case for ensemble members produced with the RegCM4 model. Nevertheless, there is still greater similarity between outputs from a single RCM forced by multiple GCMs than between results from different RCMs forced by a single GCM. RCA4 has the best apparent skill both in terms of absolute value (accumulation) accuracy and in seasonality (shape) of annual precipitation cycles. CCAM shows bias toward excessively wet summers and autumns and overly dry springs. Outputs of dynamical downscaling by RegCM4 do not show consistent seasonality, but are severely wet-biased in almost all months. Interestingly, in contrast to other potential subdomains with simpler precipitation regimes, for example, the lower Indus plains, it does not appear that a normalization procedure would change interpretation of a relative ensemble member skill in precipitation simulation. Often normalization of individual members—for example, dividing monthly precipitation values by annual totals to yield the monthly fraction of annual precipitation—can bring the ensemble into closer alignment by limiting magnitude biases inherent in model-specific precipitation algorithms. In this case, however, the ensemble members with the least monthly absolute precipitation biases are also those with the most accurate representation of precipitation seasonality, that is, the shape of the annual cycle.

For the purpose of evaluating model skill in the simulation of energy inputs (represented by near-surface air temperature), rather than proceeding by spatial means, we note how the important influence of snow and ice on the UIB water resources (through meltwater runoff generation) points to other key metrics. Specifically, the timing and magnitude of meltwater generation are influenced by the vertical position of freezing ($0°C$) isotherms and the spatial distribution of energy that is linked to lapse rates (vertical temperature gradients). As at the continental scale, within the mini-ensemble there are greater differences between the temperature-related results for the UIB (Fig. 7.5) from an individual GCM downscaled by

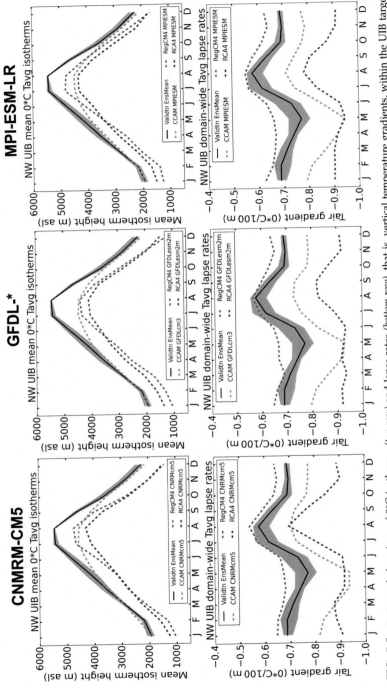

FIG. 7.5 Estimated elevations of 0°C isotherms (*top row*) and lapse rates (*bottom row*), that is, vertical temperature gradients, within the UIB target domain. Columns show various RCMs driven by a common GCM. The RCMs are differentiated by line color, that is, *red* for RegCM4, *green* for CCAM, and *blue* for RCA4. The reference climatology derived from three meteorological reanalyses are shown with a *solid black line* (mean estimate) and *gray* bounds (range from minimum to maximum estimates).

multiple RCMs than between those from an individual RCM driven by different GCMs. This points to the dominant influence of the land surface process modules in the RCMs in modulating simulated temperatures at the local scale. The land-surface process modules calculate energy exchanges at the land-atmosphere interface that are strongly affected, in the case of the UIB, by cryosphere-linked surface conditions, including albedo and soil temperature.

Considering the two temperature (energy) metrics separately provides complementary insights into the potential sources of ensemble member biases. The biases in 0C isotherm elevations fit two patterns: (1) ensemble members downscaled by RegCM4 and RCA4 show strong year-round cold biases; that is, isotherm elevations are too low and (2) CCAM-derived ensemble members show accurate (near-zero bias) isotherm elevations in cold months (October to May) but quite strong cold biases in summer months (June to September). Interestingly, the biases in lapse rates also show two patterns, but the groupings are different from those for isotherms: (1) RegCM4-derived members show relatively small lapse rate biases, specifically marginally shallow gradients, throughout the year and (2) ensemble members downscaled by CCAM and RCA4 yield vertical temperature gradients that are consistently too steep; that is, lapse rate absolute values are too large. The conclusions resulting from the combined evaluation of lapse rates and isotherm elevations are as follows:

- RegCM4 shows the nearly accurate spatial temperature gradients but exhibits a strong, consistent (i.e., year-round) cold bias.
- CCAM in cold months (October to May) shows accurate spatial aggregate temperatures over the UIB subdomain, but this is composed of cold bias at high elevations and warm bias at low elevations. In warm months (June to September) strong cold biases appear.
- RCA4 shows cold bias in isotherm elevation and overly steep temperature gradients. This implies generalized cold bias with smaller biases at low elevations and larger biases over high elevation areas.

These broadly generalized cold biases could result from a range of sources in the algorithm formulations—including excess daytime cloud cover, excess surface albedo, and so on—used by individual RCMs. Interestingly, these cold biases, which are particularly pronounced at high elevations, can be expected to result in erroneous simulated cryosphere (snow and glacier) resilience to warming because historical conditions are estimated to be further below freezing when in reality they are closer to melt. These biases could be in part removed for the purposes of impact assessment modeling by use of additional statistical downscaling. Nevertheless, at least some bias would be expected to persist due to underestimation of potential shortwave radiation (albedo) warming feedbacks resulting from transition from snow- and ice-covered to bare surface conditions.

7.7 VERTICAL PATTERNS OF PROJECTED UIB CLIMATE CHANGE FROM CORDEX-SOUTH ASIA

While it could be argued that the monthly and basin-internal variations in precipitation changes will inevitably lead to annual total runoff equivalent to the spatially and temporally aggregated values, the complexity of the UIB hydrology with its three characteristic regimes

(Archer, 2003) means that where (within the catchment) and when (during the year) precipitation occurs is of great significance for water resources in Pakistan. Given the limited existing water storage infrastructure, the volume and timing of runoff are both critical, and substantial changes in either would necessitate consequential and potentially onerous adaption measures, possibly including costly new infrastructure and drastic changes in the patterns of water use.

As with the spatial aggregates of the UIB's historical precipitation, there is substantial variation in the mini-ensemble regarding vertical patterns in projected precipitation changes (Fig. 7.6). Neither rhyme nor reason are readily apparent as ensemble consensus is rare, and both increases and decreases in precipitation are frequently projected for a given month when an individual RCM is driven by different GCMs. Furthermore, the vertical profiles of projected changes do not immediately suggest underlying physical mechanisms. In light of this uncertainty (divergence), a reasonable approach to impact assessment studies would be a prioritized scenario approach with multiple hydrological model runs driven by inputs statistically downscaled from the individual ensemble members in order from highest historical skill, for example, for precipitation starting with RCA4/MPI-ESM-LR and finishing with RegCM4/GFDL-ESM2M.

In contrast to projected precipitation change, there is substantially more frequent mini-ensemble consensus with regard to vertical patterns of projected change in monthly means of the UIB daily minimum near-surface air temperatures (Tmin). Close agreement is most common below 3000 m, and when RCM downscaling of a given GCM diverges from the other two, it is most frequently due to CCAM projecting much greater (up to 6°C more) warming than the others.

In terms of physical mechanisms driving vertical patterns of projected temperature, it is logical to consider whether any of the ensemble members simulate elevation-dependent warming (EDW). Pepin et al. (2015) reviewed both observational evidence of historical EDW and theoretical mechanisms—including shortwave radiative forcing from snow-line-related albedo change—underpinning the expectation of EDW. Palazzi et al. (2017) assessed the CMIP5 ensemble of GCMs to identify the extent to which current generation GCMs project EDW occurrence over the broader Tibetan Plateau-Himalayan region. Rather than finding projections of linearly enhanced warming as a function of elevation, their evaluation identified different change regimes separated by the 0°C isotherm. They further employed multiple regression techniques that showed albedo to be the strongest predictand of projected EDW.

In the CORDEX-South Asia mini-ensemble, EDW is most strongly projected by the members dynamically downscaled with CCAM. The pattern of vertical position of warming maxima progressing upward with the annual temperature cycle, as expected in albedo-driven EDW, appears to hold in CCAM members. Projected EDW is more limited in RCA4 members and almost absent in RegCM4 members. These differences are quite likely linked to differences in the respective land-surface process modules of the individual RCMs.

There is demonstrable pervasive cold-bias, with minor variations, throughout the mini-ensemble in representation of the UIB's historical temperature. As such there is not a strong argument of exclusion of any members from scenario formulation for impact assessment purposes. Nevertheless, since impacts are likely to be more severe with greater warming, in this case it would make sense to prioritize scenarios derived from CCAM members if computation capacity is limited (Fig. 7.7).

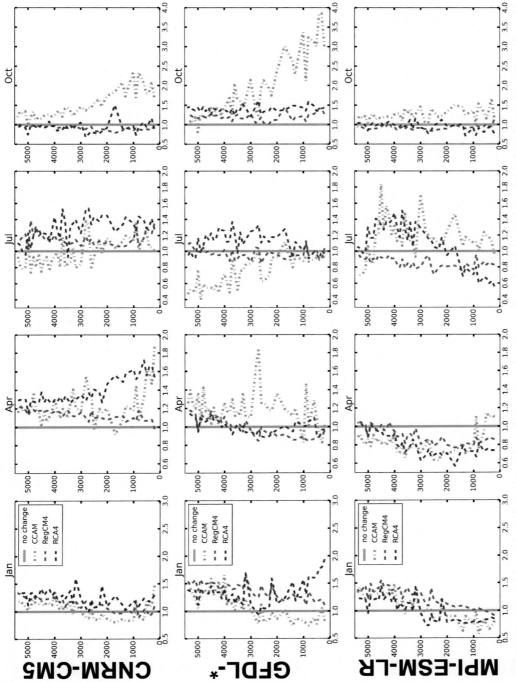

FIG. 7.6 Vertical profiles of precipitation change (multiplicative change factor). Columns show selected months throughout the year. Rows show various RCMs driven by a common GCM. The RCMs are differentiated by line color, that is, *red* for RegCM4, *green* for CCAM, and *blue* for RCA4. The broad *gray line* shows null change, that is, 1.0 multiplicative change factor.

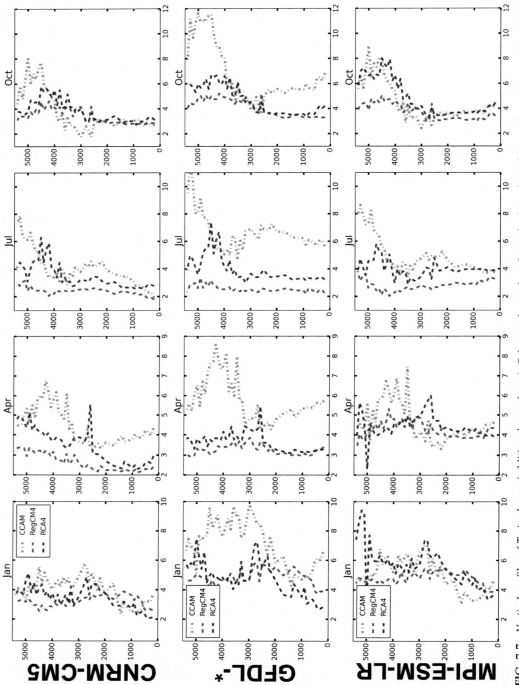

FIG. 7.7 Vertical profiles of Tmin change (additive change factor). Columns show selected months throughout the year. Rows show various RCMs driven by a common GCM. The RCMs are differentiated by line color, that is, *red* for RegCM4, *green* for CCAM, and *blue* for RCA4.

7.8 SUMMARY AND PRIORITIES FOR FURTHER RESEARCH ON THE UIB'S FUTURE CLIMATE

The model evaluation sections of this chapter highlighted substantial shortcomings of the available outputs produced with the current generation of RCMs. Mountain regions with steep topographic gradients, such as the UIB, will require substantially higher spatial resolution in order to capture topographically driven gradients in precipitation and temperature. Realistic representation of precipitation, including its spatial distribution, is necessary for accurate assessment of mass inputs to the water balance. Realistic representation of the spatial distribution of temperature is critical to accurately simulating key distinctions in the freeze/thaw state, which are important for both determination of precipitation phase (i.e., rain versus snow) and surface storage (or meltwater generation) as seasonal snowpack or glacier ice. Given the convective character of monsoonal precipitation in the pluvial regime foothill catchments of the lower reaches of the UIB it would be reasonable to use RCMs at spatial resolutions that permit convection (Chan et al., 2018). Similar spatial resolutions will be required to accurately simulate the topographic forcing on precipitation in high-elevation accumulation zones of glacial (and nival) catchments in the upper reaches of the UIB (Rasmussen et al., 2011).

Sufficient spatial resolution is not the only factor required for skilled representation of the UIB climate by RCMs. It is also crucial to accurately represent critical physical processes. The identified strong cold biases in the CORDEX-South Asia mini-ensemble point toward possible deficiencies in the current generation of models, which will need to be remedied to improve RCM simulations and thus reduce uncertainties regarding future conditions. Strong cold biases suggest serious flaws in the representation of climate processes influencing the surface energy balance. Cloud formation and property schemes strongly influence the energy balance through incoming shortwave radiation and outgoing longwave radiation. Surface albedo, which is strongly linked to the presence or absence of snow and ice, also strongly influences the surface energy balance through net shortwave absorption. Additionally, model schemes governing the thermal inertia of either the snowpack (cold content) or the near-surface soil moisture temperature can also strongly influence the energy balance through sensible and latent heat fluxes. Detailed diagnoses of the sources of cold biases in the mini-ensemble are outside the scope of this chapter. Nevertheless, such investigations will be an essential component of rigorous approaches to the development of future climate scenarios as input for impact assessment studies of the UIB.

Furthermore, specific to contexts similar to the UIB, properly skilled representation of glacier-atmosphere feedbacks will require incorporation of glacier dynamics (mass displacement)—principally avalanching and basal sliding—in RCM land-surface process modules. Without representation of these key components of cryosphere, neither the true sensitivity of glaciers to climate variability nor their influence on local climate will be correctly taken into account. Given the identified likely predominance of shortwave (albedo) effects in EDW, they will be crucial determinants in the presence and magnitude of a potential warming feedback in the upper reaches of the UIB.

The purpose of this inventory of current RCM shortcomings is not to argue that climate change impact assessment studies are presently futile due to uncertainties and biases inherent in the available future projections. Rigorous development of future climate scenarios is, at a minimum, a useful intellectual exercise. This is because it focuses attention on key factors— such as EDW mechanisms—influencing the rate of climate change in the UIB and also provides impetus for prioritizing improvements in forthcoming generations of RCMs. More pragmatically, for the purposes of water resources management planning, it is, above all, important for scientists and engineers to consider the present determinants of hydrological behavior in the three regimes—pluvial, nival, and glacial—of the UIB tributary catchments (Archer, 2003) and how these might evolve in a warming climate with ever more erratic precipitation. A logical starting point for improving the resilience of the UIB's water resources management systems and infrastructure to deal with climate change variabilities is to improve their capacity to cope with present climate variabilities. In point of fact, the current coefficient of variation, that is, the standard deviation divided by the period mean, of annual runoff in many UIB tributary catchments is similar to or greater than currently projected mean changes in annual precipitation for the region. Thus, if existing UIB infrastructure and resource allocation systems can be made to robustly handle the full range of drought and flood conditions experienced in recent decades, a very strong step will have been taken toward safeguarding Pakistan's future water security.

References

Archer, D.R., 2003. Contrasting hydrological regimes in the upper Indus Basin. J. Hydrol. 274, 198–210. https://dx.doi. org/10.1016/S0022-1694(02)00414-6.

Archer, D.R., 2004. Hydrological implications of spatial and altitudinal variation in temperature in the upper Indus basin. Nord. Hydrol. 35, 213–227.

Archer, D.R., Fowler, H.J., 2004. Spatial and temporal variations in precipitation in the upper Indus Basin, global teleconnections and hydrological implications. Hydrol. Earth Syst. Sci. 8, 47–61. https://dx.doi.org/10.5194/ hess-8-47-2004.

Archer, D.R., Fowler, H.J., 2008. Using meteorological data to forecast seasonal runoff on the river Jhelum, Pakistan. J. Hydrol. 361, 10–23. https://dx.doi.org/10.1016/j.jhydrol.2008.11.004.

Archer, D.R., Forsythe, N., Fowler, H.J., Shah, S.M., 2010. Sustainability of water resources management in the Indus Basin under changing climatic and socio economic conditions. Hydrol. Earth Syst. Sci. 14, 1669–1680. https://dx. doi.org/10.5194/hess-14-1669-.

Barnett, T.P., Adam, J.C., Lettenmaier, D.P., 2005. Potential impacts of a warming climate on water availability in snow-dominated regions. Nature 438, 303–309. https://dx.doi.org/10.1038/nature04141.

Chan, S.C., Kendon, E.J., Roberts, N., Blenkinsop, S., Fowler, H.J., 2018. Large-scale predictors for extreme hourly precipitation events in convection-permitting climate simulations. J. Clim. 31 (6), 2115–2131. https://dx.doi. org/10.1175/JCLI-D-17-0404.1.

Forsythe, N., Kilsby, C.G., Fowler, H.J., Archer, D.R., 2012a. Assessment of runoff sensitivity in the upper Indus Basin to interannual climate variability and potential change using MODIS satellite data products. Mt. Res. Dev. 32 (1), 16–29. https://dx.doi.org/10.1659/MRD-JOURNAL-D-11-00027.1.

Forsythe, N., Fowler, H.J., Kilsby, C.G., Archer, D.R., 2012b. Opportunities from remote sensing for supporting water resources management in village/valley scale catchments in the Upper Indus Basin. Water Resour. Manag. 26 (4), 845–871. https://dx.doi.org/10.1007/s11269-011-9933-8.

Forsythe, N., Hardy, A.J., Fowler, H.J., Blenkinsop, S., Kilsby, C.G., Archer, D.R., Hashmi, M.Z., 2015. A detailed cloud fraction climatology of the Upper Indus Basin and its implications for near surface air temperature. J. Clim. 28 (9), 3537–3556. https://dx.doi.org/10.1175/JCLI-D-14-00505.1.

Forsythe, N., Fowler, H.J., Li, X.-F., Blenkinsop, S., Pritchard, D., 2017. Karakoram temperature and glacial melt driven by regional atmospheric circulation variability. Nat. Clim. Chang. 7 (9), 664–670. https://dx.doi.org/ 10.1038/nclimate3361.

Fowler, H.J., Archer, D.R., 2006. Conflicting signals of climate change in the Upper Indus Basin. J. Clim. 19, 4276–4292.

Fowler, H.J., Blenkinsop, S., Tebaldi, C., 2007. Linking climate change modelling to impacts studies: recent advances in downscaling techniques for hydrological modelling. Int. J. Climatol. 27 (12), 1547–1578. https://dx.doi.org/10.1002/joc.1556.

Haarsma, R.J., Roberts, M.J., Vidale, P.L., Senior, C.A., Bellucci, A., Bao, Q., Chang, P., Corti, S., Fučkar, N.S., Guemas, V., von Hardenberg, J., Hazeleger, W., Kodama, C., Koenigk, T., Leung, L.R., Lu, J., Luo, J.-J., Mao, J., Mizielinski, M.S., Mizuta, R., Nobre, P., Satoh, M., Scoccimarro, E., Semmler, T., Small, J., von Storch, J.-S., 2016. High resolution model Intercomparison project (HighResMIP v1.0) for CMIP6. Geosci. Model Dev. 9, 4185–4208. https://dx.doi.org/10.5194/gmd-9-4185-2016.

Hewitt, K., 2005. The Karakoram anomaly? Glacier expansion and the "elevation effect", Karakoram Himalaya. Mt. Res. Dev. 25, 332–340.

Immerzeel, W.W., Wanders, N., Lutz, A.F., Shea, J.M., Bierkens, M.F.P., 2015. Reconciling high-altitude precipitation in the upper Indus basin with glacier mass balances and runoff. Hydrol. Earth Syst. Sci. 19 (11), 4673–4687. https://dx.doi.org/10.5194/hess-19-4673-2015.

Lutz, A.F., Immerzeel, W.W., Kraaijenbrink, P.D.A., Shrestha, A.B., Bierkens, M.F.P., 2016. Climate change impacts on the upper Indus hydrology: sources, shifts and extremes. PLoS One 11 (11), e0165630. https://dx.doi.org/10.1371/journal.pone.0165630.

Minora, U., Bocchiola, D., D'Agata, C., Maragno, D., Mayer, C., Lambrecht, A., Mosconi, B., Vuillermoz, E., Senese, A., Compostella, C., Smiraglia, C., Diolaiuti, G., 2013. 2001–2010 glacier changes in the Central Karakoram National Park: a contribution to evaluate the magnitude and rate of the "Karakoram anomaly". Cryosphere Discuss. 7, 2891–2941. https://dx.doi.org/10.5194/tcd-7-2891-2013.

Mölg, T., Maussion, F., Collier, E., Chiang, J.C.H., Scherer, D., 2017. Prominent mid latitude circulation signature in high Asia's surface climate during monsoon. J. Geophys. Res. Atmos. 122, 12702–12712. https://dx.doi.org/10.1002/2017JD027414.

Palazzi, E., Filippi, L., von Hardenberg, J., 2017. Insights into elevation-dependent warming in the Tibetan Plateau-Himalayas from CMIP5 model simulations. Clim. Dyn. 48 (11–12), 3991–4008. https://dx.doi.org/10.1007/s00382-016-3316-z.

Pepin, N., et al., Mountain Research Initiative EDW Working Group, 2015. Elevation-dependent warming in mountain regions of the world. Nat. Clim. Chang. 5 (5), 424–430. https://dx.doi.org/10.1038/nclimate2563.

Rasmussen, R., Liu, C., Ikeda, K., Gochis, D., Yates, D., Chen, F., Tewari, M., Barlage, M., Dudhia, J., Yu, W., Miller, K., 2011. High-resolution coupled climate runoff simulations of seasonal snowfall over Colorado: a process study of current and warmer climate. J. Clim. 24 (12), 3015–3048. https://dx.doi.org/10.1175/2010JCLI3985.1.

Rummukainen, M., 2010. State-of-the-art with regional climate models. WIREs Clim. Change 1, 82–96. https://dx.doi.org/10.1002/wcc.008.

Sharif, M., Archer, D.R., Fowler, H.J., Forsythe, N., 2013. Trends in timing and magnitude of flow in the Upper Indus Basin. Hydrol. Earth Syst. Sci. 17, 1503–1516. www.hydrol-earth-syst-sci.net/17/1503/2013/ https://doi.org/10.5194/hess-17-1503-2013.

Wake, C.P., 1989. Glaciochemical investigations as a tool to determine the spatial variation of snow accumulation in the Central Karakoram, Northern Pakistan. Ann. Glaciol. 13, 279–284.

PART III

WATER AND FOOD SECURITY OF INDUS RIVER BASIN

Transboundary Indus River Basin: Potential Threats to Its Integrity

Muhammad Jehanzeb Masud Cheema*,†,
Muhammad Uzair Qamar*

*Department of Irrigation and Drainage, University of Agriculture, Faisalabad, Pakistan †Precision Agriculture, Center for Advanced Studies in Agriculture and Food Security, University of Agriculture, Faisalabad, Pakistan

8.1 INTRODUCTION

The agricultural sector has expanded enormously over the past five decades, enhancing its production to meet the world's demand for food and fiber. Today irrigated agriculture is practiced worldwide on about 324 million hectares of land—including land in Pakistan, India, Bangladesh, China, and the United States, among others—and produces about 40% of the world's agricultural products (FAO, 2012a; Turral et al., 2011). Most of the irrigated agriculture in Pakistan and India is concentrated in the plains of the Indo-Gangetic Basin, which comprises 90% of all agriculture in the Indus Basin. It is estimated that 93% of the Indus Basin's available water resources are used for agriculture, while the rest is used in urban and industrial sectors (FAO, 2012b).

The catchment area of the basin is shared by China, Afghanistan, India (39%), and Pakistan (47%) and is mostly mountainous, with at least 40% of it located above 2000 m above sea level (m a.s.l.). The lifeline of irrigated agriculture in the basin is the Indus River, which originates in the northern Himalayas' Mount Kailash in Tibet (China), at an altitude greater than 5000 m. The river traverses from east to west through India and longitudinally through Pakistan, ultimately flowing into the Indian Ocean to the south. The Indus is fed by 24 tributaries, with eight being major tributaries. The Jhelum, Chenab, Ravi, Sutlej, and Beas rivers are the major eastern tributaries, while the Kabul and Gomal rivers are western tributaries and Gilgit river joins the Indus River at junction point of three mountains ranges near town of Juglot in the North of the basin (Fig. 8.1; FAO, 2012b).

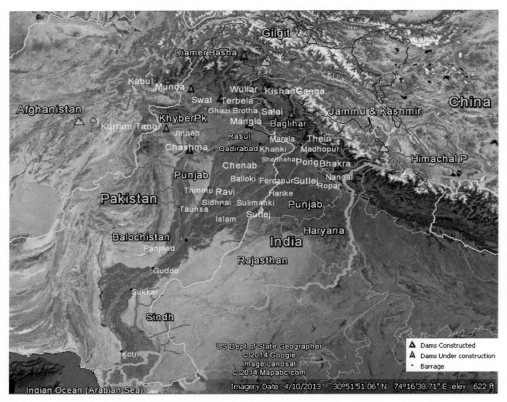

FIG. 8.1   Location of reservoirs and barrages constructed on the Indus River and its tributaries.

The mainstay of flows in the Indus and its tributaries is snowmelt from upstream glaciers of the Himalayan, Karakoram, and Hindu Kush ranges. The flows available from snowmelt contribute up to 80% of the total, while the rest is contributed by rainfall (Yu et al., 2013; FAO, 2012b). More than $200\,km^3$ of flows are available in the rivers annually (Yu et al., 2013; Sharma et al., 2008). The water flows in the rivers are regulated through reservoirs constructed on the major rivers, built after the signing of the famous Indus Waters Treaty of 1960 over the rights to river water between India and Pakistan. Eight major reservoirs (three in Pakistan and five in India) are currently utilized for hydropower generation and irrigation supplies, while the construction of new reservoirs is also in progress.

The water releases from these reservoirs are controlled by a series of barrages downstream that divert available water to 26 million hectares (mha) of agricultural land in the basin (16mha in Pakistan and 10mha in India). Surface water availability varies depending on the season. The lack of a reliable surface water supply and erratic rainfall have forced farmers to augment their water supplies by means of groundwater extraction. As a result, large numbers of tubewells have been installed on both sides of the basin to extract groundwater. A significant percentage of the irrigated area is totally dependent on groundwater, while a larger percentage uses it in conjunction with surface water supplies, otherwise known as "conjunctive use."

Because the basin is underlain by an extensive unconfined aquifer, with a surface area of 0.16 million km^2, Pakistan and India share the groundwater. The basin aquifer has large groundwater reserves with an annual replenishable potential of approximately $90\,km^3$. A high recharge rate is found in the north below the foothills of the Himalayas and reduces toward the south (Fig. 8.2; Laghari et al., 2012). Most of the groundwater flows from northeast to southwest in the basin, according to water table contour maps from 2004 (Chadha, 2008). The extensive reliance on the Indus Basin for water makes it one of the most depleted water basins in the world (Sharma et al., 2010). The dependence on the Indus Basin for water has increased to such an extent that there are times in a hydrological year when water does not drain into the sea (Molle et al., 2010). Therefore information on the water sources and sinks is considered important for efficient water management in the basin. However, the fundamental information on water flows, sources of water, and water demand in the basin is either missing or not accessible.

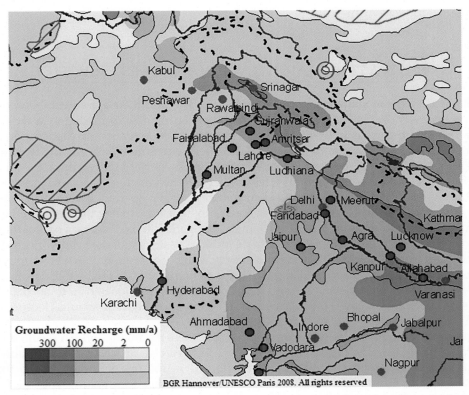

FIG. 8.2   Transboundary Indus River Basin aquifer with annual recharge information. *(Extracted from original; source: http://www.whymap.org.)*

III. WATER AND FOOD SECURITY OF INDUS RIVER BASIN

Downstream riparian countries depend on their upstream neighbors for data collection and sharing. If this does not happen, the downstream countries cannot prepare to cope with floods and droughts or generate hydropower (Zawahri, 2008). This problem is more severe in basins in developing countries, and the Indus Basin is an example. The vastness of the basin, budget constraints, political distrust, and its transboundary nature is a hindrance in transboundary river basin water resources management. More recently, the upstream riparian countries' aggressive use of water as a potential deterrent has threatened the survival of the lower riparian's water-based economy and raised regional security issues.

The geopolitical nature of the basin has made the Indus a test case for scientists and policy analysts seeking to efficiently manage its resources. India and Pakistan, major beneficiaries, are facing tremendous pressure on their water resources due to an exponential increase in population and the unconstrained use of water resources. The situation is leading to an alarming reduction in per capita water availability. Population shifts to cities and industrialization have become major competitors with the traditional water users in the agricultural sector. Changing climate and global warming threaten the spatial and temporal availability of freshwater resources. Changes in precipitation patterns and intensity are also being observed in the Upper Indus Basin, where most of the glaciers are located (Cheema and Pawar, 2015). All these problems directly affect the region's water security, which ultimately affects the region's food security.

8.2 WATER SHARING MECHANISM IN PRACTICE

India and Pakistan share major portions of the basin and their agricultural-based economies rely heavily on the river water to irrigate their farmlands. Immediately after independence from British rule in 1947, various conflicts arose over water distribution on the rivers in the Indus Basin due to the lack of an official agreement between the two countries to use and share water resources. Two major headworks, one at Madhopur on the Ravi River and the other at Ferozpur on the Sutlej River (both rivers flow eastward from India to Pakistan) went under Indian control. India, being the upper riparian, diverted all flows from the northeast to southwest flowing rivers (called the Eastern Rivers: the Ravi, the Beas, and the Sutlej) to meet its growing irrigation demand. Irrigation in the Pakistani part of the Punjab province was dependent on these headworks. India's possession of the headworks resulted in administrative problems in regulating and supply water. Consequently, water scarcity and an environmental threat were created in parts of the basin previously fed by these rivers, mostly in the eastern parts of Pakistan. The situation became a continuous barrier to normalizing relationships between the two states. Tensions thus remained high for about a decade over the use of water available from the Indus River system.

However, in 1960 the World Bank mediated to resolve the matter, which resulted in India and Pakistan signing the Indus Water Treaty (IWT). Twelve articles with eight annexures, various subsections, and subannexures are defined in the treaty. According to the IWT, the flows of the three main west-flowing rivers (the Indus, the Jhelum, and the Chenab) are available to Pakistan, while India has exclusive rights to the waters of the rivers flowing east. Pakistan has unrestricted use of all the waters of the Western Rivers, which India has an obligation to let flow. India has the right to use the water from the Western Rivers for domestic use, nonconsumptive use, agricultural use (as per IWT Annexure C), and hydropower (as per Annexure D). For smooth implementation of the treaty, a permanent Indus Water Commission (IWC) was established under Article VIII. The treaty not only channelized the rights of the Indus water system between riparian countries but also paved the way for the Indus Basin Plan in Pakistan. By outlining a systematic mechanism to resolve conflicts regarding river water rights, the Indus Basin Plan led to large-scale, water-related infrastructure development in Pakistan. A number of link canals and reservoirs were constructed in both Pakistan and India. The information on reservoirs constructed or under construction is provided in the Table 8.1.

With the World Bank as guarantor, the signed treaty survived three wars, expeditious decolonization, and disproportional geographical development. However, during 2007–17, a series of conflicts have been raised due to continuous upstream interventions. The Wullar, Baglihar, and Kishan Ganga projects are a few examples. These interventions are considered as a potential threat to the timely availability of water to downstream users, thus creating panic among the people and government of Pakistan. More recently, India's prime minister, while addressing farmers in Punjab (a state bordering Pakistan), vowed to end the treaty unilaterally. Some consider the prime minister's statement to be a result of public pressure for more water to grow crops. While in Pakistan it is considered as a tactic of fifth-generation warfare.

TABLE 8.1 Major Reservoirs Constructed in the Indus Basin

S. No.	Reservoir	River	Country	Construction Year
1	Mangla	Jhelum	Pakistan	1966
2	Chashma	Indus	Pakistan	1971
3	Tarbela	Indus	Pakistan	1976
4	Diamer-Basha	Indus	Pakistan	Under construction
5	Kurramtangi	Kurram	Pakistan	Under construction
6	Munda	Swat	Pakistan	Under construction
7	Bhakra	Sutlej	India	1963
8	Pong	Beas	India	1974
9	Pandoh	Beas	India	1977
10	Salal	Chenab	India	1995
11	Thein	Ravi	India	2001
12	Baglihar	Chenab	India	2004
13	NimooBazgo	Indus	India	Under construction
14	Chutak	Indus	India	Under construction

Source: Cheema, M.J.M., 2012. Understanding water resources conditions in data scarce river basins using intelligent pixel information – Case: Transboundary Indus Basin. Dissertation. Delft University of Technology Delft, The Netherlands. Available from: http://repository.tudelft.nl/ assets/uuid:7b569411-9934-4b23-b631-36a58f60363f/CheemaMJM_PhD_Thesis.pdf.

Because of its huge dependence on the Indus water system, Pakistan would consider defiance of its access to the water as catastrophic. Such a situation would directly affect its regional economy and security. Moreover, because both India and Pakistan possess nuclear weapons, a regional conflict could become a global tragedy.

8.3 INDUS RIVER BASIN AND REGIONAL ECONOMY

Pakistan and India are two emerging economies (Table 8.2), but they still face a number of problems, including those related to poverty, overpopulation, sanitation, and education (Cheema and Pawar, 2015). Agriculture serves as the backbone of the local economies, and water is the key input for food production. Agriculture also depends on timely occurrence of the monsoon season and a sufficient amount of annual rainfall. To overcome the uncertainties and vagaries of the monsoon season, farmers resort to various methods of irrigation, and irrigated agriculture is the biggest consumer of water in the area. Sustainable use of water for food production, human consumption, and industry therefore is currently the prime challenge. Water scarcity and stiff competition for water between different sectors have resulted in reduced water availability for irrigation. This situation has more tangible consequences at the regional scale.

Therefore management of irrigation water is important and is directly related to poverty reduction, as the agriculture sector comprises 22% of Pakistan's gross domestic product (GDP) and nearly 40% of its workforce (women comprise 30% of the total workface), with most of the residing in rural areas (FAO, 2012b). Although agriculture contributes only 15% to 17% to India's GDP, it supports the livelihood of almost half of the country's 1.2 billion people. It is sobering that the water shortage in India has contributed to 320,000 farmers committing suicide during past two decades (Merriott, 2016).

The sustainability of the Indus River Basin is of paramount importance not only for the two countries but also globally, because it plays a role both in feeding and employing the most densely populated area of the world and in maintaining global food security. Both countries are major growers and exporters of staple crops, for example, rice, wheat, and sugarcane. India and Pakistan are the first and fourth largest exporters of rice, respectively. In 2017 India and Pakistan were ranked among the nine biggest sugar exporters in the world (fourth and ninth, respectively). Moreover, both countries have very strong wheat export credentials. Therefore water-related instability in the Indus water system could have serious global repercussions.

TABLE 8.2   Economic Statistics for Pakistan and India

Statistics	Pakistan	India
Population (million—2013)	182.1	1252.0
GDP (billion US $—2013) Gross National Income Per Capita (US $—2013)	236.6	1877.0
Gross National Income Per Capita (US $—2013)	1380	1570
Irrigated land as % of total agricultural land (2010)	75.98	35.19

Source: World Bank Group, 2013. Agricultural and World Development Indicators Datasets Utilized. Available from: http://data.worldbank.org/.

However, the challenges to the IRB are no longer regional, as its regional economies are receiving support from global players. For example, the United States and China are key players in the region. With China's investment in Pakistan's economy, the local water conflict has taken on a global dimension. Upstream blockage of the Indus water system would unilaterally revoke the IWT and set a very dangerous precedent. Presently, India and China do not share a water treaty, which will allow China to construct dams and reservoirs on the tributaries thus reducing share of water coming from China.

Therefore the need for proper water allocation and regulatory mechanisms in the Indus Basin is dire due to the increasing demand for water among the competing countries, which will only increase over time. Unilateral revocation of the IWT would put regional peace at risk. However, the concerns of the signatories need to be addressed under the changing geographical and climatic conditions. The climate change issue may create a higher level of threat to sustainability of IRB water resources. Thus the production of food, fiber, fuel, and other industrial inputs with less water availability will be a major challenge for both rain-fed and irrigated agriculture in the two countries.

8.4 EMERGING THREATS TO THE INDUS RIVER BASIN

The transboundary Indus River Basin is an example of complex hydrology coupled with strained hydro-political relationships between riparian countries. The basin acts as the breadbasket for more than a billion people in the region. The growing population demands more food, but agricultural lands are shrinking due to peri-urban expansions. At the same time, the basin is facing serious challenges of physical water shortages and lower water and land productivity. This water scarcity is being experienced by both the industrial and agricultural sectors. Irrigated agriculture is suffering the most, as 93% of water in the basin is consumed by this sector alone. Qureshi (2011) estimated a 32% shortfall of water available for agriculture by 2025, thus resulting in a food shortage of 70 million tons. The climate change and siltation of reservoirs will further reduce the surface storage capacity by 30%. The combination of these factors means that by 2050 the Indus will be able to effectively feed 26 million fewer people than it does today (FAO, 2013; Immerzeel et al., 2010).

Stagnant or decreasing agricultural productivity, increasing dependence on groundwater, high risk of climatic variability, enhancing industrialization, and unplanned and un-regularized urban growth are some of the realities of the Indus River Basin. These pose a variety of challenges for water resources governance, management, and use. Groundwater overdraft, food and nutritional security, decreased fresh water availability vis-à-vis escalating demand, and water pollution are major challenges and even greater threats for the environmental security and peace of the region.

A number of studies carried out in the recent past have concluded that the future of the Indus Basin in terms of water availability is extremely uncertain in the long run. The uncertainty stems mainly from the climate change that is directly affecting the temperature and precipitation patterns of the region. The region, which has the highest population growth-rate in the world, is the region most affected by climate change according to the recent studies.

The increasing population trend is expected to put more pressure on the water resources of the basin in future, which can have serious adverse impacts on the regional peace.

This section highlights a few of the potential issues that can adversely affect the sustainability of the basin in terms of both food and internal security.

8.4.1 Growing Population

The present population of the basin is approximately 300 million people, of which about 61% live in Pakistan and 35% in India. An estimate projects that by 2050 this population will increase to as many as 383 million people (FAO, 2012b; Laghari et al., 2012). The region has a higher population density than the rest of the world due to a backward economy and strong social taboos. This increased growth rate adversely impacts the economic growth, trade balance, and any plans to limit poverty.

Rapid growth and associated settlement have already significantly increased the water demands for human and industrial consumption, and thus competition over irrigation water for agriculture has increased manifold. As mentioned earlier, by 2050 the Indus will be able to effectively feed 26 million fewer people than it does today, and the population is predicted to expand. This growing population and reduced water availability has led to the categorization of the Indus Basin as a water-scarce basin, according to World Business Council of Sustainable Development (WBCSD) data and its definition of renewable internal freshwater resources available on an annual basis (Finley et al., 2008).

8.4.2 Reduced or Fluctuating Surface Flows

The average annual flows of major rivers in the basin show decreasing trends for both west- and east-flowing rivers (Table 8.3). These flows represent the pre-IWT (1922–61) and post-IWT (1985–2002 and 2007–10) situations. The average flow of eastern rivers into Pakistan

TABLE 8.3 Average Flows in Major Rivers of the Indus Basin Before and After IWT

	River	Rim Station	Average Annual Flow (1922–61) (km^3)	Average Annual Flow (1985–2002) (km^3)	Average Annual Flow (2007–10) (km^3)
West flowing rivers	Indus	Kalabagh	114.4	94.1	101.9
	Jhelum	Mangla	28.3	23.7	19.3
	Chenab	Marala	31.9	24.5	23.9
East flowing rivers	Ravi	Below Madhopur	8.6	4.0	1.1
	Sutlej	Below Ferozepur	17.2	2.2	0.8
	Total		200.4	148.5	147.0

Source: Khan, A.R., 1999. An Analysis of the Surface Water Resources and Water Delivery Systems in the Indus Basin. Research Report 93. IWMI, Lahore, Pakistan, p. 66; Government of Pakistan, 2011. Pakistan Statistical Year Book 2011, Agricultural Data Set. Federal Bureau of Statistics, Statistics Division, Islamabad, Pakistan; IUCN, 2011. Water Resources of Pakistan: The Government's Main Objectives, in Pakistan Water Gateway. Available from: http://waterinfo.net.pk/?q=node/19.

was reduced by 75% and 92% during the years 1985–2002 and 2007–10, respectively. About 17% reduction in the average flow of the west-flowing rivers is also observed (Cheema, 2012).

The variability in river flows affects overall surface water supplies to the irrigated areas. The average canal water available at the farm gate openings for each command area in the IRB is estimated at $113 \, km^3$. However, the amount fluctuates between $136 \, km^3$ to $93 \, km^3$ depending on the variability in the river flows (UNDP, 2016).

Climate change and its variabilities as well as upstream interventions are considered as other causes of reduction in the river flows (Ahmad, 2009). Most of the irrigated agriculture in the basin is based on these flows; thus any change in availability can have severe effects on food security in the basin.

8.4.3 Reduced Environmental Flows

Political differences, mistrust, escalating tensions about surface boundaries, and disagreements about water use between the riparian countries have led to various conflicts between competing countries over water distribution on the rivers in the Indus Basin. India, being the upper riparian, has diverted all flows from the northeast to southwest flowing rivers (called the "Eastern Rivers": the Ravi, the Beas, and the Sutlej) to meet its growing irrigation demand. Consequently, it created water scarcity and an environmental threat in parts of the basin previously fed by these rivers, mostly in eastern parts of Pakistan. Table 8.3 shows decreasing flow trends for both west- and east-flowing rivers. A 92% reduction in average flow of eastern rivers entering into Pakistan was observed by 2010. No minimum environmental flows were allocated in the IWT. As a result, these rivers flow during the flood period and remain dry for almost 335 days in a year. The consequential effect is degradation of the river ecosystems and in the livelihood of local people.

This situation also resulted in reduced flows for the Lower Indus Basin. According to some studies, in order to restrict seawater intrusion and associated salinization, about 8–10 billion m^3 of water is required to flow continuously below the Kotri barrage, the last gauging station on the Indus River. However, this threshold flow cannot be obtained due to variable upstream flows. The absence of storage facilities upstream of Kotri has made it technically impossible to ensure minimum required flows. The reduced surface flows are a consistent environmental threat to flora, fauna, and aquatic life, especially in the Lower Indus Basin. The situation has created a continuing barrier to normalizing relationships between India and Pakistan.

8.4.4 Accelerated Water Scarcity

Reduced water availability and a rising population have resulted in converting the basin into a water-scarce region. Water availability in Pakistan declined from $3385 \, m^3$ per annum per capita in 1977 to $1396 \, m^3$ in 2011, while India went from $2930 \, m^3$ to $1539 \, m^3$ during the same period. Now per capita water availability has reduced to $1000 \, m^3/yr$. The reduction in water availability directly influences crop production and hence can adversely affect the food security of the region (FAO, 2012c). The rising population and urbanization are also affecting land use patterns by reducing agricultural lands even while there is mounting pressure for increased food production on available land. Moreover, both Pakistan and India are losing water in the form of virtual water trades (i.e., hidden water), as food and other commodities are traded to other countries—India and Pakistan are globally ranked first and third, respectively, in trade of rice (Konar et al., 2013). In this way, Pakistan and India jointly export around $5\,km^3$ of water each year, which equals half of the total water storage capacity of the Tarbela Dam in Pakistan.

8.4.5 Rising Water Demand

The rise in population and the growth of high delta crops (rice, sugarcane, corn, etc.) in the basin have increased the water demand and competition among various water users. It is anticipated that domestic and industry water demands will receive higher priority than the agriculture sector, which will have to compete more rigorously to get its share. The farmers are opting to grow high delta crops to get more economic benefit but at the cost of water, which is considered to be a free commodity. For example, the price of canal water in Pakistan is merely 135 Pakistani rupees (PKR), while groundwater pumpage is encouraged in both India and Pakistan as governments provide subsidies for tubewell installation and electricity tariffs. Such initiatives have given farmers the freedom to sow crops requiring a significant amount of water, which has resulted in a water shortage that has severely affected productivity.

It is foreseen that due to the slackness in developing water resources for the near future, the rise in water demand from the domestic and industrial sectors will largely be met by water previously allocated and used for agriculture (UNDP, 2016). At the same time, unchecked groundwater abstractions could adversely affect land and water productivity not only because of depleting aquifers but also because of deteriorating groundwater quality.

8.4.6 Lower Land and Water Productivity

The land and water productivity (crop yield per unit of water consumed) of various staple crops is highly variable in the region. In the case of rice, the land and water productivity vary between $2.6\,t/ha$ to $6.18\,t/ha$ and $0.20\,kg/m^3$ to $2.04\,kg/m^3$, respectively (Immerzeel et al., 2010; Cai et al., 2010). While, in the case of wheat, average water productivity is estimated at $1.0\,kg/m^3$ and $0.5\,kg/m^3$ in India and Pakistan, respectively. For comparison, in the case of wheat, the state of California produces more than $1.5\,kg/m^3$ of water while India and Pakistan produce only $1.0\,kg/m^3$ and $0.5\,kg/m^3$, respectively. The situation is similar for the other crops grown in the region. As a major user of the Indus water system, Pakistan is the least efficient water-productive country in the region. It has been estimated that the country is using water to irrigate wheat crops with 12% less efficiency than India does. Watto and Mugera (2016) examined the water productivity in Pakistan and stated that it was $0.76\,kg/m^3$ for wheat, which was 24% less than the global average of $1.0\,kg/m^3$. For rice crops the percentage is estimated to be $0.45\,kg/m^3$ compared with the Asian average of $1.0\,kg/m^3$, thus making Pakistan's water use efficiency 55% lower than the Asian average. Similarly, for cereal crops the water use efficiency in Pakistan is extremely low. It is $0.13\,kg/m^3$ in Pakistan compared to $0.39\,kg/m^3$ and $0.83\,kg/m^3$ in India and China, respectively. The low Indus water productivity is largely because of less trained manpower and poor (traditional) irrigation practices (Hussain et al., 2003). Traditional agricultural practices and poor land and water management, without consideration for integrated approaches, will not be able to meet future demand.

8.4.7 Groundwater Quantity and Quality

Variable surface supplies and the rising demand for water for agriculture, drinking water, and industry have forced an extensive pumpage of groundwater. Around 50% of agricultural water requirements in the IRB are being met through the use of groundwater. Moreover, sustainable use of groundwater is becoming difficult because of the heavy reliance on the resource for irrigation and municipal purposes. Groundwater usage contributes around $1.3 billion to the national economy of Pakistan each year. Studies have shown that due to the use of groundwater, crop yields have increased from 150% to 200% and cropping intensities have increased from 70% to 150% (Qureshi et al., 2003). The groundwater aquifer was ignored in the Indus Water Treaty of 1960 that allowed the two countries (India and Pakistan) to carry out unmetered groundwater pumpage. The potential areas of groundwater pumpage are located in Pakistani and Indian Punjab, as well as the Indian state of Haryana (Cheema et al., 2014; Fig. 8.3).

The total groundwater potential in the IRB is estimated at $85 \, km^3$, while about $68 \, km^3$ of groundwater is being pumped annually, and that amount is increasing due to annual increases in the number of tubewells (Cheema, 2012). In Pakistani Punjab, tubewell density is estimated at five to six tubewells per km^2, while in the Punjab and Haryana states of India, 27 and 14.1 tubewells were installed per km^2, respectively, in 2001, and the number is increasing (Vijay Shankar et al., 2011). This extensive pumpage has resulted in depletion of the aquifer at a rate of $31 \, km^3$, making the Indus Basin aquifer the most overstressed aquifer in the world and causing a severe decline in the water table (Gleeson et al., 2012; Richey et al., 2015; Lutz et al., 2016). The water table varies between 10 m to 200 m with an average decline rate estimated at 1.5 m/yr, depending on the hydrogeological conditions of aquifer and recharge potential.

The situation could be even worse from a regional perspective, as negative flux created by continuous abstraction at one location can disturb the natural groundwater flow paths and cause environmental degradation elsewhere. The groundwater in the Indus flows naturally from the northeastern to southwestern part of the basin. A change in flow may severely affect the already degraded environment in the middle parts of the basin, including the state of Rajasthan and the eastern parts of the Punjab province.

Due to the preceding exploitations, the issue reaches beyond the quantity of groundwater to its quality, along with the added cost of the energy needed to abstract the water. Secondary salinization associated with the extensive use of inferior groundwater is becoming a major threat to the sustainability of irrigated agriculture in the basin (Qureshi, 2011). This threat is more intense for areas in the lower Punjab and Sindh provinces of Pakistan. The Indian part of the Indus receives sufficient rainfall to replenish the groundwater, while the arid nature of the Pakistani Punjab and Sindh provinces has made them less resilient to drought, resulting in social, economic, and political problems.

It is therefore important to pay careful attention to the role of this transboundary aquifer and to consider it as being equally important as the surface water exchanges between administrative boundaries.

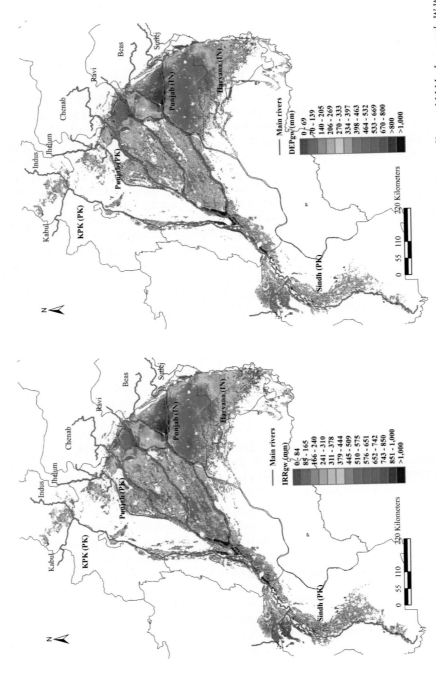

FIG. 8.3 Spatial groundwater abstraction and depletion during a single year in Indus River Basin. (*Source: Cheema, M.J.M., Immerzeel, W.W., Bastiaanssen, W.G.M., 2014. Spatial quantification of groundwater abstraction in the irrigated Indus basin. Groundwater 52 (1), 25–36.*)

8.4.8 Hydropower Generation and Energy Crisis

Apart from irrigating land, the five major tributaries of the Indus generate massive hydropower resources for India and Pakistan, although limits have been defined in the Indus Water Treaty for the upper riparian for Chenab and Sutlej. Similarly, a number of hydropower schemes are operational on the other rivers. Such projects supply electricity to the industrial sectors of the countries. With the demands of increasing industrialization and population, the development of hydropower has become a priority for the governments of the respective countries. In fact, the Ravi, Sutlej, and Beas valleys on the Indian side already have numerous power projects.

Being the upper riparian, India generates sufficient hydropower resources from the tributaries of the Indus. According to one estimate, the country is extracting 12,127 MW (megawatts) of electricity from runoff at the river hydropower projects installed at the various locations in the Upper Indus Basin. However, by receiving only 6500 MW, Pakistan fails to get maximum benefit of the hydropower potential, which is estimated to be as much as 50,000 MW. Details on the hydropower attained by the two countries from different rivers are provided in Table 8.4.

The agricultural sector of India consumes 23% of the total power supply. However, due to massive subsidies, it contributed only 7% to the India's national revenue generated during 2011–2012. The farmers in the states of Indian Punjab and Haryana, the breadbasket of India, are presently charged no money for extraction of groundwater using electrical tubewells. The provincial government pays a monthly subsidy for free power supplies. However, due to the increasing demand for electricity, the long waits for electrical connections, the frequent power outages, and the willingness of farmers to pay for an uninterrupted power supply, the Punjab State Power Corporation Limited (PSPCL) has introduced a policy for payment of tubewell connections.

For Pakistan the statistics are not encouraging because country's economic dependence on power generation is immense. For example, a power outage in 2011–2012 cost 1439 billion in Pakistani rupees, which comprised about 7% of country's GDP (Pasha et al., 2013). Based on the preceding statistics, it can be interpreted that for both counties the Indus Basin is of paramount importance. However, a number of authors have concluded that climate change will decrease 30% to 40% of the Indus Basin's flows (e.g., Briscoe and Qamar, 2007; Ahsan et al., 2016). The magnitude of the diminishing flows will adversely impact hydropower generation, thus undermining the efforts of Pakistan and India to become energy-secure countries.

TABLE 8.4 Hydropower Attained by the Two Countries From Indus River and Its Major Tributaries

S..No.	River/Tributaries of River	India (MW)	Pakistan (MW)
1	Indus	98	5352
2	Sutlej	4534	–
3	Beas	2267	–
4	Ravi	2059	–
5	Chenab	2015	–
6	Jhelum	1154	1000

8.4.9 Climate Change

With Pakistan and India competing for water from the Indus system to fulfill the requirements of their predominately agricultural-based economies, geopolitical stability is essential for the region's security. However, based on the system's location in the region, which is expected to be the area most affected by climate change, the goal of achieving consolidated regional peace appears to be very challenging (Kreft et al., 2016). The climate changes in the Indus Basin threaten access to fresh water, impair food production, and intensify the frequent occurrence of extreme climatic events (e.g., floods and droughts), resulting in loss of agricultural land. Because of the transboundary nature of the Indus Basin, it will be a major challenge to adopt a holistic approach to ensure its sustainability, which is even more problematic as the adversities related to climate change (e.g., hunger, poverty, water shortage, etc.) adhere to neither boundaries nor barriers.

The Indus Basin depends heavily on its upstream mountainous part for its downstream supply of water, and downstream demands are high. Climate change is expected to adversely impact the sources of water in the Indus Basin, including snow and glacier melt and rainfall. The latter provides 50%–80% of the flow through runoff originating in the Indus system, with the rest of the runoff contributed in the monsoon season. The catchment in the Upper Indus Basin accommodates more than 5000 glaciers covering an area of about $13,000 \, km^2$.

The future temperature-related climate projections for the Indus Basin indicate an increase of at least five degrees in temperature by the end of this century (Rasul et al., 2012). The strong correlation of rising temperature with increasing streamflow magnitude is likely to elevate the stream flows in the near term. However, with decreasing glacier volumes owing to increased melt rates, the flow magnitudes are expected to decrease in the long term (Lutz et al., 2016).

On the other hand, the future precipitation projections for the Upper Indus Basin are argued to be highly uncertain due to the complex nature of the process (Lutz et al., 2016). However, recent studies have revealed that the intensity and frequency of hydrological events in the future are going to be extreme (Lutz et al., 2016). The monsoon season, which as stated also contributes to the flows of the Indus basin, is becoming shorter and more extreme in nature and thereby making floods and droughts a normal occurrence (Turner and Annamalai, 2012). The recent past has exhibited exceptional variability in the hydrological response of the Indus Basin, with extreme drought spanning from 2000 to 2003 (Ahmad et al., 2004) and unprecedented monsoon rainfall in 2010 (Houze et al., 2011). The intense monsoon rainfall resulted in massive floods, which affected one-fifth of Pakistan's total area. Because of the floods, the fragile economy suffered a $10.056 billion loss. The increased intensity of the monsoon disturbs the natural ecosystem of the basin by accelerating the erosion rates.

It is thus evident that the climate change poses a very serious threat to the sustainability of water resources and food security in the Indus Basin.

8.4.10 Aging Treaty

Water resources in the basin states are under stress, which is straining the water-sharing mechanism between the riparian countries. The sharing mechanism was chalked out in 1960 and was given the name Indus Water Treaty (IWT). The recent advancements on optimizing the use of water resources for irrigation, food, and energy, combined with climate change concerns, strongly demand that the treaty be revisited. For example, India's construction of new dams and hydropower projects is not deemed acceptable by Pakistan. The International Court of Justice must be involved in various ongoing projects to check whether they are violating the treaty. Recent examples include the Baglihar Dam on the Chenab; the Nimoo Bazgo and Chutak on the Suru tributary of Indus, and the Kishanganga hydropower plant located upstream of the Jhelum.

Afghanistan is also planning to control the water of the Kabul River (Lashkaripour and Hussaini, 2008) with financial and technical support from India. Similar structures are proposed upstream of the Jhelum and Chenab (Khan, 2009). It is argued that, although there is a provision in the IWT to construct hydropower generation. The construction of these storage structures on western rivers will have catastrophic consequences for Pakistan, as reduced flows resulting from filling these dams during low flow seasons could destroy the rabi crops in Pakistan (PILDAT, 2010). This could escalate political tension among the riparian countries.

Moreover, the treaty is mainly focused on available surface water supplies without considering the transboundary aquifers under current climate change scenarios. Both countries are calling the IWT an ineffective forum for resolving water issues. The frequency of extreme events is expected to increase in the near future, and provisions for floodwater management strategies need immediate attention. Pakistan wants to include them in the composite dialogue; however, India doesn't agree because of political differences.

8.5 CONCLUSIONS

The basin countries continue to become overpopulated, more polluted, but less tolerant than ever before. These realities have resulted in intensification of strains on the Indus River Basin that can be seen in diminishing river flows and dropping water tables. According to a statement by Khalid Mohtadullah, former member of the Water and Power Development Authority (WAPDA) and senior advisor to the International Centre for Integrated Mountain Development (ICIMOD) of Pakistan, "We are altering our natural systems in various ways and various scales at all levels at an unprecedented rate." The major source of degradation is the agricultural sector as it claims the maximum share of water. However, the sector also employs almost 60% of the local workforce, which comprises the most underprivileged class in the region.

The agricultural products grown in the area provide food to the region and also to the international market. Therefore agricultural productivity compared with the given volume of water is an important component in the efficient fulfillment of these roles. Rather amazingly, the major consumer of the Indus water system (i.e., Pakistan) has no comprehensive agriculture policy, whereas the Indian agriculture policy is always criticized for lack of liberalization. Historically, agricultural productivity and the profit margin linked with agriculture are observed to be positively correlated with agricultural policy. Such a policy smooths the way for the introduction of modern technologies in agriculture, product commercialization, and linkage of agricultural and nonagricultural industries. In the absence of a suitable agriculture policy, the ecosystem of the region, which is known for growing high delta crops (rice, sugarcane, etc.), is at serious risk. The major share of water to irrigate these crops is extracted from groundwater resources. The lack of groundwater recharge in the face of mounting extraction is proving to be a serious threat to groundwater sustainability. Therefore a comprehensive agricultural policy safeguarding the interests of the stakeholders committed to the basin will play a significant role not only in increasing agricultural production but also in protecting the region's ecosystem.

The desire to have a comprehensive agricultural policy can be achieved only if the region has a well-thought-out regional water distribution policy. The present water distribution treaty (i.e., the IWT) has a number of ambiguities that need to be addressed to make it a comprehensive, transparent, and lasting document. The document focuses on the localized water rights of Pakistan and India without monitoring the utilities on the bordering states (e.g., China and Afghanistan). By limiting its scope, the document overlooks the water-consumption activities in China and Afghanistan, which ultimately affects flow magnitudes in the Indus waterways. Moreover, the treaty covers only the surface water component of irrigation water and completely ignores groundwater extraction. Overtime, the share of groundwater in irrigation has increased exponentially. The basic reasons for this increase are inadequacy and unreliability in the supply of canal water. These problems result from lack of proper maintenance of irrigation-related infrastructures due to lack of funding. Anthropogenic and uncontrolled extraction of groundwater reorients the energy and water fluxes between atmosphere and land, thus altering the climate system and eco-hydrological processes. In regions where groundwater is overexploited, the terrestrial water storage is depleted rapidly, resulting in unsustainable water-use practices (unequal discharge and recharge) and human-induced climate change. Therefore a comprehensive policy on groundwater extraction should be part of the IWT to create a balance between water extraction and economic growth in the riparian states and to ensure sustainable water use.

Another reason for revisiting the IWT is the mind-boggling rise in human population. The population increase in the Indus Basin is responsible for an increased pressure on water resources due to more demand for food production and industrial purposes. South Asia, which is already the most populated region of the world, is currently accommodating 237 million people in the basin. Based on medium population estimates, the population is expected to grow to 319 million in 2025 and to 383 million by 2050. Whereas, the high population estimates for 2025 and 2050 are 336 million and 438 million, respectively. The population on the Chinese side of the basin is very small compared with the other three riparian, whereas the number of people on the Indian and Pakistani sides are expected to increase substantially. Moreover, the city of Kabul, the capital of Afghanistan, is situated in the Indus Basin. Since 2001 Kabul has been considered to be the fastest growing city in the world (Lashkaripour and Hussaini, 2008; Setchell and Luther, 2009). The population of Kabul has tripled since 2001, to 4.5 million. Population is perhaps the only front that requires a regional approach. These staggering population numbers demand that, instead of quantifying the water sharing rights or serving political motives, the treaty should encourage the enhancement of the region's dangerously low level of water productivity. Unless the Indus Basin riparian countries consider making and then following a unified approach, all water policies are doomed to fail.

Climate change is the most potent threat to the water resources of the Indus Basin. The threats are mainly because of glacier melting due to high temperature, which can reduce Indus Basin flows as much as 50% the worst case future scenario. The picture gets very disturbing, once this aspect is combined with the fact that the population of the region is on the rise. It is in the interest of both India and Pakistan to revisit the treaty, assess the climate change impacts of the flows of Indus basin, and subsequently translate them into a water-sharing mechanism.

It is therefore important to overcome a host of overlapping socioeconomic, environmental, and policy pressures as the region strives to fulfill its future water needs. There is a need for persistent effort at the governmental and nongovernmental level to bring all stakeholders and interest groups to a common platform in order to facilitate better understanding of each other's concerns, clear any misgivings, and ultimately build a consensus on how to address the real, emerging issues related to the Indus Basin. Indus waterway disputes used to be a product of territorial changes. Today they are largely the product of a decrease in water flow, changes in demography and population, as well as usage issues resulting from increased irrigation and the need for drinking water. Efforts, which essentially include cultural and social paradigm shifts, that will help the country evolve to a modern society should be encouraged. The modernization of irrigation structures, encouragement of farmer training and capacity building, and the effective use of information communication technology are some of suggested strategies. Agriculture-related strategies, such as assessing the relative social and economic value of surface and groundwater irrigation, delaying rice transplanting and short duration rice varieties, and farmer training for conjunctive irrigation, are important. Decreasing water resource availability and the misuse of water in irrigation result in highly confusing situations. Better coordination between national and state governments, with context-specific modifications to achieve the larger goal of effective localized water management, is of paramount importance.

The constant threat of flooding and the socioeconomic pressure to develop these regions call for combined action from both countries. Long-term cooperative solutions may lift huge economic burdens from both nations. Discussion on the merits, demerits, and provisions for the design aspects of proposed major irrigation projects in the Indus Basin, in light of siltation and waterlogging problems, is also inevitable. There is a need for joint, impartial research that will provide alternative approaches to address the present and future challenges emanating from the Indus Waters Treaty. Formation of an Indus Waters Experts Group might be a good starting point. This group should have multidisciplinary, multi-organization researchers, and a mix of governmental and nongovernmental sector specialists. An integrated, holistic approach to transboundary river basin management is also needed, in which the basin is accepted as the logical unit of operation. A multi-sectoral, integrated system, complemented by information sharing, transparency, and wide participation, is therefore best suited to encompass all of these elements. Such an integrated system for the evaluation of interactions between the hydrological processes in the mountains, river flow generation, water retention in reservoirs, groundwater pumping, and agricultural water use in the Indus Basin is largely lacking and should be investigated.

References

Ahmad, S., 2009. Water availability in Pakistan: Paper presented by Dr. Shahid Ahmad, Member PARC. In: National Seminar on "Water Conservation, Present Situation and Future Strategy". Project Management & Policy Implementation Unit (PMPIU) of the Ministry of Water & Power, Islamabad, Pakistan, p. 114.
Ahmad, S., Hossain, Z., Sarwar, A., Majeed, R., Saleem, M., 2004. Drought Mitigation in Pakistan: Current Status and Options for Future Strategies. International Water Management, Working paper, No. 85.

Ahsan, M., Shakir, A.S., Zafar, S., Nabi, G., 2016. Assessment of climate change and variability in temperature, precipitation and flows in upper Indus basin. Int. J. Sci. Eng. Res. 7 (4), 1610–1620.

Briscoe, J., Qamar, U., 2007. Pakistan's Water Economy Running Dry. Oxford University Press, Karachi. Commissioned by World Bank, 2007.

Cai, X., Sharma, B.R., Matin, M.A., Sharma, D., Gunasinghe, S., 2010. An assessment of crop water productivity in the Indus and Ganges River Basins: Current status and scope for improvement. Research Report 140, IWMI, Colombo, Sri Lanka.

Chadha, D.K., 2008. Development, management and impact of climate change on transboundary waters management. In: Paper Presented at the 4th International Symposium on Transboundary Waters Management, 15–18 October 2008, Thessaloniki, Greece. Available from, http://www.inweb.gr/twm4/abs/CHADHA%20Devinder.pdf.

Cheema, M.J.M., 2012. Understanding water resources conditions in data scarce river basins using intelligent pixel information – Case: Transboundary Indus Basin. Dissertation, Delft University of Technology, Delft, The Netherlands. Available from, http://repository.tudelft.nl/assets/uuid:7b569411-9934-4b23-b631-36a58f60363f/CheemaMJM_PhD_Thesis.pdf.

Cheema, M.J.M., Pawar, P., 2015. Bridging the divide: Transboundary science and policy interaction in the Indus Basin. VF Report. Environmental Security Program, Stimson Center, Washington DC, USA.

Cheema, M.J.M., Immerzeel, W.W., Bastiaanssen, W.G.M., 2014. Spatial quantification of groundwater abstraction in the irrigated Indus basin. Groundwater 52 (1), 25–36.

FAO (Food and Agriculture Organization of the United Nations), 2012a. FAOSTAT Statistical Database, dataset for inputs/land utilized. Available from, http://faostat3.fao.org/download/R/RL/E.

FAO (Food and Agriculture Organization of the United Nations), 2012b. Irrigation in Southern and Eastern Asia in Figures: AQUASTAT Survey 2011, 129-140. FAO, Rome, Italy.

FAO (Food and Agriculture Organization of the United Nations), 2012c. AQUASTAT Database, dataset for country water resources utilized. Available from, http://www.fao.org/nr/water/aquastat/main/index.stm.

FAO (Food and Agriculture Organization of the United Nations), 2013. FAO Statistical Yearbook 2013: World Food and Agriculture. Rome, Italy.

Finley, T., Leathers, G., Zhang, H.X., 2008. Use of the WBCSD Global Water Tool to assess global water supply risk and gain valuable strategic perspective. In: Paper Presented at the Water Environment Federation's Annual Technical Exhibition Conference (WEFTEC) 2008, 18–22 October 2008, Chicago, IL.

Gleeson, T., Wada, Y., Bierkens, M.F.P., van Beek, L.P.H., 2012. Water balance of global aquifers revealed by groundwater footprint. Nature 488, 197–200. https://dx.doi.org/10.1038/nature11295.

Houze, R.A., Rasmussen, K.L., Medina, S., Brodzik, S.R., Romatschke, U., 2011. Anomalous atmospheric events leading to the summer 2010 floods in Pakistan. Bullet. Am. Meteorol. Soc. 92, 291–298.

Hussain, I., Sakthivadivel, R., Amarasinghe, U., Mudasser, M., Molden, D., 2003. Land and water productivity of wheat in the western Indo-Gangetic plains of India and Pakistan: A comparative analysis. Research Report 65, International Water Management Institute, Colombo, Sri Lanka.

Immerzeel, W.W., van Beek, L.P.H., Bierkens, M.F., 2010. Climate change will affect the Asian water towers. Science 328 (5984), 1382–1385.

Khan, A.F., 2009. India Building Small Dams on Indus. DAWN, Karachi.

Konar, M., Hussein, Z., Hanasaki, N., Mauzerall, D., Rodriguez-Iturbe, I., 2013. Virtual water trade flows and savings under climate change. Hydrol. Earth Syst. Sci. 17, 3219–3234.

Kreft, S., Eckstein, D., Melchior, I., 2016. Global Climate Risk Index 2017. Germanwatch e.V, Berlin, Germany.

Laghari, A.N., Vanham, D., Rauch, W., 2012. The Indus basin in the framework of current and future water resources management. J. Hydrol. Earth Syst. Sci. 16, 1063–1083.

Lashkaripour, G.R., Hussaini, S.A., 2008. Water resource management in Kabul River Basin, Eastern Afghanistan. Environmentalist 28 (3), 253–260. https://dx.doi.org/10.1007/s10669-007-9136-2.

Lutz, A.F., Immerzeel, W.W., Kraaijenbrink, P., Shrestha, A.B., Bierkens, M.F., 2016. Climate change impacts on the upper Indus hydrology: sources, shifts and extremes. PLoS One 11 (11), e0165630. https://dx.doi.org/10.1371/journal.pone.0165630.

Merriott, D., 2016. Factors associated with the farmer suicide crisis in India. J. Epidemiol. Glob. Health 6, 217–227.

Molle, F., Wester, P., Hirsch, P., 2010. River basin closure: processes, implications and responses. Agric. Water Manag. 97, 569–577.

Pasha, H.A., Pasha, A.G., Saleem, W., 2013. Economic Costs of power load shedding in Pakistan. Retrieved from, http://pdf.usaid.gov/pdf_docs/PBAAF356.pdf.

PILDAT (Pakistan Institute of Legislative Development and Transparency), 2010. Pakistan-India relations: Implementation of Indus water treaty. A Pakistani narrative. Pakistan Institute of Legislative Development and Transparency (PILDAT), Islamabad, p. 14 Background paper 6-020.

Qureshi, A.S., 2011. Water Management in the Indus Basin in Pakistan: challenges and opportunities. Mt. Res. Dev. 31 (3), 252–260.

Qureshi, A.S., Shah, T., Akhtar, M., 2003. The Groundwater Economy of Pakistan. IWMI Working Paper No. 64, International Water Management Institute (IWMI), Colombo, Sri Lanka.

Rasul, G.A., Mahmood, A., Khan, S.I., 2012. Vulnerability of the Indus delta to climate change in Pakistan. Pak. J. Meteorol. 8, 89–107.

Richey, A.S., Thomas, B.F., Lo, M.H., Reager, J.T., Famiglietti, J.S., Voss, K., 2015. Quantifying renewable groundwater stress with GRACE. Water Resour. Res. 51, 5219–5238. https://dx.doi.org/10.1002/ 2015WR017349.

Setchell, C.A., Luther, C.N., 2009. Kabul, Afghanistan: A case study in responding to urban displacement. Humanitarian Exchange Magazine 45. Retrieved from: http://www.odihpn.org/report.asp?id=3053.

Sharma, B.R., Amarasinghe, U.A., Sikka, A., 2008. Indo-Gangetic River Basins: Summary Situation Analysis. Project Report, IWMI, New Delhi, India.

Sharma, B.R., Amarasinghe, U.A., Xueliang, C., De Condappa, D., Shah, T., Mukherji, A., Bharati, L., Ambili, G., Qureshi, A., Pant, D., Xenarios, S., Singh, R., Smakhtin, V., 2010. The Indus and the Ganges: River basins under extreme pressure. Water Int. 35 (5), 493–521.

Turner, A.G., Annamalai, H., 2012. Climate change and the South Asian summer monsoon. Nat. Clim. Change 2, 587–595.

Turral, H., Burke, J., Faurès, J., 2011. Climate Change, Water and Security. Water Report No. 36, FAO, Rome, Italy. Available from, http://www.fao.org/docrep/014/i2096e/i2096e.pdf.

UNDP, 2016. Development Advocate Pakistan. United Nations Development Programme Pakistan, 4th Floor, Serena Business Complex, Khayaban-e-Suharwardy, Sector G-5/1, P. O. Box 1051, Islamabad, Pakistan. ISBN: 969-978-8736-16-3.

Vijay Shankar, P.S., Kulkarni, H., Krishnan, S., 2011. India's groundwater challenge and the way forward. Econ. Polit. Wkly. 46 (2), 37–45.

Watto, M., Mugera, A., 2016. Wheat farming system performance and irrigation efficiency in Pakistan: a bootstrapped metafrontier approach. Int. Trans. Oper. Res. 00 (2016), 1–21.

Yu, W., Yang, Y., Savitsky, A., Alford, D., Brown, C., Wescoat, J., Debowicz, D., Robinson, S., 2013. The Indus Basin of Pakistan: The Impacts of Climate Risks on Water and Agriculture, Directions in Development Sseries. World Bank, Washington, DC.

Zawahri, N.A., 2008. International rivers and national security: The Euphrates, Ganges–Brahmaputra, Indus, Tigris, and Yarmouk rivers. Nat. Res. Forum 32, 280–289.

Further Reading

Government of Pakistan, 2011. Pakistan Statistical Year Book 2011, Agricultural Data Set. Federal Bureau of Statistics, Statistics Division, Islamabad, Pakistan.

IUCN, 2011. Water Resources of Pakistan: The Government's Main Objectives, in Pakistan Water Gateway. Available from: http://waterinfo.net.pk/?q=node/19.

Khan, A.R., 1999. An Analysis of the Surface Water Resources and Water Delivery Systems in the Indus Basin. Research Report 93. IWMI, Lahore, Pakistan, p. 66.

World Bank Group, 2013. Agricultural and World Development Indicators Datasets Utilized. Available from, http:// data.worldbank.org/.

Indo-Ganges River Basin Land Use/ Land Cover (LULC) and Irrigated Area Mapping

Murali Krishna Gumma, Prasad S. Thenkabail[†],
Pardhasaradhi Teluguntla[†], Anthony M. Whitbread**

[*]International Crops Research Institute for the Semi-Arid Tropics (ICRISAT), Patancheru, India
[†]U.S. Geological Survey (USGS), Western Geographic Science Center, Flagstaff, AZ, USA

9.1 INTRODUCTION

Geo-spatial information on the distribution of irrigated areas is limited to district-level crop statistics published by state or national governments in different parts of the world. Although data that has been collected by irrigation and agriculture departments are available, there are often differences between sources in regard to the extent of the irrigated areas (Biggs et al., 2006; Gumma, 2008). The World Summit on Sustainable Development (WSSD) organized by the United Nations and held at Johannesburg, South Africa, declared water as the most critical resource in the 21st century, with increasing demands and decreasing supplies. Much of the freshwater is consumed by irrigation and evapotranspiration from different land use/land cover (LULC) classes, especially through transpiration from natural vegetation. Of the different types of LULC classes, irrigation is known to consume about 60% of the world's available freshwater resources. With ballooning global populations projected to be 8.3 billion in 2030 compared to just over 6 billion at present and with food and nutritional intake expected to increase from 2500 cal to 3000 cal per day per person (FAO, 2003), the demand for water for irrigation will only grow. This is neither feasible due to a shortage of resources nor desirable due to the, still to be fully understood, environmental impacts of large, medium, and even small irrigation schemes on natural environments along river courses.

Given the importance of irrigation to the world's food bank, a calculation of water resources that includes a detailed, accurate, and sophisticated LULC system is required. Land cover is likely to be the single most important factor of change in all river basins. It is well established that LULC changes have a significant effect on many processes in basins, including soil erosion, global warming (Penner et al., 1994), and biodiversity impacts (Chapin et al., 2000), and such changes are expected to have a greater influence on human habitability than climate change will (Skole, 2004). Through the use of the AVHRR pathfinder data sets, several global LULC maps have been produced (e.g., De Fries et al., 1998; Loveland et al., 2000).

Accurate information on the extent of river basin cropland is critical for food security assessments, water allocation decisions, and yield estimations. This information will also help decision makers monitor dynamic landscapes, such as agricultural lands, fallow croplands, and land cover such as forests, water bodies, and wetlands. Ex-ante assessments on the effect of changes in land use will facilitate sustainable land use planning, socially, economically, and ecologically. Moreover, countries' departments of agriculture and revenue will need such spatial information at the village-level in order to send advisories to farmers on timely inputs and best practices for agricultural management.

Croplands in the Indo-Ganges River Basin are frequently affected by abiotic stresses such as drought. Crop year 2015 was declared the hottest year on record by the world Meteorological Organization (WMO). Very high temperatures over both land and ocean in 2015 were accompanied by many extreme weather events, such as heatwaves, flooding, and severe drought (WMO, 2015). Several studies have been conducted on land use mapping and changes. However, the main purpose for analyzing changes in agricultural land use is to monitor cropping patterns and cropland changes (Singh, 1989; Wan et al., 2004; Yaduvanshi et al., 2015; Hao et al., 2016; Li et al., 2016; Olsson et al., 2016; Wang et al., 2016). These analyses rely heavily on agricultural statistics (e.g., extent of area). Besides discrepancies at the district, state, and provincial levels, there are discrepancies in the statistics reported by agricultural agencies and irrigation agencies. Variations in land use on such a large scale are not sufficient to fully clarify their effect on river basins. On the other hand, remote sensing with satellite imagery can give detailed maps of land use and identify where significant cropping pattern changes have occurred in response to variations in rainfall (Badhwar, 1984; Thiruvengadachari and Sakthivadivel, 1997). Remote sensing utilizing satellite imagery has been used to quantify water use and productivity in irrigation systems (Thiruvengadachari and Sakthivadivel, 1997), but is less frequently used to identify changes in irrigated command areas in response to variations in rainfall and water supply. Normalized Difference Vegetation Index (NDVI) time-series data have been used for mapping land use changes (Gumma et al., 2011b) and irrigated areas (Biggs et al., 2006; Thenkabail et al., 2009b). Time series data have also been used for detecting changes in irrigated areas in major river basins (Bhutta and Van der Velde, 1992; Gaur et al., 2008).

The present study analyzed the spatial extent of major croplands along with other LULC in the Indo-Ganges Basin. Water deficiency for rainfed crops was assessed based on low rainfall, which was reflected in the NDVI imagery from the 2013–14 period. The area under each type of land use for each year and the changes between years were estimated.

The estimates of both land use and land cover changes were compared with ground survey data and secondary sources, such as published statistics on rice systems. The focus was on the areas where significant changes had occurred in the cropping pattern.

In addition to the preceding assessments, the overarching goal of this study was to map the spatial extent of the above-mentioned croplands. The study used MODIS 250 m time-series data to map major croplands using SMTs that were first advocated for cropland mapping by Thenkabail et al. (2007) and later successfully applied in global and regional mapping of croplands (Thenkabail et al., 2007a, 2009a, 2012; Biradar et al., 2009; Pittman et al., 2010; Gray et al., 2014; Gumma et al., 2015a; Salmon et al., 2015; See et al., 2015).

9.2 STUDY AREA

The study area (see nonhatched area in the basin boundary in Fig. 9.1) covers 63% (133,071,400 ha) of the Indo-Gangetic Plain (total area=217,699,000 ha.). Crop year 2013 was a normal year and 204 areas had significant rainfall deficit in terms of amount and

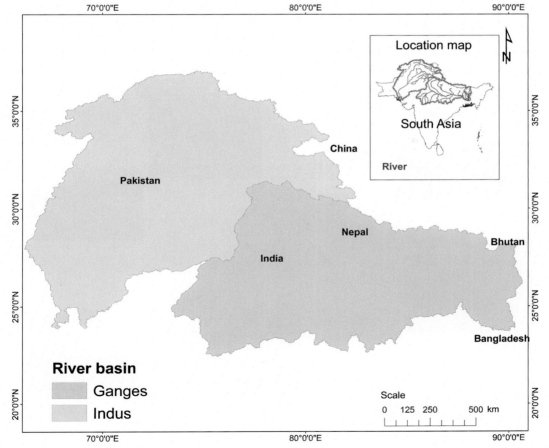

FIG. 9.1 Study area of Indo-Ganges River Basin.

distribution (Fig. 9.2). The area of the Ganges and the Indus basins falling within the three MODIS tiles (h24v06, h25v06, and h26v06; each tile is 1000 by 1000 km) was chosen as the area of study. The three tiles were mosaic tiles stacked into a single contiguous tile by running batch scripts in ERDAS Imagine 8.6 (ERDAS, 2003) from which the areas in the Ganges and Indus basins were delineated (Fig. 9.1). About 95% of the Ganges Basin (total area 95,111,154 ha.) and 37% of the Indus Basin (116,113,290 ha.) were covered by the three MODIS tiles. The characteristics of the 7-band, 8-day interval MODIS data of years 2001 and 2002 used in this study are shown in Fig. 9.1 and Table 9.1.

The origin of the Ganges River Basin is a highly fertile glacier called Gangotri, which is located in the Himalayans about 4267 m above sea level. The Ganges River Basin encompasses an area with a very high population density of about 530 persons per square kilometers, with the river flowing through 29 cities with populations of more than 100,000, 23 cities with populations between 50,000 and 100,000, and about 48 towns (Aitken, 1992; Ilich, 1996).

The source of the Indus River lies in Western Tibet in the Mount Kailas region at an altitude of 5500 m a.s.l. The Indus Basin comprises the Indus River, its five major left bank tributaries, the Jhelum, Chenab, Ravi, Beas, and Sutlej Rivers, and one major right bank tributary, the Kabul (Khan, 1999). The catchments contain some of the largest glaciers in the world

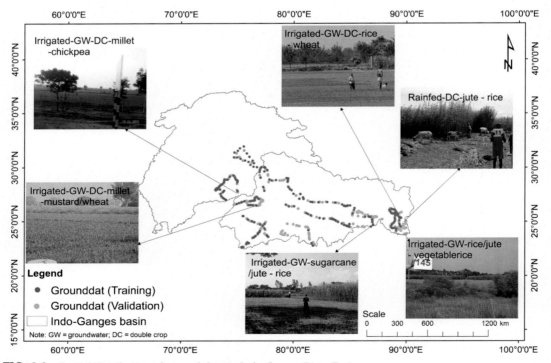

FIG. 9.2 Spatial distribution of ground data in Indo-Ganges River Basin.

TABLE 9.1 Total Geographical Area

Country	Total Geographical Area ('000 ha)	Area Covered in Indo-Ganges Region ('000 ha)	Percent of Indo-Ganges Basin (%)
Bangladesh	14,804	4273	2.0
Bhutan	4365	1816	0.8
India	345,623	123,795	56.9
Nepal	16,210	14,762	6.8
Pakistan	89,167	53,916	24.8
Afghanistan	65,200	7270	3.3
China	9,596,960	11,868	5.5
Total	1,01,32,329	217,699	

outside the polar regions (Meadows, 1999). Only 37% of the total basin area is covered in this study (Fig. 9.1), but much of this area covers the Punjab and Sindh regions, which are heavily irrigated by the Indus River.

9.3 DATA

9.3.1 Satellite Data

The MODIS data we used for the Krishna River Basin is archived at the NASA-USGS website (http://e4ftl01.cr.usgs.gov/MOLT/MOD13Q1.005). MODIS 2013–2014, recorded every 16 days (Table 9.2), and Terra sensor data were used for the present study. The format has two specific bands (band 1, red; and band 2, near infrared) that are processed for land applications as a MODIS vegetation index product (MOD13Q1). MOD13Q1 is computed from MODIS level 5 bands 1–2 (centered at 648 nm and 858 nm).

MODIS imagery was used to map the spatial extent of land use/land cover for the years 2013–14 and 2015–16. The process began with rescaling 16-day NDVI images that were later stacked into a single file data composite for each cropping year (Thenkabail et al., 2005; Dheeravath et al., 2010; Gumma et al., 2011a, 2015b). MODIS 16-day composites were converted into a NDVI monthly Maximum Value Composite (MVC) (NDVI MVC) using Eq. (9.1) (Casanova et al., 1998), where MVCi is the monthly MVC of the ith month and $i1$ and $i2$ represent all the 16-day data in a month:

$$NDVI_{MVC_i} = Max \left(NDVI_{i1,}, NDVI_{i2,} \right) \qquad (9.1)$$

In the present study, monthly NDVI MVC images were used for classification and a NDVI 16-day data set was used for identifying and labeling land use/land cover classes, including irrigated areas. The main reason for using MVC was to avoid noise (clouds) in some of the areas (Gumma et al., 2014).

TABLE 9.2   MODIS—250 m Terra Vegetation Indexes 16-Day L5 Product Used in This Study

MODIS Data Sets	Units	Band Width nm/Range	Potential Application
250 m 16 days NDVI	NDVI	−1 to +1	Vegetation conditions
250 m 16 days EVI	EVI	−1 to +1	Canopy structural variations
250 m 16 days red reflectance (Band 1)	Reflectance	620–670	Absolute land cover transformation, vegetation chlorophyll
250 m 16 days NIR reflectance (Band 2)	Reflectance	841–876	Cloud amount, vegetation land cover transformation
250 m 16 days blue reflectance (Band 3)	Reflectance	459–479	Soil/vegetation differences
250 m 16 days MIR reflectance (Band 7)	Reflectance	2105–2155	Cloud properties, land properties

9.3.2 Ground Data

Ground data was collected in 2013 from September 13 to September 26 for 227 sample points and in 2015 from September 21 through September 30 for 326 sample points covering about 8000 km of road travel in the Indo-Ganges Basin (Fig. 9.2). Ground data were collected based on pre-classified output, Google Earth imagery, and tracking GPS attached to the image processing software that captured ground survey information while in motion. Detailed information was collected for class identification and for labeling point locations. Point-specific information was collected from 250 m × 250 m plots and consisted of GPS locations, land use categories, land cover percentages, cropping patterns during different seasons (through farmer interviews), crop types, and watering methods (irrigated, rainfed). Samples were obtained within large contiguous areas of a particular LULC. Landsat 8 products were used as additional ground survey information in class identification. A stratified systematic sample design was adopted based on road network or footpath access. Where possible, a systematic location of sites was done every 5 km or 10 km along the road network by vehicle or by foot (Thenkabail et al., 2004, 2005; Gumma et al., 2011e), which is detailed in a description of the ground survey methodological approach.

9.3.3 Ideal Spectra Signatures

Ideal spectra signatures (Fig. 9.3) were generated using 16-day NDVI time-series composite and precise ground survey information that was also used for the class identification process. Ideal spectral signatures were based on 204 unique, ideal samples available from field data. This ground survey information was collected during 2013–14 that corresponded with the 2013–14 MODIS data. Ninety-four other samples were of noncroplands. The 298 samples were grouped according to their unique categories and grouped major rice systems as shown in Fig. 9.3. The samples were chosen to generate ideal spectra signatures that refer to crop intensity, crop type, and cropping systems. Each signature was generated with a group of

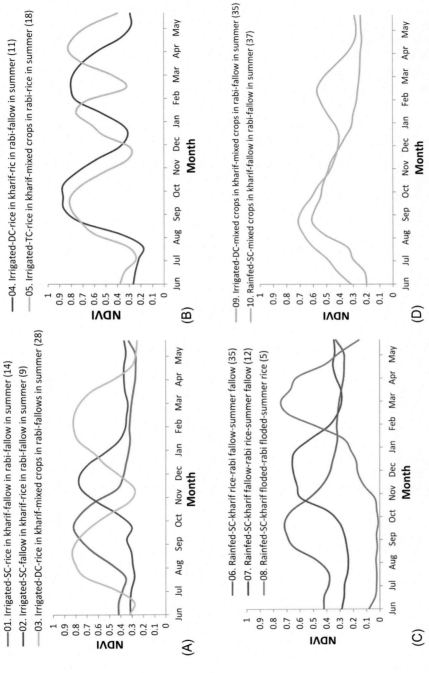

FIG. 9.3 Ideal spectral signatures of 10 cropland classes in South Asia (Gumma et al., 2016). For example, 01. Irrigated, SC, rice in kharif, fallows in rabi, fallows in summer means irrigated single cropland area with rice grown during kharif (June–October), but left fallow during rabi (November–February) and also left fallow in summer (March–May). (A) and (B) are irrigated-rice systems, (C) is rainfed-rice systems, and (D) is other major croplands.

similar samples. Take, for example, Fig. 9.3A, class 1: the "01. Irrigated-SC –rice in kharif – fallow in rabi-fallow in summer (14)" signature defines irrigated croplands during the kharif season followed by cropland fallows during the rabi season, and the cropland fallows during the summer season. This signature was generated by 14 ground survey samples that were smoothed. Overall, a total of 25 unique cropland classes that are either irrigated or rainfed have differing cropping intensities (e.g., classes 1, 2, 6, 7, 8, and 10 are single crop; classes 3, 4, and 9 are double crop; and class 5 is a triple crop), and distinct phenological cycles.

9.3.4 Mapping Land Use/Land Cover

An overview of the methods (Fig. 9.4) and details are described next. The process began with mapping land use/land cover areas using MODIS 16-day time-series data with spectral matching techniques and field-plot information.

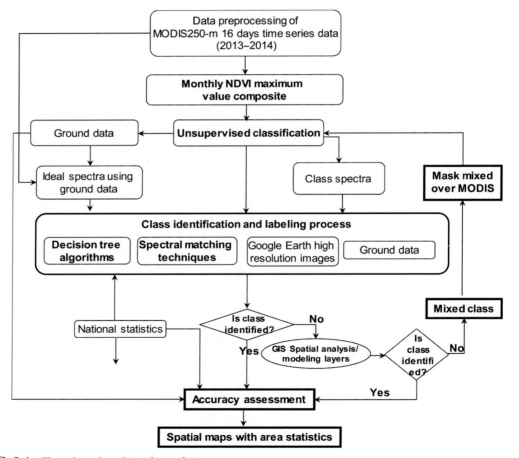

FIG. 9.4   Flow chart describing the analysis process.

MODIS 16-day time-series composite vegetation index images at 250 m resolution were obtained from June 1, 2013 to May 31, 2014 (MOD13Q1 data product). The MOD13Q1 data set are available in the public domain and are pre-calibrated (http://modis-sr.1tdri.org/html). The large scene size and daily overpass rate of MODIS make it attractive for mapping large crop areas, and NDVI images derived from MODIS have high fidelity with biophysical parameters (Huete et al., 2002). The 16-day NDVI images were stacked into a 23-band file for each crop year (two images per month). Monthly maximum value composites were created using 16-day NDVI MODIS data to minimize cloud effects (Holben, 1986).

Unsupervised classification was used to generate initial classes. The unsupervised ISOCLASS cluster algorithm (ISODATA in ERDAS Imagine 2016TM) that was run on the NDVI-MVC generated an initial 100 classes, with a maximum of 100 iterations and a convergence threshold of 0.99. Although ground survey data were available at the time of image classification, unsupervised classification was used to capture the full range of NDVI over a large area. Use of unsupervised techniques is recommended for large areas that cover a wide and unknown range of vegetation types, and where landscape heterogeneity complicates identification of homogeneous training sites (Cihlar, 2000; Biggs et al., 2006; Gumma et al., 2011e). Identification of training sites is particularly problematic for small, heterogeneous irrigated areas.

Land use/land cover classes were identified based on NDVI temporal signatures along with ground survey data. We observed crop growth stages and cropping patterns from temporal signatures, such as (1) onset of cropping season (e.g., monsoon and winter), (2) duration of cropping season (e.g., monsoon and winter), (3) magnitude of crops during different seasons and years (e.g., water-stressed and normal years), and (4) end of cropping season (Gumma et al., 2011a,e).

The process of labeling class identification was done based on spectral matching techniques (SMTs) (Thenkabail et al., 2007b; Gumma et al., 2016). Initially, 160 classes from the unsupervised classification were grouped based on spectral similarity or closeness of class signatures. Each group of classes was matched with ideal spectra signatures and ground survey data and assigned class names (Gumma et al., 2014, 2016). Classes with similar NDVI time-series and land cover were merged into a single class, and classes showing significant mixing, for example, continuous irrigated areas and forest, were masked and reclassified using the same ISOCLASS algorithm. Some continuous irrigated areas mixed with forests in the Western Ghats were separated using a 90 m digital elevation model (DEM) from the Shuttle Radar Topography Mission (SRTM) and an elevation threshold of 630 m, Landsat imagery, and ground survey data through spatial modeling techniques such as overlay matrix, recode, and proximity analysis (Tomlinson, 2003; Gumma et al., 2011d). This resulted in nine classes of LULC. While class aggregation could have been performed statistically using a Euclidean or other distance measure, we employed a user-intensive method that incorporates both ground survey data and high-resolution imagery in order to avoid lumping classes that might be spectrally similar but have distinct land cover. The NDVI of some classes differed in only one or 2 months, which would cause the classes to be merged if an automated similarity index were used.

Classification was done at 250 m spatial resolution. In the present study area, average land holding size is less than a pixel and there are different land use/land cover classes with 250 m × 250 m pixel (6.25 ha). Full pixel areas are not an accurate representation of actual cropland areas. The cropland fraction was calculated using the methodology described in (Thenkabail et al., 2007b; Gumma et al., 2011a,d, 2015a). Subpixel areas were important when a particular pixel was named as cropland but also contained other land use/land cover classes (grasses, trees, shrubs, etc.).

Ground data points were used to assess the accuracy of the classification results, based on a standard procedure (Jensen, 1996; Congalton and Green, 1999, 2008), to generate an error matrix and accuracy measures for each land use/land cover map. Error matrices and Eq. (9.2) (Farr and Kobrick, 2000) "Cohen's kappa coefficient (κ)" are commonly used for accuracy assessment. For example, these are useful when building models that predict discrete classes or when classifying imagery. κ can be used as a measure of agreement between model predictions and reality (Congalton, 1991a) or to determine if the values contained in an error matrix represent a result significantly better than random (Jensen, 1996). κ is computed as

$$\kappa = \frac{N\sum_{i=1}^{r} x_{ii} - \sum_{i=1}^{r}(x_{i+} \times x_{+i})}{N^2 - \sum_{i=1}^{r}(x_{i+} \times x_{+i})} \qquad (9.2)$$

where N is the total number of sites in the matrix, r is the number of rows in the matrix, xii is the number in row i and column i, $x+i$ is the total for row i, and $xi+$ is the total for column i (Jensen, 1996).

9.3.5 Matching Class Spectra With Ideal Spectra to Group Classes Using Spectral Matching Techniques (SMTs)

The initial 160 unsupervised classes (called class spectra) were arranged into a number of groups based on quantitative spectral matching techniques (Homayouni and Roux, 2003; Thenkabail et al., 2007a,b). Fig. 9.3 shows the grouping of some of these classes. This homogeneous class was then matched with ideal spectra (see Fig. 9.3) for preliminary class identification and labeling. Additional verification was conducted using ground data, and high-resolution imagery from Google Earth and GeoCover by overlaying minor (e.g., district) administrative boundaries in the Google Earth application. Mixed classes remained because of the large extent and diverse land use of small holdings. To resolve these mixed classes we used various other sources, such as irrigation command area boundaries, rainfall, district-level statistics, and high-resolution imagery using spatial modeling (Gumma et al., 2014). Some classes did not resolve conclusively, even after using ground survey information and other information previously mentioned. These classes were then subset, reclassified, and reanalyzed following the preceding protocols (Thenkabail et al., 2007b; Gumma et al., 2011a, 2014).

Fig. 9.5 illustrates the group of croplands in kharif-croplands in rabi-fallow in summer classes, which was matched with class 4 (see Fig. 9.2) in ideal spectra signatures. Initially classes were grouped based on decision tree algorithms and spectral similarity. This group of classes was then matched with ideal spectra (Fig. 9.5C) for preliminary class identification and labeling. Additional verification was conducted using ground survey information data, and high-resolution imagery from Google Earth and GeoCover by overlaying minor (e.g., district) administrative boundaries in the Google Earth application. In Fig. 9.5 initially 14 classes were grouped through decision tree algorithm, and finally 21 classes were closely correlation with ideal spectra class 4.

9.3.6 Subpixel Area Calculations

Full pixel areas (FPAs) are not correct representations of actual areas. For example, in each cropland class there are often thousands or millions of pixels. The proportion of area cropped in each of these thousands or millions of pixels varies significantly, even in a single cropland class. This situation results, for example, if we map all pixels with 50% or more covered in maize crops, because in a maize class, there will be pixels in which the maize proportion varies between 50% and 100%. So to get actual areas, the FPAs need to be multiplied by the cropland area fraction (CAF). The CAF will depend on the percentage of area in a pixel that actually belongs to a class. In the case of the previously discussed maize crop class, the CAF will vary between 0.5 and 1.0. Therefore the FPAs with CAFs of 0.5 will be multiplied by 0.5, FPAs with CAFs of 0.55 will be multiplied by 0.55, and so on. Overall, the actual areas are equivalent to the subpixel areas (SPAs), as shown in previous well-established studies (Thenkabailc et al., 2007). That is, each pixel in each class is assessed for its actual area as follows:

$$SPAs \text{ or actual areas} = FPAs^* CAFs$$

Extensive details of this methodology are explained in (Thenkabailc et al., 2007). SPAs or actual area calculations gain greater significance as pixel sizes become coarser. In the present study the MOD13Q1 pixel cover is 250 m per side, and its area is 6.25 ha. So for a pixel for an area only 50% cropped, a FPA-based area calculation per pixel will be 6.25 ha, whereas the SPA or actual area will be 3.125 ha (6.25 ha * 0.5). Thus unless we calculate areas based on SPA, there will be huge discrepancies in actual areas.

9.3.7 Accuracy Assessments

Accuracy assessment was based on a total of 346 independent ground samplings, as described in Section 9.3.2. These data points were not used in class identification and labeling. The accuracy of the classification results, based on (Jensen, 2004), were used to generate an error matrix and accuracy measures for final classification. The columns (x-axis) of an error matrix contain the ground survey data points, and the rows (y-axis) represent the results of the classified rice maps (Congalton, 1991b). The error matrix is a multidimensional table in which the cells contain changes from one class to another. The statistical approach of accuracy assessment consists of different multivariate statistical analyses. A frequently used measure is the Kappa measurement system, which is designed to compare results from different regions or different classifications.

9.4 RESULTS AND DISCUSSIONS

9.4.1 Spatial Distribution of Croplands in Indo-Ganges Basin

The objectives of this study and based on the methods described in above sections, distinct cropland classes of Indo-Ganges basin.

- Identify crop type\dominance in different seasons with a focus on major croplands, including source of irrigation.
- Establish the season in which croplands are cultivated and also establish the season in which croplands are fallow.

Using the preceding focus, we identified and labeled cropland classes in the Indo-Ganges Basin (see Fig. 9.4) based on the methods and approaches discussed throughout Section 9.2. Fig. 9.5 shows the spatial distribution of all the cropland areas in the Indo-Ganges Basin with 19 distinct cropland classes and two other land cover\land use (LCLU); their statistics are provided in Table 9.3.

A total of 21 LULC classes (Fig. 9.6 and Table 9.3) were mapped, showing clear spectral separateability on one or more single dates, and/or one or more multiple dates, and/or over a near-continuous time interval (Thenkabail et al., 2005). The total study area in the Ganges and Indus basins was 216 Mha (Table 9.3), of which there was a high degree of irrigation (see classes 1 through 9 in Fig. 9.6 and Table 9.3).

The LULC name or label *was* based on the predominance of a particular land cover. For example, the name for class 01 is "Irrigated-SW/GW-DC-rice-wheat." The land cover (Thenkabailc et al.) of this class was dominated by rice-wheat (rice during the kharif season and wheat during the rabi season), and the total area of the class was 18.8 Mha, with 90% of the area covered by croplands, 4% by other LULC, 3% by grasses, and 1% by shrubs. The rice area was predominant in Punjab and Haryana during kharif, and wheat was a winter crop. Class 10 was labeled "Rainfed-DC-rice-fallows-jute/rice/mixed crops" since this was an intensely cropped area class that is heavily dependent on rain. At the time of this ground data, 94% of class 10 had cropland that grew rice during the kharif and rabi seasons, with the rest as follows: 3% comprising trees, 1% comprising shrubs, 1% comprising grasses, and 1% comprising water bodies. Since we had several different types of ground data (e.g., cover percentages, digital photos, observations marked on maps and images), we were able to label classes with as close a match with reality as possible, leading to a final set of 21 LULC classes (Fig. 9.6 and Table 9.3).

The FPA of croplands in the Indo-Ganges Basin was 210,482,000 ha, with 175,670,700 (or 90%) being SPAs during kharif.

9.4.2 Spatial Distribution of Irrigated and Rainfed Croplands in the Indo-Ganges Basin

A total area of 114 Mha (FPA) was estimated in the Indus Basin, including all three major classes (i.e., irrigated croplands, rainfed croplands and other LULCs); and a total area of 102 Mha (FPA) was estimated in the Ganges Basin, including trees, shrubs, grass, water bodies, and other LULC classes (Table 9.4). After excluding the area covered by trees, shrubs, grass,

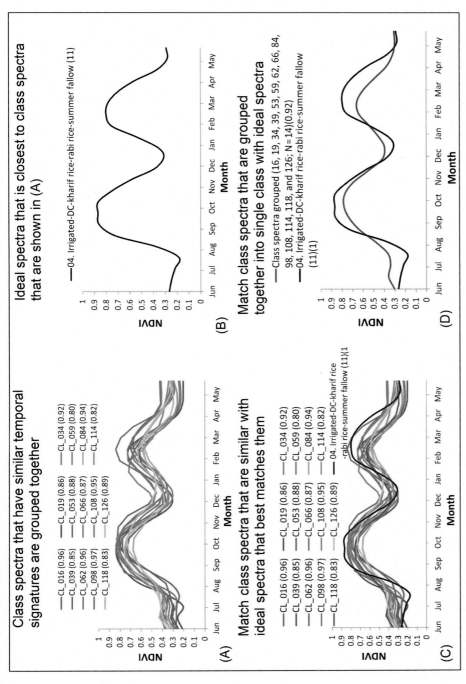

FIG. 9.5 Spectral matching techniques (SMTs) to match class spectra with ideal spectra. The process involves two steps: 1. *Grouping similar class spectra (A):* For example, of the 160 unsupervised classes, we group all classes that have similar time-series spectral signatures (C and D). (D) For example, classes 16, 34, 53, 62, 66, 84, 98, 108, and 126 group into a single class [they are highly correlated with R-square values of 0.87 or above]. 2. *Finding an ideal spectra that matches closest to class spectra (B):* From the ideal spectral library (Fig. 9.3), we selected an ideal spectra (B) that matches closest to class spectra that are grouped together (A). 3. *Matching group of similar classes with ideal spectra:* For example, the legend "CL_062(0.96)" in (D) means that class 62 has a spectral correlation similarity R-square value (SCS R-square value) of 0.96 with ideal spectra class 4 (04. irrigated, double crop, rice in kharif season, rice in rabi season, and fallow in summer season). So classes 16, 34, 53, 62, 66, 84, 98, 108, and 126 are grouped and given initial label same as ideal spectra class 4. The name is further verified using numerous other ancillary data (e.g., ground data, other published data, high-resolution imagery). 4. *Combining all similar class spectral classes to a single class (D):* Since all the 12 class spectral classes are very highly correlated to one another and in turn they are highly correlated with ideal spectra, the 14 class spectral classes are combined into a single class.

TABLE 9.3 LULC Areas in Indo-Ganges River Basin With Irrigation Source

Land Use/Land Cover	Indus (FPA) ('000ha)	Ganges (FPA) ('000ha)	Sample	TREE (%)	SHRUBS (%)	GRASS (%)	Water Bodies (%)	Other LULC (%)	Croplands (%)	Indus (King et al., 2003) ('000ha)	Ganges (King et al., 2003) ('000ha)
01. Irrigated-SW/GW-DC-rice-wheat	7983	12,920	31	2	1	3	0	4	90	7208	11,666
02. Irrigated-SW/GW-DC-rice-rice	3853	5901	47	3	1	1	1	0	94	3604	5520
03. Irrigated-SW-DC-sugarcane/rice-rice/plantations (20%)	173	917	36	2	1	0	0	1	97	167	885
04. Irrigated-SW-DC-beans/cotton-wheat	1187	8872	32	1	1	6	0	1	90	1072	8010
05. Irrigated-GW-DC-millet/sorghum/potato-wheat/mustard	6315	14,644	55	1	1	3	0	1	94	5929	13,748
06. Irrigated-DC-fallows/pulses-rice-fallow	158	259	38	3	1	2	0	1	92	146	240
07. Irrigated-GW-DC-rice-maize/chickpea	155	2640	3	4	1	2	0	2	92	142	2420
08. Irrigated-TC-rice-mixed crops-mixed crops	112	2260	39	4	1	1	0	0	94	105	2122
09. Irrigated-SW-DC-cotton/chili/maize-fallow/pulses	7	84	29	1	1	1	0	4	93	7	78
10. Rainfed-DC-rice-fallows-jute/rice/mixed crops	125	2320	30	3	1	1	1	0	94	117	2171
11. Rainfed-SC-rice-fallow/pulses	7	59	7	1	1	1	1	2	94	7	55

TABLE 9.3 LULC Areas in Indo-Ganges River Basin With Irrigation Source—cont'd

Land Use/Land Cover	Indus (FPA) ('000ha)	Ganges (FPA) ('000ha)	Sample	TREE (%)	SHRUBS (%)	GRASS (%)	Water Bodies (%)	Other LULC (%)	Croplands (%)	Indus (King et al., 2003) ('000ha)	Ganges (King et al., 2003) ('000ha)
12. Rainfed-DC-millets-chickpea/fallows	745	999	5	0	3	0	7	0	90	668	896
13. Rainfed-SC-cotton/pigeonpea/mixed crops	421	3024	178	1	2	4	0	5	89	374	2682
14. Rainfed-SC-groundnut/millets/sorghum	356	204	68	1	1	2	0	3	93	330	189
15. Rainfed-SC-pigeonpea/mixed crops	93	1432	35	2	1	2	0	1	93	86	1337
16. Rainfed-SC-millet-fallows/mixed crops	3658	3483	11	0	2	9	0	1	88	3222	3068
17. Rainfed-SC-fallow-chickpea-	18	28	21	1	1	1	0	9	89	16	25
18. Rainfed-SC-millets/fallows-LS	5140	7	6	1	1	12	0	5	81	4164	5
19. Rainfed-SC-mixed crops/plantations	1155	1624	36	8	3	2	0	0	86	997	1402
20. Shrub lands/trees/rainfed-mixedcrops30%	7	187	9	2	41	8	1	0	48	3	90
21. Other LULC	82,506	40,240	129	27	20	19	5	16	13	10,635	5187

*The table shows full-pixel area (FPA), crop area fraction (CAF), and subpixel area (King et al., 2003) or actual area. SPA = FPA * CAF.*

III. WATER AND FOOD SECURITY OF INDUS RIVER BASIN

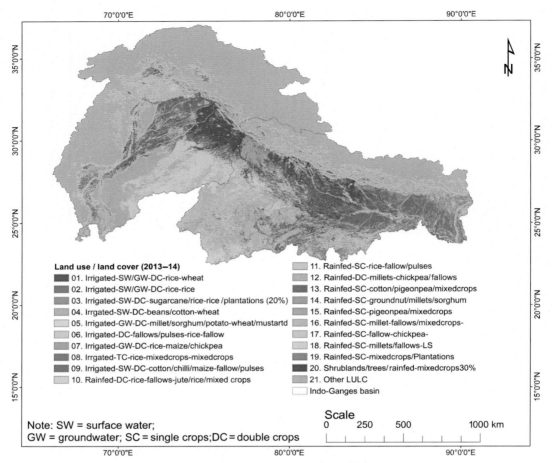

FIG. 9.6 Spatial distribution of croplands and their characteristics in the Indo-Ganges River Basin. The 21 classes show cropland classes that have single or double or triple cropping and where rice or other crops dominate. The classes also show seasonality of cropping and when croplands are left fallow.

water bodies, and other LULC classes, in all three major classes, a total area of 28.4 Mha was estimated (i.e., all the area covered by croplands in the Indus Basin and 56.6 Mha in the Ganges Basin).

A total area of 19.9 Mha (FPA) in the Indus Basin and a total area of 48.5 Mha (FPA) in the Ganges Basin was estimated with 310 samples, in which trees comprised 2%, shrubs comprised 1%, grass comprised 2%, water bodies comprised 0%, other LULC classes comprised 2%, and cropland comprised 93%. Excluding tree, shrubs, grass, water bodies, and other LULC classes, the exact area of irrigated croplands was 18.4 Mha in the Indus Basin and 44.7 Mha in the Ganges Basin (Table 9.4).

A total area of 11.7 Mha (FPA) in the Indus Basin and a total area of 13.4 Mha (FPA) in the Ganges Basin was estimated with 406 samples, in which trees comprised 2%, shrubs

TABLE 9.4 Irrigated Areas vs. Rainfed Cropland Area in Indo-Ganges River Basin With Irrigation Source

	Indus (FPA) ('000 ha)	Ganges (FPA) ('000 ha)	Sample	TREE (%)	SHRUBS (%)	GRASS (%)	Water Bodies (%)	Other LULC (%)	Croplands (%)	Indus (King et al., 2003) ('000 ha)	Ganges (King et al., 2003) ('000 ha)
01. Irrigated-croplands	19,943	48,498	310	2	1	2	0	2	93	18,379	44,689
02. Rainfed-croplands	11,725	13,367	406	2	6	3	1	2	86	9984	11,921
03. Other LULC	82,506	40,240	129	27	20	19	5	16	13	10,635	5187

*The table shows full-pixel area (FPA), crop area fraction (CAF), and subpixel area (King et al., 2003) or actual area. SPA = FPA * CAF.*

III. WATER AND FOOD SECURITY OF INDUS RIVER BASIN

comprised 6%, grass comprised 3%, water bodies comprised 1%, other LULC classes comprised 2%, and cropland comprised 86%. Excluding trees, shrubs, grass, water bodies, and other LULC classes, the exact area of rainfed croplands in the Indus Basin was 9.98 Mha and 11.9 Mha in the Ganges Basin.

A total area of 82.5 Mha (FPA) in the Indus Basin and a total area of 40.2 Mha (FPA) in the Ganges Basin was estimated with 129 samples, in which trees comprised 27%, shrubs comprised 20%, grass comprised 19%, water bodies comprised 5%, other LULC classes comprised 16%, and cropland comprised 13%. Excluding tree, shrubs, grass, water bodies, and other LULC classes, the exact area of croplands occurring in other LULC classes was 10.6 Mha in the Indus Basin and 5.2 Mha in the Ganges Basin.

An accuracy assessment was done for the final LULC map in the Indo-Ganges River Basin, and three classes were classified using ground points. A total of 209 samples, comprising 101 and 36 columns, were classed as irrigated croplands, rainfed croplands, and noncroplands (other LULCs). Based on 346 reference totals and 346 classified totals, a total of 317 correct pixels occurred in classification, with an overall classification accuracy of 91.62%; thus an overall *kappa Statistics assessment* was done with 0.84 accuracy.

A total of 218 pixels were classified for the irrigated croplands class, in which 200 pixels were classified as irrigated croplands, 13 pixels were classified as rainfed croplands, and 5 pixels were classified as noncroplands (other LULCs), with an overall reference total of 209 pixels. A total of 200 pixels were correctly classified for irrigated croplands, with a producer's accuracy of 95% and a user's accuracy of 92% and an overall kappa coefficient of 79%.

A total of 93 pixels were classified for the rainfed croplands class, in which 7 pixels were classified as irrigated croplands, 86 pixels as rainfed croplands, and zero pixels as noncroplands (other LULCs), with an overall reference total of 101 pixels. A total of 86 pixels were correctly classified for rain-fed croplands, with 85% producer's accuracy and 92% user's accuracy and an overall kappa coefficient of 89%.

A total of 36 pixels were classified for the noncroplands (other LULCs) class, in which 2 pixels were classified as irrigated croplands, 2 pixels as rainfed croplands, and 31 pixels as noncroplands (other LULCs), with an overall reference total of 36 pixels. A total of 31 pixels were correctly classified for noncroplands (other LULCs), with 86% producer's accuracy and 89% user's accuracy and an overall kappa coefficient of 87%.

9.4.3 Class Signatures and Onset-Peak-Senescence-Duration of Crops

The class signatures of NDVI (CS-NDVI) are unique spectral properties of a class that can be mapped using NDVI time-series data of a class (see Fig. 9.5A). It is not possible to have spectral signatures when single-date images or images with a few dates are used, as is often the case with LULC studies. Since near-continuous MODIS data were used in this study, a unique set of LULC class signatures were possible (see Fig. 9.3) for classes mapped in Fig. 9.6.

The threshold NDVIs and NDVI signatures over time help us determine (see Table 9.3 and Figs. 9.6 and 9.7) (1) onset of a cropping seasons (e.g., kharif and rabi), (2) duration of the cropping seasons (e.g., for kharif and rabi), (3) magnitude of the crops during different seasons and years (e.g., drought versus normal years), and (4) end of cropping season (senescence). In order to illustrate these possibilities, the MODIS CS-NDVI signatures are presented and

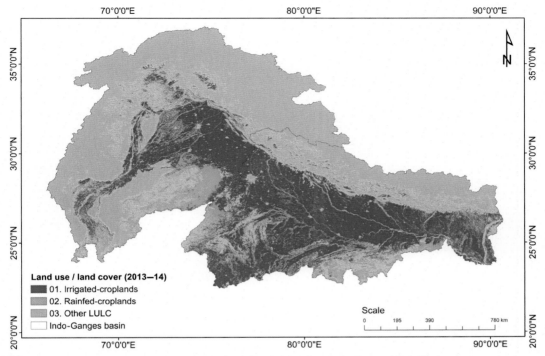

FIG. 9.7   Spatial distribution of irrigated and rainfed croplands in Indo-Ganges River Basin.

discussed for a set of distinct classes (Fig. 9.3) and thematically similar classes (see Fig. 9.5A). The NDVI of irrigated classes never falls below 0.7 on the peak growth stage during the kharif season (see Fig. 5A).

9.4.4 Accuracies and Errors

Accuracy assessment was performed on two independent classification products as shown in Figs. 9.6 and 9.7. Accuracies of the classes were established based on 346 ground sample data (see Tables 9.5 and 9.6), which provided an overall accuracy of 71% with a kappa coefficient of 0.73. The user's and producer's accuracies of most classes were above 80%. Even when they were somewhat lower, the class mix was mainly among cropland classes.

9.4.5 Comparison With District Wise Cropland Statistics

Fig. 9.6 illustrates the spatial extent of rice-wheat growing areas (see also Table 9.6) derived from MOD13Q1 time-series data with spectral matching techniques. To assess how well the spatial extent of rice-wheat (rice during kharif, wheat during rabi) were estimated, we correlated district wise statistics of rice areas obtained from national statistics, resulting in a $R2$ value of 0.84.

TABLE 9.5 Accuracy Assessment Using Error Matrix (Land Use/Land Cover Classification)

	CL_01	CL_02	CL_03	CL_04	CL_05	CL_06	CL_07	CL_08	CL_09	CL_10	CL_11	CL_12	CL_13	CL_14	CL_15	CL_16	CL_17	CL_18	CL_19	CL_20	CL_21	Row Total	Number Correct	Producer's Accuracy (%)	User's Accuracy (%)	Kappa
CL_01	20	5	0	2	2	0	0	0	0	0	2	0	0	0	0	1	0	0	0	0	3	35	20	74%	57%	54%
CL_02	0	17	1	0	0	0	2	3	0	1	0	0	0	0	0	0	0	0	0	0	1	25	17	59%	68%	65%
CL_03	1	0	8	0	0	0	0	0	0	0	0	0	0	0	0	0	0	0	0	0	0	9	8	80%	89%	89%
CL_04	0	0	0	19	2	0	0	4	0	0	0	4	5	0	0	1	0	0	0	0	0	35	19	79%	54%	51%
CL_05	3	0	0	2	45	0	0	0	0	1	5	2	0	0	0	0	0	0	0	0	0	58	45	83%	78%	73%
CL_06	3	0	0	0	0	13	0	0	0	0	0	0	0	0	0	0	0	0	0	0	1	17	13	76%	76%	75%
CL_07	0	2	1	0	0	0	5	0	0	0	0	0	0	0	0	0	0	0	0	0	0	8	5	71%	63%	62%
CL_08	0	2	0	0	1	4	0	14	0	0	0	0	0	0	0	1	0	0	0	0	0	22	14	54%	64%	61%
CL_09	0	0	0	0	0	0	0	0	6	0	0	0	0	0	0	0	0	2	0	0	0	8	6	100%	75%	75%
CL_10	0	2	0	0	1	0	0	2	0	8	0	0	0	0	0	0	0	0	0	0	0	13	8	80%	62%	60%
CL_11	0	0	0	0	0	0	0	0	0	0	7	0	0	0	0	0	0	0	0	0	0	7	7	50%	100%	100%
CL_12	0	0	0	0	0	0	0	0	0	0	0	2	0	0	0	0	0	0	0	0	0	2	2	25%	100%	100%
CL_13	0	0	0	0	0	0	0	0	0	0	0	0	5	0	0	0	0	0	0	0	0	5	5	50%	100%	100%
CL_14	0	0	0	0	0	0	0	0	0	0	0	0	0	8	2	0	0	0	0	0	0	10	8	89%	80%	79%
CL_15	0	0	0	0	0	0	0	0	0	0	0	0	0	1	7	1	0	0	0	0	0	9	7	78%	78%	77%
CL_16	0	0	0	0	0	0	0	0	0	0	0	0	0	0	0	9	0	0	1	0	0	10	9	69%	90%	90%
CL_17	0	0	0	0	0	0	0	0	0	0	0	0	0	0	0	0	4	0	0	0	0	4	4	67%	100%	100%
CL_18	0	0	0	0	0	0	0	0	0	0	0	0	0	0	0	0	2	17	0	0	0	19	17	89%	89%	89%
CL_19	0	0	0	0	0	0	0	0	0	0	0	0	0	0	0	0	0	0	13	0	0	13	13	76%	100%	100%
CL_20	0	0	0	0	0	0	0	0	0	0	0	0	0	0	0	0	0	0	2	2	0	4	2	100%	50%	50%
CL_21	0	1	0	1	3	0	0	3	0	0	0	0	0	0	0	0	0	0	1	0	24	33	24	83%	73%	70%
Column Total	27	29	10	24	54	17	7	26	6	10	14	8	10	9	9	13	6	19	17	2	29	346	253			

Overall Kappa Statistics = 0.7097

Overall Classification Accuracy = 73.12%

TABLE 9.6 LULC Areas in Indo-Ganges River Basin With Irrigation Source

Land Use/ Land Cover	01. Irrigated-Croplands	02. Rainfed-Croplands	03. Noncroplands (Other LULC)	Reference Totals	Classified Totals	Number Correct	Producers Accuracy	Users Accuracy	Kappa
01. Irrigated-croplands	200	13	5	209	218	200	96%	92%	79%
02. Rainfed-croplands	7	86	0	101	93	86	85%	92%	89%
03. Noncroplands (Other LULC)	2	2	31	36	35	31	86%	89%	87%
Column Total	209	101	36	346	346	317			
Overall Classification Accuracy = 91.62%					Overall Kappa Statistics = 0.8420				

*The table shows full-pixel area (FPA), crop area fraction (CAF), and subpixel area (King et al., 2003) or actual area. SPA = FPA * CAF.*

9.4.6 Discussion of Methods

Mapping cropland areas is useful for understanding and determining options for producing more food, which is critical for ensuring global food security. Greater food production for a growing population requires more land. Since cropland expansion is not feasible and has costly environmental and ecological impacts (Tilman, 1999; Tilman et al., 2002; Thenkabail et al., 2012; Kuemmerle et al., 2013), cropland intensification by cultivating existing fallow croplands is a possible option. This study investigated the Indo-Ganges Basin extensively using MODIS 250 m NDVI time-series data to arrive at cropland classes (see Fig. 9.6) from which two cropland classes (irrigated croplands and rainfed croplands) (see Fig. 9.7) were identified, with a total cropland area of about 84 Mha (see Table 9.2 and Fig. 9.6) as fallow cropland during the rabi season.

The present research used MOD13Q1.5 temporal data to identify major croplands and irrigated areas across the Indo-Ganges Basin. MODIS captures imagery on a daily basis. The 16-day composites from the daily acquisitions combine to make a time-series data set over a crop year or a calendar year. This type of data set provides temporal profiles of crop-growing locations to identify the start of a season, the peak growth stage, and the harvest date during each season. The value of NDVI as a function of time also helps in identifying the type of crop in an eco-region based on certain peak thresholds for that crop. This study applied a spectral matching technique that is found to be ideal in mapping irrigated areas (Thenkabail et al., 2007a) and mapping rice areas (Gumma et al., 2011a). Mapping the spatial distribution of rice fallows using a MODIS 250 m 16-day time-series and ground survey information with spectral matching techniques represents a significant, new advancement in the use of this technology. The advantage of using a spectral matching technique in this study is that we were able to selectively use the ideal spectral profiles of rice during the rainy season. The rainfed rice spectral class varies from 0.25 to 0.70 for purely rainfed rice and 0.25 to 0.85 for irrigated rice during the rainy season. The qualitative (shape) difference between ideal spectra and class spectra is narrow and represents the fallow lands accurately.

Some discrepancies were also found during the comparison between national statistics and MODIS-derived cropland areas. The mismatch occurred in the eastern part of India, where there was misclassification with irrigated areas due to similar growing conditions during the cropping season. Most of these areas were corrected using rainfall data and spatial modeling techniques.

Cropped area fractions (the proportion of an irrigated/rice area in a pixel) were assigned based on land use proportions in each class to better calibrate the MODIS pixel area to the real irrigated/rice area. Also, this method relies on ground survey information that is a truly representative sample of the fragmented rice systems. Higher resolution imagery could be used to provide a more accurate estimate of pure classes, but wall-to-wall coverage, repeat coverage during a crop's growing period, costs, and massive processing are all major issues that are hard to surmount for areas as large as the Indo-Ganges Basin. Results clearly show that present methods and MODIS time-series data have many advantages, such as capturing a large-scale cropping pattern. But to minimize errors, additional research will be needed using multi-sensor images with advanced fusion techniques (Gumma et al., 2011c).

The areas of the Indo-Ganges Basin where rice predominates are usually located in zones that receive high rainfall during the monsoon season (Satyanarayana et al., 1997; Ali et al., 2014).

Following the rice season, enough soil moisture may remain to facilitate the growth of short-duration crops. While an important reason for fallowing these lands is the scarcity of water during the rabi season, short-season crops and especially legumes, which can biologically fix N, have a higher chance of success. Accurate up-to-date spatial distribution of rice fallows and statistics are important to guide breeders, agronomists, and policymakers toward promoting short-duration crops in this region. Research is required to provide information on systems' aspects, such as the amount of water available for soil, the incidence of unseasonal rainfall, and other agronomic factors.

Cropping systems were mapped in the Indo-Ganges Basin for the years 2013–14, accounting for a total area of 85 Mha. The source of water and the crop intensity were also considered in the classification of the land cover. The net irrigated cropland area derived from MODIS was 63 Mha (18.3 Mha in the Indus River Basin and 44.7 Mha in the Ganges River Basin), and the rainfed cropland area was 21.9 Mha (10 Mha in the Indus River Basin and 11.9 Mha in the Ganges River Basin).

9.5 CONCLUSIONS

This study espoused a vegetation phenological approach using MODIS 250 m time-series data to map agriculture croplands in the Indo-Ganges River Basin. Annual average NDVI and timing of onset of greenness allowed separation of groundwater from surface water. The specific land use categories separated were as follows. The time-series NDVI phenological signatures were distinctly different in the Indo-Ganges Basin in terms of (1) spatial distribution of major croplands, (2) use of surface water for irrigation of continuous crops, and (3) use of groundwater to irrigate crops. Overall, 19 cropland classes, rice fallows 2.3 Mha (see Fig. 9.6 and Table 9.2) of subpixel areas (SPAs) or actual areas during kharif. The overall accuracy of cropland mapping was 73% with a kappa coefficient of 0.71%. The irrigated versus rainfed croplands (with rice as kharif crop) showed producer's accuracies between 96% and 85% and a user's accuracy of 92%.

References

Aitken, B., 1992. Seven Sacred Rivers. Penguin Books India, New Delhi.

Ali, M., Ghosh, P., Hazra, K., 2014. Resource conservation technologies in rice fallow. In: Ghosh, P.K., Kumar, N., Venkatesh, M.S., Hazra, K.K., Nadarajan, N. (Eds.), Resource Conservation Technology in Pulses. Scientific Publishers, ISBN: 978-81-7233-885-5, pp. 83–88. Chapter 7.

Badhwar, G.D., 1984. Automatic corn-soybean classification using landsat MSS data. I. Near-harvest crop proportion estimation. Remote Sens. Environ. 14, 15–29.

Bhutta, M.N., Van der Velde, E.J., 1992. Equity of water distribution along secondary canals in Punjab, Pakistan. Irrig. Drain. Syst. 6 (2), 161–177.

Biggs, T.W., Thenkabail, P.S., Gumma, M.K., Scott, C.A., Parthasaradhi, G.R., Turral, H.N., 2006. Irrigated area mapping in heterogeneous landscapes with MODIS time series, ground truth and census data, Krishna Basin, India. Int. J. Remote Sens. 27, 4245–4266.

Biradar, C.M., Thenkabail, P.S., Noojipady, P., Li, Y., Dheeravath, V., Turral, H., Velpuri, M., Gumma, M.K., Gangalakunta, O.R.P., Cai, X.L., Xiao, X., Schull, M.A., Alankara, R.D., Gunasinghe, S., Mohideen, S., 2009. A global map of rainfed cropland areas (GMRCA) at the end of last millennium using remote sensing. Int. J. Appl. Earth Obs. Geoinf. 11, 114–129.

Casanova, D., Epema, G., Goudriaan, J., 1998. Monitoring rice reflectance at field level for estimating biomass and LAI. Field Crop Res. 55, 83–92.

Chapin III, F.S., Zavaleta, E.S., Eviner, V.T., Naylor, R.L., Vitousek, P.M., Reynolds, H.L., Hooper, D.U., Lavorel, S., Sala, O.E., Hobbie, S.E., Mack, M.C., 2000. Consequences of changing biodiversity. Nature 405 (6783), 234.

Cihlar, J., 2000. Land cover mapping of large areas from satellites: status and research priorities. Int. J. Remote Sens. 21, 1093–1114.

Congalton, R.G., 1991a. Remote sensing and geographic information system data integration: error sources and. Photogramm. Eng. Remote Sens. 57, 677–687.

Congalton, R.G., 1991b. A review of assessing the accuracy of classifications of remotely sensed data. Remote Sens. Environ. 37, 35–46.

Congalton, R., Green, K., 1999. Assessing the Accuracy of Remotely Sensed Data: Principles and Practices. Lewis, New York.

Congalton, R.G., Green, K., 2008. Assessing the Accuracy of Remotely Sensed Data: Principles and Practices. CRC Press.

De Fries, R.S., Hansen, M., Townshend, J.R., Sohlberg, R., 1998. Global land cover classifications at 8 km spatial resolution: the use of training data derived from Landsat imagery in decision tree classifiers. Int. J. Remote Sens. 19 (16), 3141–3168.

Dheeravath, V., Thenkabail, P.S., Chandrakantha, G., Noojipady, P., Reddy, G.P.O., Biradar, C.M., Gumma, M.K., Velpuri, M., 2010. Irrigated areas of India derived using MODIS 500 m time series for the years 2001–2003. ISPRS J. Photogramm. Remote Sens. 65, 42–59.

ERDAS, 2003. ERDAS Field Guide. vol. 1. October 2003.

FAO, 2003. Water Resources Research Institute/FAO. In: 29 (Ed.).

Farr, T.G., Kobrick, M., 2000. Shuttle radar topography mission produces a wealth of data. EOS Trans. 81, 583–585.

Gaur, A., Biggs, T.W., Gumma, M.K., Parthasaradhi, G., Turral, H., 2008. Water scarcity effects on equitable water distribution and land use in major irrigation project – a case study in India. J. Irrig. Drain. Eng. 134 (1), 26–35.

Gray, J., Friedl, M., Frolking, S., Ramankutty, N., Nelson, A., Gumma, M., 2014. Mapping Asian cropping intensity with MODIS. IEEE J. Sel. Top. Appl. Earth Observ. Remote Sens. 7, 3373–3379.

Gumma, M.K., 2008. Methods and approaches for irrigated area mapping at various spatial resolutions using AVHRR, MODIS and LANDSAT ETM+ data for the Krishna river basin, India. PhD Thesis, JNTU, Hyderabad, Telangana.http://publications.iwmi.org/pdf/H042567.pdf. (Accessed 3 July 2017).

Gumma, M.K., Andrew, N., Thenkabail, P.S., Amrendra, N.S., 2011a. Mapping rice areas of South Asia using MODIS multitemporal data. J. Appl. Remote. Sens. 5, 053547.

Gumma, M.K., Gauchan, D., Nelson, A., Pandey, S., Rala, A., 2011b. Temporal changes in rice-growing area and their impact on livelihood over a decade: a case study of Nepal. Agric. Ecosyst. Environ. 142, 382–392.

Gumma, M.K., Thenkabail, P.S., Hideto, F., Nelson, A., Dheeravath, V., Busia, D., Rala, A., 2011c. Mapping irrigated areas of Ghana using fusion of 30 m and 250 m resolution remote-sensing data. Remote Sens. 3, 816–835.

Gumma, M.K., Thenkabail, P.S., Muralikrishna, I.V., Velpuri, M.N., Gangadhararao, P.T., Dheeravath, V., Biradar, C.M., Acharya Nalan, S., Gaur, A., 2011d. Changes in agricultural cropland areas between a water-surplus year and a water-deficit year impacting food security, determined using MODIS 250 m time-series data and spectral matching techniques, in the Krishna River basin (India). Int. J. Remote Sens. 32, 3495–3520.

Gumma, M.K., Thenkabail, P.S., Nelson, A., 2011e. Mapping irrigated areas using MODIS 250 meter time-series data: a study on Krishna River basin (India). Water 3, 113–131.

Gumma, M.K., Thenkabail, P.S., Maunahan, A., Islam, S., Nelson, A., 2014. Mapping seasonal rice cropland extent and area in the high cropping intensity environment of Bangladesh using MODIS 500m data for the year 2010. ISPRS J. Photogramm. Remote Sens. 91, 98–113.

Gumma, M.K., Kajisa, K., Mohammed, I.A., Whitbread, A.M., Nelson, A., Rala, A., Palanisami, K., 2015a. Temporal change in land use by irrigation source in Tamil Nadu and management implications. Environ. Monit. Assess. 187, 1–17.

Gumma, M.K., Mohanty, S., Nelson, A., Arnel, R., Mohammed, I.A., Das, S.R., 2015b. Remote sensing based change analysis of rice environments in Odisha, India. J. Environ. Manag. 148, 31–41.

Gumma, M.K., Thenkabail, P.S., Teluguntla, P., Rao, M.N., Mohammed, I.A., Whitbread, A.M., 2016. Mapping rice-fallow cropland areas for short-season grain legumes intensification in South Asia using MODIS 250 m time-series data. Int. J. Digit. Earth 9, 981–1003.

Hao, Z., Hao, F., Singh, V.P., 2016. A general framework for multivariate multi-index drought prediction based on multivariate ensemble streamflow prediction (MESP). J. Hydrol. 539, 1–10.

Holben, B.N., 1986. Characteristics of maximum-value composite images from temporal AVHRR data. Int. J. Remote Sens. 7, 417–1434.

Homayouni, S., Roux, M., 2003. Material mapping from hyperspectral images using spectral matching in urban area. In: Landgrebe, P. (Ed.), IEEE Workshop in honour of Prof. Landgrebe, Washington DC, USA, October 2003.

Huete, A., Didan, K., Miura, T., Rodriguez, E.P., Gao, X., Ferreira, L.G., 2002. Overview of the radiometric and biophysical performance of the MODIS vegetation indices. Remote Sens. Environ. 83, 195–213.

Ilich, N., 1996. Ganges River Basin Water Allocation Modelling Study. Consultant Report to the World Bank, Calgary.

Jensen, J.R., 1996. Introductory Digital Image Processing: A Remote Sensing Perspective. Prentice Hall, Upper Saddle, NJ.

Jensen, J.R., 2004. Introductory Digital Image Processing: A Remote Sensing Perspectivee, third ed. Prentice Hall, Upper Saddle River, NJ. 544 p.

Khan, A.R., 1999. An analysis of surface water resources and water delivery systems in the Indus Basin. International Water Management Institute (IWMI), Lahore, Pakistan.

King, M.D., Closs, J., Spangler, S., 2003. EOS Data Products Handbook Version 1. NASA Goddard Space Flight Center, Greenbelt, MD.

Kuemmerle, T., Erb, K., Meyfroidt, P., Müller, D., Verburg, P.H., Estel, S., Haberl, H., Hostert, P., Jepsen, M.R., Kastner, T., 2013. Challenges and opportunities in mapping land use intensity globally. Curr. Opin. Environ. Sustain. 5, 484–493.

Li, Z., Hao, Z., Shi, X., Déry, S.J., Li, J., Chen, S., Li, Y., 2016. An agricultural drought index to incorporate the irrigation process and reservoir operations: a case study in the Tarim River basin. Glob. Planet. Chang. 143, 10–20.

Loveland, T.R., Reed, B.C., Brown, J.F., Ohlen, D.O., Zhu, Z., Yang, L.W., Merchant, J.W., 2000. Development of a global land cover characteristics database and IGBP DISCover from 1 km AVHRR data. Int. J. Remote Sens. 21 (6–7), 1303–1330.

Meadows, P., 1999. The Indus River: Biodiversity, Resources, Humankind. Oxford University Press for the Linnean Society of London, ISBN: 0195779053, p. 441.

Olsson, P.-O., Lindström, J., Eklundh, L., 2016. Near real-time monitoring of insect induced defoliation in subalpine birch forests with MODIS derived NDVI. Remote Sens. Environ. 181, 42–53.

Penner, J.E., Charlson, R.J., Schwartz, S.E., Hales, J.M., Laulainen, N.S., Travis, L., Leifer, R., Novakov, T., Ogren, J., Radke, L.F., 1994. Quantifying and minimizing uncertainty of climate forcing by anthropogenic aerosols. Bull. Am. Meteorol. Soc. 75 (3), 375–400.

Pittman, K., Hansen, M.C., Becker-Reshef, I., Potapov, P.V., Justice, C.O., 2010. Estimating global cropland extent with multi-year MODIS data. Remote Sens. 2, 1844–1863.

Salmon, J.M., Friedl, M.A., Frolking, S., Wisser, D., Douglas, E.M., 2015. Global rain-fed, irrigated, and paddy croplands: a new high resolution map derived from remote sensing, crop inventories and climate data. Int. J. Appl. Earth Obs. Geoinf. 38, 321–334.

Satyanarayana, A., Seenaiah, P., Sudhakara Babu, K., Prasada Rao, M., 1997. Extending pulses area and production in rice fallow. In: Asthana, A.N., Ali, M. (Eds.), Recent Advances in Pulses Research. Indian Society of Pulses Research and Development, Indian Institute of Pulses Research, Kanpur, India, pp. 569–580.

See, L., Fritz, S., You, L., Ramankutty, N., Herrero, M., Justice, C., Becker-Reshef, I., Thornton, P., Erb, K., Gong, P., 2015. Improved global cropland data as an essential ingredient for food security. Glob. Food Secur. 4, 37–45.

Singh, A., 1989. Review article digital change detection techniques using remotely-sensed data. Int. J. Remote Sens. 10, 989–1003.

Skole, D.L., 2004. Geography as a great intellectual melting pot and the preeminent environmental discipline. Ann. Assoc. Am. Geogr. 94 (4), 739–743.

Thenkabail, P.S., Stucky, N., Griscom, B.W., Ashton, M.S., Diels, J., Van Der Meer, B., Enclona, E., 2004. Biomass estimations and carbon stock calculations in the oil palm plantations of African derived savannas using IKONOS data. Int. J. Remote Sens. 25, 5447–5472.

Thenkabail, P.S., Schull, M., Turral, H., 2005. Ganges and Indus river basin land use/land cover (LULC) and irrigated area mapping using continuous streams of MODIS data. Remote Sens. Environ. 95, 317–341.

Thenkabail, P.S., Biradar, C.M., Noojipady, P., Cai, X., Dheeravath, V., Li, Y., Velpuri, M., Gumma, M., Pandey, S., 2007. Sub-pixel area calculation methods for estimating irrigated areas. Sensors 7, 2519–2538.

Thenkabail, P., GangadharaRao, P., Biggs, T., Gumma, M., Turral, H., 2007a. Spectral matching techniques to determine historical land-use/land-cover (LULC) and irrigated areas using time-series 0.1-degree AVHRR pathfinder datasets. Photogramm. Eng. Remote Sens. 73, 1029–1040.

Thenkabail, P.S., GangadharaRao, P., Biggs, T., Gumma, M.K., Turral, H., 2007b. Spectral matching techniques to determine historical land use/land cover (LULC) and irrigated areas using time-series AVHRR pathfinder datasets in the Krishna River Basin, India. Photogramm. Eng. Remote. Sens. 73, 1029–1040.

Thenkabail, P., Biradar, C., Noojipady, P., Dheeravath, V., Li, Y., Velpuri, M., Gumma, M., Reddy, G., Turral, H., Cai, X., Vithanage, J., Schull, M., Dutta, R., 2009a. Global irrigated area map (GIAM) for the end of the last millennium derived from remote sensing. Int. J. Remote Sens. 30 (14), 3679–3733.

Thenkabail, P.S., Biradar, C.M., Noojipady, P., Dheeravath, V., Li, Y., Velpuri, M., Gumma, M., Gangalakunta, O.R.P., Turral, H., Cai, X., Vithanage, J., Schull, M.A., Dutta, R., 2009b. Global irrigated area map (GIAM), derived from remote sensing, for the end of the last millennium. Int. J. Remote Sens. 30, 3679–3733.

Thenkabail, P.S., Knox, J.W., Ozdogan, M., Gumma, M.K., Congalton, R.G., Wu, Z., Milesi, C., Finkral, A., Marshall, M., Mariotto, I., 2012. Assessing future risks to agricultural productivity, water resources and food security: how can remote sensing help? Photogramm. Eng. Remote. Sens. 78, 773–782.

Thiruvengadachari, S., Sakthivadivel, R., 1997. Satellite Remote Sensing for Assessment of Irrigation System Performance: A Case Study in India. Research Report 9 International Irrigation Management Institute, Colombo, Sri Lanka.

Tilman, D., 1999. Global environmental impacts of agricultural expansion: the need for sustainable and efficient practices. Proc. Natl. Acad. Sci. 96, 5995–6000.

Tilman, D., Cassman, K.G., Matson, P.A., Naylor, R., Polasky, S., 2002. Agricultural sustainability and intensive production practices. Nature 418, 671–677.

Tomlinson, R., 2003. Thinking about Geographic Information Systems Planning for Managers. ESRI Press, p. 283.

Wan, Z., Wang, P., Li, X., 2004. Using MODIS land surface temperature and normalized difference vegetation index products for monitoring drought in the southern Great Plains, USA. Int. J. Remote Sens. 25, 61–72.

Wang, J., Ling, Z., Wang, Y., Zeng, H., 2016. Improving spatial representation of soil moisture by integration of microwave observations and the temperature–vegetation–drought index derived from MODIS products. ISPRS J. Photogramm. Remote Sens. 113, 144–154.

WMO, 2015. WMO (World Meteorological Organization) Press Release, 25-01-16. (Accessed 21 Aug 2015).

Yaduvanshi, A., Srivastava, P.K., Pandey, A.C., 2015. Integrating TRMM and MODIS satellite with socio-economic vulnerability for monitoring drought risk over a tropical region of India. Phys. Chem. Earth Parts A/B/C 83–84, 14–27.

10

Increasing Water Productivity in the Agricultural Sector

Asad Sarwar Qureshi

International Center for Biosaline Agriculture (ICBA), Dubai, United Arab Emirates

10.1 INTRODUCTION

The world population is expected to increase to more than 9 billion by 2050 (FAO, 2011). Demand for food and fiber will also increase as incomes and standards of nutrition rise and as people consume more land- and water-intensive diets (i.e., consumption of more meat and dairy products). Current estimates suggest that by 2050 annual cereal production will rise to about three billion tonnes and meat production to more than 200 million tonnes (FAO, 2006). To produce this amount of food, irrigated land must increase by 35%, and 20% more water must be diverted to agriculture. However, due to increasing inter-sectoral competition for water and decreasing investment in irrigation development, expansion of irrigated lands will be no more than 5%, and water diversions for irrigation are projected to decrease by 8% because of climate change and other management issues. Thus in the future irrigation's contribution to food security will have to come from improving existing systems and increasing the productivity of available water resources. Under the "business as usual" scenario (i.e., continuing with the current agricultural practices and water productivity levels), $4500 \, \text{km}^3 \, \text{yr}^{-1}$ more water will be needed to feed the world's population by 2050 (Falkenmark et al., 2009). This is roughly twice the amount of water presently used in irrigation (Kijne et al., 2009).

Alarmingly, most of this population increase is expected in developing countries where access to water for agriculture and domestic purposes is already under stress and where 1.2 billion people lack access to safe drinking water. Physical water scarcity is already affecting food production in the arid parts of the world, for example, in North Africa and the Middle East. Although there are varying opinions on the degree and severity of water scarcity in Asia and Africa, there is broad agreement that increasing water scarcity will turn water into a key limiting factor in food production and the livelihoods of people who are impoverished throughout rural Asia and most of Africa, with particularly severe water scarcity in the breadbaskets of Northwest India and Northern China.

Water distribution across regions is also not equitable. Asia is particularly hard hit, with just 36% of the world's water resources supporting 60% of the world's population (Qureshi and Shoaib, 2015). Africa has a better balance, 11% of the world's available fresh water with 13% of the world's population. The population of the Arab world is 3% of the world's population, but it receives only 1% of the world's renewable water resources. For Arab countries in the Gulf, per capita water availability is very low; that is, except for Egypt, which has $1120\,m^3$, the per capita water availability ranges between $107\,m^3$ in Kuwait and $148\,m^3$ in Saudi Arabia. Due to population growth, it is expected that per capita water availability will reduce to half by 2025. Increasing demand and decreasing quality have put enormous pressure on the agricultural sector to reduce its use of water. Currently, more than 90% of the available water in Asia and 88% in the Arab world is used for agriculture (Qureshi et al., 2016). Despite this high-water allocation for agriculture, more than 50% of their food requirements are imported, and this gap is likely to double in the next two decades.

The water demand is driven by four major factors: population growth; increased living standards; water allocation between agriculture, industry, and domestic sectors; and efficiency of water use and distribution systems. The issue of water demand must be addressed because, if per capita water allocations are not modified, economies will be destabilized and food security will be threatened. Land expansion is no more a viable option to increase agricultural production due to increasing land degradation and urbanization problems (Cai et al., 2011). As opportunities for development of new water resources are limited and costs are rising, increasing the productivity of existing water resources becomes a more attractive alternative. Increasing the *productivity* of existing water resources is central to produce more food, to fight poverty, to reduce competition for water, and to ensure that there is enough water for nature.

It must be realized that it is not just the volume of water delivered but also the way it is delivered that controls the effective use of resources. Increasing water productivity means using less water for crops while maintaining or even increasing total crop productivity (unrelated to the economic value of water) by enhancing the efficient use of water through improved management and advanced irrigation technologies. Several strategies can be considered as a part of these activities, such as redesigning total irrigation systems for higher efficiency, reusing marginal waters, reducing evaporation losses, practicing deficit irrigations, and minimizing water losses to unrecoverable sinks. Replacing high-water-consuming crops with lower-water-consuming crops can also increase the economic productivity of water if the high-value crop has a shorter growing season and the land is cropped only once a year. Increasing on-farm irrigation efficiencies may not save water; however, improved irrigation systems make farm operations more efficient and competitive.

Management of the demand for water requires increased water productivity in agriculture, rational water allocations for different sectors, and reduced water loss at all levels. Improvements in water productivity can generate higher agricultural outputs, increase farm incomes, and reduce poverty. Averting a water crisis is a massive undertaking that will require a combination of conservation, improved efficiency, and increased cooperation among competing interests. This chapter discusses concepts of water productivity at different scales and targets different strategies, with the aim of enabling policies and institutions to successfully adopt different interventions that can help increase water productivity at different scales.

10.2 CONCEPTS OF WATER PRODUCTIVITY

The term water productivity is defined as the amount of beneficial output per unit of water depleted. In a broader sense, it is related to "crop production per unit of water used" (Molden et al., 2010; Kijne et al., 2009). Improving agricultural water productivity is related not only to increased production of rain-fed and irrigated crops but also to maximized production of fish, trees, and livestock. The used water includes "green" water (effective rainfall) for rain-fed areas and both "green" and "blue" water (diverted water from surface and groundwater systems) for irrigated areas.

$$\text{Water Productivity (WP)} = \frac{\text{Crop produced}}{\text{Water consumed}}$$

In defining water productivity, we need to express explicitly which *production* (biomass or yield) and which *water consumption* (transpiration or evapotranspiration) is being considered (Perry et al., 2009). Water productivity can be expressed in physical or economic terms. Physical water productivity is the quantity of the product divided by the quantity of the input and can be expressed in terms of mass (kg), or even in monitory terms ($) to compare different crops (Molden et al., 2007). The physical production may include crop yield, biomass, fish, and livestock production, which are expressed in unit of kilogram; it can also be stated in economic values ($) like market value of grain and/or biomass or nutritional values (kilocalories). Water input can be gross inflow, net inflow, available water, irrigation, and actual evapotranspiration.

The physical production of a crop per unit of consumed water is denoted by different indicators. The water productivity is expressed as the biomass or yield (kg) per cubic meter (m^3) of evapotranspiration (ET) or crop transpiration (T). Evapotranspiration is the sum of evaporation (E) and transpiration (T) of soil water through plant systems and into the atmosphere. Transpiration is the flow of water vapor from stomates of leaves that causes liquid water to move from soil to roots, through stems, and on to leaves. Water vapor exits through the same stomates that carbon dioxide enters. The water vapor lost by transpiration in exchange for carbon dioxide is the primary process for plant growth and development. Other nutrients are delivered to crops from the soil by the water used in transpiration. Evaporation is the direct conversion of water into water vapor when wet leaves or soil are exposed to drier air and radiant heat.

Generally, a linear relationship is observed between crop biomass and transpiration for a given situation; however; the slope of this relationship may vary for diverse conditions (Howell, 1990). Since crop transpiration is usually considered a direct measure of the physiological performance of a crop, water productivity is expressed as crop yield/biomass per unit of transpiration ($WP_T = Y/T$). The inevitable loss of water due to soil evaporation negatively affects the water productivity from WP_T to WP_{ET}, which is expressed in terms of yield/biomass per unit of ET. WP_{ET} represents the total amount of water used in crop production.

WP_T is used for farm-level analysis whereas for the scales above farm level, WP_{ET} is widely used. Relatively lower WP_{ET} values stress the need to reduce soil evaporation through different management measures. When water is a limiting resource, actual yield per unit of applied water (i.e., irrigation and precipitation) ($WP_{AW} = Y_{act}/AW_{I+P}$) also becomes important

(Molden, 2007). Economic productivity ($/m^3) uses valuation techniques to derive the value of water, income obtained from water use and benefits derived from water or increased welfare. The introduction of water productivity measures makes it possible to undertake a holistic and integrated performance assessment by.

- including all types of water uses in a system;
- including a wide variety of outputs;
- integrating measures of technical and allocative efficiency;
- incorporating multiple use and sequential reuse as the water cascades through the basin;
- including multiple sources of water; and
- integrating nonwater factors that affect productivity.

Water productivity can be measured at different scales to meet the needs of different stakeholders in the system, from farmers to policymakers. Farmers are interested in increasing WP at field scale to reduce water charges and generate more income. For irrigation managers, increasing WP at the system scale is important, whereas for policymakers, maximizing outputs from efficient use of all available water is the key issue. As pursuit of "maximum outputs from the system" is often the research focus, balancing the benefits of all stakeholders is always important to ensure sustainable development. This is achieved by defining the inputs of water and outputs in units appropriate to the users' indicator needs.

The output derived from water use can be defined in the following ways:

- Physical output, which can be total biomass or harvestable product
- Economic output (the cash value of output) either gross benefit or net benefit
- Water input, which can be specified as volume (m^3) or as the value of water expressed as the highest opportunity cost in alternative uses of the water

Different indicators used to assess water productivity for a cropping system at different scales are briefly discussed next (Cook et al., 2009):

- *Plant scale*: At this scale crop physiologists need to assess how efficient a crop or cultivar of a crop is in converting water into biomass or crop yield. The output can be quantified either as total biomass or as crop yield (harvestable produce), whereas the relevant water input is the water used in transpiration.
- *Field scale*: This scale is central in assessing how efficiently a cropping system converts water into beneficial output. At this scale the output can be quantified as total biomass or crop yield (kg) and the water inputs as the amount of water that was used in transpiration (m^3).
- *Farm scale*: This scale is of interest to farmers, agronomists, and water specialists in assessing the opportunities of saving water lost through non-beneficial use. At this scale the output can be quantified as total biomass, crop yield (kg), or crop value ($), while the water input is the amount of water depleted from the system through (1) evaporation, (2) flows to sinks that are not recoverable, (3) pollution to levels that render it unfit for use, and (4) incorporation into the product.
- *Irrigation system scale*: At this scale it is important to evaluate how productively the water available to the irrigation system is being used. The irrigation manager considers both the amount of water depleted and the amount recaptured for reuse downstream. At this scale the output can be quantified in physical and economic terms, and the water can be accounted for in either volume or value terms.

- *Basin scale*: This scale is important for assessing options for increasing productivity of the renewable water that enters the basin, mainly as rainfall. The output includes all the benefits derived by, for example, biomass or harvestable produce, landscape, and could even include the value of near-shore marine life. The water input becomes the net inflow, which is difficult to value in monetary terms and therefore generally is assessed as volume. This approach is particularly useful in assessing the opportunities for investing in water infrastructure.

10.3 AGRICULTURAL WATER PRODUCTIVITY ACROSS THE GLOBE

The major objective of improving agricultural water productivity is to increase crop production in irrigated and rain-fed areas and the associated economic, social, environmental, and ecological benefits per unit of water used (Molden et al., 2007). This also includes enhancing benefits from livestock, agroforestry, and aquaculture sectors with minimum water use (Rockström and Barron, 2007; Cai et al., 2011). The concepts of water use efficiency and water productivity are often used in the same context of increasing crop production by using least water resources, although they vary in definitions. Water use efficiency considers water input, whereas water productivity is usually calculated based on the amount of water consumed.

There is significant variation in water productivity across different cropping systems, under both irrigated and rain-fed conditions. The lower water productivities are usually found in sub-Saharan Africa and regions in South Asia due to poor access to irrigation water and items such as seeds, fertilizer, and pesticides. Therefore increasing productivity of water in these regions would result in greater food security and improved livelihood for the millions of rural people living in poverty in these areas (Rockström et al., 2010). It has been estimated that 80% of the required additional food can be obtained by increasing the productivity of low-yield farming systems (Molden, 2007; Cai et al., 2011). The estimated water productivity values of wheat crops for different countries are given in Table 10.1.

The differences in WP values can be linked to yield gaps in different countries. In addition to irrigation water availability, crop yields are also dependent on non-water factors such as rate of fertilizer use, seed quality, disease, and pest management (FAO, 2009; Kingwell et al., 2016). For example, wheat yields are ranging between 1.0 and 7.0 tha^{-1} at different locations in Iran. Hussain et al. (2003) reported a yield of 4.5 and 4.1 tha^{-1} for wheat production in India (Bhakra region) and Pakistan (Punjab region), respectively. The crop yields obtained in sub-Saharan Africa are about half of the yields achieved in parts of South Asia (Kadigi et al., 2012). The yield gaps (difference between the average national yield and the average potential yields) are the highest in developing countries and most notably in sub-Saharan Africa (World Bank, 2007). The potential yield of a crop is the yield that is attained without any water, nutrients, and biotic stress during crop growth (van Ittersum et al., 2013). However, actual obtained yields are usually lower than the potential yields, resulting in yield gaps. In addition to low use of fertilizer and poor extension services, yield gaps can also be caused by volatile weather conditions resulting in frequent crop failure such as in Russia and other Central Asian countries (Schierhorn et al., 2014) (Fig. 10.1).

TABLE 10.1   Estimated Water Productivity Values of Wheat Crop for Selected Countries

Countries	Water Productivity (kg/m³)	Reference
Syria	0.48–1.10	Oweis et al. (2000)
Iran	0.86–2.50	Abbasi and Mehrpour (2004)
India	1.28–1.82	Zhang et al. (2003)
China	1.23–1.49	Hussain et al. (2003)
Morocco	0.32–1.06	Mrabet (2002)
Pakistan	1.08–1.62	Hussain et al. (2003)
Turkey	1.33–1.45	Sezen and Yazar (1996)
Uzbekistan	0.44–1.02	Kamilov et al. (2003)

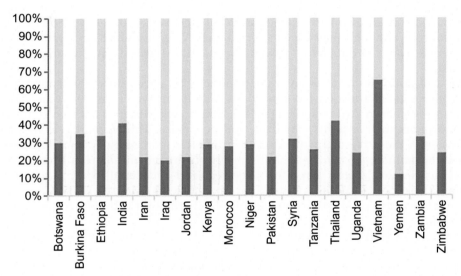

FIG. 10.1   Yield gaps for major grains in different countries (*blue* columns denote actual obtained yields as a percentage of potential yields). (*Modified from Rockström, J., Karlberg, L., Wani, S.P., Barron, J., Hatibu, N., Oweis, T., Bruggeman, A., Farahani, J., Qiang, Z., 2010. Managing water in rainfed agriculture: the need for a paradigm shift. Agric. Water Manag. 97, 543–550. doi:10.1016/j.agwat.2009.09.009; Qureshi, A.S., 2014. Reducing carbon emissions through improved irrigation management: a case study from Pakistan. Irrig. Drain. 63, 132–138.*)

10.4 OPPORTUNITIES FOR INCREASING WATER PRODUCTIVITY IN AGRICULTURE SECTOR

Increasing demand and decreasing water quality and quantity have put enormous pressure on the agriculture sector to increase productivity of the available water resources in order to save water for industrial, environmental, and domestic needs. Improvements in water productivity can be achieved by choosing and adapting water-efficient crops, reducing unproductive water losses, and maintaining ideal agronomic conditions for crop production (Kijne et al., 2003; Rockström and Barron, 2007). The adoption of improved agronomic practices can reduce unproductive water losses thereby increasing water use efficiency and crop production. However, it is important to realize that higher crop water productivities also require management of other soil and plant stresses, such as nutrient deficiencies and management of weeds and crop diseases (Bouman, 2007). Other interventions can also contribute significantly in increasing crop water productivity, such as supplemental and deficit irrigation, precision agriculture, soil fertility management, soil moisture conservation, and the use of drought- and disease-resistant crop varieties (Oweis and Hachum, 2009; de Vries et al., 2010; Mzezewa et al., 2011).

In water-scarce basins it is inevitable that water use in one sector or at one scale will have a direct impact on water quality and quantity in other sectors of the same basin. Effective management of basin water for production requires a better understanding of a complex set of water-related interactions that occur across spatial and temporal scales, and within various locations in the same basin. Effective integrated management of basin water resources is complicated by the fact that the use of water and land at one location affects how water is used at other locations, often in complex ways. Misunderstanding can lead to policies that adversely affect one set of users while trying to improve conditions for other users. There are at least two dimensions to this—one is the consequence of upstream use on downstream availability, and the other is how actions taken at one scale affect users at another scale. For example, the degree to which field-scale interventions improve water productivity at the basin scale or the degree to which policies affecting basin allocation affect farm and community practices are often not clear.

The key levels usually considered for water resources in agriculture are for an individual plant, a field, an irrigation scheme, or a river basin. As we move from one scale to another scale, key processes change. For example, for a rain-fed agricultural field, infiltration, transpiration, and evaporation are important parameters. However, for at basin scale, stream flow and water allocation among multiple users and environmental flows get priority. Therefore there is a strong need to develop common water accounting procedures for analyzing the use, depletion, and productivity of water at the plant, field, irrigation system, and basin scale (Falkenmark et al., 1989). These procedures are necessary to evaluate the impact of alternative water management interventions on water productivity at different scales.

At each scale different actions are needed to enhance water productivity. The range of options to increase water productivity at different scales is discussed next. However, successful implementation of these options requires more in-depth analysis and evaluation of the cost of various interventions for increasing productivity. Implementing these options may be constrained by the lack of incentives of local farmers and irrigation system managers. Devolution of responsibility to local user groups may be effective in increasing irrigation efficiency and water productivity in some situations.

10.4.1 Plant-Scale Interventions

In water-scarce environments about half of yield improvements are attained due to improvements in the crop genotypes and the other half due to better agronomic and water management practices (Morison et al., 2008). Plant breeding and genetic improvement of plants have delivered substantial productivity gains over recent decades. Genetic modifications have helped in increasing susceptibility of crops to pests and diseases, producing more grains, reducing the growing period, and increasing the capacity of plants to tolerate more heat and water stress. Crops that withstand disease and pest attacks increase water productivity by reducing crop failure and unproductive water consumption. Similarly, drought tolerance (that is, capacity of a plant to survive severe water stress and continue to maturity) also increases production from low and unreliable rainfall.

Genetically improved crop varieties with more efficient photosynthetic processes may produce more biomass for the same amount of transpiration (Perry et al., 2009). Significant progress has been made for the genetic enhancement of dryland crops with a view toward increasing crop productivity (Gowda et al., 2009). More drought-tolerant varieties of pearl millet, sorghum, cowpeas, and barley are now available for dry areas of Asia and Africa. The successful plantation of these varieties in sub-Saharan Africa is relatively lower than in Asia due to the unavailability of nonwater inputs such as good nitrogen and disease-free seed and lack of pest management. Moreover, in this region many "orphan crops" (e.g., cassava, millet, yams, and teff) are grown. The productivities of these crops are low due to the lack of physiologically advanced varieties that can withstand water and heat and that are salt-resistant (Mueller et al., 2012).

Water productivity improvements can also be achieved by using crop varieties that are most adaptable to local conditions and by growing the crops in seasons and/or areas that have low atmospheric evaporative demands. The adoption of improved, early maturing, and high-yielding crop varieties during the past 25 years has increased the average yields of many cereal crops and reduced the growth period. The improvements in plant breeding have shorten the growth cycle of many crops, which helps in reducing soil evaporation. The introduction of high-yielding and short-duration rice varieties has increased average yields from 2 to 3 tha^{-1} to 5–6 tha^{-1} and reduced the crop duration time from 140 days to 110 days (Hobbs and Gupta, 2003). The availability of hybrid varieties can produce 15% to 20% more rice than inbred high-yielding varieties with comparable maturity periods. Better nutrient management offers another opportunity to increase rice yields with the same amount of water consumption.

10.4.2 Field-Scale Interventions

Water productivity at the field scale can be increased by (1) reducing evaporation, especially during early growth stages of a crop, (2) reducing seepage and percolation losses during irrigation events, (3) choosing suitable soils for different crops, especially for rice, and (4) minimizing non-beneficial use of water by weeds and trees. At the field scale interventions should aim at increasing water productivity by reducing non-beneficial evaporation and/or increasing productive transpiration (Sadler et al., 2007). Practicing night-time irrigation, better weed control, and mulching of fields in high-temperature areas and seasons are also good options for controlling excessive evaporation. Options for capturing more water from the soil include the use of varieties with deeper rooting systems and practices that control the moisture of soil. Applying irrigation according to a crop's actual water requirements can help reduce unproductive soil evaporation. Since crops at different growth stages react differently to water stress, ensuring full water deliveries at crops' critical growth stages can have a large impact on crop yields (Perry et al., 2009).

Commonly used flooding and basin irrigation methods are less expensive and easy to operate but are considered less efficient because much of the applied water is wasted through evaporation, runoff, and percolation below the root zone. Improvements in water productivity at the field scale can be attained by adopting improved irrigation methods and deficit irrigation concepts. Water conservation measures such as precision land leveling, zero tillage, and bed planting can reduce water use considerably at the field scale. Bed planting uses 40% less water than the basin irrigation method does (Qureshi, 2014). These technologies have the capacity to improve water productivity at the field level if farmers do not at the same time expand their cultivated area or increase their cropping intensity (Ahmad et al., 2007). However, these methods require precise grading of the topography, high instant flow rates, and relatively high levels of automation and management. Zero tillage technology is now widely practiced in many countries, including the United States, Brazil, Argentina, Pakistan, India, and Zimbabwe.

The bed planting in rice-wheat cropping systems used in India, Pakistan, and Uzbekistan have shown significant savings in water (Hobbs and Gupta, 2003; Mollah et al., 2009). Improved irrigation management for rice crops can yield maximum water savings, as nearly 55% of all the rice and wheat produced in the world comes from irrigated areas (Qureshi et al., 2004). Different irrigation regimens have been developed to reduce rice crops' consumption of water. These regimens include a shallow water layer with wetting and drying (SWD), alternate wetting and drying (AWD), and semi-dry cultivation (SDC). In China these irrigation regimens have shown water savings of 318%, 7% to 25%, and 20% to 50% under SWD, AWAD, and SDC, respectively (Tuong et al., 2009).

The alternate wetting and drying (AWD) method for rice has become popular in many countries, including the Philippines, Bangladesh, and Vietnam. With the AWD regimen farmers allow ponded water to disappear from the field and infiltrate for several days until the perched field's water table reaches a depth of 15–20 cm. In Bangladesh, for example, where 90% of the irrigation water is used for rice production, changing from flooding to the AWD irrigation method reduces the water demand by 20% to 30%, amounting to $73 million worth of irrigation costs (Mollah et al., 2009).

The available evidence indicates that higher water productivity can be obtained using the AWD method for rice (Bouman, 2007; Tuong et al., 2009). This technique may also boost concentrations of essential nutrients, particularly zinc, in harvested rice (Price et al., 2013). Fereres and Soriano (2006) estimated that annual water diversions can be reduced by as much as 40% from full ET under a deficit irrigation technique (deliberately under-irrigating in less sensitive crop growth periods and ensuring full deliveries at critical growth periods), depending on the crop. However, the impact of all these techniques on different agroecological conditions needs to be carefully evaluated, as reduced irrigation applications may increase the threat of soil salinization.

In irrigated areas the efficiency of the irrigation can be increased up to 95% by using pressurized systems, that is, drip and sprinkler systems, because they reduce seepage losses in the conveyance and distribution systems. At present 15% of the total irrigated area in the world (44 million ha) is equipped with pressurized systems (35 million ha with sprinklers and

9 million ha with drip systems), and most of these are concentrated in Europe and America (Kulkarni, 2011). These irrigation methods provide unique agronomic as well as water and energy conservation benefits. Farmers prefer these systems to achieve higher water use efficiencies and to expand their irrigated area with the same diverted volume of water. The major disadvantages of these systems, especially for smallholder farmers, are their capital and operational costs. Manufacturing these systems locally along with making improvements to reduce labor via automation may help reduce costs and achieve higher irrigation efficiencies.

Different irrigation technologies involve different trade-offs between crop yield, water use, and capital cost of equipment and structures. Modern irrigation methods such as drip or trickle irrigation, surge irrigation, and sprinkler systems are more expensive but usually offer greater potential to minimize runoff, drainage, and soil evaporation. Therefore choice of appropriate irrigation method based on soil type, crops grown, crop water demand, and cost are of paramount importance for managing water for irrigation.

10.4.3 Irrigation System-Scale Interventions

At the irrigation system level water efficiency depends on the management of water runoff and control of seepage and percolation losses in both the water delivery system and on-farm independent, interactive systems. System water losses (the amount of water that leaves the system without contributing productively) caused by interacting problems may be quite serious in certain situations. To increase water productivity at the irrigation system level, we need to improve water allocation and distribution so that all using the system get their fair share. In situations where allocation and distribution are not equitable, farmers getting less than their share are underproductive, and those who get extra water usually tend to waste it through over-irrigation and/or bad management.

A typical example of such a system is the Indus Basin Irrigation System (IBIS) in Pakistan, which irrigates about 16 million ha of land (Qureshi, 2014). This is a fixed rotational distribution system, where each farmer can take an entire flow of the outlet once every seven days and for a period proportional to the size of his land holding. The supplied amount of water is usually not sufficient to irrigate the entire farm in one irrigation turn, and the farmers must decide whether to under-irrigate all land or leave a fraction unirrigated (Qureshi, 2014). The fixed rotational system favors head-end farmers and discriminates against tail-ender farmers. The tail-end farmers get 20% less water than middle farmers, who in turn get 20% less water than head-end farmers. Similar trends are seen in the crop water productivity of head-end, middle, and tail-end farmers (Latif and Tariq, 2009). Equitable water distribution among all farmers therefore remains a challenge in traditional irrigation systems.

For improving water productivity at the irrigation system-scale, managers must provide irrigation water in an equitable, predictable, and timely manner to all water users. However, this task is done less and less satisfactorily due to monopoly, discretion, and negligence in the water sector. This results in inequitable distribution of water, poor technical performance, and a pervasive environment of mistrust and conflict between water managers and water users. The managers should explore hydrometeorological networks, databases, and information systems that can support reservoir and canal operations, provide information on the initiation of floods and droughts, and help farmers with suitable irrigation schedules for different crops (Pereira et al., 2002). The managers should also use tools to develop water allocation and delivery operation rules to ensure equitable and timely delivery of irrigation water to all farmers regardless of their location and land holding (Rossi et al., 2002). Developing small farm reservoirs to harvest rainwater and spate irrigation could be a useful strategy for increasing water supplies (Oweis and Hachum, 2009).

The distribution of irrigation water can be improved from simple syphon tubes for field water use to sophisticated canal automation and telemetry (Kulkarni, 2011). The water administration system used by water-user associations in South Africa to manage their water accounts and supply to clients has reduced water loss up to 20% through improved water releases in canals (ICID, 2008). An upgrade of the traditional irrigation system used in the Mula region in the southeastern part of Spain has led to a fivefold reduction in annual water loss. This has resulted in less exploitation of groundwater with significant savings in pumping energy and an increase in crop productivity and the quality of fruits (Kulkarni, 2011). Automatic hydrodynamic gates on irrigation canals are already in place in France, Morocco, Iran, and Pakistan. In Iran water is delivered to farmers based on actual crop water requirements. The downstream control water canals in the northern province of Pakistan are providing water on a crop-demand basis, thereby reducing releases from reservoirs.

Reuse of drainage water and wastewater offers an attractive way to increase water use efficiency and productivity of an irrigation system. The conjunctive use of groundwater with surface water in the rice-wheat irrigation systems in Pakistan and India is of special significance. Seepage and percolation from rice fields and irrigation networks become a recharge to shallow unconfined aquifers. The water stored in the aquifer can be used again to supplement canal irrigation supplies at the time of need. However, this practice will increase production costs, as energy for pumping groundwater is becoming expensive.

The possibility of recycling does not negate the need to conserve water on-farm. Water recycling and the conjunctive use of groundwater and surface water are happening as a desperate response from farmers where surface water supplies are limited. Therefore the cost-effectiveness of recycling and conjunctive use of ground water and surface water need to be carefully compared with other conservation measures, such as canal lining to reduce seepage and percolation losses in irrigation systems. The recycled waters are mostly of inferior quality; therefore their effects on soil, crops, and the environment need to be carefully evaluated.

To reduce non-beneficial evaporation in an irrigation system, fallow lands and free surface areas should be reduced. Development and adoption of new irrigation schedules for preparing land using early season rainfalls could make it possible to conserve water in reservoirs, allowing more area to be irrigated in the dry season. The adoption of the above-mentioned interventions usually requires more resources, such as labor, capital, and management skills. Therefore an economic analysis of the adoption of alternative interventions for raising water productivity is very essential to increase the effectiveness of these strategies.

10.4.4 Basin-Scale Interventions

Water productivity improvement at the basin level relates to interventions at the level of field, farm, and irrigation systems. From the basin perspective water that would otherwise be lost to sink, percolate to saline groundwater, or flow to the sea can be of value. Therefore such losses need to be controlled in order to increase water productivity at the basin scale.

The reuse of drainage and wastewater is of immense importance for increasing water productivity at the basin level. However, making it feasible to use this water for agricultural use requires maintaining the quality of the drainage and wastewater.

Basin-level water productivity can also be increased by shifting from low-value crops to higher value crops. For example, in the face of declining returns for rice and wheat, diversification to higher valued crops has been encouraged in many countries. However, often it is done without an assured water supply and support for research, extension, and marketing services needed for success. This is perhaps the major reason for farmers' lack of interest in diversified cropping. To promote successful diversification, an assured water supply together with proper market access and extension services need to be ensured. Otherwise, farmers will continue growing traditional crops without caring about the value of water.

For better management of basin water resources, an effective institutional and managerial system is inevitable. A study of 15 irrigation systems in South and Southeast Asia has shown that without addressing institutional and managerial capacity, it is not possible to increase the performance of a basin (Murray-Rust and Snellen, 1993). Basin-level management should have the capacity to effectively operate and maintain different sub-basins, including water allocation and distribution, and to coordinate the entire system through effective feedback and communication. Management should also take important actions like constructing new storages to preserve extra amounts of water.

To understand interactions of different users in a basin and their shared consequences, simulation models may be the most appropriate tools to use. Models can help determine the current situation and processes and, through scenario calculations, can predict what might happen in the future at different scales with the adoption of alternative water-management strategies.

For basin-scale analysis it is not sufficient to have models that look only at the impact on one sector or one class of users because the potential benefits identified by using such models may be outweighed by their greater cost to other users or sectors. For example, if based on field-scale model simulations, we choose to change cropping patterns to improve water productivity and farm profitability in one part of the basin, its effects on other users in other parts of the basin must also be studied using a different model that can simulate a larger domain.

Currently, a wealth of simulation models for different scales and different purposes are available and can be used for up-scaling purposes. For successful application of integrated modeling at field, farm, irrigation systems, and basin scales, four sequential steps are involved:

(1) Data collection and establishment of a structured database
(2) Data analysis using range of appropriate tools
(3) Model development/or application to understand current conditions
(4) Scenario development and assessment leading to conclusions and recommendations

Different water users in a basin are affected not only by the flow of water as inputs and outputs of the different sub-systems but also by the spatial and temporal relationships among the users. Spatial relationships are important in terms of both water quantity and water quality. Irrigation activities at the upstream affect the downstream in two ways: water deliveries may be increased due to augmentation of return flow, but the quality of these return flows may create an adverse effect on downstream users. This is particularly true in basins where extensive irrigated agriculture is practiced at the upstream part of the basin. As a result, salinity inevitably increases at the downstream part of the basin and directly effects the production potential of downstream areas. The situation becomes even worse when there are water quality requirements for the downstream users to fulfill environmental and ecosystem demands.

From the temporal perspective we may find that short-term "losses" may turn out to be useful water gains in the long term. Groundwater is important in this regard. Aquifers act as reservoirs to capture seepages from canals, percolation from rivers and irrigated areas, and recharges from rainfall. Although these supplies are not immediately available, they can be tapped at later stages when no other water resources are available, such as periods of low rainfall or incidences of drought. It is generally argued that improvement in productivity primarily depends on better matching irrigation supplies with crop demands. Therefore a more flexible irrigation scheduling system capable of distributing water on demand is necessary for the optimization of crop production.

10.5 CONCLUSIONS

Global water resources are under stress due to quality degradation and overuse, whereas the growing population's demand for food is increasing. Therefore increasing the productivity of existing lands with the same amount of available water is critical to produce more food, to fight poverty, to reduce competition for water, and to ensure that there is enough water for the environment. Water productivity can be improved by (1) improving the management capacity of farmers through the adoption of innovative irrigation technologies and water delivery systems and (2) optimizing water applications using scientific irrigation schedules for different crops to avoid non-beneficial water losses.

Using water conservation techniques, introducing water-efficient cropping systems, and using deep-rooted crops to maximize the utilization of stored soil water and nutrients are verified ways to increase water productivity. The management of irrigation water is linked to the maintenance of soil water through improved timing of irrigation to minimize negative effects of water deficits on yields and quality. Therefore changes to the physical and managerial aspects of water delivery and to the design of farm irrigation systems will be required to enable farmers to irrigate their fields with the right amount of water at the right time.

The efficiency of water use can be improved several ways. These include building small pits or basins (mini reservoirs) to store water accessed during high flow seasons and then using it during low flow periods. Reducing non-beneficial losses through night-time irrigation can reduce loss due to evaporation. Better weed control and mulching of fields in high-temperature areas and seasons can also control field losses and improve water productivity.

Despite their high installation and maintenance costs, pressurized irrigation systems (such as drip and sprinkler systems) offer greater potential for minimizing runoff, drainage, and soil evaporation. The disadvantage of these methods is that they require intensive supervision; therefore management tools such as decision support systems are needed to provide accurate predictions of crop water requirements, application efficiencies, and timing of chemical applications to control pest outbreaks. System-level management is needed to reduce conveyance losses, field runoff, and leaching and for equitable water distribution to all water users. Poorly managed advanced irrigation systems can be as inefficient and unproductive as poorly managed traditional systems. Therefore the choice of an appropriate irrigation method, based on soil type, crops grown, crop water demand, and cost, is of paramount importance for improving the productivity of water.

To increase basin-level water productivity, water losses to sinks and saline groundwater should be controlled, and a paradigm shift is needed to move from low-value crops to higher value crops. For example, to address the issue of water-deprived basins, many countries have encouraged replacing rice with higher valued crops. However, this requires an assured water supply and support for research, extension, and marketing services. Otherwise, farmers will continue to practice their traditional cropping patterns without caring about the value of water.

References

Abbasi, N., Mehrpour, M., 2004. Effect of Water Deficit Stress on Physiological Properties of Promising Lines of Wheat. Research Report No. 33, Agricultural Engineering Research Institute, Iran.

Ahmad, M.D., Turral, H., Masih, I., Giordano, M., Masood, Z., 2007. Water saving technologies: myths and realities revealed in Pakistan's rice-wheat systems. International Water Management Institute, Colombo, Sri Lanka. (IWMI Research Report 108): 44 pp.

Bouman, B., 2007. A conceptual framework for the improvement of crop water productivity at different spatial scales. Agric. Syst. 93, 43–60.

Cai, X., Molden, D., Mainuddin, M., Sharma, B., Ahmad, M., Karimi, P., 2011. Producing more food with less water in a changing world: assessment of water productivity in 10 major river basins. Water Int. 36, 42–62. https://dx.doi.org/10.1080/02508060.2011.542403.

Cook, S.E., Fisher, M.J., Andersson, M.S., Rubiano, J., Giordano, M., 2009. Water, food and livelihoods in river basins. Water Int. 34, 13–29.

de Vries, M.E., Rodenburg, J., Bado, B.V., Sow, A., Leffenaar, P., Giller, K.E., 2010. Rice production with less irrigation water is possible in a Sahelian environment. Field Crops Res. 116, 154–164. https://dx.doi.org/10.1016/j.fcr.2009.12.006.

Falkenmark, M., Lundquist, J., Widstrand, C., 1989. Macro-scale water scarcity requires micro-scale approaches: aspects of vulnerability in semi-arid development. Nat. Resour. Forum 13, 258–267.

Falkenmark, M., Karlberg, L., Rockström, J., 2009. Present and future water requirements for feeding humanity. Food Security 1, 59–69.

FAO, 2006. World Agriculture: Towards 2030/2050. Interim Report. Prospects for Food, Nutrition, Agriculture and Major Commodity Groups. FAO, Rome. Available at http://www.fao.org/fileadmin/user_upload/esag/docs/Interim_report_AT2050web.pdf.

FAO, 2009. FAO Water Unit | Water News: water scarcity. http://www.fao.org/ (Accessed 12 March 2009).

FAO, 2011. The State of the World's Land and Water Resources for Food and Agriculture (SOLAW) – Managing Systems at Risk. Food and Agriculture Organization of the United Nations/Earthscan, Rome/London.

Fereres, E., Soriano, A., 2006. Deficit irrigation for reducing agricultural water use. J. Exp. Bot. 58 (2), 147–159. https://dx.doi.org/10.1093/jxb/erl165.

Gowda, C.L.L., Serraj, R., Srinivasan, G., Chauham, Y.S., 2009. Opportunities for improving crop water productivity through genetic enhancement of dryland crops. In: Wani, S.P., et al. (Eds), Rainfed Agriculture: unlocking the potential. CABI International, Wallingford, UK, pp. 133–163.

Hobbs, P.R., Gupta, R.K., 2003. Resource-conserving technologies for wheat in the rice-wheat system. In: Improving Productivity and Sustainability of Rice-Wheat Systems: Issues and Impact. Am. Soc. Agron. Spec. Publ. 65, 149–171.

Howell, T.A., 1990. Relationships between crop production and transpiration, evapotranspiration, and irrigation. Irrigation of agricultural crops. Agron. Monogr No. 30.

Hussain, I., Sakthivadivel, R., Amarasinghe, U., 2003. Land and water productivity of wheat in the Western Indo-Gangetic plains of India and Pakistan: a comparative analysis. In: Kijne, J.W., Barker, R., Molden, D. (Eds.), Water Productivity in Agriculture: Limits and Opportunities for Improvement. In: Comprehensive Assessment of Water Management in Agriculture Series, vol. 1. CAB International, Wallingford, UK, pp. 251–271. In Association with International Water Management Institute (IWMI), Colombo.

ICID, 2008. Water Saving in Agriculture. International Commission on Irrigation and Drainage, New Delhi. Publication No. 95.

Kadigi, R.M.J., Tesfay, G., Bizoza, A., Zinabo, G., 2012. Irrigation water Use Efficiency in Sub-Saharan Africa. Briefing Paper No. 4, Agriculture Policy Series, The Global Development Network (GDN), Oxford.

Kamilov, B., Ibragimov, N., Esanbekov, Y., Evett, S., Heng, L., 2003. Irrigation scheduling study of drip irrigated cotton by use of soil moisture neutron probe. In: Proceedings of the UNCGRI/IAEA National Workshop in Optimization of Water and Fertilizer use for Major Crops of Cotton Rotation, Tashkent, Uzbekistan, 24–25 December, 2002.

Kijne, J.W., Barker, R., Molden, D. (Eds.), 2003. Water Productivity in Agriculture: Limits and Opportunities for Improvement. In: Comprehensive Assessment of Water Management in Agriculture Series, vol. 1. CABI International/In Association with International Water Management Institute (IWMI), UK/Colombo.

Kijne, J., Barron, J., Hoff, H., Rockström, j., Karlberg, L., Gowing, J., Wani, S.P., Wichelns, D., 2009. Opportunities to Increase Water Productivity in Agriculture with Special Reference to Africa and Asia. Project Report, Stockholm Environment Institute, Sweden. 39 pp.

Kingwell, R., Carter, C., Elliott, M.P., White, P., 2016. Russia's Wheat Industry: Implications for Australia. Aegic.

Kulkarni, S., 2011. Innovative technologies for water saving in irrigated agriculture. Int. J. Water Resour. Arid Environ. 1 (3), 226–231.

Latif, M., Tariq, J.A., 2009. Performance assessment of irrigation management transfer from government to farmer-managed irrigation system: a case study. Irrig. Drain. 58, 275–286.

Molden, D. (Ed.), 2007. Water for Food, Water for Life: Comprehensive Assessment of Water Management in Agriculture. Earthscan/In Association with International Water Management Institute (IWMI), London/Colombo.

Molden, D., Oweis, T.Y., Steduto, P., Kijne, J.W., Hanjra, M.A., Bindraban, P.S., 2007. Pathways for increasing agricultural water productivity. In: Molden, D. (Ed.), Water for Food, Water for Life: Comprehensive Assessment of Water Management in Agriculture. Earthscan/in Association with International Water Management Institute (IWMI), London/Colombo, pp. 279–310.

Molden, D., Oweis, T., Steduto, P., Bindraban, P., Hanjra, M.A., Kijne, J., 2010. Improving agricultural water productivity: between optimism and caution. Agric. Water Manag. 97, 528–535. https://dx.doi.org/10.1016/j.agwat.2009.03.023.

Mollah, M.I.U., Bhuyia, M.S.U., Kabir, M.H., 2009. Bed planting – a new crop establishment method for wheat in Rice-wheat cropping system. J. Agric. Rural Dev. 7 (1&2), 23–31.

Morison, J.I.L., Baker, N.R., Mullineaux, P.M., Davies, W.J., 2008. Improving water use in crop production. Phil. Trans. Roy. Soc. B 363, 639–658.

Mrabet, R., 2002. Wheat yield and water use efficiency under contrasting residue and tillage management systems in a semi-arid area of Morocco. Exp. Agric. 38, 237–248.

Mueller, N.D., Gerber, J.S., Johnston, M., Ray, D.K., Ramankutty, N., Foley, J.A., 2012. Closing yield gaps through nutrient and water management. Nature 490 (7419), 254–257.

Murray-Rust, M., Snellen, W.B., 1993. Irrigation-water management for sustainable development in Iran. In: Irrigation/Water Management for Sustainable Agricultural Development. Report of the Expert Consultation of the Asian Network on Irrigation/Water Management, 25–27 August 1992, Bangkok, Thailand, pp. 216–221 RAPA Publication 1994/24.

Mzezewa, J., Gwata, E.T., van Rensburg, L.D., 2011. Yield and seasonal water productivity of sunflower as affected by tillage and cropping systems under dryland conditions in the Limpopo Province of South Africa. Agric. Water Manag. 98, 1641–1648.

Oweis, T., Hachum, A., 2009. Water harvesting for improved rainfed agriculture in the dry environments. In: Wani S.P., Rockström, J., Oweis, T. (Eds.), Rainfed Agriculture: Unlocking the Potential. In: Comprehensive Assessment of Water Management in Agriculture Series, vol. 7. CAB International/In Association with International Crops Research Institute for the Semi-Arid Tropics (ICRISAT)/International Water Management Institute (IWMI), Wallingford, UK/Patancheru, Andhra Pradesh, India/Colombo, pp. 164–181.

Oweis, S., Zhang, H., Pala, M., 2000. Water use efficiency of rainfed and irrigated bread wheat in a Mediterranean environment. Agron. J. 92, 231–238.

Pereira, L.S., Oweis, T., Zairi, A., 2002. Irrigation management under water scarcity. Agric. Water Manag. 57, 175–206.

Perry, C., Steduto, P., Allen, R.G., Burt, C.M., 2009. Increasing productivity in irrigated agriculture: agronomic constraints and hydrological realities. Agric. Water Manag. 98, 1641–1648. https://dx.doi.org/10.1016/j.agwat.2009.05.005.

Price, A.H., Norton, G.J., Salt, D.E., Oliver Ebenhoeh, O., Meharg, A.A., Meharg, C., Islam, M.R., Sarma, R.N., Dasgupta, T., Ismail, A.M., McNally, K.L., Zhang, H., Dodd, I.C., Davies, W.J., 2013. Alternate wetting and drying irrigation for rice in Bangladesh: is it sustainable and has plant breeding something to offer? Food Energy Security 2, 120–129.

Qureshi, A.S., 2014. Reducing carbon emissions through improved irrigation management: a case study from Pakistan. Irrig. Drain. 63, 132–138.

Qureshi, A.S., Shoaib, I., 2015. Evaluating benefits and risks of using treated municipal wastewater for a agricultural production under desert conditions. In: Paper presented in the DT12 Conference, held in Cairo, November 16–19, 2015.

Qureshi, A.S., Masih, I., Turral, H., 2004. Furrow-Bed technologies: Option for improving water use efficiency in the rice-wheat cropping systems. In: Paper Presented in the workshop on Water and Wastewater Development for Development Countries (WAMDEC), July 28–30, Zimbabwe.

Qureshi, A.S., Iftikhar, M.I., Shoaib, I., Khan, Q.M., 2016. Evaluating heavy metal accumulation and potential health risks in vegetables irrigated with treated wastewater. Chemosphere 163, 54–61. https://dx.doi.org/10.1016/j.chemosphere.2016.07.073.

Rockström, J., Barron, J., 2007. Water productivity in rainfed systems: overview of challenges and analysis of opportunities in water scarcity prone savannahs. Irrig. Sci. 25, 299–311.

Rockström, J., Karlberg, L., Wani, S.P., Barron, J., Hatibu, N., Oweis, T., Bruggeman, A., Farahani, J., Qiang, Z., 2010. Managing water in rainfed agriculture: the need for a paradigm shift. Agric. Water Manag. 97, 543–550. https://dx.doi.org/10.1016/j.agwat.2009.09.009.

Rossi, G., Cancellieri, A., Preira, S.L., Oweis, T., Shatanawi, M., Zairi, A. (Eds.), 2002. Tools for Drought Mitigation in Mediterranean Regions. Kluwer Academic Publishers, Dordrecht, The Netherlands.

Sadler, E.J., Camp, C.R., Evans, R.G., 2007. In: Lascano, R.J., Sojka, R.E. (Eds.), New and future technology in irrigation of agricultural crops. American Society of Agronomy, Madison, WI, pp. 609–626. Agronomy Monograph 30 (2).

Schierhorn, F., Faramarzi, M., Prishchepov, A., Koch, V., Müller, D.F., 2014. Quantifying yield gaps in wheat production in Russia. Environ. Res. Lett. 9, 84–97.

Sezen, S.M., Yazar, A., 1996. Determination of water-yield relationship of wheat under Cukurova conditions. J. Agric. Forest. 20, 41–48 (in Turkish, with English abstract).

Tuong, T.P., Bouman, B.A., Lampayan, R., 2009. A Simple Tool to Effectively Implement Water Saving Alternate Wetting and Drying Irrigation for Rice. ICID Newsletter, Issue 2009/4.

van Ittersum, M.K., Cassman, K.G., Grassini, P., Wolf, J., Tittonell, P., Hochman, Z., 2013. Yield gap analysis with local to global relevance—a review. Field Crops Res. 143, 4–17.

World Bank, 2007. Agriculture for Development. World Development Report 2008. World Bank, Washington, DC.

Zhang, X., Pei, D., Hu, C., 2003. Conserving groundwater for the North China Plain. Irrig. Sci. 21, 159–166.

Hydrological Cycle Over the Indus Basin at Monsoon Margins: Present and Future

Shabeh ul Hasson[*,†], *Jürgen Böhner*[*]

[*]Centre for Earth System Research and Sustainability (CEN), Institute of Geography, University of Hamburg, Hamburg, Germany [†]Department of Space Sciences, Institute of Space Technology, Islamabad, Pakistan

11.1 INTRODUCTION

The Indus Basin has intricate cryo-hydrologic and climate regimes, which are defined by distinct large-scale circulation systems, latitudinal extent ranging from hyper-arid to humid zones, altitudinal range from zero mean sea level delta to the second highest peak in the Karakoram-2 (K2), and a high degree of continentality, such as the heavily concentrated Himalayan cryosphere, which is the largest cryosphere outside the polar regions; the Indian Ocean; the Thar Desert; and a vast region of irrigated plains. The Indus River system (IRS) comprises the main Indus trunk and its six major tributaries—the Upper Indus Basin and the Jhelum, Chenab, Ravi, and Sutlej rivers—which support the world's largest contiguous irrigation system and thus also the food security and socioeconomic well-being of approximately 260 million people who live in the basin (CIESIN, 2005). The Indus trunk originates in the Tibetan Plateau and flows approximately 3600 km southwest through India, Afghanistan, and Pakistan before converging with the Arabian Sea. The major tributaries originate mainly in the highly glaciated Hindu Kush-Karakoram-Himalayan (HKH) headwaters. At high altitudes the tributaries depend on snowmelt and glacier melt and then at relatively low altitudes on the monsoon waters from the western Himalayas, eastern Hindu Kush, and adjacent plains. Lying at the margins of the South Asian summer monsoon system and the western disturbances, the headwaters of the Indus and its tributaries exhibit a delicate hydro-climatology that is nourished and influenced by the associated precipitation regimes and the interactions and modulations of the complex HKH terrain and its aboding cryosphere.

The westerly precipitation regime (WPR) is associated with year-round western disturbances that bring moisture from the Atlantic Ocean and the Caspian Sea and contribute mainly to solid precipitation in the upper part of the basin during the spring and winter. The monsoonal precipitation regime (MPR) is associated with the South Asian summer monsoon system that brings moisture from the Bay of Bengal and the Arabian Sea to the western Himalayas, their foothills, and the Punjab plains from late-June to September. The MPR largely coincides with the mid-to-late melt season, during which the HKH glaciers contribute meltwater to the IRS. The early-melt season spans from March to late-June, during which the main contribution comes from melting of snow accumulated during spring and winter under WPR. In contrast to the glacial and nival melt regimes active in the far north and the pluvial regimes dominating around the middle of the basin with a clear east-west gradient, the Lower Indus Basin is arid and hyper-arid. Such heterogeneously distributed precipitation—spatially and across the seasons—amounts to approximately 430 ± 87 mm yr^{-1} when aggregated for the entire basin (Table 11.1).

The total surface water received from the IRS is equivalent to approximately 210 mm of runoff, 80% of which is available during the high flow period in late-March to September

TABLE 11.1   Observational Uncertainty of Annual Precipitation Over the Indus Basin

S. No.	Observed Datasets	Acronym	Reference	ONDJF	MAM	Westerly	Monsoon	Annual
1	APHRODITE VR1101	APHRO	Yatagai et al. (2012)	71	77	148	160	331 ± 49
2	CMAP	CMAP	Xie and Arkin (1997)	157	122	279	193	502 ± 71
3	CPC-Unified	CPCUNI	Xie et al. (2010)	63	67	130	142	292 ± 88
4	CRU TS3.2	CRU	Harris et al. (2014)	89	96	185	195	409 ± 61
5	EWEMBI1	EWEMBI	Frieler et al. (2017)	103	104	207	201	434 ± 70
6	GPCC	GPCC	Schneider et al. (2014)	92	93	185	204	415 ± 68
7	GPCP V2.3	GPCP	Adler et al. (2003)	104	99	203	230	465 ± 71
8	MSWEP V1.2	MSWEP	Beck et al. (2017)	148	156	304	266	608 ± 75
9	PERSIANN-CDR	PERSIAN	Ashouri et al. (2015)	114	103	217	238	488 ± 99
10	PREC-L	PREC-L	Chen et al. (2002)	84	90	174	173	370 ± 59
11	UDEL	UDEL	Willmott and Matsuura (2001)	102	100	202	188	417 ± 60
	Ensemble Stats			102 ± 29	101 ± 23	203 ± 51	199 ± 35	430 ± 87

(Hasson et al., 2013, 2014a). Since the Indus is among highly irrigated basins, more than 80% of its annual waters are diverted to and consumed by irrigation (210 mmyr^{-1}—Laghari et al., 2012). The mean annual discharge into the Arabian Sea, which is measured at the last gauging site (located at Kotri), is around 40 mm in runoff equivalent. Pakistan is the major beneficiary of the IRS flows with an exclusive control over the western tributaries of the main Indus River and the Jhelum and Chenab rivers under the 1960 Indus Water Treaty (Mehta, 1988). The mean annual surface water received in Pakistan through the IRS at its river inflow measurement (RIM) stations—the Kabul at Nowshera, the Indus at Tarbela, the Jhelum at Mangla, the Chenab at Marala, the Ravi at Balloki, and the Sutlej at Sulemanki—amounts to 142 million acre feet (Ali et al., 2009). The largest contribution comes from the Indus at Tarbela, which amounts to about 45% of the annual total (Table 11.2).

To summarize, the Indus River flows are generated and collected from distinct sources: (1) nival and glacial melt regimes of the HKH (March to September), which are nourished mainly by solid precipitation received during the winter and spring seasons through year-round western disturbances and (2) the South Asian summer monsoonal rainfall along the Himalayan foothills and over adjacent plains from July to August (Hasson et al., 2014b). Since climate-induced pressures on such distinct hydrological regimes and on the forms and strengths of their nourishing precipitation regimes at their extreme margins can substantially imbalance the water demand and supply, an assessment of prevailing and projected changes in the hydrological cycle and subsequent water availability under a changing climate is crucial to support an informed decision on basin-scale future planning and development. However, in view of the transboundary nature of the Indus Basin (Fig. 11.1), a basin-scale assessment is difficult when using typical watershed models under perturbed forcing and also because of the extensive diversions engineered across the IRS. Thus present-day, state-of-the-art global and regional climate models (GCMs/RCMs) are primary choices.

GCMs/RCMs can simulate the climate system of the Earth or a region including hydrological cycle, and their changes under a suite of probabilistic warmer climates as anticipated under anthropogenic perturbed forcing, for instance, use of the latest generation of the global climate modeling experiments participating in the Coupled Model Intercomparison Project Phase 5 (CMIP5; Taylor et al., 2012). However, the robustness of hydroclimatic changes projected by these experiments depends largely on the degree of representation of relevant physical processes in the employed models and the reproducibility of dependently observed

TABLE 11.2 Surface Water Availability From the IRS (1961–2001)

S. No.	Tributary	Apr–Sep %	Oct–Mar %	% of Total (142 MAF)
1	Indus at Tarbela	86	14	44
2	Jhelum at Mangla	78	22	16
3	Chenab at Maralla	83	17	19
4	Kabul at Nowshera	82	18	16
5	Eastern rivers			05

Data from Ali, G., Hasson, S., Khan, A.M., 2009. Climate Change: Implications and Adaptation of Water Resources in Pakistan. Global Change Impact Studies Centre (GCISC), GCISC RR-13 Islamabad, Pakistan.

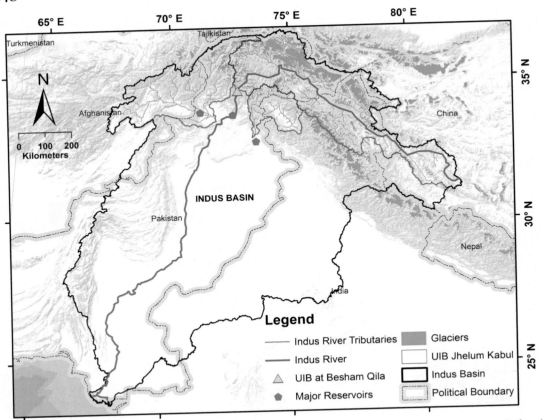

FIG. 11.1   Broad view of the Indus Basin and the sub-watersheds that constitute the Upper Indus. *(Source: Authors.)*

phenomenon, collectively called fidelity (Hasson, 2016a). Thus it is crucial to analyze the fidelity of CMIP5 experiments prior to assessing their projected changes in the hydrological cycle. In terms of water balance closure and reproducibility of statistical properties of major hydrological cycle components, fidelity assessment of previous-generation CMIP3 experiments reveals that these experiments exhibited serious difficulty in simulating the hydro-climatology of the Indus Basin (Hasson et al., 2013, 2014a), while the latest-generation experiments and their dynamically downscaled invariants do not improve for the realism (Boos and Hurley, 2013; Levine et al., 2013; Hasson, 2016a; Hasson et al., 2016a, 2018). These experiments therefore must be investigated for their suitability to assess robust changes in future water budgets.

With this background, this chapter analyzes 36 CMIP5 experiments and presents changes in 21st-century water budgets based on moderate-to-high fidelity experiments. Future water budget changes are assessed under the climate change scenario of the Representative Concentration Pathway 4.5 (RCP4.5), which assumes a consistent $4.5\,\mathrm{Wm}^{-2}$ increase in radiative forcing by 2100. Because of the remarkable spread in the simulated quantities of the selected experiments, their projected mean ensemble changes are interpreted primarily in connection with a qualitative agreement among the individual experiments on the direction of change. Before presenting the future water budgets, the chapter characterizes prevailing precipitation regimes from a comprehensive database of 11 proxy gridded observations, in order to better quantify the observational uncertainty while interpreting the realism of CMIP5 experiments.

11.2 DATA AND METHODS

11.2.1 Data

The observed precipitation variability was assessed from a broad database employing 11 gridded observational data sets. Among these, six data sets were constructed by statistically interpolating station observations, while five are satellite-based remotely sensed observations, which are further adjusted either to the rain gauges or to the reanalysis data sets (Table 11.1). The outlet discharge (D) from the Indus Basin, measured at its last gauging site located at Kotri, was collected for the period 1977–2000 from the Water and Power Authority (WAPDA) in Pakistan.

To assess water balance consistency, monthly precipitation (p), evaporation (e), and total runoff (r) were obtained for the historical period (1960–2000) from the entire set of latest-generation coupled climate modeling experiments participating in the CMIP5 data archive. Such historical data were available from 36 experiments and their many realizations; however, only the first realization was considered for analysis. For the future change assessment, p, e, and r were obtained for the entire 21st century from the CMIP5 experiments simulated under the RCP4.5 scenario. The RCP scenarios, presenting a broad range of probabilistic future climates, refer to the combined anthropogenic effect on greenhouse gas emissions, measured in terms of radiative forcing (Moss et al., 2010). RCP4.5, typically taken as a mid-level scenario, assumes that the total radiative forcing will rise to $4.5\,\mathrm{Wm}^{-2}$ by 2100 and then will immediately stabilize starting from the 22nd century (Masui et al., 2011; Thomson et al., 2011).

11.2.2 Methods

Interannual and intra-annual variability of precipitation—a relatively well-observed component of the hydrological cycle—is presented from 11 gridded observational data sets to better quantify the observational uncertainty for robust fidelity assessment of climate modeling experiments. Following Hasson (2016a) the fidelity of CMIP5 experiments was assessed in two aspects: physical consistency and realism. For a robust water budget assessment, physical consistency of the climate modeling experiments requires that their simulated water cycles are closed. Therefore water balance closure was taken as a principle criterion for the experiment selection. The realism of the climate modeling experiments was taken as their skill in reproducing the observed annual water budget and the bimodal distribution of prevailing precipitation regimes.

For physical consistency, annual quantities of simulated p, e, and r for the 1961–2000 period were precisely integrated over the Indus Basin within its natural boundary, in order to avoid the noise from the coarse resolution climate model grid cells that partially reside outside the basin; of particular concern were deltaic grid cells that might include part of the ocean where the water budget remains consistently negative. Details on the employed integration procedure are reported in Hasson et al. (2013, 2014a, 2016a). Going forward the climatological basin integrated quantities of p, e, and r are denoted as P, E, and R, respectively. Given the fact that the long-term annual water storage over land is negligible at any given location, the hydrological quantities satisfy p-e~r (Peixoto and Oort, 1992; Karim and Veizer, 2002), and when integrated over the whole basin, these hydrological quantities satisfy P-E ~ R~ D, where D is the long-term observed basin outlet discharge (Hasson et al., 2013). Since IRS flows are heavily diverted for irrigation, the basin outlet discharge at Kotri ($40\,\mathrm{mm\,yr^{-1}}$) was not directly comparable to the basin-integrated P-E or R from the CMIP5 experiments, which do not implement irrigation. Thus assuming no diversion within the basin, a "natural" discharge amounting to $210\,\mathrm{mm\,yr^{-1}}$ was computed by adding the total annual diversion of 170 mm (Laghari et al., 2012) to the observed discharge at Kotri (Hasson et al., 2013). The observed water budget (P-E) was computed from the balance Eq. P-E~D that features the same observational uncertainty as does P. The water balance for a particular experiment was considered closed if the magnitude of its simulated P-E was comparable to that of R within their statistical uncertainties, here taken as a 95% confidence interval around the mean. Further, the CMIP5 experiments that simulated the magnitude of P-E close to observation were considered as high skill experiments. Simulated and observed annual cycles of P were further compared to assess whether the CMIP5 experiments reproduced bimodal precipitation distribution.

Based on high-fidelity CMIP5 experiments, mean future changes in P-E were presented under the RCP4.5 scenario for three 21st century climates (2011–2040, 2041–2070, and 2071–2100) for the complete hydrological year (October–September) as well as for three distinct seasons: October–February (ONDJF), March–May (MAM), and June–September (JJAS). JJAS refers to the summer monsoon season while ONDJF and MAM are accumulation seasons associated with WPR. The spring season (MMA) is believed to exhibit marked hydroclimatic changes (Hasson et al., 2017), thus the season was considered separately. Since the multi-member ensemble mean can be biased to a few wettest or driest experiments, the majority agreement was additionally computed as a measure of robustness on the projected direction of change in P-E. The majority agreement was computed as the difference between the number of experiments projecting wet and dry future.

11.3 RESULTS AND DISCUSSION

11.3.1 Observed Precipitation

Owing to uneven, sparse, and largely inaccessible transboundary point observations, the limited skill of interpolation techniques as well as difficulties in remote sensing of solid moisture in the complex Himalayan terrain, the constructed spatially complete gridded observations are expected to feature distinct skills and uncertainties. Thus the mean annual precipitation estimates computed from 11 different observational data sets feature a remarkable spread, ranging from 292 mm from CPCUNI to 608 mm from MSWEP (Table 11.1). Typically, the multisource satellite-based merged products are relatively wetter over the Indus Basin as compared to their interpolated counterparts. Fig. 11.2 shows that the interpolated

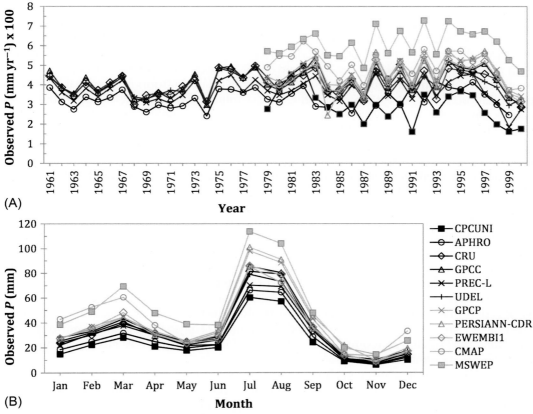

FIG. 11.2 Basin-integrated observed precipitation from 11 gridded observational data sets: (A) interannual variability and (B) annual cycle. Purely interpolated data sets are shown in *solid black lines*, while the multisource remotely sensed merged data sets are shown in *solid gray lines*. Gray- and *black*-filled markers indicate the wettest and driest data sets, respectively. *(Source: Authors.)*

data sets feature high agreement from 1961 until the 1980s, except for APHRO that consistently suggests lower estimates by around $100 \, \text{mm} \, \text{yr}^{-1}$.

For the 1979–2000 period, the lowest precipitation is suggested by CPCUNI followed by APHRO that amounts to 292 mm and 331 mm, respectively. Overall, interpolated data sets agree well with each other on a mean annual precipitation of around 370 mm (400 mm excluding the driest CPCUNI and APHRO data sets) as well as on the interannual variability that ranges between 60 mm and 100 mm. From remotely sensed observations, MSWEP features the highest precipitation followed by CMAP dataset. Overall, remotely sensed data sets suggest the mean annual precipitation around 500 mm (460 mm excluding the wettest MSWEP) while all data sets agree on the interannual variability of around 70 mm, except for PERSIAN that features higher interannual variability of 100 mm. Consistent lower estimates by interpolated data sets against their remotely sensed counterparts indicate their systematic underestimation of the solid precipitation over the Himalayan watersheds. Important to note that regardless of the "wetness" scatter of gridded observations, around all data sets correctly identify wet and dry years and the drought period over the basin, starting 1998.

On intra-annual scale, again remotely sensed data sets are wetter than interpolated data sets throughout the year (Fig. 11.2). All data sets show a bimodal distribution of precipitation with peaks in March and July for WPR and MPR, respectively. The scatter of peak westerly precipitation is around 40 mm while such scatter is around 60 mm for the peak monsoonal precipitation. CPCUNI estimates are the lowest followed by APHRO through the annual course. On the other hand, MSWEP is the wettest data set across the year, followed by CMAP during WPR, but by GPCP and PERSIAN during MPR. The observed precipitation in the wettest data sets is almost twice the estimates observed in the driest data sets. Besides such marked differences in the precipitation magnitudes, the contributions from westerly and monsoonal precipitations to the mean annual totals are found to be roughly the same for all data sets (Table 11.1).

On a spatial scale, Fig. 11.3 shows that precipitation over the Indus is heterogeneously distributed. The precipitation occurs mainly over the Himalayas, along its foothills and adjacent monsoon-dominated plains; whereas the Lower Indus Basin is primarily arid and hyper-arid. WPR mostly influences the upper part of the Indus, consisting of Hindu Kush, western Himalaya, and western and central Karakoram ranges and foothills. Along a narrow strip in the Upper Indus, the mean ensemble rate of westerly precipitation is around $1 \, \text{mm} \, \text{day}^{-1}$ during ONDJF and up to $4 \, \text{mm} \, \text{day}^{-1}$ over a wider area during MAM. On the other hand, MPR is dominant over the western Himalayas, along its foothills and adjacent Punjab plains. The rate of mean maximum monsoonal precipitation (122 days) is more than $8 \, \text{mm} \, \text{day}^{-1}$ over and along the Himalayan foothills, while the rest of the Indus receives meager amounts. It is important to mention that for all seasons the mean ensemble precipitation rate is lower than the extent of disagreement on such rate by the individual ensemble members (second row, Fig. 11.3), whereas such disagreement is high over the areas of maximum precipitation. Regardless of such marked observational uncertainty, all observational data sets agree well on the interannual variability, bimodal distribution of precipitation, and ratio of contributions from westerly and monsoonal precipitation to the annual total. Analyzing a subset of the observations considered here, Hasson (2016a) also concludes that irrespective of substantial differences in their suggested seasonal or annual magnitudes, the observational data sets are skillful in representing the statistical properties of the annual cycle, such as monsoonal onset and retreat timings, a sharp linear growth in between, and overall seasonality of WPR and MPR.

11.3.2 Physical Consistency and Realism of CMIP5 Experiments

In Fig. 11.4 P-E is plotted against R along with the scatter plot of E versus P for the historical period (1961–2000). Dashed lines indicate the amount of observed discharge into the Arabian Sea while thin solid lines indicate approximated "natural" discharge assuming no irrigation in the Indus Basin. All models lying over the diagonal line suggest that their water balance over the Indus Basin is strictly closed. The models clustering around the crossing of thin solid lines are able to reproduce the observed water budget.

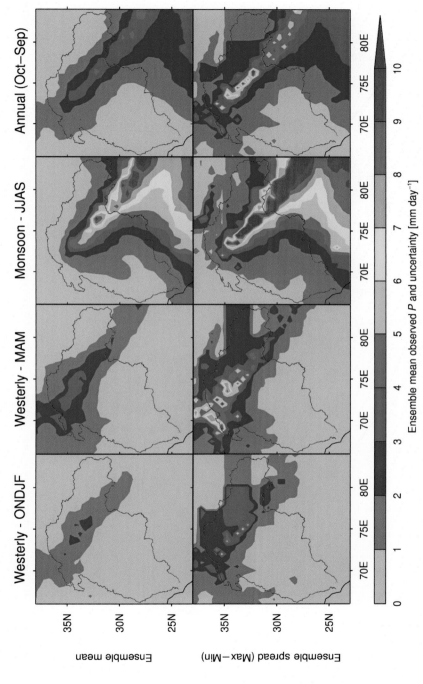

FIG. 11.3 Observed mean ensemble climatology of seasonal and annual precipitation along with their uncertainties taken as the spread (maximum minus minimum) among the 11 observational data sets in mm day^{-1}. (*Source: Authors.*)

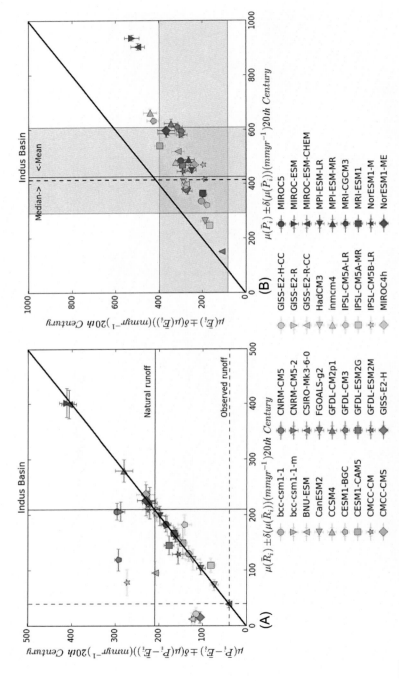

FIG. 11.4 (A) Climatological mean annual basin-integrated water budget (P-E) against the climatological mean annual basin-integrated runoff (R) along with their 95% confidence intervals. *Dashed lines* show the observed outlet discharge into the Arabian Sea while thin *solid lines* indicate the approximated "natural" discharge (absent irrigation) in equivalent areal height units; (B) E versus P along with their 95% confidence intervals along the mean and their observational uncertainty. (*Source: Authors.*)

Fig. 11.4 shows that while the majority of experiments are able to conserve water up to a reasonable degree of approximation, surprisingly 13 experiments fail such a physical consistency test. Among physically inconsistent experiments, 11 simulate P-E higher than R, indicating the loss of water to unphysical sinks by the land-surface modules of the employed climate models. On the other hand, two experiments (IPSL-CMA-LR and IPSL-CMA-MR) simulate R higher than P-E, suggesting that the land-surface routines implemented in the employed climate models are subject to unphysical sources of water. Such gain or loss of water introduces unsystematic biases in the dynamics and thermodynamics of the simulated regional climate and cannot be disentangled from the climate change signal. Therefore the projected changes from the experiments featuring inconsistent water balance are not reliable and must be excluded from further applications.

Among the physically consistent experiments, nine (BCC-CSM1–1-m, CESM1-BGC, CMCC-CMS, CMCC-CMS, MIROC4h, MIROC5, MPI-ESM-LR, NorESM1-M, NorESM1-ME) are able to reproduce the observed water budget amounting to 210 mm yr^{-1}. Among the remaining 14 experiments, 11 markedly underestimate and three overestimate the observed water budget. CSIRO-Mk3-6-0 experiments simulate the lowest water budgets, and MIROC-ESM experiments suggest the highest water budgets. An examination of the individual components of water budgets reveals that experiments simulated under similar forcing largely differ from each other on their simulated P and E quantities. The spread of P is even higher than that of E. Yet most of the models either cluster around the mean and median of observed P or lie within its observational uncertainty owing to its marked spread. These factors emphasize the need to employ a large observational database while validating climate model simulations, as an underestimation reported for one data set can be interpreted contrarily to another data set (Hasson et al., 2016b). Fig. 11.4 also suggests that the realism of physically consistent experiments depends on the magnitude of their simulated P. The driest model simulates the lowest while the wettest model simulates the highest water budgets. It is worth noting that there is no linear relationship between the physical consistency and the realism; most of the physically inconsistent experiments are close to reality, while few show a strictly closed water balance overestimate P.

Because the Indus Basin experiences two distinct precipitation regimes forming bimodal distribution, it is important to investigate the realism of physically consistent models in terms of reproducing such a distribution. Fig. 11.5 shows that with moderate inter-dataset agreement, CMIP5 experiments are largely able to reproduce the WPR within the observational uncertainty. With only slight differences, CanESM2 and CSIRO-Mk3-6-0 underestimate while MIROC5 overestimates. Most of the experiments simulate peak westerly P during March as observed, while only three experiments (MIROC-ESM, MIROC-ESM-CHEM, and MPI-ESM-MR) suggest a peak in April. Analyzing earlier generations of similar experiments participating in the CMIP3 archive (Hasson et al., 2014a), also reported their satisfactory skill in representing WPR.

In contrast to WPR, CMIP5 experiments feature diversified skill for the precipitation regime associated with the South Asian summer monsoon. Surprisingly, nine experiments (BCC-CSM1-1, CSIRO-Mk3-6-0, GISS-E2-H, GISS-E2-H-CC, GISS-E2-R, GISS-E2-R-CC, IPSL-CMA-LR, IPSL-CMA-MR, and IPSL-CMB-LR) do not reveal the monsoon over the Indus Basin. The absence of monsoonal P explains why the physically consistent CSIRO-Mk3-6-0 experiment is the driest with the lowest simulated P-E. Note that all experiments

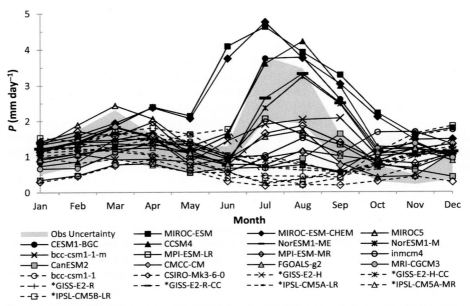

FIG. 11.5   Climatological mean annual cycle of basin-integrated simulated precipitation from physically consistent CMIP5 experiments *(shown with solid lines)* plus those that do not simulate the South Asian summer monsoon at all over the Indus Basin *(shown with dashed lines)*. Among the experiments with absent monsoon, the names of those that breach the water balance closure are preceded by (*). The spread shown in *gray* refers to the observational uncertainty, which is computed as the differences between the maximum and the minimum monthly estimates among 11 observational data sets. *(Source: Authors.)*

with absent monsoonal P do not feature a closed water balance, with the exception of two (CSIRO-Mk3-6-0 and BCC-CSM1-1). The rest of the physically consistent experiments feature a remarkable spread of their simulated P. Relating this spread to the skill of identifying accurate monsoonal onset and retreat timings and seasonality, Hasson et al. (2016a) reported a delayed onset/retreat and an underestimation of seasonality by CMIP5 experiments. Such skill has been found to be even poorer in the dynamically downscaled high-resolution invariants of the CMIP5 experiments, emphasizing the need for more accurate representation of the physical processes and feedbacks for improving realism (Hasson, 2016a).

Fig. 11.5 similarly shows either delayed onset or retreat, or both. Only the wettest experiments (MIROC-ESM and MIROC-ESM-CHEM) simulate peak monsoonal P in July while the rest delay it by a month or two. Only 10 experiments (bcc-csm1-1-m, CanESM2, CCSM4, CESM1-BGC, MIROC-ESM, MIROC-ESM-CHEM, MPI-ESM-LR, MPI-ESM-MR, NorESM1-M, and NorESM1-ME) are close to the observed ratio of roughly equal contributions from westerly and monsoonal P, while the rest indicate a weaker monsoon.

The weak monsoon over the Indus shown by several experiments is mainly related to their difficulty in propagating the monsoonal system to its western extremity. Fig. 11.6 shows the distinct skill of 36 CMIP5 experiments in simulating both the magnitude and the spatial extent of MPR at its extreme margins over the Indus Basin. Possible causes of this remarkable spread

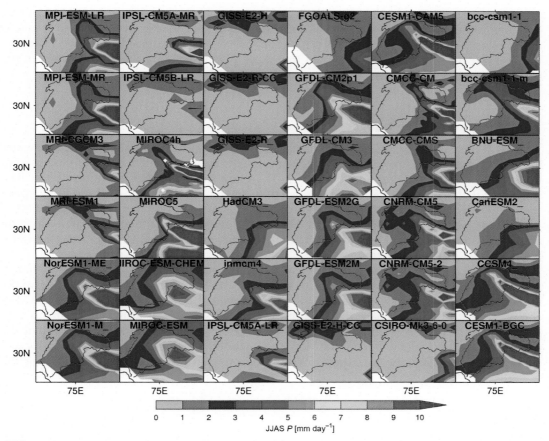

FIG. 11.6 Spatial maps of the monsoonal precipitation from 36 CMIP5 experiments. *(Source: Authors.)*

include (1) underrepresentation of the Karakoram topography, (2) missing irrigation over the Punjab plains, and (3) cold bias in the north Arabian Sea surface temperatures.

The underestimated real topography along 60–80°E in coarse resolution GCMs let the dry mid-latitude air enter into the region, which negatively influences the monsoonal thermal regime and suppresses moist convection (Boos and Hurley, 2013). Consequently, the monsoonal low-level jet and the associated precipitation regime are too weakened to reach far into the observed northwestern area (Chakraborty et al., 2006). The missing irrigation in CMIP5 models results in similar effects. In reality, annually approximately 170 mm of water is diverted to and evaporated from the irrigated fields (Laghari et al., 2012). This behavior thermodynamically affects the sub-regional climate, resulting in evaporative cooling of the surface and subsequently a lower-than-observed land-sea thermal contrast, which reduces the penetration of the westerlies but provides a conducive environment for the monsoonal currents to approach the Upper Indus Basin (Saeed et al., 2009). Further, an anomalously cooler north Arabian Sea surface during winter, spring, and summer negatively affects evaporation and subsequently the amount of moisture in the low-level monsoonal jet that suppresses the monsoon convection; whereas the low-level convergence over land weakens the monsoonal flow and associated precipitation, preventing its extension northwest over the Indus Basin (Levine et al., 2013). Given that these phenomena are well represented in the present-day climate models, the observed extent of MPR can be well reproduced.

11.3.3 Future Water Budget Changes

Most of the analyzed CMIP5 experiments are found physically consistent in terms of water balance, and a significant number of the experiments are moderately close to the observations, particularly for the annual quantities. Therefore changes projected under the RCP4.5 scenario for the 21st century are presented from only 14 moderate-to-high fidelity experiments. In view of the marked ensemble spread, the mean ensemble changes from the selected experiments are further supported by their majority agreement on their simulated direction of change. Such agreement provides a qualitative measure of the robustness of mean ensemble change in future water budgets.

Basin integrated analysis suggests that most of the experiments simulate a negative change in the water budget for the westerly seasons (ONDJF and MAM) and a positive change for the monsoon season (Fig. 11.7). Changes in the westerly water budget are higher during MAM as compared with ONDJF where the maximum negative change of around 0.4 mm day^{-1} is projected by MIROC-ESM-CHEM. During the monsoon season, MIROC5 projects that the maximum water budget will increase. Since the extent of negative change in the westerly water budget is either higher or equal in magnitude to that of the monsoon season for the

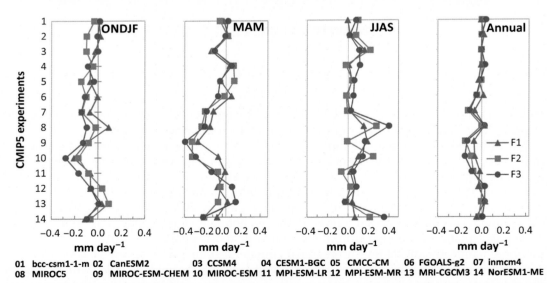

01 bcc-csm1-1-m 02 CanESM2   03 CCSM4   04 CESM1-BGC 05 CMCC-CM   06 FGOALS-g2 07 inmcm4
08 MIROC5   09 MIROC-ESM-CHEM 10 MIROC-ESM 11 MPI-ESM-LR 12 MPI-ESM-MR 13 MRI-CGCM3 14 NorESM1-ME

FIG. 11.7 Basin integrated changes in the seasonal and annual water budgets from three 21st century climates under the RCP4.5 scenario from 14 selected experiments. F1, F2, and F3 refer to 2011–2040, 2041–2070, and 2071–2100 climates. *(Source: Authors.)*

individual experiments, their annual water budgets are projected to experience minor changes. It is noteworthy that the pattern of the water budget changes remains consistent across the future climates studied, although the changes will become much greater in far future. For ONDJF, although two experiments (MIROC5 and MRI-CGCM3) initially suggest an increase in the water budget, they agree that later it will decrease. Exceptions for MAM seasons are CMCC-CM and FGOALS-g2. In contrast, three experiments (CanESM2, CCSM4, and CMCC-CM) project a decrease in the ONDJF water budget by mid-century climate, but they suggest either no change or a positive change by the end of the century. Similar experiments for the MAM season are MPI-ESM-MR and MRI-CGCM3. For the monsoon season, about half of the selected experiments project either no or a slight negative change by the middle or end of the century, indicating only medium-level agreement on the projected magnitude of the changes and their direction.

On a spatial scale, again the patterns of seasonal and annual water budget changes as projected under the near-future climate remain largely the same, though they intensify by the end of century (Fig. 11.8). For the annual water budget, the mean ensemble change indicates a slight decrease over the Upper Indus Basin but increase over the Lower Indus Basin (up to 0.2 and 0.1 mm day^{-1}, respectively). On an intra-annual scale, the mean ensemble water budget is projected to increase up to 1 mm day^{-1} for MPR over its dominated area, which also depicts its northwestward extension under mid-century and end-of-century climates. Hasson et al. (2016a) also showed such extension in the monsoonal spatial domain over Northwest Pakistan, using the CMIP5 experiments under the RCP8.5 scenario. Nevertheless, there was not good agreement among the individual experiments about the increase in the annual and seasonal water budgets throughout the century, as the coefficients of variation are above unity (indicated by stippling in Fig. 11.8), implying that the disagreement on mean ensemble change among individual experiments (standard deviation) is higher than the change itself. This fact is reinforced by the majority agreement, which suggests no clear majority on a positive direction for the annual and intra-annual water budgets throughout the 21st century. In fact, the mean ensemble increases in the water budgets are influenced by few wet experiments. This is, however, not the case with a decrease, as most of the individual experiments agree well on the mean ensemble decrease in the ONDJF water budget by up to 0.1 and 0.2 mm day^{-1} over the Lower and Upper Indus, with the decrease over the upper basin intensifying throughout the 21st century. As compared with ONDJF, MAM features a marked decrease in the water budget over the Upper Indus, which continues under warmer climates, reaching up to 1 mm day^{-1} by the end of the century.

Negative westerly (positive monsoonal) water budgets are mainly related to weakening (strengthening) of prevailing precipitation regimes. Analyzing the CMIP3 experiments, Hasson et al. (2014a) also reported a decrease in the westerly precipitation and an increase in the monsoonal precipitations and their water budgets. Decreasing westerly precipitation and water budgets are further in agreement with the reports of a growing number of dry spring days, as analyzed from the observations as well as from the CMIP5 experiments (Hasson et al., 2016a, 2017). Such changes in the westerly precipitation can be related to the northward shift of the western disturbances (Fu and Lin, 2011). Since the frozen water resources of the Upper Indus are nourished by westerly precipitation, its negative changes projected under warmer climates imply less accumulation and enhanced warming, resulting initially in increased but later in decreased meltwater contributions from the existing

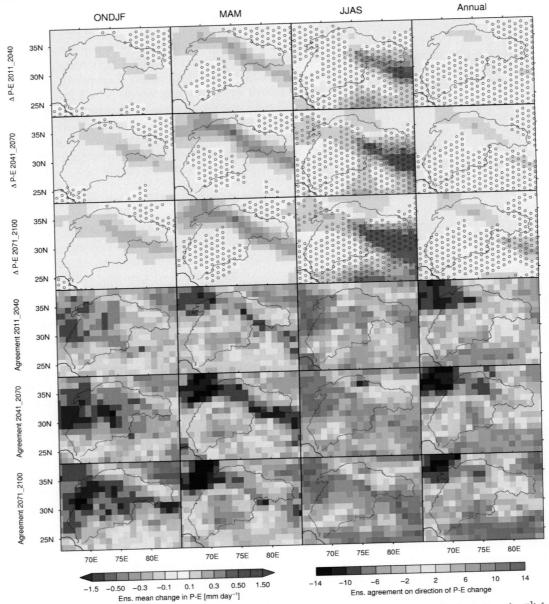

FIG. 11.8   Spatial maps of the mean ensemble changes in the seasonal and annual water budgets (mm day^{-1}) for three 21st century climates under the RCP4.5 scenario from 14 selected experiments along with their majority agreement on such changes. Majority agreement is computed as the difference between the number of experiments projecting wet and dry future. Stippling shown by hollow circles identifies locations where the coefficient of variation (ratio of standard deviation to mean) of the P-E change ensemble is above unity. *(Source: Authors.)*

cryosphere. Coupling of the hydrological model with the extreme-end warming scenario of RCP8.5 from the fine-scaled CMIP5 experiments over the Upper Indus also suggests consistently increasing water availability due to enhanced contributions from the existing cryosphere until it reaches a tipping point or ultimately vanishes (Hasson, 2016b). This scenario suggests more dependence on the erratic and short-spanned monsoon waters in the far future. Although the skill of the CMIP5 experiments in reproducing the monsoonal regime is limited due to unrealistic representation of few region-specific but important features and their associated processes, medium-level agreement among these experiments suggests an increase in the monsoonal water budgets, mainly due to the intensification of monsoonal P under warmer climates. Analyzing changes under the RCP8.5, Hasson et al. (2016a) additionally reported an increase in monsoonal precipitation, in its seasonality, and in its spatial domain, which are linked to increased risks of floods and droughts and related disasters at the monsoonal margins.

11.4 CONCLUSIONS

Because the Indus Basin hydro-climatology, nourished by distinct precipitation regimes of the southwestern summer monsoon and year-round western disturbances at their extreme margins, can suffer from a substantial imbalance between water demand and water supply due to climate-induced pressures, basin-scale assessment of the prevailing and projected changes in the hydrological cycle under warmer climates is of utmost importance for making informed decisions on the planning and management of water resources and subsequently of their dependent sectors, particularly the transboundary irrigated agrarian economies. Therefore this chapter characterizes the prevailing precipitation regimes and analyzes 21st century changes in basin-scale water budgets using the full CMIP5 data archive under the RCP4.5 scenario.

To assess robust changes in future water budgets, the CMIP5 experiments were first analyzed for their fidelity, which was assessed in terms of physical consistency of closed water balance and the reproducibility of observed precipitation distribution and water budget, by employing a comprehensive observational database. Results suggest that approximately 13 experiments failed such a physical consistency test, while around one fourth did not see the monsoonal precipitation regime over the Indus Basin. These low-fidelity experiments must be excluded from any further application for the Indus Basin as the unsystematic dynamic and thermodynamic biases introduced due to unphysical water gain and loss cannot be disentangled from the simulated regional climate and its change signal. On the other hand, bias correcting the magnitudes and the statistical properties of a nonexistent monsoonal precipitation regime does not seem appropriate. This further emphasizes that the considered fidelity assessment procedure should be taken as a selection principle for regional and global climate modeling experiments for the Indus Basin.

Regardless of the abovementioned difficulties in simulating the intricate hydroclimatology of the Indus Basin, the CMIP5 experiments still serve as a primary source on the future assessment of water resources for a transboundary basin under warmer climates in the foreseeable future. Thus changes in the water budgets are presented from 14 selected experiments featuring moderate-to-high fidelity. In view of the remarkable spread in the simulated quantities of these selected experiments, their projected mean ensemble changes are mostly interpreted in connection with the qualitative agreement among the individual experiments. Results reveal high agreement among individual experiments on their projected mean ensemble decrease in the water budget under WPR, which dominates over the Upper Indus Basin during the spring, throughout the 21st century. In contrast, a moderate agreement suggests an intensification of MPR and an increase in the water budget over an extended area. The overall change in the annual water budget is small, suggesting a slight decrease over the Upper Indus Basin but an increase over the Lower Indus Basin. Such projections under warmer climates in the future postulate increasing pressure on the existing cryosphere of the Upper Indus Basin until a tipping point is reached, after which there will be more dependence on monsoonal precipitation.

It is important to mention that the changes presented in this chapter on the future water budgets under the RCP4.5 scenario offer only one of many possible futures. It is therefore warranted that changes under low- and high-end climate change scenarios of RCP2.6 and RCP8.5 are assessed in order to present the entire uncertainty spectrum. Further, the first-order water budget changes discussed need to be assessed in detail, particularly for the melt-dominated Upper Indus, by incorporating the spatio-temporal dynamics of the existing cryosphere. This can be achieved by coupling high-fidelity climate modeling experiments with sophisticated hydrological models that can simulate pluviometric, nival, and glacial dynamics. Moreover, the increasing availability of remotely sensed observational data sets provides a great opportunity to better quantify the observational uncertainty over the poorly monitored Indus Basin for the robust fidelity assessment of climate modeling experiments; however, such observational data sets need to be validated against the independent station observations, particularly for the Upper Indus Basin.

Acknowledgments

The authors acknowledge the World Climate Research Programme's Working Group on Coupled Modeling, which is responsible for CMIP, and the climate modeling groups for producing and making available their model outputs through the Earth System Grid Federation infrastructure, an international effort led by the US Department of Energy's Program for Climate Model Diagnosis and Intercomparison, the European Network for Earth System Modeling, and other partners in the Global Organization for Earth System Science Portals. The authors also acknowledge the groups developing and making available the gridded observational data sets, the provision of discharge data from the Water and Power Development Authority (WAPDA) in Pakistan, and the support of Cluster of Excellence "CliSAP" (EXC177), Universität Hamburg, funded through the German Research Foundation (DFG).

References

Adler, R.F., Huffman, G.J., Chang, A., Ferraro, R., Xie, P.P., Janowiak, J., Rudolf, B., Schneider, U., Curtis, S., Bolvin, D., Gruber, A., 2003. The version-2 global precipitation climatology project (GPCP) monthly precipitation analysis (1979–present). J. Hydrometeorol. 4 (6), 1147–1167.

Ali, G., Hasson, S., Khan, A.M., 2009. Climate Change: Implications and Adaptation of Water Resources in Pakistan. Global Change Impact Studies Centre (GCISC), GCISC RR-13, Islamabad, Pakistan.

Ashouri, H., Hsu, K.L., Sorooshian, S., Braithwaite, D.K., Knapp, K.R., Cecil, L.D., Nelson, B.R., Prat, O.P., 2015. PERSIANN-CDR: daily precipitation climate data record from multisatellite observations for hydrological and climate studies. Bull. Am. Meteorol. Soc. 96 (1), 69–83.

Beck, H.E., van Dijk, A.I., Levizzani, V., Schellekens, J., Miralles, D.G., Martens, B., de Roo, A., 2017. MSWEP: 3-hourly 0.25 global gridded precipitation (1979-2015) by merging gauge, satellite, and reanalysis data. Hydrol. Earth Syst. Sci. 21 (1), 589.

Boos, W.R., Hurley, J.V., 2013. Thermodynamic bias in the multimodel mean boreal summer monsoon. J. Clim. 26 (7), 2279–2287.

Chakraborty, A., Nanjundiah, R.S., Srinivasan, J., 2006. Theoretical aspects of the onset of Indian summer monsoon from perturbed orography simulations in a GCM. Ann. Geophys. 24 (8), 2075–2089.

Chen, M., Xie, P., Janowiak, J.E., Arkin, P.A., 2002. Global land precipitation: a 50-yr monthly analysis based on gauge observations. J. Hydrometeorol. 3 (3), 249–266.

CIESIN, 2005. Center for International Earth Science Information Network: Columbia University, and Centro Internacional de Agricultura Tropical (CIAT), Gridded Population of the World, Version 3 (GPWv3): Population Density Grid, Future Estimates.

Frieler, K., Lange, S., Piontek, F., Reyer, C.P., Schewe, J., Warszawski, L., Zhao, F., Chini, L., Denvil, S., Emanuel, K., Geiger, T., 2017. Assessing the impacts of 1.5 C global warming–simulation protocol of the inter-sectoral impact model Intercomparison project (ISIMIP2b). Geosci. Model Dev. 10 (12), 4321.

Fu, Q., Lin, P., 2011. Poleward shift of subtropical jets inferred from satellite-observed lower-stratospheric temperatures. J. Clim. 24 (21), 5597–5603.

Harris, I.P.D.J., Jones, P.D., Osborn, T.J., Lister, D.H., 2014. Updated high-resolution grids of monthly climatic observations –the CRU TS3 10 dataset. Int. J. Climatol. 34 (3), 623–642.

Hasson, S., 2016a. Seasonality of precipitation over Himalayan watersheds in CORDEX South Asia and their driving CMIP5 experiments. Atmosphere 7 (10), 123.

Hasson, S., 2016b. Future water availability from Hindukush-Karakoram-Himalaya upper Indus Basin under conflicting climate change scenarios. Climate 4 (3), 40.

Hasson, S., Lucarini, V., Pascale, S., 2013. Hydrological cycle over south and southeast Asian river basins as simulated by PCMDI/CMIP3 experiments. Earth Syst. Dyn. 4 (2), 199–217.

Hasson, S., Lucarini, V., Pascale, S., Böhner, J., 2014a. Seasonality of the hydrological cycle in major south and southeast Asian river basins as simulated by PCMDI/CMIP3 experiments. Earth Syst. Dyn. 5 (1), 67–87.

Hasson, S., Lucarini, V., Khan, M.R., Petitta, M., Bolch, T., Gioli, G., 2014b. Early 21st century snow cover state over the western river basins of the Indus River system. Hydrol. Earth Syst. Sci. 18 (10), 4077–4100.

Hasson, S., Pascale, S., Lucarini, V., Böhner, J., 2016a. Seasonal cycle of precipitation over major river basins in south and Southeast Asia: a review of the CMIP5 climate models data for present climate and future climate projections. Atmos. Res. 180, 42–63.

Hasson, S., Gerlitz, L., Schickhoff, U., Scholten, T., Böhner, J., 2016b. Recent climate change over high Asia. In: Climate Change, Glacier Response, and Vegetation Dynamics in the Himalaya. Springer, Cham, pp. 29–48.

Hasson, S., Böhner, J., Lucarini, V., 2017. Prevailing climatic trends and runoff response from Hindukush–Karakoram–Himalaya, upper Indus Basin. Earth Syst. Dyn. 8 (2), 337.

Hasson, S., Böhner, J., Chishtie, F., 2018. Low fidelity of CORDEX and their driving experiments indicates future climatic uncertainty over Himalayan watersheds of Indus basin. Clim. Dyn., 1–22.

Karim, A., Veizer, J., 2002. Water balance of the Indus River Basin and moisture source in the Karakoram and western Himalayas: implications from hydrogen and oxygen isotopes in river water. J. Geophys. Res. Atmos. 107 (D18), 1–12.

Laghari, A.N., Vanham, D., Rauch, W., 2012. The Indus basin in the framework of current and future water resources management. Hydrol. Earth Syst. Sci. 16 (4), 1063–1083.

Levine, R.C., Turner, A.G., Marathayil, D., Martin, G.M., 2013. The role of northern Arabian Sea surface temperature biases in CMIP5 model simulations and future projections of Indian summer monsoon rainfall. Clim. Dyn. 41 (1), 155–172.

Masui, T., Matsumoto, K., Hijioka, Y., Kinoshita, T., Nozawa, T., Ishiwatari, S., Kato, E., Shukla, P.R., Yamagata, Y., Kainuma, M., 2011. An emission pathway for stabilization at 6 Wm- 2 radiative forcing. Clim. Chang. 109 (1–2), 59.

Mehta, J.S., 1988. The Indus water treaty: a case study in the resolution of an international river basin conflict. Nat. Resour. Forum 12 (1), 69–77.

Moss, R.H., Edmonds, J.A., Hibbard, K.A., Manning, M.R., Rose, S.K., Van Vuuren, D.P., Carter, T.R., Emori, S., Kainuma, M., Kram, T., Meehl, G.A., 2010. The next generation of scenarios for climate change research and assessment. Nature 463 (7282), 747.

Peixoto, J.P., Oort, A.H., 1992. Physics of Climate. American Institute of Physics Press, New York.

Saeed, F., Hagemann, S., Jacob, D., 2009. Impact of irrigation on the south Asian summer monsoon. Geophys. Res. Lett. 36(20).

Schneider, U., Becker, A., Finger, P., Meyer-Christoffer, A., Ziese, M., Rudolf, B., 2014. GPCC's new land surface precipitation climatology based on quality-controlled in situ data and its role in quantifying the global water cycle. Theor. Appl. Climatol. 115 (1–2), 15–40.

Taylor, K.E., Stouffer, R.J., Meehl, G.A., 2012. An overview of CMIP5 and the experiment design. Bull. Am. Meteorol. Soc. 93 (4), 485–498.

Thomson, A.M., Calvin, K.V., Smith, S.J., Kyle, G.P., Volke, A., Patel, P., Delgado-Arias, S., Bond-Lamberty, B., Wise, M.A., Clarke, L.E., Edmonds, J.A., 2011. RCP4.5: a pathway for stabilization of radiative forcing by 2100. Clim. Chang. 109 (1–2), 77.

Willmott, C.J., Matsuura, K., 2001. Terrestrial Air Temperature and Precipitation: Monthly and Annual Time Series (1950–1999). http://climate.geog.udel.edu/~climate/html_pages/README.ghcn_ts2.html.

Xie, P., Arkin, P.A., 1997. Global precipitation: a 17-year monthly analysis based on gauge observations, satellite estimates, and numerical model outputs. Bull. Am. Meteorol. Soc. 78 (11), 2539–2558.

Xie, P., Chen, M., Shi, W., 2010. CPC unified gauge-based analysis of global daily precipitation. In: Preprints, 24th Conference on Hydrology, Atlanta, GA, American Meteorological Society.

Yatagai, A., Kamiguchi, K., Arakawa, O., Hamada, A., Yasutomi, N., Kitoh, A., 2012. APHRODITE: Constructing a long-term daily gridded precipitation dataset for Asia based on a dense network of rain gauges. Bull. Am. Meteorol. Soc. 93 (9), 1401–1415.

PART IV

WATER EXTREMES IN INDUS RIVER BASIN

Water Resources Forecasting Within the Indus River Basin: A Call for Comprehensive Modeling

Thomas E. Adams III

TerraPredictions, LLC, Blacksburg, VA, United States

12.1 INTRODUCTION

The hydrometeorological setting of the Indus River Basin (Fig. 12.1) is complex. Details are discussed in Section 12.1.2. These complexities are derived from both natural influences and human activities. From a water resources management perspective these complexities clearly demonstrate the necessity for the development of a comprehensive hydrometeorological modeling system for the Indus River Basin that spans forecast horizons from short lead times (hours) to climate time scales (months, seasons, interannual). Giupponi and Sgobbi (2013) emphasize the need for approaching water resources management using integrated water resources management tools and methods to improve natural resources management in a sustainable way. Such systems should include the capabilities of an early warning system for floods as well as meeting the needs for longer lead times for water resources planning and management. Importantly, significant flow contributions from *headwater* regions of the Indus River originate in Afghanistan, China, and India, which leaves Pakistan vulnerable and dependent on the need to obtain real-time, observed, and forecast streamflow information from neighboring countries. Unfortunately, with the contentious political conditions in the region, the likelihood of development and implementation of a comprehensive, fully cooperative *transboundary* hydrometeorological modeling system is unlikely. Consequently, this implies that any meaningful discussion of the development of a comprehensive hydrometeorological modeling system for the Indus River Basin should begin with the assumption that very limited cooperation from neighboring countries will be substantially forthcoming. Contributions by other authors in this book discuss issues related to the geopolitics of the

267

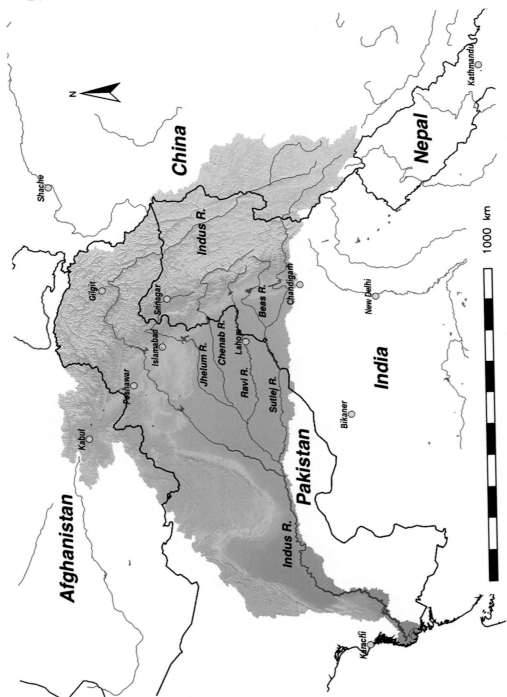

FIG. 12.1 Map of the Indus River Basin with major tributaries (*blue lines*). Country borders for Pakistan, India, China, and Afghanistan are shown as *heavy dark gray* lines. Prominent reservoirs are shown as *red areas*, and some major cities are identified for reference as *yellow circles*. The *Arabian Sea* lies to the left of the map. (*From Lehner, B., Liermann, C.R., Revenga, C., Vörösmarty, C., Fekete, B., Crouzet, P., Döll, P., Endejan, M., Frenken, K., Magome, J., Nilsson, C., Robertson, J.C., Rödel, R., Sindorf, N., Wisser, D., 2011. High-resolution mapping of the world's reservoirs and dams for sustainable river-flow management. Front. Ecol. Environ. 9 (9), 494–502.*)

IV. WATER EXTREMES IN INDUS RIVER BASIN

region and implications to water security and sustainability within the Indus River Basin. The aim of the current chapter is to

1. explain the necessity for hydrological services (Section 12.1.1);
2. illustrate the complexity of the Indus River Basin hydro-climatology (Section 12.1.2);
3. catalog the recent history and geographic occurrence of flooding within the basin (Section 12.1.3);
4. identify regions of irrigation and underscore the importance of water resources within the basin (Section 12.1.4);
5. report on the current status of hydrometeorological monitoring within the basin (Section 12.2);
6. describe the status of hydrological modeling and forecasting within the basin for each of the countries, Afghanistan, China, India, and Pakistan (Section 12.3);
7. identify needed improvements and benefits to hydrological modeling and forecasting within the basin, particularly in Pakistan (Section 12.4); and
8. present the challenges to making the needed changes to hydrological services in Pakistan (Section 12.5).

12.1.1 Necessity for Hydrological Services

A hydrological service (HS) is an entity whose central activity is to provide information to the public, governmental bodies, and other end-users on the status and trends of a country's water resources (UNISDR, 2012). These activities include real-time hydrologic forecasts and outlooks, ranging from hourly to interannual time scales, and development and management studies with planning horizons. Most typically, the activity of an HS is focused on the assessment of water resources related to water management for water supply, irrigation, hydropower generation, including drought monitoring and outlooks, as well as flood forecasting and warnings. HSs typically coordinate directly with disaster risk management (DRM) agencies at regional and national levels.

The World Meteorological Organization (WMO, 2006) reports that in most countries, the functions of HSs are typically served by several related water agencies, often with overlapping responsibilities. HS activities include systematic water resources monitoring, including data collection, processing, storage, and archiving, the production and dissemination of related water resources data and information, and hydrological forecasting spanning a range of time scales, with the issuance of flood alerts and warnings as needed (WMO, 2011; Adams and Thomas, 2016). Water resources cannot be properly managed without quantification of their current and potential future physical distribution, quality, and variability. Data from hydrological networks are used by public and private sectors for a variety of applications, including reservoir management, hydropower generation, water supply, water allocations for irrigation, and DRM. Recognizing that there is no substitute for water as a natural resource and that it is essential for survival and development, the United Nations and World Bank Group (2016b) called for a comprehensive and coordinated approach to water in a *joint statement* from the *high-level panel on water*, emphasizing the need for increased attention to and investment in water-related services.

Herold and Rudari (2013) have found that insufficient hydrometeorological monitoring, modeling, and forecasting capabilities exist in low- and medium-income countries within

National Hydrological Services (NHSs) programs. Recently WMO, WBG, and GFDRR (2018) undertook a global assessment to identify obstacles confronting the establishment and sustainability of NHSs. The initiative drew from global databases and reports, national assessments in Africa, and many other regional and national studies. The global assessment report details the current status of hydrological information and service delivery systems, focusing on conditions in low- and middle-income countries. The study shows that HSs in most countries cannot meet the needs of their populations. The identified problem areas include

1. fragmented and narrowly focused policy environments;
2. insufficient budgets;
3. inability to attract, train, and retain qualified staff;
4. limited and often deteriorating hydrometeorological monitoring networks;
5. insufficient maintenance of hydrological infrastructure;
6. inadequate data management systems;
7. insufficient integration between hydrological and meteorological services;
8. poor connection with end-users; and
9. unsatisfactory service delivery.

Importantly, the study findings emphasize the benefits derived from modernizing and maintaining robust HSs. Fortunately, recent World Bank (WB)-funded projects to address HS modernization in Afghanistan, India, and Pakistan are either underway or are planned (WBG, 2011, 2016, 2017). National approaches to flood warning and water resources management systems are needed because intermittently funded and isolated projects lead to fragmentation and disarray of monitoring and modeling systems that deteriorate and fall into disuse over time.

Modernization of HSs in Pakistan is critical. Kugelman and Hathaway (2009) and contributors discuss the worsening water crisis in Pakistan, which includes diminished water availability, from about 5000 m^3 per capita in the early 1950s to less than 1500 m^3 per capita in 2009. Approximately 90% of Pakistan's water is allocated for irrigation and other agricultural uses. Khan (2009) traces many of the water problems in Pakistan to governance issues, mismanagement, and suspect decision making.

The water problems facing Pakistan and neighboring Indus River Basin countries are common in low- and middle-income countries. The United Nations and World Bank Group *high-level panel on water* (United Nations and World Bank Group, 2016a) underscore the critical need for comprehensive water management within a robust NHS, identifying that:

- Water scarcity affects more than 40% of the global population and is projected to rise.
- Floods and other water-related disasters account for 70% of all deaths related to natural disasters.
- Over 1.7 billion people are currently living in river basins where water use exceeds recharge.
- More than 80% of wastewater resulting from human activities is discharged into rivers or seas without any pollution removal.
- At least 663 million people lack access to safe drinking water.
- 2.4 billion people lack access to basic sanitation services, such as toilets or latrines.
- Each day nearly 1000 children die due to preventable water and sanitation-related diarrheal diseases.

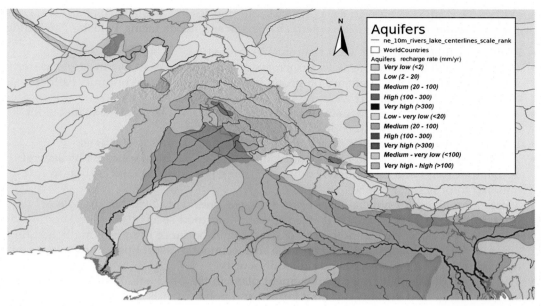

FIG. 12.2   Map showing geographic extent and recharge rates (mm/year) of groundwater aquifers. *Blue shading* indicates *major unconfined alluvial aquifers; pink shading* indicates *complex hydrogeologic structures; yellow to red shading* indicates *local and shallow aquifers. (From WHYMAP GWR BGR and UNESCO 2015—Bundesanstalt für Geowissenschaften und Rohstoffe (BGR) and UNESCO, 2008. Groundwater Resources of the World 1:25000000. Hannover, Paris. Available from: https://www.whymap.org/whymap/EN/Home/whymap-node.html.)*

The complexity of water resources management in the Indus River Basin has grown in recent decades by the use of groundwater for irrigation (Briscoe and Qamar, 2006). Fig. 12.2 shows major groundwater aquifers in the Indus River Basin region, indicating mean annual recharge rates. Increased groundwater use in Pakistan for irrigation has led to water-logging in many areas and decreased water quality due to increased salinity from irrigation. In other regions groundwater withdrawals have led to reduced water table levels and greater pumping costs (Shams-ul Mulk, 2009; Yu et al., 2013).

Related to flood forecasting needs, Girons et al. (2017) emphasize the benefits derived from early warning systems for flood loss mitigation, stating that they "found that efforts to promote and preserve social preparedness may help to reduce disaster-induced losses by almost one half…."

12.1.1.1 Pakistan's Needs

Significant hydrometeorological infrastructure improvements are needed in the Indus River Basin to manage water resources, principally for irrigation, water supply, and flood prediction. Yu et al. (2013) describe the influence of the complex hydro-climatology of the Indus River Basin in shaping water resources availability and decision making by the central government of Pakistan based on current and possible future climatic conditions. A study by Briscoe and Qamar (2006) conveys heightened awareness of Indus River Basin water resources issues in Pakistan.

Within the context of current water stresses and policies in Pakistan, Yu et al. (2013) identify several water resources management challenges and needs facing Pakistan, including:

- Multiyear storage in the Indus basin remains limited.
- Water and food demands are likely to increase on a per capita basis and in aggregate terms, as population increases. Reliance on groundwater resources will continue. Falling water tables and increased salinity in many places may worsen.
- An array of allocation entitlements economically constrains the waters available for agricultural production and coping with climatic risks.
- Low water-use efficiencies and agriculture productivities are top concerns.
- A common set of water and agricultural policy challenges is complicated by several dynamic stresses and institutional shifts, including constitutional devolution from national to provincial levels.
- Most national and provincial development plans continue to focus on the role of infrastructure in addressing challenges of water and food security.
- Recent policy documents highlight the increasing importance of improving irrigation efficiency, of improving yields, and of the socioeconomic distribution of development opportunities and benefits, including food security.
- The important role that water management plays in the productivity of the agriculture sector is recognized in many different forums and policy reports. *However, these linkages are not always comprehensively addressed (with systems-based models) in federal and provincial planning documents and budgets.*

Emphasis is added to the latter point, as it underscores the need for a comprehensive modeling-based system for water resources management, including flooding-related impacts. While annual flood control benefits are estimated to be significantly smaller compared with benefits accrued from irrigation or hydropower generation, flooding impacts are often significant, and are discussed in Section 12.1.3. An example of the economic impact of major flooding was seen in 2010. The Asian Development Bank and World Bank (2010) lists key issues related to flood response and management: (1) the deferred maintenance of flood embankments resulting in structural failures, (2) insufficient storage capacity to absorb flood peaks, (3) lack of response mechanisms to early warnings, (4) need for expanding flood early warning systems (FEWSs), and (5) encroachment into flood plains and riverine areas. The Asian Development Bank and World Bank (2010) also observes that postflood assessments emphasize the need for both nonstructural and structural measures to be implemented for comprehensive flood management.

12.1.2 Hydro-Climatology

The hydro-climatology of the Indus River Basin is complex. Much of the basin falls within an arid zone that extends from West Africa to eastern India between about 10°N to 40°N latitude, as illustrated in the map showing the Köppen-Geiger climate classification (Köppen, 1936) globally in Fig. 12.3 (refer to Table 12.1, Kottek et al., 2006). A detailed map of the Köppen-Geiger climate classification for the basin region is shown in Fig. 12.4. The climate of South Asia is dominated by the summer monsoonal circulation system. The Himalayas commonly receive excessive rainfall, particularly over south-facing slopes, which

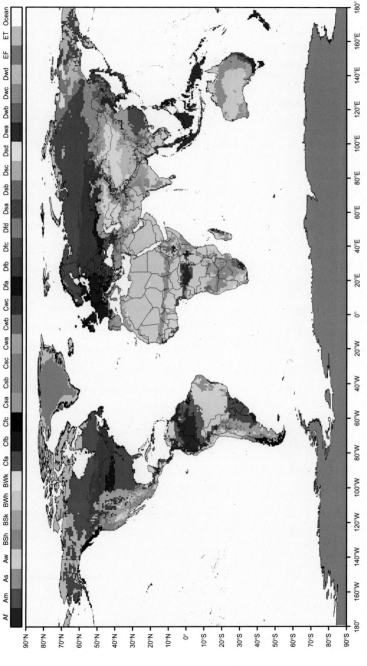

FIG. 12.3 Map showing global Köppen-Geiger climate classification; refer to Table 12.1 for key and description details. *(From Kottek, M., Grieser, J., Beck, C., Rudolf, B., Rubel, F., 2006. World map of the Köppen-Geiger climate classification updated. Meteorol. Z. 15 (3), 259–263.)*

IV. WATER EXTREMES IN INDUS RIVER BASIN

TABLE 12.1   Köppen-Geiger Climate Classification Key and Description (Köppen, 1936)

ID	Code	Description
1	Af	Tropical/rainforest
2	Am	Tropical/monsoon
3	Aw	Tropical/savana
4	BWh	Arid/desert/hot
5	BWk	Arid/desert/cold
6	BSh	Arid/steppe/hot
7	BSk	Arid/steppe/cold
8	Csa	Temperate/dry summer/hot summer
9	Csb	Temperate/dry summer/warm summer
10	Csc	Temperate/dry summer/cold summer
11	Cwa	Temperate/dry winter/hot summer
12	Cwb	Temperate/dry winter/warm summer
13	Cwc	Temperate/dry winter/cold summer
14	Cfa	Temperate/without dry season/hot summer
15	Cfb	Temperate/without dry season/warm summer
16	Cfc	Temperate/without dry season/cold summer
17	Dsa	Cold/dry summer/hot summer
18	Dsb	Cold/dry summer/warm summer
19	Dsc	Cold/dry summer/cold summer
20	Dsd	Cold/dry summer/very cold winter
21	Dwa	Cold/dry winter/hot summer
22	Dwb	Cold/dry winter/warm summer
23	Dwc	Cold/dry winter/cold summer
24	Dwd	Cold/dry winter/very cold winter
25	Dfa	Cold/without dry season/hot summer
26	Dfb	Cold/without dry season/warm summer
27	Dfc	Cold/without dry season/cold summer
28	Dfd	Cold/without dry season/very cold winter
29	Et	Polar/tundra
30	Ef	Polar/frost

IV. WATER EXTREMES IN INDUS RIVER BASIN

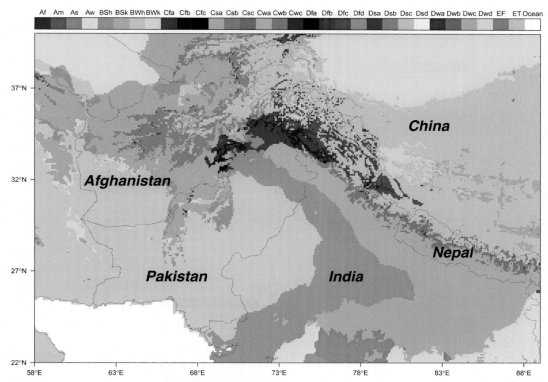

Af Am As Aw BSh BSk BWh BWk Cfa Cfb Cfc Csa Csb Csc Cwa Cwb Cwc Dfa Dfb Dfc Dfd Dsa Dsb Dsc Dsd Dwa Dwb Dwc Dwd EF ET Ocean

FIG. 12.4 Map showing Köppen-Geiger climate classification for the Indus River Basin region; refer to Table 12.1 for key and description details. *(From Kottek, M., Grieser, J., Beck, C., Rudolf, B., Rubel, F., 2006. World map of the Köppen-Geiger climate classification updated. Meteorol. Z. 15 (3), 259–263.)*

leads to flooding episodes during the summer monsoons (June-July-August-September, JJAS), often producing devastating landslides (Bhatt and Nakamura, 2005). It is widely observed that orographic effects of major orogenic belts, such as the Himalayas, control the occurrence, enhanced intensity, and distribution of precipitation due to their prominent relief (Andermann et al., 2011). Fig. 12.5A depicts the fraction of mean annual precipitation that occurred during the monsoon season from 1981 to 2010. The fraction of mean annual precipitation falling during the winter season (December-January-February, DJF) is shown in Fig. 12.5B. Fig. 12.5 shows the dominance of monsoonal precipitation on the annual water budget for the basin and particularly as it relates to major regions of groundwater recharge (shown in Fig. 12.2), significant flood deaths (Fig. 12.10), and more humid climatic conditions. However, with high evapotranspiration (ET) and low rainfall, spatially averaged mean annual runoff is negligible over much of the basin as Fig. 12.5D shows. MODIS satellite derived ET estimates from Running et al. (2017, 2018), shown in Fig. 12.6, provide a sense of the variability and magnitude of ET estimates over the region. Bastiaanssen et al. (2012) have developed techniques to estimate the surface energy balance and actual ET of the transboundary Indus basin region from satellite measurements and the recently developed ETLook model. These estimates are substantially different from Running et al. (2017, 2018).

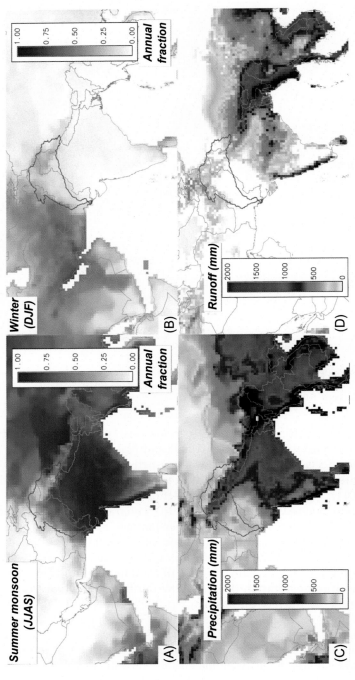

FIG. 12.5 Map showing the fraction of mean annual precipitation, 1981–2010, for the (A) summer monsoon season (June-July-August-September, JJAS), (B) winter season (December-January-February, DJF), (C) mean annual precipitation (Becker et al., 2013; Schneider et al., 2015), and (D) mean annual runoff (1950–2000) (Fekete et al., 2002) for the Indus River Basin region. The Indus R basin is identified by a *red* outline.

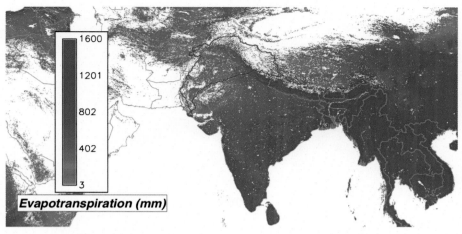

FIG. 12.6   Map of MODIS estimated mean annual *evapotranspiration*, 2000–13, for the South Asia region, showing the Indus River Basin as a *red* outline. *White areas* (other than oceans), on land areas, are *barren, sparse vegetation (rock, tundra, desert);* perennial *snow, ice; permanent* wetlands/inundated marshland; urban. *(From Running, S., Mu, Q., Zhao, M., 2018. MOD16A3 MODIS/Terra Net Evapotranspiration Yearly L4 Global 500m SIN Grid V006 [Data set]. https://dx.doi.org/10.5067/MODIS/MOD16A3.006. Available from: http://files.ntsg.umt.edu/data/NTSG_Products/ MOD16/MOD16A3.105_MERRAGMAO/Geotiff/MOD16A3_ET_2000_to_2013_mean.tif.)*

Global spatial datasets are beneficial in identifying broad features and trends over *small scales* and are particularly useful for decision makers. For example, the Consultative Group on International Agricultural Research (CGIAR), Consortium for Spatial Information (CSI) (http://www.cgiar-csi.org) developed an Aridity Index (AI) (Eq. 12.1) (from UNEP, 1997), which is shown for the Indus basin in Fig. 12.7A. Using Table 12.2, generalized AI classes can be identified that illustrate a predominant aridity, ranging from semiarid to hyperarid (with very high PET values, shown in Fig. 12.8), for much of the basin south of the mountainous Hindu Kush and Karakoram region and in the Tibetan Plateau (Fig. 12.7B). Only the foothills of the Hindu Kush and Karakoram fall within the Dry subhumid to Humid AI zones. It is clear from Fig. 12.9, which shows dominant vegetation classes in the Indus basin, how the climatic conditions influence the geographic distribution of plant species, both natural crops and crops that are sustainable with irrigation. These data form the basis for hydrologic model parameter estimation related to ET estimation and energy balance.

The AI is given by

$$AI = \frac{MAP}{MAPE} \tag{12.1}$$

where MAP is the mean annual precipitation (from, e.g., Fig. 12.5C) and MAPE is the mean annual PET (see, e.g., Fig. 12.8).

Monthly average PET (mm/month), using the Hargreaves method (Hargreaves et al., 1985; Hargreaves and Allen, 2003), is given by

$$PET = 0.0023 \cdot RA \cdot (T_{mean} + 17.8) \cdot TD^{0.5} \tag{12.2}$$

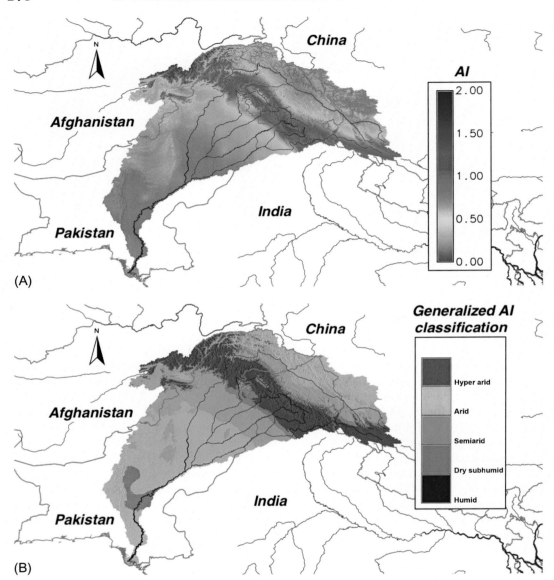

FIG. 12.7 Indus River Basin (A) Aridity Index (Eq. 12.1) and (B) *generalized AI classification* from Table 12.2, 1950–2000, showing major tributaries as *blue lines*. Country borders for Pakistan, India, China, and Afghanistan are shown as *heavy dark gray* lines. *(From UNEP, 1997. World Atlas of Desertification. Technical Report. United Nations Environment Programme (UNEP), London, UK.)*

TABLE 12.2   Aridity Index (Eq. 12.1)

Value	Climate Class
<0.03	Hyper arid
0.03–0.20	Arid
0.20–0.50	Semiarid
0.50–0.65	Dry subhumid
>0.65	Humid

Source: Classification From UNEP, 1997. World Atlas of Desertification. Technical Report. United Nations Environment Programme (UNEP), London, UK.

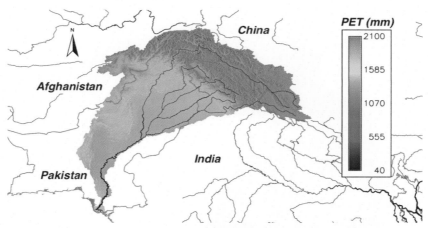

FIG. 12.8   Mean annual potential evapotranspiration (PET, Eq. 12.2), for the Indus River Basin, showing major tributaries as *blue lines*. Country borders for Pakistan, India, China, and Afghanistan are shown as *heavy dark gray* lines. *(From UNEP, 1997. World Atlas of Desertification. Technical Report. United Nations Environment Programme (UNEP), London, UK.)*

The method requires monthly average geo-data sets of (1) mean temperature (T_{mean}, °C), (2) daily temperature range (TD, °C), and (3) extraterrestrial radiation (RA, radiation on top of atmosphere expressed in millimeter per month as equivalent of evaporation).

For modeling purposes, greater spatial and temporal hydrometeorological detail is needed to produce flood warnings and alerts and accurate predictions for water resources management, such as discussed in Adams and Thomas (2016) related to regional and national flood forecasting systems.

12.1.3 Flooding

Referring to Table 12.3, the Dartmouth Flood Observatory (Brakenridge and Albert, 1985) has cataloged the occurrence of floods globally, including the Indus River Basin, since 1985. The number of deaths from these floods is shown in Fig. 12.10. Note that the Dartmouth Flood Observatory flood *severity classes* are defined as Class 1, Class 1.5, and Class 2:

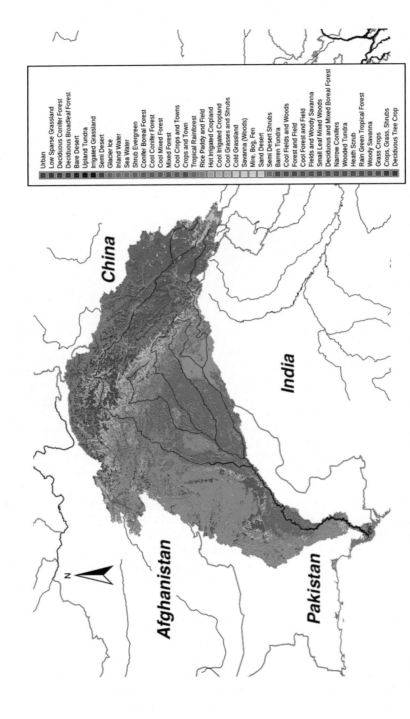

FIG. 12.9 Dominant vegetation types for the Indus River Basin, showing major tributaries as *blue lines*. Country borders for Pakistan, India, China, and Afghanistan are shown as *heavy dark gray lines*. (*From Loveland, T.R., Reed, B.C., Brown, J.F., Ohlen, D.O., Zhu, Z., Yang, L., Merchant, J.W., 2000. Development of a global land cover characteristics database and IGBP DISCover from 1 km AVHRR data. Int. J. Remote Sensing 21 (6/7), 1303–1330.*)

Continued

TABLE 12.3 Indus River Basin Floods, 1985–2017, Listing Country of Occurrence, Location, Deaths, Displaced Population, the Main Cause of the Flooding, and Flood Severity

ID	Country	Latitude	Longitude	Area (km²)	Began	Ended	Deaths	Displaced	Main Cause	Severity
33	India	76.9103	32.8353	117,441.17	7/18/1985	7/30/1985	340	20,000	Heavy rain	1
205	Pakistan	72.1319	31.8882	220,520.86	7/18/1988	8/5/1988	158	163,000	Monsoonal rain	1
234	Pakistan	68.0427	25.6834	33,007.62	8/18/1988	8/25/1988	0	200,000	Torrential rain	1
253	India	75.0459	30.9339	268,528.55	9/21/1988	10/8/1988	731	1,250,000	Heavy rain	2
401	India	76.2592	33.8036	90,818.98	3/21/1990	3/22/1990	17	0	Torrential rain	1
412	Pakistan	72.2929	34.4927	21,727.03	5/16/1990	5/18/1990	4	1200	Heavy rain	1
435	Pakistan	72.2699	31.0394	190,865.32	7/10/1990	7/12/1990	0	6000	Heavy rain	1
523	Pakistan	72.1321	34.0792	79,582.00	6/11/1991	6/12/1991	39	2400	Heavy rain	1
690	Afghanistan	69.8855	35.0968	47,999.00	9/3/1992	9/3/1992	3000	4000	Monsoonal rain	1
858	India	77.2520	31.7869	52,338.52	7/3/1994	7/12/1994	60	0	Monsoonal rain	1
880	India	76.9177	31.7821	16,795.70	8/8/1994	8/10/1994	11	0	Torrential rain	1
883	Pakistan	68.1903	27.7849	318,187.81	8/9/1994	8/15/1994	26	40,000	Monsoonal rain	1
887	India	77.2700	31.7668	14,450.04	8/20/1994	8/23/1994	9	0	Torrential rain	1
888	Pakistan	67.8824	29.6075	17,316.85	8/22/1994	8/27/1994	24	0	Torrential rain	1
894	India	77.0555	31.5294	5275.13	8/31/1994	9/5/1994	20	0	Monsoonal rain	1
973	Pakistan	68.5019	29.8869	672,265.35	7/19/1995	8/10/1995	600	600,000	Monsoonal rain	1
1108	Pakistan	72.4557	30.8497	202,950.18	9/2/1996	9/7/1996	119	100,000	Heavy rain	2
1172	India	77.3409	31.7905	57,530.65	8/8/1997	8/14/1997	200	0	Monsoonal rain	1
1181	Pakistan	72.5263	32.3559	276,909.38	8/12/1997	9/3/1997	165	836,300	Monsoonal rain	2
1182	Afghanistan	70.2244	34.8425	6428.10	9/7/1997	9/8/1997	30	600	Heavy rain	1
1251	Pakistan	74.5774	34.9725	5802.05	7/25/1997	7/29/1997	16	0	Monsoonal rain	1
1252	Pakistan	72.2365	32.2874	127,909.46	7/25/1997	7/28/1997	31	0	Monsoonal rain	1

TABLE 12.3 Indus River Basin Floods, 1985–2017, Listing Country of Occurrence, Location, Deaths, Displaced Population, the Main Cause of the Flooding, and Flood Severity—cont'd

ID	Country	Latitude	Longitude	Area (km²)	Began	Ended	Deaths	Displaced	Main Cause	Severity
1265	India	74.8298	33.9398	40,014.04	8/22/1997	8/26/1997	26	0	Monsoonal rain	1
1387	Afghanistan	70.1175	34.4397	11,603.78	6/12/1998	6/15/1998	30	100	Heavy rain	1
1616	India	77.6925	31.8060	12,415.81	7/31/2000	8/4/2000	140	4000	Brief torrential rain	1
1757	Pakistan	73.2608	34.9888	111,090.80	7/22/2001	7/25/2001	230	40,000	Brief torrential rain	1
1774	Afghanistan	69.1549	34.6188	1282.21	8/8/2001	8/8/2001	2	0	Brief torrential rain	1
1779	India	76.1481	32.6061	3009.11	8/14/2001	8/15/2001	16	0	Heavy rain	1
1780	Pakistan	70.7261	30.0691	3031.91	8/16/2001	8/16/2001	6	0	Brief torrential rain	1
2017	Pakistan	71.2064	30.0167	1352.07	8/3/2002	8/7/2002	3	800	Heavy rain	1
2040	Pakistan	72.0269	34.4920	4902.73	8/22/2002	8/28/2002	22	3000	Heavy rain	1
2160	Pakistan	68.0677	27.5247	433,469.94	2/16/2003	2/22/2003	20	3000	Heavy rain	1
2163	Pakistan	73.7749	34.4037	26,796.79	2/15/2003	2/24/2003	27	0	Heavy rain	1
2211	Afghanistan	69.2724	34.9580	1769.35	4/18/2003	4/20/2003	2	800	Heavy rain	1
2215	Afghanistan	68.8736	33.9512	1449.13	4/23/2003	4/24/2003	0	2000	Heavy rain	1
2272	Afghanistan	69.1603	34.2175	1496.24	7/5/2003	7/6/2003	2	280	Brief torrential rain	1
2280	Afghanistan	69.5093	33.4920	33,247.73	7/15/2003	7/18/2003	9	0	Heavy rain	1
2343	Pakistan	72.0521	35.2235	3519.30	9/5/2003	9/6/2003	30	0	Brief torrential rain	1
2357	Afghanistan	71.0404	34.9865	1785.32	9/19/2003	9/22/2003	3	0	Heavy rain	1
2641	Pakistan	68.3724	29.5182	75,648.02	3/20/2005	3/24/2005	20	3500	Heavy rain	1
2675	Pakistan	71.5076	34.2720	28,630.77	6/21/2005	8/5/2005	5	50,000	Snowmelt	2

Continued

2678	India	78.4423	31.6718	6530.74	6/26/2005	6/29/2005	6	5000	Heavy rain	1
2689	Pakistan	73.5637	30.9230	433,454.63	7/5/2005	8/14/2005	40	452,000	Monsoonal rain	1
2727	Afghanistan	71.0849	34.9305	3208.66	9/11/2005	9/12/2005	12	0	Heavy rain	1
2908	Pakistan	72.6179	35.3523	1429.80	7/3/2006	7/4/2006	22	0	Heavy rain	1
2916	Pakistan	69.9486	33.0562	6984.07	7/9/2006	7/13/2006	23	50	Heavy rain	1
2917	Pakistan	74.1578	35.2679	456.86	7/11/2006	7/12/2006	4	0	Heavy rain	1
2931	Pakistan	73.1998	34.6098	182,706.86	7/24/2006	8/22/2006	248	0	Monsoonal rain	1
2936	Afghanistan	69.7467	34.0789	63,167.25	7/30/2006	8/9/2006	35	0	Heavy rain	1
2958	India	73.8842	32.8864	141,449.13	8/31/2006	9/11/2006	39	3000	Monsoonal rain	1
2992	Afghanistan	70.7909	34.2199	1385.12	11/10/2006	11/12/2006	9	0	Heavy rain	1
3003	Pakistan	66.5167	28.0363	28,938.65	12/4/2006	12/6/2006	5	0	Heavy rain	1
3042	India	74.2156	33.4819	68,107.39	3/20/2007	3/31/2007	12	2000	Heavy rain	1
3055	Afghanistan	69.6682	35.2702	40,765.63	4/17/2007	4/18/2007	6	0	Heavy rain	1
3101	Pakistan	72.2258	34.9758	12,117.23	6/16/2007	6/20/2007	22	500	Brief torrential rain	1
3109	Afghanistan	71.1407	35.2782	62,335.91	6/24/2007	7/3/2007	113	4000	Heavy rain	1
3112	Pakistan	67.9025	25.2563	115,766.52	6/26/2007	7/20/2007	280	400,000	Tropical cyclone	1
3116	Pakistan	71.8699	34.2749	29,411.27	6/28/2007	7/22/2007	130	2000	Heavy rain	1
3164	India	76.5268	31.2388	191,449.95	8/12/2007	8/17/2007	76	15,000	Monsoonal rain	1
3165	Pakistan	74.3382	35.9513	16,541.18	8/12/2007	8/14/2007	22	0	Heavy rain	1
3347	Pakistan	71.7026	33.9163	32,493.08	8/2/2008	8/4/2008	35	200,000	Torrential rain	2
3359	Pakistan	72.6347	30.7385	165,918.21	8/9/2008	8/20/2008	37	90,752	Heavy monsoonal rains	1.5
3526	Pakistan	72.2471	33.9207	30,560.13	8/15/2009	8/17/2009	27	1200	Torrential rain	1.5
3647	Pakistan	73.3855	35.5882	6706.99	5/8/2010	5/23/2010	0	0	Heavy rain	1
3658	Pakistan	67.9052	25.1175	59,028.73	6/6/2010	6/7/2010	23	4000	Tropical cyclone Phet	1

IV. WATER EXTREMES IN INDUS RIVER BASIN

TABLE 12.3 Indus River Basin Floods, 1985–2017, Listing Country of Occurrence, Location, Deaths, Displaced Population, the Main Cause of the Flooding, and Flood Severity—cont'd

ID	Country	Latitude	Longitude	Area (km²)	Began	Ended	Deaths	Displaced	Main Cause	Severity
3666	Pakistan	72.0792	36.0270	9325.99	6/22/2010	6/24/2010	46	0	Torrential rain	1
3696	Pakistan	73.2579	33.3634	129,691.63	7/27/2010	11/15/2010	1750	10,000,000	Monsoonal rain	2
3697	Afghanistan	68.8594	35.2106	176,799.06	7/27/2010	8/3/2010	65	180	Torrential rain	1
3700	India	76.1256	33.5267	145,020.41	8/6/2010	8/8/2010	150	180	Torrential rain	1.5
3851	Afghanistan	70.5453	35.2213	27,920.61	8/5/2011	8/9/2011	6	20	Monsoonal rain	1
3974	Pakistan	73.5222	35.0305	98,264.36	8/23/2012	8/29/2012	26	1200	Monsoonal rain	1.5
3983	India	76.8569	31.3063	37,710.76	9/16/2012	9/18/2012	45	200	Torrential rain	2
3984	Pakistan	68.7756	28.0445	23,036.11	9/10/2012	10/29/2012	400	742,000	Monsoonal rain	1.5
4065	India	76.6135	32.7135	131,743.41	6/12/2013	6/27/2013	5748	75,000	Monsoonal rain	1.5
4078	Pakistan	70.6377	32.8144	549,425.38	8/1/2013	8/7/2013	135	81,000	Heavy rain	1.5
4082	Pakistan	69.1065	28.7404	462,763.09	8/7/2013	8/21/2013	70	12,000	Monsoon rain	1.5
4179	Pakistan	75.3419	33.9449	253,686.93	9/1/2014	10/11/2014	300	30,000	Monsoonal rain	2
4239	India	76.4058	33.1767	70,287.98	3/20/2015	3/31/2015	44	2907	Heavy rain	1.5
4245	Pakistan	71.8901	33.5131	21,148.15	4/26/2015	4/28/2015	44	4800	Torrential rain	1.5
4270	Pakistan	72.8428	35.0129	73,485.14	6/25/2015	6/29/2015	10	0	Torrential rain	1.5
4272	Pakistan	72.2737	34.3432	137,039.96	7/15/2015	8/19/2015	166	803,000	Monsoonal rain	1.5
4339	Pakistan	71.7788	33.0545	111,748.39	3/12/2016	4/13/2016	121	2400	Heavy rain	1.5
4372	Pakistan	72.7933	36.1038	31,431.28	7/3/2016	7/6/2016	30	80	Heavy rain	1.5
4489	Pakistan	72.3232	33.4014	22,2047.34	6/25/2017	7/3/2017	11	1000	Monsoonal rain	1
4501	Pakistan	74.0057	35.4339	42,978.95	8/3/2017	8/10/2017	5	0	Torrential rain and snowmelt	1.5

Source: Dartmouth Flood Observatory, http://floodobservatory.colorado.edu.

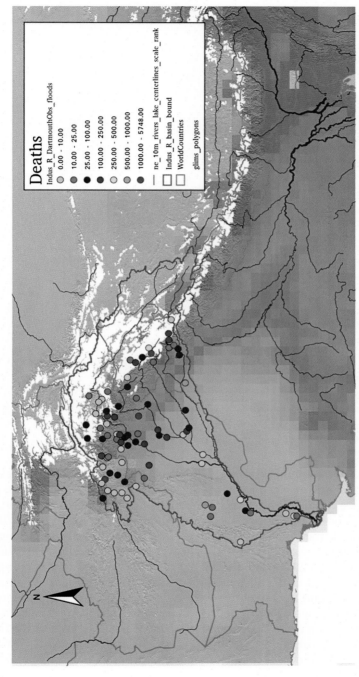

FIG. 12.10 Map showing the location and number of fatalities from 85 floods in the Indus River Basin (*red outline*), 1985–2017, listed in Table 12.3. Major tributaries are colored as *blue lines* and glaciers are colored *white*. Mean annual precipitation is shown as the background shading (refer to Fig. 12.13). (*From GLIMS and NSIDC, 2005. Global Land Ice Measurements From Space Glacier Database. GLIMS and NSIDC, Boulder, CO; Raup, B.H., Racoviteanu, A., Khalsa, S.J.S., Helm, C., Armstrong, R., Arnaud, Y., 2007. The GLIMS geospatial glacier database: a new tool for studying glacier change. Glob. Planet. Chang. 56, 101–110. Dartmouth Flood Observatory, http://floodobservatory.colorado.edu.*)

IV. WATER EXTREMES IN INDUS RIVER BASIN

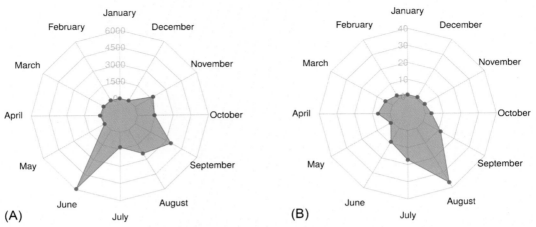

FIG. 12.11   Graphic showing the number of (A) fatalities from 85 floods in the Indus River Basin and (B) flood events by month, 1985–2017, based on the *end* of the flood period listed in Table 12.3. *(Dartmouth Flood Observatory, http://floodobservatory.colorado.edu.)*

SEVERITY CLASS: ASSESSMENT IS ON 1–2 SCALE

Floods are divided into three classes.

- *Class 1.* Large flood events: significant damage to structures or agriculture; fatalities; and/or one to two decades-long reported interval since the last similar event.
- *Class 1.5.* Very large events: with a greater than two decades but <100-year estimated recurrence interval, and/or a local recurrence interval of at one to two decades and affecting a large geographic region (>5000 km^2).
- *Class 2.* Extreme events: with an estimated recurrence interval >100 years.

Inspection of Table 12.3 reveals that nearly all flood-related deaths and the largest population displacements occur during the summer monsoon season. Interestingly, the peak of flood occurrences (August) and flood deaths (September) do not coincide (see Fig. 12.11). The greatest number of flooding deaths occurred from a single event, June 12–27, 2013 in India, in a glacially dominated region of the Himalayas. It is also noteworthy that population displacements can be massive, such as the 10,000,000 displaced in Pakistan in 2010 or the 742,000 displaced in Pakistan in 2012 from monsoonal rains and subsequent flooding. Many of the significant flooding events with substantial deaths occurred upstream of dams, where the dams offer no protection and in glacial regions where the threat of glacial lake outburst floods (GLOFs) is significant (Fig. 12.10). Table 12.3 also shows that many *flash flood*-related deaths occur, with *torrential rain* or *brief torrential rain* indicated as the cause of the flooding. Some aid in forecasting the threat of flash floods will be provided by the WMO Flash Flood Guidance System (FFGS) that is currently being implemented operationally for the South Asia region, including Pakistan. The WMO FFGS is discussed more fully in Section 12.3.1.

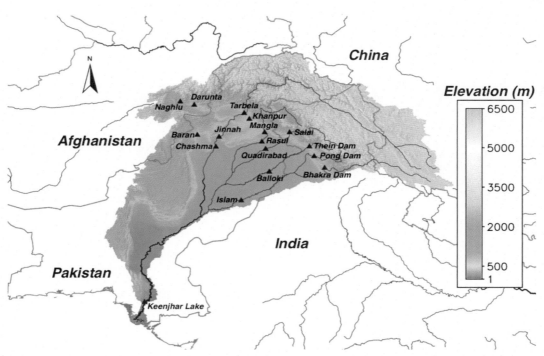

FIG. 12.12 Location of major dams (with storage ≥ 20 million m³) *(From UNEP, 1997. World Atlas of Desertification. Technical Report. United Nations Environment Programme (UNEP), London, UK.)* in the Indus River Basin *(From Lehner, B., Liermann, C.R., Revenga, C., Vörösmarty, C., Fekete, B., Crouzet, P.,Döll, P., Endejan, M., Frenken, K., Magome, J., Nilsson, C., Robertson, J.C., Rödel, R., Sindorf, N., Wisser, D., 2011. Highresolution mapping of the world's reservoirs and dams for sustainable river-flow management. Front. Ecol. Environ. 9 (9), 494–502)*, indicated by *blue* triangles, showing major tributaries as *blue lines*. Country borders for Pakistan, India, China, and Afghanistan are shown as *heavy dark gray* lines.

12.1.4 Irrigation and Water Resources

The use of water resources to support irrigation in the Indus basin is critical. Shams-ul Mulk (2009) describes the history of the development of the irrigation system in Pakistan. Briscoe and Qamar (2006) report that 90% of Pakistan's water resources are allocated to irrigation and meeting agricultural requirements. The importance of irrigation is further indicated in that irrigation is the principal purpose for the five largest dams in the basin and six of the seven largest dams, listed in Table 12.4 and shown in Fig. 12.12 with other major dams. Reservoir use for water supply and hydroelectric power generation is important as well. The major irrigation regions, shown in Fig. 12.13, in the Indus basin are located downstream of the major reservoirs. Management of water resources in Pakistan for water allocation to the Provincial Irrigation Departments is administered at the national level, with decision making based on river-flow monitoring by the Pakistan Indus River System Authority (IRSA). The location of the major irrigation regions correspond to AI (Eq. 12.1) zones identified as semiarid to arid in

TABLE 12.4 Major Dams in the Indus River Basin With Storage $\geq$20 Million m³

Name	River	Country	Year	Capacity	Depth	Elevation	Basin Area	Main Use	Longitude	Latitude
Tarbela	Indus	Pakistan	1976	13,940.0	66.9	342	199,690	Irrigation	72.691250	34.091250
Bhakra Dam	Sutluj	India	1963	9621.0	104.7	503	52,623	Irrigation	76.434583	31.412083
Pong Dam	Beas	India	1974	8570.0	45.2	411	12,614	Irrigation	75.948750	31.974583
Mangla	Jhelum	Pakistan	1967	7300.0	41.7	334	33,509	Irrigation	73.642917	33.145417
Thein Dam	Ravi	India	2000	3670.0	85.2	471	6141	Irrigation	75.729583	32.449583
Keenjhar Lake	local	Pakistan	1200	2526.5	25.3	15	1804	Water supply	68.050417	24.899583
Chashma	Indus	Pakistan	1971	1073.0	9.5	186	345,486	Irrigation	71.375417	32.437083
Naghlu	Kabul	Afghanistan		500.0	49.5	1070	26,279	Hydroelectric	69.712917	34.645417
Salal	Chinab	India	1986	285.0	42.5	466	21,812	Hydroelectric	74.808750	33.145417
Balloki	Ravi	Pakistan		259.7	27.9	186	17,906		73.857083	31.224583
Jinnah	Indus	Pakistan		246.3	28.0	211	317,969		71.525417	32.920417
Qadirabad	Chenab	Pakistan	1968	178.0	17.5	209	34,828	Irrigation	73.685417	32.327083
Islam	Sutlej	Pakistan		148.8	28.6	139	98,869		72.550417	29.827917
Rasul	Jhelum	Pakistan	1967	132.3	2.4	208	38,280	Irrigation	73.512083	32.685417
Khanpur	Haro	Pakistan	1983	132.0	35.7	588	770	Irrigation	72.929583	33.803750
Darunta	Kabul	Afghanistan	1963	125.0	11.0	613	34,705	Hydroelectric	70.362917	34.484583
Baran	Baran	Pakistan	1962	121.0	18.1	427	426	Irrigation	70.504583	33.016250
Tanda	Kohat Toi	Pakistan	1967	96.7	53.7	516	43	Irrigation	71.396250	33.574583
Marala	Chenab	Pakistan	1968	93.1	5.0	242	25,938	Irrigation	74.487917	32.699583
Warsak	Kabul	Pakistan	1960	76.5	8.9	405	68,573	Hydroelectric	71.354583	34.163750
Rawal	Korang	Pakistan	1962	58.6	13.0	524	266	Water supply	73.121250	33.695417
Ferozepur	Sutlej	Pakistan		45.2	30.1	193	85,472		74.553750	30.995417
Nangal Dam	Satluj	India	1954	43.1	13.9	351	52,699	Irrigation	76.367083	31.382917
Simly	Soan	Pakistan	1982	35.5	29.6	715	159	Water supply	73.342083	33.724583
Namal	Golar	Pakistan	1913	27.6	12.5	360	447	Water supply	71.804583	32.670417

Notes: Reservoir capacity is millions m³; depth in m; elevation in m; basin area in km². The Year for Keenjhar Lake is estimated.
Source: From UNEP, 1997. World Atlas of Desertification. Technical Report. United Nations Environment Programme (UNEP), London, UK.

IV. WATER EXTREMES IN INDUS RIVER BASIN

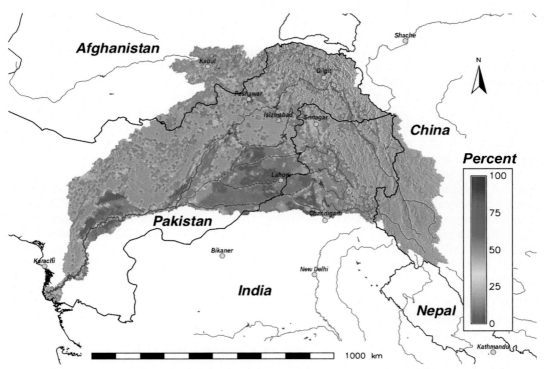

FIG. 12.13 Indus River Basin showing *percent irrigation* with major tributaries highlighted as *blue lines*. Country borders for Pakistan, India, China, and Afghanistan are shown as *heavy dark gray* lines. Prominent reservoirs are shown as *blue* areas and some major cities are identified for reference as *yellow circles*. The Arabian Sea lies to the *left* of the map. *(From Siebert, S., Henrich, V., Frenken, K., Burke, J., 2013. Global Map of Irrigation Areas version 5. Rheinische Friedrich-Wilhelms-University, Bonn, Germany/Food and Agriculture Organization (FAO) of the United Nations, Rome, Italy. Available from: http://www.fao.org/nr/water/aquastat/irrigationmap/index10.stm.)*

Fig. 12.7 in the Punjab region and hyperarid in the Sindh Province in southern Pakistan. This implies that water losses due to ET will be greatest in these areas (see Fig. 12.6) due to readily available sources of irrigation water and high PET, Eq. (12.2), shown in Fig. 12.8.

Bengali (2009) states that with Pakistan's total land area of 80 million hectares (ha), 21 million ha are cultivated of which 18 million ha are irrigated. About 12 million ha of Pakistan's irrigated land lies within the Indus River system. During major flood events, such as the major 2010 flood, significant risk of damage or failure is placed on *barrage* canal control structures. Consequently, water is diverted to the canal system to reduce damage to the barrages, which exacerbates flooding in irrigated areas.

Kugelman and Hathaway (2009); Bengali (2009); Khan (2009); Shams-ul Mulk (2009) discuss the competing interest for scarce water resources in Pakistan, recognizing that too little of the available water resources are allocated for water supply purposes.

12.2 STATUS OF HYDROMETEOROLOGICAL MONITORING

Some level of hydrometeorological monitoring and modeling exists in each of the four countries, Pakistan, India, Afghanistan, and China, that comprise Indus River Basin drainage. But these systems are not integrated, and observational and forecast data sharing is minimal. Recent interest in funding hydrometeorological infrastructure development projects by the World Bank Group point to the need for improved hydrometeorological monitoring (WBG, 2011, 2016). To the extent data are shared, data sharing does not adequately form the foundation of a robust and responsive hydrologic forecast system that serves flood and water resources forecasting purposes within the basin.

Specifically, in Pakistan, the Pakistan Meteorological Department (PMD) maintains a network of 32 automatic weather stations (AWS) and 116 daily stations. There is also an independent AWS called the flood automatic warning system (FAWS) consisting of a network of nine stations. The PMD Flood Forecasting Division (FFD), located in Lahore, utilizes streamflow observations collected by the Pakistan Water and Power Development Authority (WAPDA).

Independent hydrometerological observation monitoring networks are found in the following:

1. PMD/FFD: meteorological observations;
2. Punjab Irrigation Department: canal data, including design discharge, observed head gauge, and discharge;
3. Sindh Irrigation Department: canal data, including design discharge, observed head gauge, and discharge;
4. IRSA: streamflow monitoring at ~25 locations, 7 sites with rating tables;
5. Pakistan WAPDA, including Glacial Monitoring Research Center (GMRC): streamflow monitoring/AWS (44, ~30 operational) and about 197 manual, daily reporting stations from observers; and
6. Pakistan Agricultural Research Council (primarily for research and climate monitoring): <20 stations.

However, it remains unclear if some of these networks may actually overlap or not, with at least some monitoring stations being counted by multiple agencies. It is unclear, to a degree, where FFD, WAPDA, IRSA, and the irrigation districts responsibilities begin and end in terms of hydromet data collection. A major problem with the 44 WAPDA telemetered stations is that they are restricted geographically to the central region of Pakistan, of which only about 30 are functional.

The decline of hydrological monitoring networks is a global phenomenon and is not restricted to the Indus River Basin countries, as discussed by Hellen and Fry (2006) and others. Mishra and Coulibaly (2009) review methodologies used in the design of hydrometeorological monitoring networks. Zemadim et al. (2013) discuss the role of and need for local participation in the success of hydrometeorological monitoring in developing countries, identifying that the most commonly reported problems are

1. installation of equipment in catchments where little is known about the catchment characteristics;

2. theft and vandalism;
3. postinstallation damage due to floods or other natural events; and
4. institutional and policy barriers that hinder operation and maintenance of monitoring stations.

Zemadim et al. (2013) say that problems arise, in part, due to the lack of local stakeholder involvement in the establishment and operation of hydrological monitoring systems.

12.3 STATUS OF HYDROLOGICAL MODELING AND FORECASTING

12.3.1 Pakistan

The PMD FFD currently utilizes two hydrological modeling systems, namely, the Integrated Flood Analysis System (IFAS) and the Deltares-based FEWS, which utilizes the US National Oceanic and Atmospheric Administration (NOAA), National Weather Service (NWS) Sacramento Soil Moisture Accounting (SAC-SMA) (Burnash et al., 1973; Burnash, 1995) hydrologic model and the Deltares SOBEK 1D unsteady flow, hydrodynamic model. Both modeling systems are currently utilized only during the monsoonal flood season. Flood forecasting continues to rely on empirical regression relationships that translate observed upstream streamflows to downstream locations with a 24-h lead time. This technique is similar to the method used by Kohler (1944) in the NOAA/NWS in the 1940s.

The FFD FEWS implementation was aided with help from both Deltares and National Engineering Services Pakistan, principally for SOBEK model implementation in 2007. The PMD FEWS implementation does not include inflows upstream of Tarbela Dam or from the Kabul River; no model improvements have been made since that time.

IFAS is a large-scale hydrologic modeling system developed by the International Center for Water Hazard and Risk Management (ICHARM) in Japan and used internationally for flood analysis. IFAS provides interfaces to input both satellite-based and ground-based rainfall data. It includes Geographic and Information Systems (GIS) functions to create river channel networks and to estimate parameters of a default runoff analysis engine, the Public Works Research Institute distributed hydrologic model, and interfaces to display output results. The rainfall runoff forecasting model was customized for the Indus River Basin under the project "Strategic Strengthening of Flood Warning and Management Capacity of Pakistan" (2012–14). The project was undertaken by UNESCO with the help of government of Japan in collaboration with the PMD. The modeling system has the capability of forecasting flood wave propagation. An additional benefit of the modeling system is that it covers upstream of Tarbela Dam from Skardu and the Kabul River Basin downstream to Kotri, which the FEWS implementation currently lacks. The model was delivered to FFD/PMD in March 2014 and ran on a trial basis during the 2014 flood season for its calibration and validation. The IFAS includes a Rainfall-Runoff-Inundation component to estimate flood water inundation.

The PMD Standard Operating Procedures (SOPs) outline current flood forecasting operational procedures but does not include the use of either FEWS or IFAS.

The PMD FFD has full responsibility for providing flood and Indus River and main tributary streamflow forecasts to the major stakeholders. Hydrologic modeling and forecasts are made during the monsoon flood season, June-October, otherwise FFD monitors the weather and compiles streamflow observations, which it publishes in its Bulletins A and B on a daily basis, which are distributed to a variety of major stakeholders. Bulletin A statements are qualitative discussions of current and forecasted hydrometeorological conditions. Bulletin B presents quantitative 24-h lead-time flow forecasts, with no indication of forecast river levels or impacted areas. Flood forecasts are submitted to the Pakistan National Disaster Management Authority (NDMA), which issues the public flood and weather warnings.

Working with the Japan International Cooperation Agency (JICA) and PMD, Islamabad has recently installed a flash flood warning system consisting of six AWS and a Danish Hydraulics Institute (DHI)-based MIKE hydrologic modeling system to produce local urban and flash flood warnings for the Lai Nullah basin in Islamabad. Warnings are tied to a local siren system.

A major deficiency in Pakistan and with PMD/FFD is the lack of capability to provide flash flood warnings. However, with the aid of the WMO, the WMO FFGS was implemented for PMD operationally in late 2017. Clark et al. (2014) discuss important findings related to various FFGS approaches. The WMO FFGS is based on the Hydrologic Research Center (located in San Diego, California) FFGS (http://www.wmo.int/pages/prog/hwrp/flood/ffgs/index_en.php). The FFGS is designed to operate remotely at HRC, utilizing both satellite-based and ground-based precipitation observations. FFGS products will be transmitted directly to PMD for issuing flash flood warnings. Extensive training is included in the FFGS implementation by the WMO and HRC.

12.3.2 India

The Central Water Commission (CWC) has national responsibility for monitoring flood water levels and discharges along the major rivers in India during the annual monsoon period, flood forecasts are issued to local administrative bodies, project authorities, state governments, and the Home Ministry of India. The functions of the Flood Forecast Monitoring Directorate of the CWC are

1. Nodal Directorate for flood forecast-related matters;
2. monitoring of flood forecasting in the country;
3. operation of Central Flood Control Room during monsoon season on a 24/7 basis, in keeping with the NDMA of India's SOP;
4. collection and compilation of flood forecast data from field offices of CWC and preparation/issue of daily flood bulletins during monsoon season;
5. coordination with field offices of CWC regarding flood situation and flood forecasting;
6. preparation of Annual Flood Forecasting Appraisal Reports;
7. matters related to installation, operation, and maintenance of wireless/telemetry network of CWC;
8. operation and maintenance of video conferencing facility linked with NDMA and states and State Disaster Management Authority on flood matters; and
9. liaison with India Meteorological Department.

Responsibility covers 148 broad, low-lying areas, cities, and towns. Additionally, inflows and outflows of 28 reservoirs in the country are monitored and forecasted. The network encompasses the following major river systems:

1. Ganga River and its tributaries
2. Indus/Jhelum rivers
3. Brahmaputra River and its tributaries
4. Barak River
5. Eastern Rivers
6. Mahanadi River
7. Godavari River
8. Krishna and the west-flowing rivers.

The flood forecasting and monitoring system (discussed in detail here: Central Water Commission, 2018) covers 72 river subbasins, spanning 18 states, including Andhra Pradesh, Assam, Bihar, Chhattisgarh, Gujarat, Haryana, Jammu and Kashmir, Jharkhand, Karnataka, Madhya Pradesh, Maharashtra, Orissa, Telangana, Tripura, Uttarakhand, Uttar Pradesh, and West Bengal, one Union Territory of Dadra and Nagar Haveli and the National Capital Territory of Delhi. Flood forecasting and advance warning for 148 low-lying areas and towns and 28 reservoirs help other government agencies with decision making to implement flood mitigation measures, such as the evacuation of residents and shifting their movable property to safer locations. Inflow forecasts at 28 reservoirs are used by dam authorities to manage reservoir gate operation for the release of discharges downstream. Reservoir inflow forecasts are also used to ensure adequate storage in the reservoirs to meet irrigation and hydropower generation demands during nonmonsoonal periods. Such services are available annually during the flood period, from May through December. The forecast is disseminated using a range of communication pathways, such as fax, wireless, phone, mob, SMS, email, electronic media, print media, social media, website, etc. Annually, over 6000 flood forecasts and advance warnings are issued by CWC regional offices across the country to government agencies during floods.

Hydrologic modeling utilizes statistical correlations and computer-based *mathematical models*, such as the Swedish Meteorological and Hydrological Institute HYPE, US Army Corps of Engineers HEC-HMS, and DHI MIKE-11 FF models for selected river basins. CWC follows the following forecasting schedule:

- Major rivers (travel time >24 h): Forecasts based on 0800/0900 hours water-level data and issued once a day at 1000 hours with advance warning time from 24 to 36 h.
- Medium rivers (travel time 12–24 h): Forecasts based on 0600 and 1800 hours water-level data and issued twice a day at 0700 and 1900 hours with advance warning time from 12 to 24 h.
- Flashy rivers (travel time <12 h): Forecasts based on hourly water-level data and issued multiple times (more than twice) a day with advance warning time less than 12 h.

The CWC has created four flood categories used in flood monitoring to identify warning, danger, and highest flood levels (HFLs) in the country's flood forecasting network. These are low, moderate, high, and unprecedented, depending on observed river water level:

- *Low flood*: A river is said to be in a low flood state at any flood forecasting site when the water level of the river reaches or crosses the warning level, but remains below the danger level of the forecasting site. The color *yellow* has been assigned to this category.
- *Moderate flood*: If the water level of the river reaches or crosses its danger level, but remains 0.50 m below the highest flood level of the site (commonly known as HFL), the flood state is called a moderate flood. The color *pink* has been assigned to this category.
- *High flood*: If the water level of the river at the forecasting site is below the HFL of the forecasting site but still within 0.50 m of the HFL, the flood state is designated as a high flood. The color *orange* has been assigned to this category. In *high flood situations* a special *orange bulletin* is issued by the CWC to the end-user agencies contains a *special flood message* related to the high flood.
- *Unprecedented flood*: A flood is identified as unprecedented when the water level of the river reaches or crosses the HFL recorded previously, during the historical observation period. The color *red* has been assigned to this category. In *unprecedented flood situations* a special *red bulletin* is issued by the CWC to end-user agencies that contains the *special flood message* related to the unprecedented flood.

The CWC maintains approximately 226 flood forecasting stations, consisting of 166 hourly level forecasting stations for towns and important villages and 60 inflow forecasting stations for dams and reservoirs. Rainfall observations consist of twice daily, 3-hourly, and hourly stations, which are transmitted to divisional headquarters wirelessly, via telemetry, and by telephone. Worldwide Web-based graphics allow the color of point symbols or graphics to signify the basic station characteristics. The color *green* is used to identify flood-level forecast stations important to towns and villages. The color *blue* is for inflow forecast stations, such as dams, reservoirs, barrages, and weirs. These colors change dynamically to reflect conditions at the flood-level forecast stations available during the annual flood period, May through December.

India, with aid from the World Bank (WB) and International Bank for Reconstruction and Development, is proposing funding of the National Hydrology Project (NHP). The NHP has two main potential benefits:

1. reduced damages from flooding and
2. improved (dynamic- and model-based) reservoir management, implying increased hydropower generation, enhanced canal water releases for irrigation, increased drinking water supplies, and improved water supply for industrial production.

The WB states:

> [It is] assumed that these benefits are unlikely to occur if the individual states acted on their own and without the help of the NHP. This is not only because water resources are shared across states—and so many concerns (such as flood management) can only be dealt with jointly—but also because high-quality and large-scale data collection and data analysis (including modeling) are necessary to generate sufficient confidence in forecasts, maps, and other information products to change the planning, design, and operations of water and other infrastructure projects. A case in point is the reservoir operation schedule for each dam that has been in place from commissioning and continues to be adhered to (as far as possible) with little analytical basis because of the fear of potential catastrophes. Another example, going beyond water infrastructure, is the information base for decisions by civil authorities to evacuate citizens from a town or area to mitigate the effects of floods.

12.3.3 China

Little information is available on water resources activities in China. However, Liu (2016) discusses recent flood forecasting development activities at a national level in China, and additional insight on water resources modeling and assessments in China is provided by, for example, Zhijia et al. (2008) and Wang et al. (2016). Water resources modeling in China tends to be done at the provincial and river basin authority levels. Data sharing within China is limited across provincial and river basin authority boundaries.

12.3.4 Afghanistan

The Afghanistan Meteorological Department (AMD) falls under the Civil Aviation Authority of the Islamic Republic of Afghanistan (http://www.amd.gov.af). The Afghanistan Early Warning Project was developed as a joint effort between the WMO and the US Agency for International Development/Office of US Foreign Disaster Assistance (USAID/OFDA). The project has been assisting AMD with capacity building, with the aim of providing severe weather forecasts and early warnings for floods and flash floods. Afghanistan is very susceptible to hydrometeorological hazards. According to the International Disaster Database, Center for Research on the Epidemiology of Disasters, from 1980 to 2015, close to 15,000 deaths occurred, incurring an estimated USD 396 million in economic losses. Many of these floods have occurred within the Indus River Basin drainage as indicated in Table 12.3 and Fig. 12.14.

USAID reports that AMD issued its first flood warning 1 week following reception of the meteorological satellite data. Through the project, a data visualization and processing station has been installed in Afghanistan. This allows AMD to receive satellite images and products from the Meteosat-8 EUMETSAT operational geostationary satellite located over the Indian Ocean. The EUMETCast Severe Weather and Convection Detection product provides AMD with the capability to forecast flood-producing heavy precipitation.

The USAID project provides extensive capacity and infrastructure building, which includes renovation of AMD facilities, including the addition of an uninterrupted power supply, dedicated forecasting room, and IT equipment with local computing capability for data processing and visualization. Local and wide area networks provide the Internet bandwidth necessary for obtaining meteorological data.

A competency-based training program was created, focused on developing sustainable local capacity for providing hydrometeorological services. AMD staff training included use of the WMO FFGS, satellite meteorology, observation techniques, weather analysis and forecasting, hydrometeorology, meteorological instruments maintenance and calibration, and use of the Common Alerting Protocol. Additionally, USAID maintains a Spatial Data Center for Afghanistan (http://asdc.immap.org, which produces graphical, web-based hydromet hazards products, such as that shown in Fig. 12.14.

12.4 NEEDED IMPROVEMENTS AND BENEFITS

Modernization of infrastructure for National Meteorological and Hydrological Services DRM is critical. Improvements are needed to support

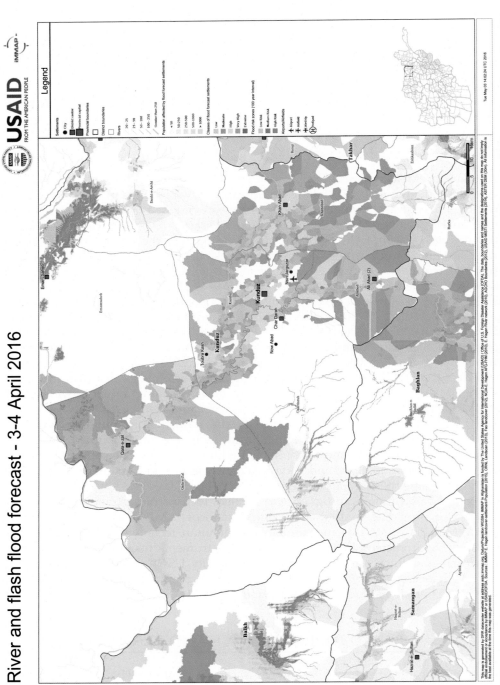

FIG. 12.14 Example USAID Spatial Data Center, Kunduz Province flood risk forecast, river and flash flood forecast, April 3–4, 2016. (*From http://asdc. immap.org/documents/4983.*)

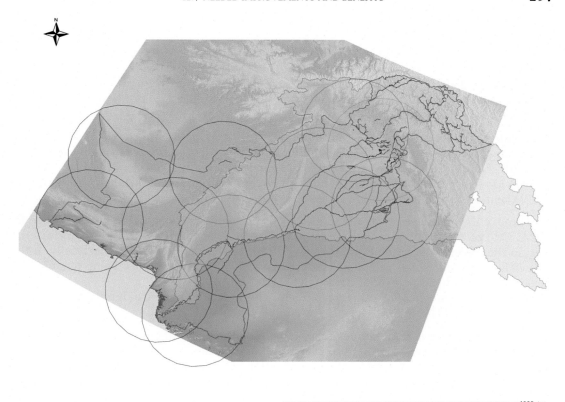

FIG. 12.15 Hypothetical configuration of 13 S-band radars to provide nearly complete coverage for all but the highly mountainous Hindu Kush and Karakoram regions of Pakistan.

- river and flash flood forecasting and warnings;
- issues related to meteorological forecasting, including, for example, modernized and expanded AWS network, including *all* hydrometeorological variables;
- precipitation estimation from radars;
- telecommunication infrastructure modernization;
- real-time and archive database implementation; and
- timely and accurate forecast delivery improvements.

12.4.1 Infrastructure Improvements

A spatially dense, well-maintained, and quality controlled hydromet network is the backbone of any hydrological forecasting system. Without quality model forcing data, hydrological forecasts will suffer significantly. Hydromet network expansion and improvements in data transmission, processing, and storage in Pakistan are critical. One aspect of this is the radar coverage, which, from a hydrologic perspective, is essentially nonexistent. PMD has proposed expansion to 18–21 radars, including the use of primarily S-band radars, but also X-band radars in the extreme northern, highly mountainous Hindu Kush and Karakoram regions of Pakistan. No radar coverage maps are available from PMD. Consequently, based on an expected 230 km range, a simple analysis provides a hypothetical arrangement of 13 S-band radars positioned within Pakistan (shown in Fig. 12.15) to provide maximum coverage within the context of

1. an S-band radar expected 230 km range;
2. placing radars in population centers (cities and towns) where there is access to existing utilities and infrastructure needed for operations and maintenance; and
3. providing significant coverage of the Indus River Basin drainage system outside of Pakistan and with overlapping coverage within Pakistan.

Three of the *replacement* S-band radars are currently under construction with aid from the JICA. One of the 13 hypothetical radar locations was placed in Nok Kundi, Balochistan, a region that receives precipitation averaging 35.3 mm/year annually. So, potentially, if this site is ignored, only nine additional radars beyond the three currently being installed are needed to provide S-band radar coverage for Pakistan.

X-band radar coverage in Pakistan (and elsewhere) is problematic in highly mountainous areas (Savina, 2011) due to

• weather radar located in Alpine areas being highly dependent on the radar visibility and on the height above the ground of the sampling volume;
• creation of a visibility model for the identification of areas systematically affected by radar-visibility-related errors such as ground clutter, side lobes, beam blocking, and anomalous propagation;
• need for antenna heating which cannot always prevent ice deposition, which is shown to produce very strong signal attenuation; and
• accessibility for maintenance.

Consequently, deployment of X-band radars does not appear to be efficacious.

Many overlapping hydromet networks exist in Pakistan, and it appears there is minimal integration of the networks and data sharing. As an example, PMD has established nearly 15 permanent conventional observatories in the northern, high mountainous areas of Pakistan to monitor weather and climatology. Recently, with the advancement in monitoring equipment technology, PMD has enhanced its meteorological monitoring network with the introduction of AWSs that monitor continuously at high altitudes. So far five AWSs are installed at Baltoro, Passu, and Batura glaciers, with the installation of additional stations at other meteorologically strategic important locations planned. Apart from meteorological monitoring, PMD is involved in situ measurement of the characteristics of glaciers, including melting rates, depth, debris cover, and velocities of different glaciers and their mass balance.

Additionally, the WAPDA GMRC maintains a network of 20 stations in the high Hindu Kush and Karakoram region of Pakistan. It does not appear PMD/FFD utilizes these stations as input to the IFAS modeling system. A notable concern to PMD/FFD is significantly improved monitoring for early warning of GLOFs.

12.4.2 Improved Capacity Building and Coordination

Generally, in Pakistan resources are not shared sufficiently between agencies. Improved interagency collaboration is needed. This deficiency includes inadequate sharing of operational requirements, data, and technical capabilities, such as with remote sensing and GIS expertise and hydrologic modeling. That is, with the limited resources of Pakistan, there is too little coordination between agencies. A clear example lies with the limited relationship PMD has with the Pakistan Space and Upper Atmosphere Research Commission (SUPARCO), Space Application Center for Response in Emergency and Disasters (SACRED), which has considerable remote sensing and GIS capability. SUPARCO-SACRED partnered with ICHARM in the development and implementation of IFAS for the Indus River Basin in Pakistan. Activities of SUPARCO-SACRED include monitoring of droughts, floods, landslides, earthquakes, and avalanches through satellite remote sensing (Emerton et al., 2016).

Improved capacity building, including fundamental science and technology education, is needed. As part of this training, methods for hydrologic and hydraulic model implementation, historical data analysis, and model calibration at FFD must be included. This need goes beyond the basic systems-related training that should be expected with modernized NHSs. With this, there must be the development of a new concept of operations (CONOPS). Current modeling capability lacks hydraulic modeling of barrages, snow accumulation/snowmelt modeling, and reservoir modeling, which are a requirement for producing monthly to seasonal hydrologic outlooks for water resources planning and management purposes. Comprehensive Indus River Basin modeling must include the entire basin, including the Kabul River into Afghanistan, areas in India, and basin areas above the Tarbela Dam and the other major reservoirs. Some additional points:

- PMD and FFD lack sufficient GIS knowledge and the needed hardware and software to support their mission.
- The office operational working environment needs to be modernized with power backup, UPS, air conditioning (at least for cooling of computer systems).
- Use Open source software solutions as much as possible to both save cost and with capacity building.
- Deal with the many streamflow monitoring stations that lack rating curves.
- Routine annual coordination meetings between PMD and FFD with the major stakeholders are needed; quarterly or at least annual meetings at lower ranking staff levels to resolve technical issues and for sharing new developments would be highly beneficial. This coordination must include on-site visits.

12.4.3 Hydrological Modeling Improvements

Hydrologic modeling improvements in the Indus basin are a necessity. A complete restructuring of the water resources enterprise in the Indus basin is needed, beginning with hydrometeorological data collection. Critically, this must happen in a fully integrated structure that should include the following:

1. an expanded, well-maintained, real-time (hourly) AWS network;

2. an expanded, well-maintained, real-time (hourly) network of stream gauging stations that measure, minimally, water level (stage), flow, temperature, and suspended sediment concentration;
3. adoption of a universal data format to be used in data transmission;
4. real-time telemetry from these stations to a centralized data collection facility;
5. raw hydrometeorological data storage and backup within a modern relational database management system (RDMS) for structured query language access;
6. real-time data processing and quality control (QC);
7. real-time and archived storage of QC'ed data;
8. open access to data by outside users to facilitate research;
9. advanced precipitation processing system estimation techniques (Smith and Krajewski, 1991; Seo, 1998; Fulton et al., 1998; Seo et al., 1999; Fulton, 2002; Kitzmiller et al., 2011; Cocks et al., 2016; Zhang et al., 2016), coupled with real-time AWS measurements for bias removal. Data from the individual radar should be mosaicked into a seamless spatially gridded field with full coverage of the entire Indus River Basin. In regions lacking radar coverage, remote sensing from satellite coverage can be optimally merged with AWS data using data fusion processes (Turlapaty et al., 2010; Wang et al., 2011; Wang, 2012; Livneh et al., 2015);
10. a comprehensive hydrologic modeling system that includes rainfall runoff, snow and glacial runoff, reservoir simulation, real-time hydrodynamic streamflow routing models, including the capability for real-time flood inundation modeling;
11. implementation of ensemble hydrologic forecasting methodologies due to the highly uncertain nature of hydrologic forecasts, which stem from the uncertainty in precipitation forecasting (Dawdy and Bergmann, 1969; Ebert and McBride, 2000; Damrath et al., 2000; Ebert, 2001; Ebert et al., 2003). This uncertainty must be conveyed to end-users of forecasts using ensemble modeling techniques (Day, 1985; Franz et al., 2003; Schaake et al., 2007; Adams and Ostrowski, 2010; Demargne et al., 2014);
12. implementation of procedures for and commitment to model calibration (Duan et al., 1992, 1993, 1994; Anderson, 2002; Koren et al., 2003);
12. establishment of a comprehensive hydrologic forecast verification system, which necessitates access to historical data archives in an RDMS (Welles et al., 2007; Demargne et al., 2009, 2010; Brown et al., 2010);
14. integration with national and regional/provincial decision support systems (DSS) for DRM, especially with national and provincial/regional disaster management authorities (DMAs);
15. improvements in the process of disseminating alerts and warnings to the public and other stakeholders is paramount; the process must be structured to meet the differing needs of end-users, understanding that alert and warning messages must meet the needs of different users;
16. end-user education to promote understanding and awareness of the *alert* and *warning* process;
17. incorporation of situational awareness procedures and tools in the forecasting workflow;

18. creation of comprehensive Concept of Operations (CONOPS) that spans national and regional/provincial agencies and major stakeholders;
19. dramatically improved science and engineering education of staff;
20. staff capacity building on all levels; and
21. establishment of programs to attract technical graduates from universities, including the use of internships and increased wages.

The establishment of national and large regional flood forecasting systems, many of which are integrated into more comprehensive water resources management systems, is discussed in Adams and Thomas (2016). WMO (2011) discusses recommended features and techniques that should be found in modern flood forecasting and warning systems. The motivation for many of the recommendations is found in, for example, McEnery et al. (2005) and National Research Council (2006a, b).

Global hydrologic modeling systems that could provide some benefits in the Indus River Basin have been developed (Ward et al., 2013; Herold and Rudari, 2013; Sampson et al., 2015), but they are based on coarsely gridded (~0.1 degrees horizontal resolution or greater) distributed modeling structures. These global and continental scale models have not shown sufficient forecast skill to produce actionable short lead-time alerts and warnings. For example, Yossef et al. (2012) report forecast skill relative to climatology for a global hydrological model in reproducing the occurrence of monthly flow extremes. Yossef et al. (2013) investigate the forecast skill of the global seasonal streamflow forecasting system FEWS-World. Sood and Smakhtin (2015) examine the structure, strengths, and weaknesses of several monthly and seasonal global simulation models and address model uncertainty, data scarcity, spatial resolution, and integration with remote sensing data. Voisin et al. (2011) evaluate the performance of the Variable Infiltration Capacity model at a daily time-step relative to a simulated reference forecast, based on model forcings from the downscaled European Center for Medium-Range Weather Forecasts analysis temperatures and winds and Tropical Rainfall Measuring Mission Multisatellite Precipitation Analysis precipitation using up to the day of forecast for the period 2002–07. Yuan et al. (2015) evaluate an experimental global seasonal hydrologic forecasting system, based on coupled climate forecast models participating in the North American Multimodel Ensemble project. The system is evaluated over major Global Energy and Water Cycle Experiment, Regional Hydro-Climate Projects river basins by comparison with ensemble streamflow predictions (ESPs). These studies serve as examples for the promise of continental and global scale ensemble hydrologic models in providing long lead-time monthly and seasonal outlooks, but are not useful for short lead-time flood forecasting.

Moreover, Smith et al. (2004b, 2012) show that significant hurdles remain with distributed model performance relative to legacy lumped parameter hydrologic models. However, Smith et al. (2004a) and Koren et al. (2004) have produced encouraging results with a reformulation of the NOAA/NWS SAC-SMA model (Burnash et al., 1973; Burnash, 1995) within the Research Distributed Hydrologic Model model framework. The Pakistan FFD currently uses the SAC-SMA model in its local FEWS modeling system.

12.5 CONCLUSIONS

The major challenges facing Indus River Basin countries, particularly Pakistan (Qureshi, 2011; Michel et al., 2013), come from (1) the need to make institutional changes where a comprehensive (Bengali, 2009), fully integrated NHS can flourish and (2) increased water stress due to a changing climate and reduced precipitation (Yu et al., 2013; Lutz et al., 2014; Rajbhandari et al., 2015; Kraaijenbrink et al., 2017) and expected increased populations and energy demands. Taken together water availability to meet water supply and agricultural needs (irrigation) will be problematic.

An initial step toward addressing shared water resources problems in the Indus River Basin region was taken by Cheema and Prakashkiran (2015) to *formulate a proposal for a practical, cooperative research project that could be implemented by Indian and Pakistani scientists toward building a shared knowledge base for water resources management in the Indus basin.* The collaboration by Cheema and Prakashkiran (2015) suggests:

- development of a comprehensive knowledge base (temporally and spatially consistent) on various hydrological processes without political interference;
- formulation of an integrated model framework that will be smart enough to establish relationships between land use, anthropogenic activities, climate, and socioeconomics and provide policy guidelines;
- exploration of alternative irrigation scenarios to improve water/land productivity in view of the changing climate; and
- investigation of the temporal and spatial extent of groundwater exploitation, considering transboundary perspectives and the development of retrofit measures.

The latter point is significant as it refers to an extensive unconfined aquifer that underlies the Indus River Basin, most of which is shared between India and Pakistan, comprising a surface area of 0.16 million km^2. The unconfined aquifer has the potential for large groundwater storage with an annual replenishable capacity of approximately 90 km^3. A high recharge rate is found in the North, below the foothills of the Himalayas, which declines toward the South (see Fig. 12.2; Laghari et al., 2012). The aquifer is well developed, with an unconfined to semiconfined porous alluvial surficial formation with the capacity to store and transmit water to deeper geologic formations. Groundwater flow is predominantly from northeast to southwest (Chadha, 2008). The water in the aquifer is used both as a supplement and as an alternative source for irrigation. Farmers of both India and Pakistan have been forced to use groundwater as a supplemental source of water for irrigation because of inadequate and variable surface water supplies, resulting in an annual contribution of more than 50% of total irrigation requirements. Consequently, local, readily available groundwater is more productive compared to surface water sources for irrigation. Qureshi (2011) reports that a large number of irrigation wells have been added annually in India and Pakistan, which has resulted in a 20%–30% increase in groundwater abstractions from the alluvial aquifer over the past 20 years.

In an attempt to address the looming transboundary water resources management issues, Cheema and Prakashkiran (2015) proposed a detailed water resources management roadmap for the Indus basin, including creating a shared knowledge base, which has the following elements:

- formation of executive committees for science, policy, and governmental administration;

- identification of the scientific research, socioeconomic investigations, regulatory, and policy requirements of the region;
- unification of the scientific facts and policy issues, and their transformation into easy messages for the general public;
- formulation of public hearing committees;
- public hearings of a sufficient duration, at an appropriate time;
- execution of identified science-policy projects;
- training, capacity building, and awareness generation activities;
- presentation of results to governmental administration committees;
- review of facts, figures, and relevant expectations of riparian countries; and
- negotiation and finalization of terms.

Initiatives such as the one proposed by Cheema and Prakashkiran (2015) are necessary for advancements in meeting the challenges of the shared regional water resources issues in the Indus basin as a sustainable long-term solution. Giupponi and Sgobbi (2013) identify a critical issue that underlies efforts to modernize HSs and DSSs, stating:

> …priority efforts should not focus on developing the tools, but rather on improving the effectiveness and applicability of integrated water resources management legislative and planning frameworks, training and capacity building, networking and cooperation, harmonization of transnational data infrastructures and, very importantly, learning from past experiences and adopting enhanced protocols for DSS development.

References

Adams, T., Ostrowski, J., 2010. Short lead-time hydrologic ensemble forecasts from numerical weather prediction model ensembles. In: Proceedings World Environmental and Water Resources Congress 2010, EWRI, Providence, RI.

Adams, T.E., Thomas, C., 2016. Flood Forecasting: A Global Perspective, first ed. Elsevier/Academic Press, New York, NY.

Andermann, C., Bonnet, S., Gloaguen, R., 2011. Evaluation of precipitation data sets along the Himalayan front. Geochem. Geophys. Geosyst. 12 (7), Q07023.

Anderson, E.A., 2002. Calibration of conceptual hydrologic models for use in river forecasting. U.S. National Weather Service, Office of Hydrology, Hydrology Laboratory. Technical Report.

Asian Development Bank and World Bank, 2010. Pakistan floods 2010 damage and needs assessment. Pakistan Development Forum, Islamabad. Technical Report.

Bastiaanssen, W.G.M., Cheema, M.J.M., Immerzeel, W.W., Miltenburg, I.J., Pelgrum, H., 2012. Surface energy balance and actual evapotranspiration of the transboundary Indus Basin estimated from satellite measurements and the ETlook model. Water Resour. Res. 48 (11), W11512.

Becker, A., Finger, P., Meyer-Christoffer, A., Rudolf, B., Schamm, K., Schneider, U., Ziese, M., 2013. A description of the global land-surface precipitation data products of the global precipitation climatology centre with sample applications including centennial (trend) analysis from 1901-present. Earth Syst. Sci. Data 5 (1), 71–99.

Bengali, K., 2009. Water management under constraints: the need for a paradigm shift. In: Kugelman, M., Hathaway, R.M. (Eds.), Running on Empty: Pakistan's Water Crisis. Woodrow Wilson Center for Scholars, Washington, DC.

Bhatt, B.C., Nakamura, K., 2005. Characteristics of monsoon rainfall around the Himalayas revealed by TRMM precipitation radar. Mon. Weather Rev. 133 (1), 149–165.

Brakenridge, R., Albert, K., 1985. Dartmouth Flood Observatory, Space-Based Measurement, Mapping, and Modeling of Surface Water for Research, Humanitarian, and Water Management Applications. INSTAAR, University of Colorado, Boulder, CO, p. 7.

Briscoe, J., Qamar, U., 2006. Pakistan's water economy: running dry. The World Bank, Washington, DC. http://documents.worldbank.org/curated/en/989891468059352743/pdf/443750PUB0PK0W1Box0327398B01PUBLIC1.pdf. Technical Report.

Brown, J.D., Demargne, J., Seo, D.J., Liu, Y., 2010. The Ensemble Verification System (EVS): a software tool for verifying ensemble forecasts of hydrometeorological and hydrologic variables at discrete locations. Environ. Model. Softw. 25 (7), 854–872.

Burnash, R.J., 1995. The NWS River Forecast System—Catchment Model, first ed. Water Resources Publications, Highlands Ranch, CO.

Burnash, R.J., Ferral, R.L., McGuire, R.A., 1973. A generalized streamflow simulation system: conceptual modeling for digital computers. US Department of Commerce National Weather Service and State of California Department of Water Resources. Technical Report.

Central Water Commission, 2018. India Water Resources Information System (WRIS). Central Water Commission, New Delhi. http://www.india-wris.nrsc.gov.in/wrpinfo/index.php?title=CWC_National_Flood_Forecasting_Network.

Chadha, D.K., 2008. Development, management and impact of climate change on transboundary aquifers of Indus Basin. In: Ganoulis, J., Aureli, A., Fried, J. (Eds.), 4th International Symposium on Transboundary Water Management, Aristotle University of Thessaloniki, Thessaloniki, Greece, p. 10.

Cheema, M.J.M., Prakashkiran, P., 2015. Bridging the divide: transboundary science & policy interaction in the Indus Basin. Stimson Center, Environmental Security Program, Visiting Fellows Program. https://www.stimson.org/sites/default/files/file-attachments/EnvSecVFPaperMarch2015-FINAL.pdf. Technical Report.

Clark, R.A., Jonathan, J.G., Zachary, L., Flamig, Y.H., Edward, C., 2014. Conus-wide evaluation of National Weather Service flash flood guidance products. Weather Forecast. 29 (2), 377–392.

Cocks, S.B., Martinaitis, S.M., Kaney, B., Zhang, J., Howard, K., 2016. MRMS QPE performance during the 2013/14 cool season. J. Hydrometeor. 17, 791–810. https://dx.doi.org/10.1175/JHM-D-15-0095.1.

Damrath, U., Doms, G., Fruehwald, D., Heise, E., Richter, B., Steppeler, J., 2000. Operational quantitative precipitation forecasting at the German Weather Service. J. Hydrology 239, 260–285.

Dawdy, D.R., Bergmann, J.M., 1969. Effect of rainfall variability on streamflow simulation. Water Resour. Res. 5 (5), 140–158.

Day, G.N., 1985. Extended streamflow forecasting using NWSRFS. ASCE J. Water Resour. Plann. Manag. 3, 157–170.

Demargne, J., Mulluski, M., Werner, K., Adams, T., Lindsey, S., Schwein, N., Marosi, W., Welles, E., 2009. Application of forecast verification science to operational river forecasting in the U.S. National Weather Service. Bull. Am. Meteorol. Soc. 90 (6), 779–784.

Demargne, J., Brown, J.D., Seo, D.J., Wu, L., Toth, Z., Zhu, Y., 2010. Diagnostic verification of hydrometeorological and hydrologic ensembles. Atmos. Sci. Lett. 11 (2), 114–122.

Demargne, J., Limin, W., Satish, K.R., James, D.B., Haksu, L., Minxue, H., Dong-Jun, S., Robert, H., Henry, D.H., Mark, F., John, S., Yuejian, Z., 2014. The science of NOAA's operational hydrologic ensemble forecast service. Bull. Amer. Meteor. Soc. 95, 79–98.

Duan, Q., Sorooshian, S., Gupta, V.K., 1992. Effective and efficient global optimization for conceptual rainfall-runoff models. Water Resour. Res. 28 (4), 1015–1031.

Duan, Q.A., Gupta, V.K., Sorooshian, S., 1993. Shuffled complex evolution approach for effective and efficient global minimization. J. Optim. Theory Appl. 76 (3), 501–521.

Duan, Q.A., Sorooshian, S., Gupta, V.K., 1994. Optimal use of the SCE-UA global optimization method for calibrating watershed models. J. Hydrol. 158 (5), 265–284.

Ebert, E.E., 2001. Ability of a poor man's ensemble to predict the probability and distribution of precipitation. Mon. Weather Rev. 129, 2461–2480.

Ebert, E.E., McBride, J.L., 2000. Verification of precipitation in weather systems: determination of systematic errors. J. Hydrol. 239, 179–202.

Ebert, E.E., Damrath, U., Wergen, W., McBride, J.L., 2003. The WGNE assessment of short-term quantitative precipitation forecasts. Bull. Am. Meteorol. Soc. 84 (4), 481–492.

Emerton, R.E., Elisabeth, M., Stephens, F.P., Thomas, C.P., Albrecht, H.W., Andy, W.W., Peter, S., James, D.B., Niclas, H., Chantal, D., Calum, A.B., Hannah, L.C., 2016. Continental and global scale flood forecasting systems. WIREs Water 3, 391–418.

Fekete, B.M., Charles, J.V., Wolfgang, G., 2002. High-resolution fields of global runoff combining observed river discharge and simulated water balances. Glob. Biogeochem. Cycles 16 (3), 15-1–15-10.

Franz, K.J., Holly, C.H., Soroosh, S., Roger, B., 2003. Verification of National Weather Service ensemble streamflow predictions for water supply forecasting in the Colorado River Basin. J. Hydrometeorol. 4 (12), 1105–1118.

Fulton, R.A., 2002. Activities to improveWSR-88D radar rainfall estimation in the National Weather Service. In: Proceedings of the Second Federal Interagency Hydrologic Modeling Conference. Las Vegas, NV.

Fulton, R., Breidenbach, J., Seo, D.J., Miller, D., 1998. The WSR-88D rainfall algorithm. Weather Forecast. 13, 377–395.

Girons, L.M., Di Baldassarre, G., Seibert, J., 2017. Impact of social preparedness on flood early warning systems. Water Resour. Res. 53 (1), 522–534.

Giupponi, C., Sgobbi, A., 2013. Decision support systems for water resources management in developing countries: learning from experiences in Africa. Water 5 (2), 798–818.

Hargreaves, G.H., Allen, R.G., 2003. History and evaluation of Hargreaves evapotranspiration equation. J. Irrig. Drain. Eng. 129 (1), 53–63.

Hargreaves, G.L., Hargreaves, G.H., Riley, J.P., 1985. Irrigation water requirements for Senegal River Basin. J. Irrig. Drain. Eng. 111 (3), 265–275.

Hellen, H.C., Fry, M., 2006. The decline of hydrological data collection for the development of integrated water resource management tools in Southern Africa. In: Proceedings of the Fifth FRIEND World Conference on Climate Variability and Change—Hydrological Impacts, Havana, Cuba, vol. 308, IAHS.

Herold, C., Rudari, R., 2013. Improvement of the global flood model for the GAR 2013 and 2015. Background paper prepared for the global assessment report on disaster risk reduction. World Bank and United Nations Office for Disaster, Risk Reduction (UNISDR), Geneva, Switzerland. Technical Report.

Khan, F., 2009. Water, governance, and corruption in Pakistan. In: Kugelman, M., Hathaway, R.M. (Eds.), Running on Empty: Pakistan's Water Crisis. Woodrow Wilson Center for Scholars, Washington, DC.

Kitzmiller, D., Suzanne, V.C., Feng, D., Kenneth, H., Carrie, L., Jian, Z., Heather, M., Yu, Z., Jonathan, J.G., Dongsoo, K., David, R., 2011. Evolving multisensor precipitation estimation methods: their impacts on flow prediction using a distributed hydrologic model. J. Hydrometeorol. 12, 1414–1431.

Kohler, M.A., 1944. The use of crest stage relations in forecasting the rise and fall of the flood hydrograph. US Weather Bureau. Technical Report (mimeo).

Köppen, W., 1936. Das geographisca system der klimate. In: Köppen, W., Geiger, G. (Eds.), Handbuch der Klimatologie. vol. 1. Verlag von Gebrüder Borntraeger, Berlin.

Koren, V., Smith, M., Duan, Q., 2003. Use of a Priori Parameter Estimates in the Derivation of Spatially Consistent Parameter Sets of Rainfall-Runoff Models. In: American Geophysical Union, Washington, DC.

Koren, V., Seann, R., Michael, S., Ziya, Z., Seo, D.J., 2004. Hydrology laboratory research modeling system (HL-RMS) of the US National Weather Service. J. Hydrol. 291, 297–318.

Kottek, M., Grieser, J., Beck, C., Rudolf, B., Rubel, F., 2006. World map of the Köppen-Geiger climate classification updated. Meteorol. Z. 15 (3), 259–263.

Kraaijenbrink, P.D.A., Bierkens, M.F.P., Lutz, A.F., Immerzeel, W.W., 2017. Impact of a global temperature rise of 1.5 degrees Celsius on Asia's glaciers. Nature. 549(257). https://www.nature.com/articles/nature23878#supplementary-information.

Kugelman, M., Hathaway, R.M., 2009. Running on Empty: Pakistan's Water Crisis. Woodrow Wilson Center for Scholars, Washington, DC.

Laghari, A.N., Vanham, D., Rauch, W., 2012. The Indus Basin in the framework of current and future water resources management. Hydrol. Earth Syst. Sci. 16 (4), 1063–1083.

Liu, Z., 2016. The development and recent advances of flood forecasting activities in China. In: Adams, T.E., Pagano, T.C. (Eds.), Flood Forecasting: A Global Perspective. Elsevier/Academic Press, New York, NY, pp. 67–86.

Livneh, B., Pierce, D.W., Francisco, M.-A., Nijssen, B., Vose, R., Cayan, D.R., Brekke, L., 2015. A spatially comprehensive, hydrometeorological data set for Mexico, the U.S., and Southern Canada 1950–2013. Scientific Data 2 (150042).

Lutz, A.F., Immerzeel, W.W., Shrestha, A.B., Bierkens, M.F.P., 2014. Consistent increase in high Asia's runoff due to increasing glacier melt and precipitation. Nat. Clim. Chang. 4(587). https://www.nature.com/articles/nclimate2237#supplementary-information.

McEnery, J., Ingram, J., Duan, Q., Adams, T., Anderson, L., 2005. NOAA's advanced hydrologic prediction service: building pathways for better science in water forecasting. Bull. Am. Meteorol. Soc. 24 (3), 375–385.

Michel, D., Powell, L., Ramay, S., 2013. Connecting the drops: an Indus Basin roadmap for cross-border water research, data sharing, and policy coordination. Stimson, Indus Basin Working Group. https://www.stimson.org/sites/default/files/file-attachments/connecting_the_drops_stimson_1.pdf. Technical Report.

Mishra, A.K., Coulibaly, P., 2009. Developments in hydrometric network design: a review. Rev. Geophys. 47 (2), RG2001.

National Research Council, 2006a. Toward a new advanced hydrologic prediction service AHPS. Committee to Assess the National Weather Service Advanced Hydrologic Prediction Service Initiative, Water Science and Technology Board, Washington, DC. Technical Report.

IV. WATER EXTREMES IN INDUS RIVER BASIN

National Research Council, 2006b. Completing the forecast: characterizing and communicating uncertainty for better decisions using weather and climate forecasts. Committee on Estimating and Communicating Uncertainty in Weather and Climate Forecasts, Washington, DC. Technical Report.

Qureshi, A.S., 2011. Water management in the Indus Basin in Pakistan: challenges and opportunities. Mt. Res. Dev. 31 (3), 252–260.

Rajbhandari, R., Shrestha, A.B., Kulkarni, A., Patwardhan, S.K., Bajracharya, S.R., 2015. Projected changes in climate over the Indus River Basin using a high resolution regional climate model (precis). Clim. Dyn. 44 (1), 339–357.

Running, S.W., Mu, Q., Zhao, M., Moreno, A., 2017. User's guide: MODIS global terrestrial evapotranspiration (ET) product (NASA MOD16A2/A3). Version 1.5 For Collection 6, NASA Earth Observing System MODIS Land Algorithm.

Running, S., Mu, Q., Zhao, M., 2018. MOD16A3 MODIS/Terra Net Evapotranspiration Yearly L4 Global 500m SIN Grid V006 [Data set]. https://dx.doi.org/10.5067/MODIS/MOD16A3.006 http://files.ntsg.umt.edu/data/NTSG_Products/MOD16/MOD16A3.105_MERRAGMAO/Geotiff/MOD16A3_ET_2000_to_2013_mean.tif.

Sampson, C.C., Smith, A.M., Bates, P.D., Neal, J.C., Alfieri, L., Freer, J.E., 2015. A high-resolution global flood hazard model. Water Resour. Res. 51 (9), 7358–7381.

Savina, M., 2011. The Use of a Cost-Effective X-Band Weather Radar in Alpine Region (Ph.D. thesis). ETH, ETH Zurich, Diss. ETH No. 20141.

Schaake, J., Hamill, T.M., Buizza, R., 2007. hEPEX: the hydrological ensemble prediction experiment. Bull. Am. Meteorol. Soc. 88, 1541–1547.

Schneider, U., Becker, A., Finger, P., Anja, M.-C., Rudolf, B., Ziese, M., 2015. GPCC Full Data Reanalysis Version 7.0 at 0.5°: Monthly Land-Surface Precipitation from Rain-Gauges built on GTS-based and Historic Data. https://dx.doi.org/10.5676/DWD_GPCC/FD_M_V7_050.

Seo, D.J., 1998. Real-time estimation of rainfall fields using radar rainfall and rain gauge data. J. Hydrol. 208, 37–52.

Seo, D.J., Breidenbach, J., Johnson, E., 1999. Real-time estimation of mean field bias in radar rainfall data. J. Hydrol. 223, 131–147.

Mulk, Shams-ul, 2009. Pakistan's water economy, the Indus River system and its development infrastructure, and the relentless struggle for sustainability. In: Kugelman, M., Hathaway, R.M. (Eds.), Running on Empty: Pakistan's Water Crisis. Woodrow Wilson Center for Scholars, Washington, DC.

Smith, J.A., Krajewski, W.F., 1991. Estimation of the mean field bias of radar rainfall estimates. J. Appl. Meteorol. 30, 397–412.

Smith, M., Koren, V., Zhang, Z., Reed, S., Seo, D., Moreda, F., Kuzmin, V., Cui, Z., Anderson, R., 2004. NOAA NWS distributed hydrologic modeling research and development. Department of Commerce, NOAA/NWS. Technical Report. NOAA NWS Technical Report NWS 51.

Smith, M.B., Seo, D.J., Koren, V.I., Reed, S., Zhang, Z., Duan, Q.Y., Moreda, F., Cong, S., 2004. The distributed model intercomparison project, DMIP—motivation and experiment design. J. Hydrol. 298 (1), 4–26.

Smith, M.B., Koren, V., Reed, S., Zhang, Z., Zhang, Y., Moreda, F., Cui, Z., Mizukami, N., Anderson, E.A., Cosgrove, B.A., 2012. The distributed model intercomparison project—Phase 2. Motivation and design of the Oklahoma experiments. J. Hydrol. 418, 3–16.

Sood, A., Smakhtin, V., 2015. Global hydrological models: a review. Hydrol. Sci. J. 60 (4), 549–565.

Turlapaty, A.C., Anantharaj, V.G., Younan, N.H., Turk, F.J., 2010. Precipitation data fusion using vector space transformation and artificial neural networks. Pattern Recogn. Lett. 31 (10), 1184–1200.

UNEP, 1997. World Atlas of Desertification. United Nations Environment Programme (UNEP), London, UK. Technical Report.

United Nations and World Bank Group, 2016a. Action plan. High level panel on water. United Nations and World Bank Group, New York, NY.

United Nations and World Bank Group, 2016b. Joint statement of the high level panel on water. In: High Level Panel on Water. New York, NY, https://sustainabledevelopment.un.org/content/documents/11296Joint%20Statement%20HLPW.pdf.

UNISDR, 2012. The role of hydrometeorological services in disaster risk management. Proceedings from the Joint Workshop, World Bank, the United Nations International Strategy for Disaster Reduction, and the World Meteorological Organization, Printing & Multimedia Services, World Bank Group, Washington, DC, p. 3.

Voisin, N., Pappenberger, F., Lettenmaier, D.P., Buizza, R., Schaake, J.C., 2011. Application of a medium-range global hydrologic probabilistic forecast scheme to the Ohio river basin. Weather Forecast. 26 (4), 425–446.

Wang, S., 2012. Assessment of Multiscale Precipitation Data Fusion and Soil Moisture Data Assimilation and Their Roles in Hydrological Forecasts (Ph.D. thesis). University of Pittsburgh, Unpublished.

Wang, S., Liang, X., Nan, Z., 2011. How much improvement can precipitation data fusion achieve with a multiscale Kalman smoother-based framework? Water Resour. Res. 47 (3), W00H12.

Wang, B., Liu, L., Huang, G.H., 2016. Forecast-based analysis for regional water supply and demand relationship by hybrid Markov chain models: a case study of Urumqi, China. J. Hydroinf. 18 (5), 905–918. https://dx.doi.org/10.2166/hydro.2016.202.

Ward, P.J., Jongman, B., Weiland, F.S., Bouwman, A., van Beek, R., Bierkens, M.F.P., Ligtvoet, W., Winsemius, H.C., 2013. Assessing flood risk at the global scale: model setup, results, and sensitivity. Environ. Res. Lett. 8 (044019), 10.

WBG, 2011. Afghanistan—Irrigation Restoration and Development (IRD) Project. National Center for Atmospheric Research, Boulder, CO. http://documents.worldbank.org/curated/en/598691468198013735/Afghanistan-Irrigation-Restoration-and-Development-IRD-Project.

WBG, 2016. National Hydrology Project for the Republic of India. World Bank Group (WBG) Project Appraisal Document (PAD), Washington, DC. http://documents.worldbank.org/curated/en/954111490207555730/pdf/India-National-Hydrology-PAD-02242017.pdf.

WBG, 2017. Pakistan Hydromet and DRM Services Project (PHDSP). World Bank Group (WBG) Project, Pakistan/South Asia-P163924, Washington, DC. http://projects.worldbank.org/P163924?lang=en.

Welles, E., Sorooshian, S., Carter, G., Olsen, B., 2007. Hydrologic verification: a call for action and collaboration. Bull. Am. Meteorol. Soc. 88, 503–511.

WMO, 2006. Guidelines on the role, operation and management of national hydrological services (WMO-No. 1003). Madrid Conference Statement and Action Plan, Madrid, Spain. Technical report. World Meteorological Organization.

WMO, 2011. Manual on flood forecasting and warning (WMO-No. 1072). World Meteorological Organization, Hydrology and Water Resources Programme, Geneva, Switzerland. Technical report.

WMO, WBG, and GFDRR, 2018. Global Assessment of the State of Hydrological Services. World Bank Group (WBG), Water Partnership Program (WPP), World Meteorological Organization (WMO), and Global Facility for Disaster Reduction and Recovery (GFDRR), Geneva, Switzerland. http://hydroconference.wmo.int/en/partners/data-management/global-assessment-state-hydrological-services.

Yossef, N.C., van Beek, L.P.H., Kwadijk, J.C.J., Bierkens, M.F.P., 2012. Assessment of the potential forecasting skill of a global hydrological model in reproducing the occurrence of monthly flow extremes. Hydrol. Earth Syst. Sci. 16 (11), 4233–4246.

Yossef, N.C., Winsemius, H., Weerts, A., van Beek, R., Bierkens, M.F.P., 2013. Skill of a global seasonal streamflow forecasting system, relative roles of initial conditions and meteorological forcing. Water Resour. Res. 49 (8), 4687–4699.

Yu, W., Yang, Y.C., Savitsky, A., Alford, D., Brown, C., Wescoat, J., Debowicz, D., Robinson, S., 2013. The Indus Basin of Pakistan: the impacts of climate risks on water and agriculture. International Bank for Reconstruction and Development/The World Bank, Washington, DC. Technical report.

Yuan, X., Roundy, J.K., Wood, E.F., Sheffield, J., 2015. Seasonal forecasting of global hydrologic extremes: system development and evaluation over GEWEX basins. Bull. Am. Meteorol. Soc. 96 (11), 1895–1912.

Zemadim, B., McCartney, M., Langan, S., Sharma, B., 2013. A participatory approach for hydrometeorological monitoring in the blue Nile River Basin of Ethiopia. International Water Management Institute, Colombo, Sri Lanka. Technical Report 155.

Zhang, J., Howard, K., Langston, C., Kaney, B., Qi, Y., Tang, L., Grams, H., Wang, Y., Cocks, S., Martinaitis, S., Arthur, A., Cooper, K., Brogden, J., Kitzmiller, D., 2016. Multi-radar multi-sensor (MRMS) quantitative precipitation estimation: initial operating capabilities. Bull. Am. Meteorol. Soc. 97 (4), 621–638.

Zhijia, L., Lili, W., Hongjun, B., Yu, S., Zhongbo, Y., 2008. Rainfall-runoff simulation and flood forecasting for Huaihe basin. Water Sci. Eng. 1 (3), 24–35.

Further Reading

Bundesanstalt für Geowissenschaften und Rohstoffe (BGR) and UNESCO, 2008. Groundwater Resources of the World 1:25000000. Hannover, Paris. https://www.whymap.org/whymap/EN/Home/whymap_node.html.

GLIMS and NSIDC, 2005. Global Land Ice Measurements From Space Glacier Database. GLIMS and NSIDC, Boulder, CO.

Koren, V.I., Smith, M., Wang, D., Zhang, Z., 2000. Use of soil property data in the derivation of conceptual rainfall-runoff model parameters. In: Conference on Hydrology. AMS, Long Beach, CA.

Lehner, B., Liermann, C.R., Revenga, C., Vörösmarty, C., Fekete, B., Crouzet, P., Döll, P., Endejan, M., Frenken, K., Magome, J., Nilsson, C., Robertson, J.C., Rödel, R., Sindorf, N., Wisser, D., 2011. High-resolution mapping of the world's reservoirs and dams for sustainable river-flow management. Front. Ecol. Environ. 9 (9), 494–502.

Loveland, T.R., Reed, B.C., Brown, J.F., Ohlen, D.O., Zhu, Z., Yang, L., Merchant, J.W., 2000. Development of a global land cover characteristics database and IGBP DISCover from 1 km AVHRR data. Int. J. Remote Sensing 21 (6/7), 1303–1330.

Raup, B.H., Racoviteanu, A., Khalsa, S.J.S., Helm, C., Armstrong, R., Arnaud, Y., 2007. The GLIMS geospatial glacier database: a new tool for studying glacier change. Glob. Planet. Chang. 56, 101–110.

Siebert, S., Henrich, V., Frenken, K., Burke, J., 2013. Global Map of Irrigation Areas version 5. Rheinische Friedrich-Wilhelms-University, Bonn, Germany/Food and Agriculture Organization (FAO) of the United Nations, Rome, Italy. http://www.fao.org/nr/water/aquastat/irrigationmap/index10.stm.

13

Review of Hydrometeorological Monitoring and Forecasting System for Floods in the Indus Basin in Pakistan

Mandira Singh Shrestha*, Muhammad Riaz Khan[†],
Nisha Wagle*, Zaheer Ahmad Babar[†], Vijay Ratan Khadgi*,
Shahzad Sultan[†]

*International Centre for Integrated Mountain Development, Kathmandu, Nepal
[†]Flood Forecasting Division, Pakistan Meteorological Department, Lahore, Pakistan

13.1 INTRODUCTION

The Indus Basin is a transboundary river basin shared by four countries, Pakistan (47%), India (39%), China (8%), and Afghanistan (6%), with a total area of 1.12 million square kilometer (FAO, 2012). The Indus River comprises 27 major tributaries, out of which six of the most significant branches, the Chenab, Ravi, Sutlej, Jhelum, Beas, and Indus rivers, flow westward through India; while the Kabul River, originating in Afghanistan flows eastward into Pakistan (Iqbal, 2013). The annual water runoff of the Indus River is about 200 cubic kilometer (Ali, 2013; Laghari et al., 2012). The Upper Indus River, which is upstream at Tarbela, carries an annual flow consisting of 70% meltwater of which 26% is contributed by glacial melt and 44% by snowmelt (Mukhopadhyay and Khan, 2015). The Indus River's flow varies, particularly due to the impact of climatic changes on water resources (Khan and Hassan, 2015).

The Indus River Basin plays an important role in the socioeconomic development of both Pakistan and India, as at least 300 million people are directly or indirectly dependent on the basin (FAO, 2012). Therefore Indus River flows play a major role in sustaining the lives and livelihoods of the inhabitants, meeting their needs for drinking water, irrigation, and hydropower, as well as the needs of the natural ecosystem. On the other hand, flooding from the river and its tributaries is a major threat in the area (Sardar et al., 2008).

13.2 FLOODS IN THE INDUS BASIN

Flooding is the greatest natural hazard in Pakistan, where 90% of the population is at risk from flooding events. According to an estimate by Leadership for Environment and Development (LEAD) in Pakistan (2015), 88 districts out of 145 can be impacted by floods. The National Disaster Management Authority (NDMA) estimates that 50 districts in Pakistan are at risk of flash floods, which occur during the monsoon season (Ahmad, 2015).

Indus floods occur during the summer from July to September as a result of intense monsoon rainfalls (Ali, 2013; Awan, 2003) supplemented by glacial meltwater as well as from occasional outburst floods due to the failure of natural dams (Ahmad et al., 2012). Monsoon weather systems originate in the Bay of Bengal, where the development of low pressure systems gives rise to heavy rain in the Himalayan foothills. The weather systems from the Arabian Sea (seasonal lows) and the Mediterranean (westerly waves) also occasionally produce destructive floods in the basin (Ali, 2013). Tributaries of the Indus Basin—the Jhelum, Chenab, and Kabul rivers and the upper and lower parts of the Indus River—are the major source of flooding (Khan et al., 2011).

Flooding in the region is strongly influenced by climate; hydrology; sediment transport characteristics; physiographic, demographic, and catchment characteristics; and socioeconomic factors (Gaurav et al., 2011; Laghari et al., 2012; Tariq and van de Giesen, 2012). Population growth, rapid urbanization, continuous degradation of ecosystem services, and climate change have resulted in increased flood risks, which are further exacerbated by inadequate flood planning and management (Ali, 2013).

Both flash floods and riverine floods occur frequently. Various districts and urban centers along the river banks are at risk of riverine floods, flash floods, and urban floods, particularly in the Punjab and Sindh provinces (FFC, 2016). Riverine floods in Pakistan are caused by heavy rainfall in the upper catchments (FFC, 2016) and are considered to be most devastating in the Indus Plain due to the flat terrain and high population density (Laghari et al., 2012). Heavy rainfall combined with cloudbursts, thunderstorms, and melting snow cause flash floods of high intensity and magnitude (Rahman and Shaw, 2015). Flash flooding mainly occurs in the mountainous region of Pakistan (Arslan et al., 2013; Rahman and Shaw, 2015) in the tributaries of the Indus, Jhelum, and Chenab rivers, causing damage to infrastructure and settlements and loss of human and animal lives (Tariq and van de Giesen, 2012). The Jhelum and Chenab rivers, which are snow-fed rivers, experience increased flooding when an early monsoon combines with peak snowmelting. Accelerated warming in the higher elevations increases glacial melt and increases the formation of glacial lakes, which give rise to increased occurrences of glacial lake outburst floods (Asraf et al., 2012).

Between 1985 and 2017 Pakistan experienced 75 major and minor flooding events, accounting for 48% of the total disasters in Pakistan (EM-DAT, 2017). It has been estimated that flood disasters alone have killed 11,087 people. It has been reported that damage from disasters has affected more than 65 million people through an economic loss of more than $19 billion, with the biggest share of the damage resulting from natural flooding disasters (Fig. 13.1). Various flood reduction approaches have been made to help develop more security in the area.

Ten major floods have occurred since 1985 in the Indus Basin and its tributaries—in 1992, 1994, 1995, 1998, 2005, 2010, 2011, 2012, 2013, and 2014—each killing more than 300 people and affecting the lives and property of millions of people, as shown in Table 13.1.

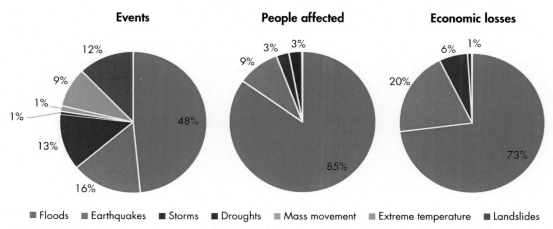

FIG. 13.1 Disaster damage showing percentage of events and people affected and the proportionate dollar impact in Pakistan between 1985 and 2017. *Source: EM-DAT.*

TABLE 13.1   Major Floods in Indus Basin in Pakistan

S. No.	Date	Location	Disaster Subtype	Total Death	Total Affected	Total Damage ('000 $)
1	Sep. 1992	Azad Kashmir and Punjab	Riverine flood[a]	1008[b]	6,655,450[a]	
2	Jul.–Sep. 1994	Murree, Risalpur, Karachi, Peshawar, Lahore, Sialkot, Multan, Bahawalpur, Shorkot, Quetta, Rawalpindi, Islamabad	Riverine flood[a]	431[b]	840,016[a]	92,000[a]
3	Jul.–Aug. 1995	Punjab, Sindh, Baluchistan, Khyber Pakhtunkwa	Riverine flood[a]	591[b]	600,000[a]	
4	Mar. 1998	Kech Valley (Balochistan province)	–	1000[a]	200,000[a]	
5	Feb. 2005	Several districts of Balochistan, Balochistan, Khyber Pakhtunkwa	Riverine flood[a]	520[a]	7,000,450[a]	30,000[a]
6	Jul.–Aug. 2010	Barkhan, Bolan, Kohlu, Nasirabad, Sibi districts (Balochistan province), Mohmand, Khyber agency, Khyber Pakhtunkhwa, Punjab province, Sindh province	Flash flood[a]	1985[b]	20,359,496[a]	9,500,000[a]
7	Aug.–Nov. 2011	Badin, Dadu, Ghotki, Hyderabad, Jacobabad, Karachi (central, east, south, west), Khaipur, Larkana, Malir, Mirpur Khas, Naushahro, Feroze, Nawabshah, Sanghar, Shikarpur, Sukkur, Tharparkar, Thatta, Umer kot districts (Sindh province)	Riverine flood[a]	516[b]	5,400,755[a]	2,500,000[a]

Continued

TABLE 13.1   Major Floods in Indus Basin in Pakistan—cont'd

S. No.	Date	Location	Disaster Subtype	Total Death	Total Affected	Total Damage ('000 $)
8	Aug.–Oct. 2012	Jaffarabad, Jhal magsi, Nasirabad districts (Balochistan province), D G khan, Rajanpur districts (Punjab province), Dadu Ghotki, Jacobabad, Larkana (Sindh Province)	Riverine flood[a]	571[b]	5,049,364[a]	2,500,000[a]
9	Aug. 2013	Balochistan, Khyber Pakhtunkhwa province, Punjab, Sindh province	Riverine flood[a]	333[b]	1,497,725[a]	1,500,000[a]
10	Sep.–Oct. 2014	Gilgit Baltistan, Azad Jammu and Kashmir (AJK), Punjab province	Riverine flood[a]	367[b]	2,530,673[a]	2,000,000[a]

[a] *EM-DAT, The International disaster database.*
[b] *Annual Flood Report (FFC, 2016).*

The 2010 Indus flood was the worst natural disaster in recorded history in Pakistan. In general, the flooding was caused by the cumulative effects of heavy rainfall in July, saturation of soil, and steady rain in August. One-fifth of the country was under water, and 20 million people were affected (9.4% of all globally reported disaster victims) (Guha-sapir et al., 2011; Khalil and Khan, 2015; Laghari et al., 2012; Lau and Kim, 2012). The United Nations estimated that the humanitarian crisis during the 2010 floods was much greater than the combined effects of the world's three worst natural disasters: the Asian tsunami and the Kashmir and Haiti earthquakes (Gaurav et al., 2011). The 2010 floods killed more than 1985 people (FFC, 2016), with an estimated economic loss of more than $10 billion, and affected at least 28 major cities and hundreds of towns and villages (Ali, 2013; Arslan et al., 2016; Lau and Kim, 2012). Similar rainfall and flooding were observed in 1976, but with less devastating effects than those in the 2010 floods, according to (Arslan et al., 2013). The difference may have been due in part to less populated settlements on the flood plains. It has been noted that the 2010 floods resulted from (1) heavy rainfall in the upstream (Gaurav et al., 2011), (2) other factors, such as weakness in flood defenses due to improper maintenance of existing flood-protection structures and distorted natural drainage networks (FFC, 2016), and (3) national decisions on water use and policy failure (Arslan et al., 2016; Mustafa and Wrathall, 2011).

In 2011 floods in the Sindh province killed 516 people and affected about 5.5 million people (EM-DAT, 2017), with half of Lower Sindh submerged during the flood events (Khan and Hassan, 2015). Flooding in 2012 killed at least 571 people and affected about 5 million people (EM-DAT, 2017) in Sindh, Balochistan and southern part of Punjab province, (FFC, 2016). In 2014 heavy rain and flooding of the Chenab, Ravi, Sutlej, and Jhelum rivers caused flash floods in Punjab, Azad Jammu and Kashmir, and the Gilgit-Baltistan province of Pakistan. The flood killed at least 367 people and affected about 2.5 million people (FFC, 2016; NDRM, 2014).

Naheed and Rasul (2011) studied the rainfall variability over Pakistan for an approximate 50-year period, from 1960 to 2009, and found variability between seasons. According to the interseasonal analysis, variability was found to be highest in post-monsoon and pre-monsoon seasons as compared with the winter and monsoon seasons. The study cautioned that the climate variability embedded with extreme (wet/dry) precipitation episodes will be hard for practitioners (people dependent on agriculture, irrigation, energy in Pakistan) to manage.

Increases in the frequency and intensity of extreme precipitation events are projected for the Hindu Kush-Himalayan region. In the Indus Basin the mean annual precipitation is projected to increase in the mid and late 21st century as compared with the 20th century, particularly in high altitudes (Forsythe et al., 2014; Immerzeel et al., 2010; Lutz et al., 2014; Su et al., 2016). Su et al. (2016) and Kulkarni et al. (2013) projected a variation in seasonal precipitation with increases during the monsoon. Similarly, Rajbhandari et al. (2014) projected an increase in rainfall intensity over the Upper Indus Basin and the Lower Indus Basin in the 2020s, 2050s, and 2080s. The increase in precipitation will likely increase runoff from the rivers, which will increase floods and flash floods in the basin.

13.3 HYDROMETEOROLOGICAL OBSERVATION/MONITORING SYSTEM OF PAKISTAN

Hydrometeorological information is important for planning, operation, and management of water resources and for flood protection (Walker, 2000). Hydrometeorological observations include precipitation, temperature, wind speed, humidity, evaporation, water level, and discharge data. Early flood warnings to enhance preparedness are largely dependent on the timely availability and quality of hydrometeorological observations (Shrestha et al., 2015). Collection of hydrometeorological data in real time requires a range of sophisticated sensors together with professional competence for management and operation. Data for flood forecasting should be transmitted in real time. For flood forecasting purposes, hydrometeorological data from the entire basin is most valuable if it reaches the forecaster in near real time.

In Pakistan a number of organizations are collecting hydrometeorological data, such as the Pakistan Meteorological Department (PMD), the Water and Power Development Authority (WAPDA), and the Irrigation Department, which are described in this section.

13.3.1 Hydrometeorological Observation by PMD

PMD is responsible for monitoring local and regional weather on a regular basis and provides flood-forecasting services. PMD maintains a network of meteorological stations to collect surface weather data and upper-air data for weather and flood forecasting. PMD has established

(i) station networks to generate meteorological information;
(ii) telecommunication systems for speedy dissemination of data;
(iii) units to analyze data for forecast and warning issuance;
(iv) weather surveillance radars;
(v) automatic weather stations; and
(vi) data processing units (PMD, 2017a).

TABLE 13.2   Number of Rainfall Stations Operated by PMD

Province	No. of Rainfall Gauging Stations
Punjab	83
Khyber Pakhtunkhwa	19
Sindh	24
Balochistan	16
Gilgit-Baltistan and Azad Kashmir	12

Source: FFD Staff.

Additionally, PMD receives weather images from China's FY-2E weather satellite.

13.3.1.1 Manual Rainfall Observations

PMD operates 154 rainfall gauging stations in various provinces (Table 13.2 shows their distribution). Daily rainfall data are recorded from 8 a.m. to 8 p.m. Pakistan Standard Time, according to World Meteorological Organization (WMO) procedures, and the data are archived at the PMD website (http://www.pmd.gov.pk/FFD/index_files/daily/rainfall_data.htm). Observational data from other stations are collected by different government organizations, such as provinces' agriculture departments, on a monthly basis and then forwarded to PMD by mail, telephone, or fax (WMO, 2010) (Personal communication with staff of FFD, Lahore).

13.3.1.2 Automatic Rainfall Observations

The floods of 2010 damaged many hydrometeorological stations. With support from the International Center for Integrated Mountain Development (ICIMOD) and WMO, as part of the Hindu Kush—Himalayan Hydrological Cycle Observing System (HKH—HYCOS) program, PMD upgraded three meteorological stations with automatic weather stations (AWSs) in Gupis, Kalam, and Lower Dir (PMD, 2016); and WAPDA installed five automatic water-level sensors, the details of which are provided in Table 13.3.

TABLE 13.3   Hydrometric Monitoring Network Upgraded Through the HKH—HYCOS in Pakistan (Shrestha et al., 2015)

Station Type	Quantity	Observation	Function	Installed and Operated
Radar sensor	5	Water level	Monitoring water level in real time	WAPDA
Tipping bucket rain gauge	8	Rainfall	Monitoring intensity and volume of rainfall in real time as an input to model to generate runoff	PMD (3); WAPDA (5)
Automated weather station	3	Rainfall, temperature, wind speed, wind direction, humidity, and pressure	Monitoring meteorological parameters in real time	PMD

TABLE 13.4   List of AWSS Installed by FFD

S. No.	Location of Stations	Organization
1	Fort Munro, D.G. Khan	FFD
2	Mari, D.G. Khan	
3	Vehova, D.G. Khan	
4	Zain Sanghar, D.G. Khan	
5	Bhandu Wala, Rajanpur	
6	Munawar Tawi, Barnala, Azad Kashmir	
7	Gupis	FFD with support from HKH—HYCOS
8	Kalam	
9	Lower Dir	

Source: FFD.

The Flood Forecasting Division (FFD) at Lahore has further installed six AWSs for obtaining real-time meteorological data from remote areas for flood forecasting purposes (according to PMD). The AWS networks installed in the catchment areas of the Dera Ghazi Khan hill torrents help forecast flash floods in these areas. In most cases, based on the AWSs the FFD has been able to issue timely warnings and alerts by accessing the amount of rainfall received on a real-time basis. The stations with AWSs are listed in Table 13.4. The locations of the AWSs are shown in Fig. 13.2. Till November 2018, FFD has installed 19 new AWS and 4 more are in the process of being installed that are developed with support from United Nations Educational, Scientific and Cultural Organization (UNESCO). The PMD's Research and Development Division has developed an Automatic Data Acquisition System and an Integrated Database Management System to provide real-time, ground-based data on rainfall and other meteorological parameters for hydrological modeling and research purposes. These stations transmit real-time data at a predefined interval to support flood forecasting. PMD has a density of one station per 75 square kilometer, which is much lower than the WMO recommendation of one station per 36 square kilometer, indicating the need for expanding the hydrometeorological network.

National Drought Monitoring and Early warning center and the Tropical cyclone monitoring center installed 45 AWSs in Pakistan. The data are transmitted to PMD through GSM network (PMD, 2017a; WMO, 2010).

13.3.1.3 *Upper-Air Observations*

There are eight upper-air observation stations in Pakistan, of which two stations, located in Karachi and Lahore, are operational. A radiosonde measurement is taken once daily (WMO, 2010). The upper-air observations are used for weather forecasting.

13.3.1.4 *Weather Radar*

PMD operates seven meteorological radars in Islamabad, Karachi, D.I. Khan, Rahim Yar Khan, Sialkot, Lahore, and Mangla. However, the radars operate continuously only during

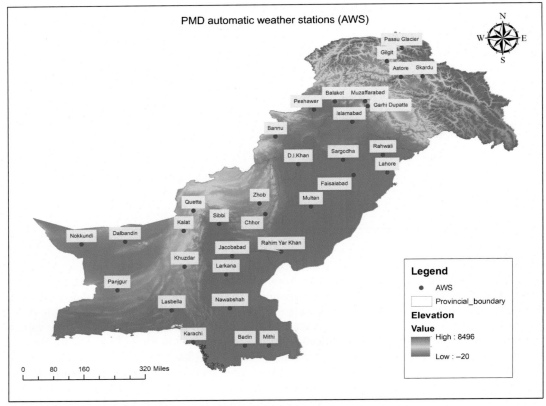

FIG. 13.2   PMD automatic weather station network. *Source: PMD.*

TABLE 13.5   Radar Facilities of PMD (Tariq and van de Giesen, 2012; WMO, 2010) (FFD Staff)

Radar Type	Location	Functions	Range
10 cm Doppler Weather radar	Lahore and Mangla	Quantitative precipitation estimates	480 km
5 cm Weather Surveillance radars	Sialkot, Karachi, D.I. Khan, Islamabad, Rahim Yar Khan	Detects the position of clouds and precipitation, quantitative precipitation estimates	280 km

the monsoon season in order to conserve energy and reduce electricity costs. For the latter reason, the PMD office collects only raw radar data; that is, the data are not yet calibrated, and the resolution of the radar images is very low (WMO, 2010). PMD is in the process of upgrading its radar infrastructure in Islamabad and Karachi (Bacha, 2016; FFD staff) and increasing the utility of the data. Table 13.5 shows the radars currently installed in Pakistan. The Sialkot radar is particularly helpful in accessing real-time information for forecasting of floods. Output from the Sialkot radar has been used effectively to calculate the amount of rainfall received in Nullahs' catchment areas, which feed the Chenab and Ravi rivers. In 2014 the radar was used to issue timely weather alerts and warnings to the concerned agencies and authorities. In the absence of real-time rainfall observations and because the rivers are transboundry, the Sialkot radar plays a pivotal role in flood forecasting.

13.3.2 Hydrometeorological Stations of WAPDA and Other Organizations

WAPDA, which was established in 1959 under the Ministry of Water and Power, implements and manages major water-resource and energy projects in Pakistan and is responsible for the operation of the Mangla and Tarbela reservoirs during floods.

WAPDA has 45 telemetric rainfall stations installed in the catchment of five major rivers, as shown in Fig. 13.3. The station data are transmitted daily to FFD through a telemetric network. As of November 2018, 26 out of 45 stations are providing real-time data to FFD. The location and number of rainfall stations are shown in Table 13.6.

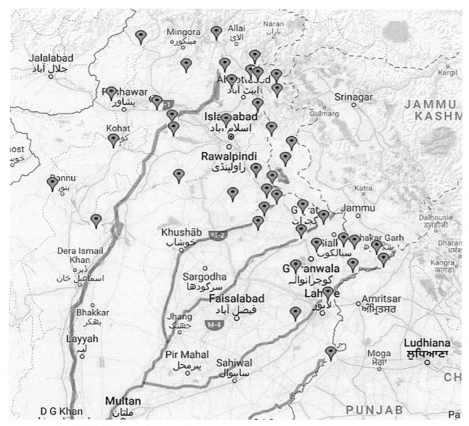

FIG. 13.3 Location of telemetry stations of WAPDA.

TABLE 13.6   Telemetric Stations (WMO, 2010) (FFD Staff)

S. No.	Location	No. of Telemetric Rainfall Stations	Functional Station
1	Indus River Basin	15	9
2	Jhelum River Basin	15	10
3	Chenab River Basin	8	4
4	Ravi River Basin	6	3
5	Sutlej River Basin	1	1

TABLE 13.7   Hydrological Station Network of the Irrigation Department

S. No.	River	Location
1	Ravi	Kot Naina, Jassar, Ravi Syphon, Shahdara, Balloki, Sidhnai bein Nullah at Chak, Amru bein nullah at Shakargrah, Deg Nullat at Q.S.Sungh, Bassantar Nullah
2	Sutlej	G.S.Wala, Bakarke, Sulemanki, Islam, Melsi Syphon
3	Chenab	Marala, Khanki, Qadirabad, Chinot Bridge, Trimmu, Rawaz Bridge, Punjnad, Aik Nullah at Ura, Palku Nullah
4	Jhelum	New Rasul, Khushab Bridge
5	Indus	Kalabagh, Taunsa, Mithankot, Ghazi Ghar, Chachran Shraif

Under WAPDA, rim stations were set up along the Indus that measure snowmelt and inflow through the stations. Instruments are installed at Bishma, Ogi, Phulra, Tarbela, and Daggar to measure the flow and rainfall rates. The Irrigation Department has hydrological station networks in several locations, as shown in Table 13.7. Each canal's discharge data are collected through its own network, using facilities of the police departments. The irrigation departments of the various provinces provide downstream discharge data. The data from both organizations are sent to the FFD.

13.3.3 Hydrometeorological Network and System in the Indus Operated by India

The India Meteorological Department (IMD), Central Water Commission (CWC), and Indian Space Research Organization (ISRO) record meteorological parameters in India.

The IMD operates 140 meteorological stations in the Indus Basin. Fifty-three stations are located in Himachal Pradesh, 47 in Punjab, 21 in Jammu and Kashmir, 12 in Haryana, and 7 in Rajasthan. The CWC has 18 meteorological stations and 26 hydromet observation sites in the basin located in Himachal Pradesh and Jammu and Kashmir. The ISRO operates 18 AWS in the basin, 17 in Punjab, and 1 in Himachal Pradesh (CWC and NRSC, 2014). The total number of stations in the Indus Basin in India is shown in Table 13.8.

The responsibility of the Indus Commission is implementation of the Indus Treaty between India and Pakistan (CWC, 2017), with the main responsibility being observation of hydrological and hydromet data and flood forecasting in Himachal Pradesh and Jammu and Kashmir.

TABLE 13.8   Meteorological Stations in the Indus Basin in India (CWC and NRSC, 2014)

S. No.	Organization	Type	Number of Stations
1	IMD	Meteorological	140
2	CWC	Meteorological	18
3	ISRO	Meteorological	18
4	CWC	Hydromet	26

13.3.4 Mechanisms for Collecting and Processing Hydrometeorological Data

The Climate Data Processing Center (CDPC) and the National Meteorological Communication Centre (NMCC) are the units of PMD responsible for storage, processing, retrieval, and distribution of meteorological data to local and international agencies. River discharges and other flood-related data are also stored in FFD.

The CDPC's main center is at Karachi, and there are four regional centers located at Karachi, Peshawar, Quetta, and Lahore that are responsible for quality control of the available data. The meteorological times-series data are updated and processed and extreme events are recorded. The NMCC collects meteorological data from various sources and transmits the data in real time for flood forecasting. The rapid dissemination of data used by NMCC comprises (PMD, 2017c)

i) telecommunication system (see Fig. 13.4),
ii) Internet service;
iii) leased line circuit of 64 kbps and 50 bauds (New Delhi & Tehran Circuit); and
iv) tsunami watch information

The telecommunication network of Pakistan in the past consisted of SSB and TP. Now most of the communication is done through leased line circuits. The Tashkent RTH (Regional Telecommunication Hub) is no longer operational. Most of the data communication has been shifted to the internet. While some remote observatories are communicating synoptic data by using Short Messaging Services (SMS) of cellular networks.

13.3.4.1 Processing and Release of Meteorological Data

Data on numerical forecasts collected from more than 100 meteorological stations are converted into WMO standards and accessed through the Global Telecommunication System (GTS). The synoptic data are put to simulation under the ICOsahedral Non-hydrostatic model (ICON), which captures the regional meteorological features of the country. The model reproduces forecasts for various meteorological parameters (temperature (min and max), precipitation, wind, relative humidity, and cloud cover). The forecast is updated twice daily and generates forecasts for 72 h (PMD, 2017b). The PMD processes the raw data and issues meteorological forecasts and warnings for use by various user agencies. The PMD extends its meteorological services to the Federal Flood Commission (FFC) on a regular basis. The meteorological data are provided through telephone or fax to the press, TV, and other information media, and the National and International aeronautical forecasts are sent in written form to the user agencies.

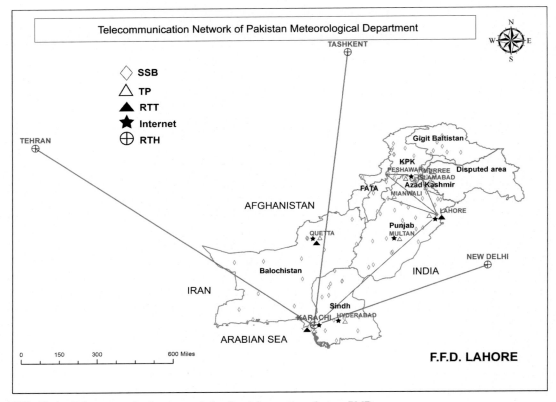

FIG. 13.4   Telecommunication network for flood forecasting. *Source: PMD.*

13.4 FLOOD FORECASTING IN PAKISTAN

Flood forecasting, warning, and dissemination services are provided by FFD of PMD. The FFD obtains hydrometeorological data from the various national and international sources, which is analyzed to produce flood forecasts and warnings and streamflow forecasting for water management at dams (PMD, 2017b).

The FFD collects data from (1) satellite imagery, (2) real-time hydrometeorological data, (3) real-time and prognostic weather charts (PMD NWF Centre and global products), (4) numerical weather prediction models (PMD NWF Centre and global products), and (5) weather radar networks (WMO, 2010). These data are used in various hydrological and statistical models for flood forecasting. Although the overall accuracy of flood forecast issued by FFD Lahore during the flood seasons is satisfactory, because of the transboundary nature of floods, water management practices and upstream operations pose a challenge for accurate flood forecasting in some of the rivers (Tariq and van de Giesen, 2012).

Most of the major tributaries of the Indus River in Pakistan originate in India. Consequently, there is mutual agreement in accordance to the Indus Water Treaty that requires India to share the river flow data through the respective Indus water commissioners

(IWCs). According to a recent report, the IWCs of India and Pakistan have agreed to share information regarding potential risks of heavy rains and floods. According to the agreement, India provides information on the Sutlej River at Rupar, Hanke, Ferozepur, Ravi, and Madhopur (The Express Tribune, 2017). Information from India includes flow of Chenab at Akhnoor for 75,000 cubic feet per second and above, for Jammu Tawi at Jammu for 20,000 cubic feet per second and above, and for Ravi below Madhopur for 30,000 cubic feet per second and above.

In addition, arrangements have been made with India to provide advance information about flood flows in the Sutlej River (below Rupar, below Harike, and below Ferozpur) and the Ravi River (below Madhopur) through hourly, three hourly, and six hourly phone calls, depending on the flood stages. Releases from the Bhakra Dam on the Sutlej River and the Pong Dam on the Beas River are also communicated daily by telephone to Pakistan at 10:30 a.m. Pakistan Standard Time (PST), along with base flow data for the Sutlej and Ravi rivers. According to *The Express Tribune* (2017) flood flow reports are also broadcast by Jammu radio stations at 9 a.m., 2:30 p.m. and 10:40 p.m. PST. However, when the information is received through the Indus Commission of Pakistan, the flow is close to the border, allowing only a short lead time for flood forecasting.

13.4.1 Numerical Weather Prediction (NWP)

The PMD's Research and Development Division is working on a number of NWP models that are used for weather and flood forecasting as well as for research (Table 13.9). Since 2007 the PMD also has been operating the High Resolution Regional Model (HRM) of the German Weather Service (DWD), which runs with 22 km resolution on a High Performance

TABLE 13.9 Weather Forecast Models Used by PMD for Operational and Research Purposes

S. No.	Model Name	Developer	Resolution
1	Global Forecast System (GFS)	National Oceanic and Atmospheric Administration (NOAA)	25 km
2	Weather Research and Forecasting (WRF)	National Center for Atmospheric Research (NCAR), NOAA, Forecast Systems Laboratory (FSL), the Air Force Weather Agency (AFWA), the Naval Research Laboratory (NRL), the University of Oklahoma (OU), and the Federal Aviation Administration (FAA)	Meters to 1000s of km
3	High-resolution Regional Model (HRM)	DWD	
4	ICOsahedral non-hydrostatic (ICON)	DWD and Max-Planck Institute for Meteorology (MPI-M)	13 km horizontal grid resolution and 40 atmospheric levels
5	Global Spectral model (GSM)	China Meteorological Administrations (CMA)	30 km
6	ECMWRF web-based products	European Center for Medium Range Weather Forecast	HRS 8 km, ENS controlled 16 km

Computing Cluster (WMO, 2010). This model was upgraded in 2010 to run with 11 km resolution. However, given the complex mountainous topography of Pakistan, since March 2015 the PMD has run the ICON model using a higher resolution.

The ICON model installed on a High Performance Computing Cluster system is driven by initial conditions from the GME (Global Model of DWD, Germany). The ICON model has 13 km horizontal grid resolution and 40 atmospheric levels, which operate twice daily for 7-day forecasts at 0:00 UTC and 12:00 UTC (PMD, 2016). Pakistan is also using the Global Spectral Model (GSM) of the China Meteorological Administration (CMA). GSM (T639) provides short-and medium-range weather forecasts for 1 week at almost 30 km horizontal resolution. The model output is updated daily at 0:00 UTC (PMD, 2016; Research and Development Division, 2017).

The Weather Research and Forecasting (WRF) modeling system has been deployed on a High Performance Cluster Computing system for operational weather forecasts up to 72 h at a finer resolution of 7 km (PMD, 2016). Likewise, 7-day 3-hourly Global Forecast System (GFS) are extracted by PMD for the Indus Basin.

Assessing the outputs from these various numerical weather forecast products (Fig. 13.5), FFD issues qualitative forecast at least 24–36 h before rainfall, while quantitative forecast is issued 12 h in advance of actual peak (WMO, 2010). The forecast products include precipitation, total cloud cover, temperature (mean, min, max), mean sea level pressure, wind, vorticity, snow depth, snow density and Convective Available Potential Energy (CAPE).

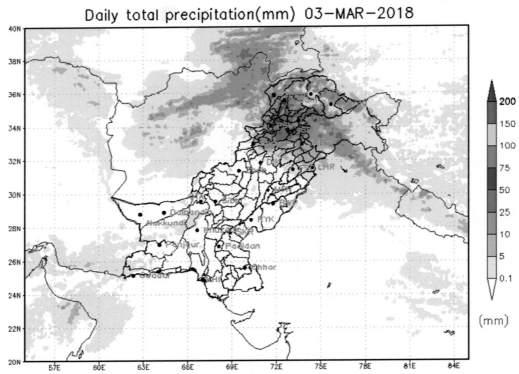

FIG. 13.5   Numerical weather prediction products for weather forecasting. *Source: PMD.*

TABLE 13.10   Summary of Models Used for Flood Forecasting in the Indus Basin

S. No.	Model Used	Data Used	Implementation	Performance	References
1	IFAS	Soil hydraulics data, satellite-based rainfall estimates, discharge, rainfall data	Research: JICA-funded UNESCO, International Center for Water Hazard and Risk Management (ICHARM)	The flood peaks calculated by the IFAS show good synchronization with the observed ones for the 2010 flood	Aziz and Tanaka (2010) and Sugiura et al. (2016)
	Distributed rainfall-runoff model mounted in IFAS	Precipitation, discharge, latent heat fluxes	Research: part of the "Strategic Strengthening Flood Forecasting and Management capacity in Pakistan" implemented by UNESCO	Model performance is dependent on the availability and quantity of measured discharge data as precipitation data are not adequate	Sugiura et al. (2013)
2	FEWS using SAMO and SOBEK routing model	Flood extent, 50 year inundation map, ArcGIS online data and global administrative area website data	Delft FEWS for operational use	Degree of accuracy of FEWS 50-year inundation maps was greater in areas where no flash floods or levee breaches were experienced	Afsar et al. (2013)
		Rainfall and snowmelt	Working paper		Werner and Dijk (2005)
3	SOBEK, Sacramento watershed model	Discharge data, rainfall-runoff, radar		Peak discharge of the river flow is calculated accurately	Awan (2003)
4	Constrained Linear System (CLS) (statistical model)		Has been replaced by Sacramento watershed model		Awan (2003)
5	Mike 11	Rainfall, discharge, evaporation, DEM	Nallah Lai Basin, operational	Good indication of floods	
6	Flash Flood Guidance system	Hydroestimator and local in situ data where available	WMO and HRC		

13.4.2 Flood Forecasting Models Used by FFD

For flood forecasting, the FFD uses the Flood Early Warning System (FEWS) and Indus-Integrated Flood Analysis System (IFAS) (Table 13.10). In regard to urban areas, flood forecasting and early warning systems have been established at Nallah Lai basin (Ahmad, 2015; Mustafa et al., 2015) extending to the twin cities of Islamabad and Rawalpindi (see Box 13.1) (WMO, 2010).

<hr>

BOX 13.1

LAI NULLAH FLOOD FORECASTING SYSTEM

On March 20, 2007, heavy rainfall over the catchments of Lai Nullah and the consequent flooding was correctly modeled by the Lai Nullah flood forecasting system established by the PMD (Hayat, 2007). The accurate alerts and early warning from the system increased the confidence of the people on the system and proved to be a useful tool for flash flood mitigation. The system supported by the Japan International Coorperation Agency (JICA) has real-time rainfall gauging stations at six locations, data are transferred using radio systems in real-time. In addition, two real-time river water level stations are in operation under this system. Water levels are observed on an hourly basis and data are transferred to WAPDA and FFD through a Meteor Burst Communication (MBC) system together with rainfall observation data (WMO, 2010). According to the PMD, the flood forecasting system for the Lai Nullah Basin is highly robust in terms of real-time monitoring of storm precipitation intensities (10 min, 30 min, 60 min) and for combining runoff and point rainfall recorded data to simulate streamflow and accumulation forecasts using Mike-11 software every 10 min for both meteorological and hydrological purposes. The system can be accessed on a real-time basis through the PMD website.

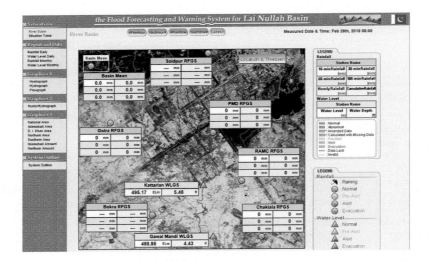

Since 2007 the FFD has been using the FEWS platform developed by Delft Hydraulic in the Netherlands for flood forecasting. The FEWS consists of (1) the Sacramento Soil Moisture Accounting (SAC-SMA) rainfall-runoff model and (2) the SOBEK, a hydraulic flow routing model (WMO, 2010). Data from 44 telemetry stations provided by the WAPDA are included in the model. The output from the rainfall-runoff model, as illustrated in Fig. 13.6, forms the input into the routing model (Awan, 2003). However, the FEWS platform's current flood forecasting system has limited geographical coverage, and rainfall-runoff information and downstream routing of the Tarbela and Kabul rivers are not incorporated into the model. The tributaries incorporated into the FEWS are not gauged according to the level and discharge rate needed for accurate flood forecasting (Jan et al., 2013; WMO, 2010). Therefore the WMO recommends upgrading the FEWS model or replacing it with advanced models in order to attain better performance and provide accurate flood forecasts.

The FFD is also implementing the IFAS developed by ICHARM with support from UNESCO under the Strategic Strengthening of Flood Warning and Management Capacity of Pakistan project and the Japan Aerospace Exploration Agency (JAXA). More than 39 districts in Pakistan are covered with flood forecasting and IFAS early warning systems (Ahmad, 2015; Mustafa et al., 2015), while 32 districts in the lower basin are served by new flood maps generated by the Rainfall–Runoff Inundation (RRI) model. A simplified structure of the IFAS model is shown in Fig. 13.7. The satellite rainfall estimates from JAXA's Global Satellite Mapping of Precipitation Near Real Time (GSMaP_NRT) are used in the IFAS to conduct rainfall-runoff modeling (Aziz, 2014; Aziz and Tanaka, 2010). Aziz and Tanaka (2010) reported that flood peaks calculated by the IFAS using GSMaP showed good synchronization with the observed peak during the 2010 flood. Aziz (2014) applied the IFAS to the Kabul River Basin using the corrected GSMaP_NRT for rainfall-runoff modeling, which showed promising results for generating sufficient lead time to the downstream communities. The discharge calculated by the GSMaP_NRT (corrected) satellite has shown good agreement with the observed discharge values from the ground discharge measuring station in terms of both flood duration and flood peak. Thus the IFAS model is being further improved, validated, and extended by the PMD before using it for operational forecasting.

The Flash Flood Guidance system is being implemented in collaboration with the WMO with the support of the Hydrologic Research Center (HRC) in the United States.

The accuracy of the flood forecasts issued by the PMD are reported to be improving, as was observed during the 2014 floods (Fakhruddin, 2014). However, there are uncertainties in the flood forecasts due to uncertainties in rainfall forecasts (Cloke and Pappenberger, 2009; Dietrich et al., 2009; Van Steenbergen and Willems, 2014), along with uncertainties in (1) estimation of model parameters, (2) initial conditions, and (3) forecasted hydromet input (Zappa et al., 2011). In the Indus Basin there are only a few gauge networks at high altitudes (Immerzeel et al., 2012; Pellicciotti et al., 2012), which leads to uneven and nonhomogenous observation data (Palazzi et al., 2013). Gauge networks are located mainly at the valley bottoms, which don't capture snowfall (Immerzeel et al., 2012; Palazzi et al., 2013), leading to underestimations of precipitation (Palazzi et al., 2013). Thus data inputs and modeling approaches are important for accurate flood forecasting (Pellicciotti et al., 2012).

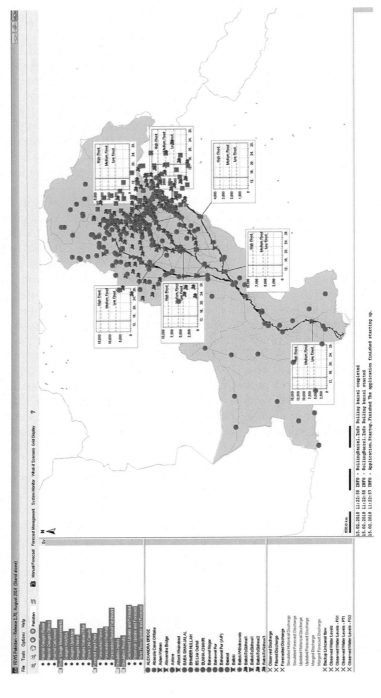

FIG. 13.6   FEWS Pakistan interface. *Source: FFD.*

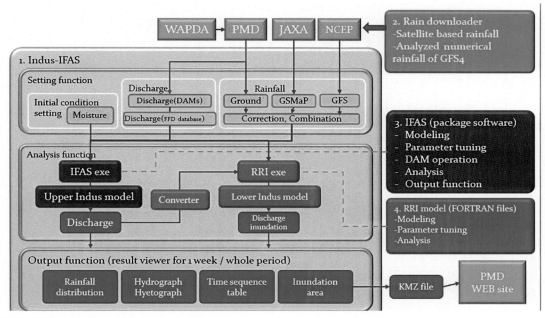

FIG. 13.7 Structure of Indus-IFAS. *Source: FFD.*

13.4.3 Real-Time Flood Monitoring and Dissemination

The PMD issues daily, fortnightly, weekly, and seasonal weather forecasts. It provides qualitative forecasts at a minimum of 24 to 36h in advance of actual precipitation and quantitative forecasts about 12h in advance of the actual peak (WMO, 2010).

The FFD analyzes meteorological and hydrological conditions from data it receives from various sources and model outputs. Data and model outputs from rim stations are considered as well. A statistical and empirical technique is applied to prepare a flood bulletin that is agreed upon by a group of meteorologists and hydrologists. The mechanism of preparation and dissemination of flood forecasts is shown in Fig. 13.8.

The FFD issues flood bulletins providing information on river situations and qualitative flood forecasts. The Bulletin "A" provides the hydrological situation in major rivers, which includes the meteorological features, including weather forecasts for next 24h and the rainfall recorded in the past 24h (FFD, 2017). The Bulletin "B" provides quantitative flood forecasts of gauging stations. The FFD also provides integrated weather, climate, and hydrological forecasting information for use by various civil organizations responsible for disaster preparedness and mitigation. The flood forecasts are provided through various mediums, such as web services, fax/phone/mobile SMS, as well as through electronic and print media. After the issuance of a flood warning by the FFD at the federal and provincial levels, the NDMA/Provincial Disaster Management Authority (PDMA) is responsible for issuing warnings through TV, radio, and tehsil (Mustafa et al., 2015).

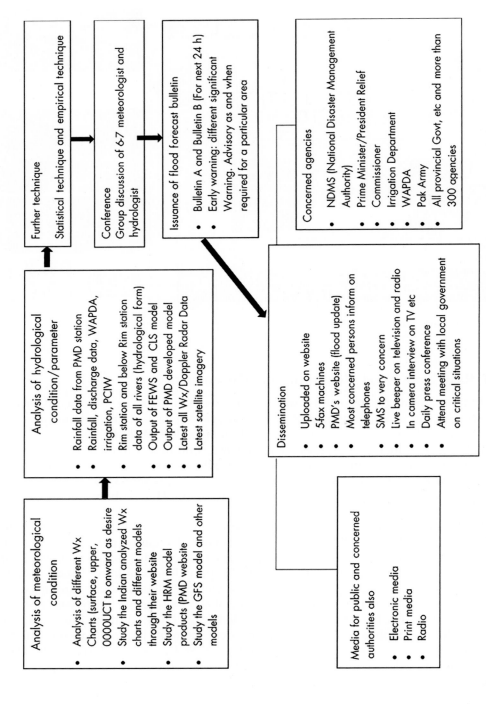

FIG. 13.8   Preparation and dissemination of flood forecasts in Pakistan. *Source: PMD.*

BOX 13.2

GLACIAL LAKE OUTBURST FLOOD MONITORING

The UNDP, with the support of the government of Pakistan and the Adaptation Fund, is supporting efforts to reduce the risks of flash floods from glacial lake outbursts in the high risk areas of the Gilgit and Chitral districts. Efficient monitoring of GLOF risks was made possible by the creation of site maps and the installation of observatories and automated weather stations in the pilot sites.

These factors were effective in reducing risks from GLOFs, as the PMD issued early warnings and alerted communities to potential dangers. Community-based disaster risk management training equipped people with techniques to protect themselves during future flash floods and other climate change-related disasters (UNDP, 2017).

A WMO mission report indicated that the 2010 flood was accurately forecasted and timely transmitted to the concerned authorities, which helped reduce the loss of lives and property. The report indicated the challenge in having adequate real-time hydromet data and a robust dissemination system and the need for public awareness and education. To address glacier lake outburst flood (GLOF) risk issues, the United Nation Development Programme (UNDP) has piloted community-based disaster risk management programs to minimize the adverse effects of flash floods (see Box 13.2).

PMD in collaboration with University of Engineering Technology, Lahore has already launched a mobile based weather information service, covering 139 districts of Pakistan to provide weather services to the users. This service will update the user about weather as well as provide advanced information about critical weather conditions, such as droughts, floods (PMD, 2016).

13.5 LOOKING AHEAD

Recent floods have indicated the lack of effective coordination among institutions involved in flood management, caused in part by limitations in the existing technical capacities, for example, in the dissemination of early warnings, in disaster preparedness measures, in emergency response, and in structural measures for flood mitigation (FFC, 2016). Further improvement of flood forecasting and warning systems is essential to reduce the impacts of future floods. While Pakistan's flood forecasting and early warning systems have demonstrated their usefulness, the system's predictive capacity is still limited. The transboundary nature of floods in the Indus Basin calls for sharing real-time hydrometeorological data from upstream countries to help improve the accuracy of flood forecasts and increase the lead time. There is also an urgent need to extend the existing FEWS coverage to the Upper Indus reach, to the Swat and Kabul rivers, and to the major hill torrents.

The PMD is seeking investments to put in place state-of-the art infrastructure and modeling tools to enhance the flood forecasting capacity. Calibration and fine-tuning of new flood forecasting models can be a time-consuming process given the complex nature of the terrain and the limited availability of transboundry data. Some experimental studies have shown that if weather stations are positioned at optimal locations, only a very few are required to model runoff with a performance that is as good or better performance than when a high-density network is used (Arsenault and Brissette, 2014). Since upgrading the density of stations requires both time and financial resources, an experiment for selecting locations from an optimizing point of view may offer a short-term strategy for Pakistan.

The Regional Specialized Meteorological Centre (RSMC) for South Asia hosted by IMD, under the WMO and the Economic and Social Commission for Asia and the Pacific (ESCAP), provides a platform for regional weather advisories on tropical cyclones. An example is the Nilofar cyclone that occurred in October 2014 when timely bulletins were shared to prepare for the cyclone between India and Pakistan (Bhatta, 2014). Such regional platforms and mechanisms could be used for promoting regional cooperation for flood preparedness and response.

Last-mile connectivity is also important for minimizing losses due to floods. The local communities that are the first responders are said not to have enough disaster preparedness information (FFC, 2016). Improving communication at all levels is an important aspect of reducing risks and increasing preparedness. Efforts must be undertaken to translate the various forecasts into information that the public can understand and that can be used effectively at local levels. Currently, there is a gap in general awareness raising and in sensitivity to and education of the communities regularly affected by floods that needs to be addressed. This gap can be tackled by providing the above to the entire community, but particularly to women and children in terms of the risks involved. The communities involved need understandable and actionable warnings, with a special focus on populations residing in the active flood plains along major, secondary, and tertiary rivers. Such actions can help minimize the adverse effects of floods in the future.

References

Afsar, M.M., Aslam, K., Atif, S., 2013. Comparative analysis of flood early warning system comparative analysis of flood early warning system (FEWS)' flood inundation maps & actual extents of 2010 floods in Pakistan. In: 5th World Engineering Congress (WEC2013). https://doi.org/10.13140/2.1.4253.0085.

Ahmad, N., 2015. Early Warning Systems and Disaster Risk Information. LEAD House, Islamabad, Pakistan. http://www.lead.org.pk/lead/attachments/briefings/LPNB7.pdf.

Ahmad, Z., Hafeez, M., Ahmad, I., 2012. Hydrology of mountainous areas in the upper Indus Basin, Northern Pakistan with the perspective of climate change. Environ. Monit. Assess. 184 (9), 5255–5274.

Ali, A., 2013. Indus Basin Floods Mechanisms, Impacts, and Management. Asian Development Bank, Mandaluyong City, Philippines. https://www.adb.org/publications/indus-basin-floods-mechanisms-impacts-and-management.

Arsenault, R., Brissette, F., 2014. Determining the optimal spatial distribution of weather station networks for hydrological modeling purposes using RCM datasets: an experimental approach. J. Hydrometeorol. 15, 517–526. https://dx.doi.org/10.1175/JHM-D-13-088.1.

Arslan, M., Tauseef, M., Gull, M., Baqir, M., Ahmad, I., Ashraf, U., Al-Tawabini, B., 2013. Unusual rainfall shift during monsoon period of 2010 in Pakistan: flash flooding in Northern Pakistan and riverine flooding in Southern Pakistan. Afr. J. Environ. Sci. Technol. 7 (9), 882–890.

Arslan, M., Ullah, I., Baqir, M., Shahid, N., 2016. Evolution of flood management policies of Pakistan and causes of flooding in year 2010. Bull. Environ. Stud. 1 (1), 29–35.

Asraf, A., Naz, R., Roohi, R., 2012. Monitoring and estimation of glacial resource of Azad Jammu and Kashmir using remote sensing and GIS techniques geographical setup and physiography. Pakistan J. Meteorol. 8 (16), 31–41.

Awan, S.A., 2003. Flood forecasting and management in Pakistan. In: Water Resources Systems—Hydrological Risk, Management and Development (Proceedings of Symposium IIS02b held during IUGG2003 at Sapporo, July 2003), pp. 90–98. IAHS Publ no. 281.

Aziz, A., 2014. Rainfall-runoff modeling of the trans-boundary Kabul River Basin using integrated flood analysis system (IFAS). Pakistan J. Meteorol. 10 (20), 75–81.

Aziz, A., Tanaka, S., 2010. Regional parameterization and applicability of integrated flood analysis system (IFAS) for flood forecasting of upper-middle Indus River classification of hydrological models. Pakistan J. Meterorol. 8 (15), 21–38.

Bacha, A., 2016. Country Report of Pakistan. Visiting Researcher Program. Asian Disaster Reduction Centre (ADRC), Kobe, Japan.

Bhatta, A., 2014. India, Pakistan Share Data on Cyclone Nilofar. Sci Dev, Net South Asia. https://www.scidev.net/south-asia/environment/news/india-pakistan-share-data-on-cyclone-nilofar.html. (Accessed 27 December 2017).

Cloke, H.L., Pappenberger, F., 2009. Ensemble flood forecasting: a review. J. Hydrol. 375, 613–626.

CWC, 2017. Central Water Commission. Indus Basin Organization. http://www.cwc.nic.in/regionaloffice/chandigarh/welcome.html. (Accessed 12 December 2017).

CWC, NRSC, 2014. Indus Basin. Central Water Commission & National Remote Sensing Centre Ministry of Water Resources, Government of India. Version 2.0.

Dietrich, J., Schumann, A.H., Redetzky, M., Walther, J., Denhard, M., Wang, Y., Pfützner, B., Büttner, U., 2009. Assessing uncertainties in flood forecasts for decision making: prototype of an operational flood management system integrating ensemble predictions. Nat. Hazards Earth Syst. Sci. 9, 1529–1540.

EM-DAT, 2017. The International Disaster Database. http://www.emdat.be. (Accessed 21 April 2017).

Fakhruddin, B.S., 2014. Pakistan Flood 2014—Lesson Identified Never Learned. https://hepex.irstea.fr/pakistan-flood-2014-lesson-identified-never-learned/. (Accessed 28 February 2018).

FAO, 2012. Irrigation in Southern and Eastern Asia in Figures, AQUASTAT Survey—FAO Water Report 37. Food and Agriculture Organization of the United Nations, Rome. http://www.fao.org/docrep/016/i2809e/i2809e.pdf.

FFC, 2016. Annual Flood Report 2016. Office of the Chief Engineering Advisor & Chairman Federal Flood Commission, Federal Flood Commission, Ministry of Water & Power Government of Pakistan, Islamabad, pp. 1–51.

FFD, 2017. Flood Forecasting Division, Flood Page. http://www.pmd.gov.pk/ffd/cp/floodpage.htm.

Forsythe, N., Fowler, H.J., Blenkinsop, S., Burton, A., Kilsby, C.G., Archer, D.R., Harpham, C., Hashmi, M.Z., 2014. Application of a stochastic weather generator to assess climate change impacts in a semi-arid climate: the Upper Indus Basin. J. Hydrol. 517, 1019–1034.

Gaurav, K., Sinha, R., Panda, P.K., 2011. The Indus flood of 2010 in Pakistan: a perspective analysis using remote sensing data. Nat. Hazards 59 (3), 1815–1826. https://dx.doi.org/10.1007/s11069-011-9869-6.

Guha-sapir, D., Hoyois, P., Below, R., Ponserre, S., 2011. Annual Disaster Statistical Review 2010: The Numbers and Trends. Ciaco Imprimerie, Louvain-la-Neuve, Brussels, Belgium. http://www.cred.be/sites/default/files/ADSR_2010.pdf.

Hayat, A., 2007. Flash flood forecasting system for Lai Nullah Basin (A case study of March 19, 2007 rainfall event). Pakistan J. Meteorol. 4 (7), 75–84.

Immerzeel, W.W., van Beek, L.P.H., Bierkens, M.F.P., 2010. Climate change will affect the Asian water towers. Science 328 (5984), 1382–1385. https://dx.doi.org/10.1126/science.1183188.

Immerzeel, W.W., Pellicciotti, F., Shrestha, A.B., 2012. Glaciers as a proxy to quantify the spatial distribution of precipitation in the Hunza Basin. Mt. Res. Dev. 32, 30–38.

Iqbal, A.R., 2013. Environmental Issues of Indus River Basin: An Analysis. ISSRA Papers. Institute for Strategic Studies, Research & Analysis, National Defence University, Islamabad, pp. 89–112.

Jan, D., Khan, I., Ullah, K., 2013. Analyzing the Shortcoming of Flood Early Warning System in Pakistan. Department of Civil Engineering, Gandhara Institute of Science and Technology, Peshawar. https://www.slideshare.net/engrdaud/flood-early-warning-system-in-pakistan.

Khalil, U., Khan, N.M., 2015. Assessment of flood using geospatial technique for Indus River Reach: Chashma-Taunsa. Sci. Int. J. 27 (3), 1985–1991.

IV. WATER EXTREMES IN INDUS RIVER BASIN

Khan, H., Hassan, S., 2015. Stochastic river flow modelling and forecasting of Upper Indus Basin. J. Basic Appl. Sci. 11, 630–636. https://dx.doi.org/10.6000/1927-5129.2015.11.84.

Khan, B., Iqbal, M.J., Yosufzai, M.A.K., 2011. Flood risk assessment of river Indus of Pakistan. Arab. J. Geosci. 4, 115–122.

Kulkarni, A., Patwardhan, S., Kumar, K.K., Ashok, K., Krishnan, R., 2013. Projected climate change in the Hindu Kush-Himalayan region by using the high-resolution regional climate model PRECIS. Mt. Res. Dev. 33, 142–151. https://dx.doi.org/10.1659/MRD-JOURNAL-D-12-00027.1.

Laghari, A.N., Vanham, D., Rauch, W., 2012. The Indus basin in the framework of current and future water resources management. Hydrol. Earth Syst. Sci. 16, 1063–1083. https://dx.doi.org/10.5194/hess-16-1063-2012.

Lau, W.K.M., Kim, K.-M., 2012. The 2010 Pakistan flood and Russian heat wave: teleconnection of hydrometeorological extremes. J. Hydrometeorol. 13 (1), 392–403. https://dx.doi.org/10.1175/JHM-D-11-016.1.

Lutz, A.F., Immerzeel, W.W., Shrestha, A.B., Bierkens, M.F.P., 2014. Consistent increase in High Asia's runoff due to increasing glacier melt and precipitation. Nat. Clim. Chang. 4, 587–592. https://dx.doi.org/10.1038/nclimate2237.

Mukhopadhyay, B., Khan, A., 2015. Boltzmann–Shannon entropy and river flow stability within Upper Indus Basin in a changing climate. Int. J. River Basin Manag. 13 (1), 87–95.

Mustafa, D., Wrathall, D., 2011. Indus Basin floods of 2010: souring of a Faustian Bargain? Water Altern. 4 (1), 72–85.

Mustafa, D., Gioli, G., Qazi, S., Waraich, R., Rehman, A., Zahoor, R., 2015. Gendering flood early warning systems: the case of Pakistan. Environ. Hazards 14 (4), 312–328.

Naheed, G., Rasul, G., 2011. Investigation of rainfall variability for Pakistan. Pakistan J. Meteorol. 7, 25–32.

NDRM, 2014. Pakistan Floods 2014: Recovery Needs Assessment and Action Framework 2014–16. National Disaster Management Authority, Pakistan.

Palazzi, E., Von Hardenberg, J., Provenzale, A., 2013. Precipitation in the Hindu-Kush Karakoram Himalaya: observations and future scenarios. J. Geophys. Res. Atmos. 118, 85–100.

Pellicciotti, F., Buergi, C., Immerzeel, W.W., Konz, M., Shrestha, A.B., 2012. Challenges and uncertainties in hydrological modeling of remote Hindu Kush–Karakoram–Himalayan (HKH) Basins: suggestions for calibration strategies. Mt. Res. Dev. 32, 39–50.

PMD, 2016. Country Report of Pakistan (2015–2016) for 43rd Session of WMO/ESCAP Panel on Tropical. Pakistan Meteorological Department, Islamabad, Pakistan.https://www.wmo.int/pages/prog/www/tcp/documents/PTC43_Pakistan_CountryReport.pdf. (Accessed 22 December 2017).

PMD, 2017a. Pakistan Meteorological Department, National Meteorological Communication Center. http://www.pmdnmcc.net/aboutPMD.aspx. (Accessed 27 July 2017).

PMD, 2017b. Pakistan Meteorological Department, Government of Pakistan. http://www.pmd.gov.pk/PMD/pmdinfo.html. (Accessed 11 October 2017).

PMD, 2017c. Pakistan Meteorological Department, National Meteorological Communication Center. http://www.pmdnmcc.net/aboutNMCC.aspx. (Accessed 31 July 2017).

Rahman, A.U., Shaw, R., 2015. Flood risk and reduction approaches in Pakistan. In: Disaster Risk Reduction Approaches in Pakistan. Springer, Japan, pp. 77–100.

Rajbhandari, R., Shrestha, A.B., Kulkarni, A., Patwardhan, S.K., Bajracharya, S.R., 2014. Projected changes in climate over the Indus river basin using a high resolution regional climate model (PRECIS). Clim. Dyn. 44, 339–357. https://dx.doi.org/10.1007/s00382-014-2183-8.

Research and Development Division, 2017. Research and Development Division. Numerical Weather Products, Pakistan Meteorological Department.http://www.pmd.gov.pk/rnd/rndweb/rnd_new/gsm.php.

Sardar, M.S., Tahir, M.A., Zafar, M.I., 2008. Poverty in riverine areas: vulnerabilities, social gaps and flood damages. Pakistan J. Life Soc. Sci. 6, 25–31.

Shrestha, M.S., Grabs, W.E., Khadgi, V.R., 2015. Establishment of a regional flood information system in the Hindu Kush Himalayas: challenges and opportunities. Int. J. Water Resour. Dev. 627, 1–15. https://dx.doi.org/10.1080/07900627.2015.1023891.

Su, B., Huang, J., Gemmer, M., Jian, D., Tao, H., Jiang, T., Zhao, C., 2016. Statistical downscaling of CMIP5 multimodel ensemble for projected changes of climate in the Indus River Basin. Atmos. Res. 178–179, 138–149. https://dx.doi.org/10.1016/j.atmosres.2016.03.023.

Sugiura, A., Fujioka, S., Nabesaka, S., Sayama, T., Iwami, Y., Fukami, K., Tanaka, S., Takeuchi, K., 2013. Challenges on modelling a large river basin with scarce data: a case study of the Indus upper catchment. In: 20th International Congress on Modelling and Simulation, pp. 1–6. https://www.mssanz.org.au/modsim2013/L1/sugiura.pdf.

Sugiura, A., Fujioka, S., Nabesaka, S., Tsuda, M., Iwami, Y., 2016. Development of a flood forecasting system on the upper Indus catchment using IFAS. J. Flood Risk Manag. 9, 265–277.

Tariq, M.A.U.R., van de Giesen, N., 2012. Floods and flood management in Pakistan. Phys. Chem. Earth 47–48, 11–20. https://dx.doi.org/10.1016/j.pce.2011.08.014.

The Express Tribune, 2017. Pakistan, India to Share Information on Potential Heavy Rains, Floods. https://tribune.com.pk/story/1457782/pakistan-india-share-information-potential-heavy-rains-floods/. (Accessed 22 December 2017).

UNDP, 2017. Communities Work Together to Reduce the Risk of Flash Floods in Northern Pakistan. http://www.pk.undp.org/content/pakistan/en/home/ourwork/environmentandenergy/successstories/glof.html. (Accessed 28 December 2017).

Van Steenbergen, N., Willems, P., 2014. Rainfall uncertainty in flood forecasting: Belgian case study of rivierbeek. J. Hydrol. Eng. 19, 1–6.

Walker, S., 2000. The value of hydrometric information in water resources management and flood control. Meteorol. Appl. 7 (4), 387–397.

Werner, M., Dijk, M.V., 2005. Developing flood forecasting systems: examples from the UK, Europe, and Pakistan. In: International Conference on Innovation Advances and Implementation of Flood Forecasting Technology. ACTIF/FloodMan/FLoodRelief, Tromsø, Norway, pp. 1–11.

WMO, 2010. WMO Fact-Finding and Needs-Assessment Mission to Pakistan: Mission Report. World Meteorological Organization, Islamabad.

Zappa, M., Jaun, S., Germann, U., Walser, A., Fundel, F., 2011. Superposition of three sources of uncertainties in operational flood forecasting chains. Atmos. Res. 100, 246–262.

14

Flood Monitoring System Using Distributed Hydrologic Modeling for Indus River Basin

Sadiq I. Khan*, Zachary Flamig†, Yang Hong‡

*University Corporation for Atmospheric Research (UCAR) at NOAA National Water Center, Tuscaloosa, AL, United States †Center for Data Intensive Science, University of Chicago, Chicago, IL, United States ‡School of Civil Engineering and Environmental Sciences, The University of Oklahoma, Norman, OK, United States

14.1 INTRODUCTION

Floods and droughts are ranked at the top of the list of natural disasters in terms of mortalities and direct and long-term economic damages. On average, floods cause more than 20,000 deaths and adversely affect about 140 million people yearly around the globe (Adhikari et al., 2010). Throughout the world, floods and droughts have been the subject of rigorous research efforts because these extreme events produce the most devastating effects on lives and infrastructure. We understand that hydrologic and land surface processes influence floods, yet we are unable to accurately monitor these natural disasters. To some extent, this is attributable to population increases, shifts into higher risk areas, and an increase in wealth (Changnon et al., 2000; Kunkel et al., 1999). Forecasting these natural hazards is a challenging task especially in regions where in situ observations of hydrometeorological variables are sparse or nonexistent. Regardless, it is clear that more needs to be done to assist communities in responding to these water extremes.

In the Indus River Basin (IRB) floods represent one of the most recurrent hydrological disasters with devastating impacts on human life, food security, the economy, and livelihoods. This region has the highest annual average number of people physically exposed to floods and landslides globally, which mainly occur seasonally from storm systems during the monsoon from July to September. Over the Indus Basin the monsoon system occurs mainly due to land-sea temperature differences inciting a low-pressure circulation promoting moisture-laden fluxes from the Bay of Bengal and the Arabian Sea and resulting in convection. The moist air moves northward, passing through the Lower Indus Plain, and encounters the relief of the Hindu Kush—Himalayas resulting in enhanced orographic effects and high rainfall rates. Every summer these heavy downpours produce severe floods and landslides that have devastating impacts on lives, livelihoods, and infrastructure. The summer monsoon comprises 50%–75% of Pakistan's annual rainfall with a long, typical track stretching from the ocean to the Himalayan foothills. The water extremes are also prominent as approximately 70% of the river flows in the Upper Indus Basin occur in just the summer monsoon months of the year (July–September) (Khan, 1993; Wang et al., 2011a; Khan et al., 2014).

The 2010 extreme monsoon over the IRB resulted in the worst recorded flooding in the area, with destruction throughout the low-lying regions of the Indus River Valley that stretches from the Himalayas to the Arabian Sea. Anomalously heavy precipitation fell during late July and early August causing one of the worst hydrometeorological disasters in history and resulting in human casualties and heavy economic destruction. By the end of August, 20% of Pakistan was submerged under the flood waters, and there were close to 2000 fatalities with more than $40 billion worth of damage (WMO, 2011). This was not an isolated incident; Pakistan is one of the top five countries prone to disaster with nearly 80% of the total population exposed to recurring flood risks every year. Therefore an operational flood monitoring component is indispensable in the overall disaster early warning system for the Indus River Basin.

Precipitation is the key variable that links the atmospheric and land surface processes of Earth's hydrologic cycle. Accurate estimates of precipitation variability both in space and time are critical for better understanding weather and water extremes such as floods and droughts, and can have various scientific and societal applications. Significant progress has been made in understanding the process and contribution of precipitation in the water budget over land surfaces; however, accurate measurement of precipitation at high spatial and temporal scales is still a challenge. Although the systematic observation of water budget components started a century ago, the information sensing related to the water cycle and understating of the hydrologic processes are still limited. Conventional, dense ground-based instruments such as rain gauges provide the best available local measurements of precipitation; however they suffer from insufficient spatial coverage as well as limited ability to capture precipitation variability.

In this chapter we discuss and demonstrate implications of geospatial data from spaceborne sensors and geospatial modeling for hydrologic applications in data-scarce environments. Specifically, the focus is to look into some of the critical questions, such as the following: What are the major uncertainties in big geospatial data for hydrologic assessments? Given that uncertainties are inherent in the hydrometeorological data, is it possible to quantify these errors and reduce them? Can improvements be made on geophysical models, for example, assimilating multisource data through supercomputing and numerical model optimization techniques? How do we establish linkages between distributed hydrologic modeling and multi-sensor remote sensing data collection to develop operational decision support systems? This study touched on some of these research questions and focused on an applied research project on flood monitoring and forecasting for the Indus River Basin.

This chapter is divided into the four sections, with Section 14.2 providing an overview on flood modeling research developed for operational flood monitoring and prediction.

Section 14.3 details data sets and methodology used for a hydrologic modeling architecture. Here we look at the advance flood modeling paradigm that assimilates the hydrometeorological forcing data sets, such as quantitative precipitation estimation into physically based numerical modeling for flood hazard assessment. Section 14.4 focuses on a demonstration project for a next-generation flood hazard assessment system for the Indus River Basin. At the end of the chapter we present the lessons learned over the course of this demonstration study as well as future plans for improvements in flood forecasting system over this hydrologically complex region.

14.2 ADVANCES IN FLOOD FORECASTING SYSTEMS

Recent advances in flood forecasting include the availability of observations of areal precipitation using a dense network of rain gauge measurements, ground- and space-based radar technologies, and storm-scale Numerical Weather Prediction (NWP) systems. For flood forecasting, these high spatio-temporal resolution estimates have been tested for hydrologic modeling approaches specifically for flood modeling applications (Vieux and Bedient, 2004). In addition to the use of these observation systems, flood modeling has seen advances in two-dimensional, distributed hydrologic modeling approaches (Hunter et al., 2008) and representation of sub-grid effects in the hydraulic models (Soares-Frazão et al., 2008). The combination of recent advances has established a paradigm for integrating robust data acquisition and hydrologic modeling in flood forecasting. However, key advances are still needed, including greater spatial coverage of accurate precipitation data, more detailed terrain databases to parameterize friction effects at the sub-grid scale, and better understanding of cause–effect relationships among weather variables and flood forecasts. Moreover, other aspects include advancing flood model execution speed coupled with incorporation of uncertainty analysis, and developing broader understanding of the effectiveness of flood controls in terms of resiliency and sustainability.

(Adams and Pagano, 2016; Sene, 2012) provide a comprehensive review of existing developments in flood forecasting and flood risk management using experiences written by experts from around the world. Here we briefly discuss some of the ongoing efforts on an operational hydrological model implemented at the regional scale in the European Flood Awareness System (EFAS; Ramos et al., 2007; Thielen et al., 2009; Bartholmes et al., 2009). This medium-range streamflow flood forecasting system was developed by the European Commission (EC) Joint Research Centre (JRC) in collaboration with other institutions and became operational in 2012. EFAS was developed as a complementary system for existing national and regional flood forecasting. The EFAS is a geospatial modeling framework that employs a distributed hydrological model that includes a one-dimensional (1-D) channel routing model. Detailed descriptions of the system are listed in (Van Der Knijff et al., 2010; Bartholmes et al., 2009; Pappenberger et al., 2008). The operational hydrological model used within the EFAS is called LISFLOOD and was developed and implemented by (Pappenberger et al., 2008; Thielen et al., 2009; Van Der Knijff et al., 2010).

Currently, another such effort on a continental scale in terms of supporting operational flood monitoring that utilizes high-resolution geospatial data such as digital terrain models, radar rainfall, and other hydrometeorological estimates is the United States-based National Oceanic and Atmospheric Administration (NOAA)-developed Flooded Locations and Simulated Hydrographs Project (FLASH) system. The FLASH project (Gourley et al., 2017) was implemented by the NOAA National Severe Storms Laboratory (NSSL) in collaboration with the University of Oklahoma and consists of the Ensemble Framework For Flash Flood Forecasting (EF5; Clark et al., 2017; Flamig et al., 2015), which contains the Coupled Routing and Excess Storage (CREST), a distributed hydrologic model (Wang et al., 2011b). EF5's modular architecture also enables frequent updates to the software and allows the modeling framework to incorporate new scientific advances as necessary (Fig. 14.1). The model is forced with precipitation estimates from NSSL's Multi-Radar, Multi-Sensor (MRMS) system, running at 10 min and 1 km resolution over the conterminous United States (Vergara et al., 2016; Gourley et al., 2014). The hydrologic model structure includes the water budget module of CREST linked with the kinematic wave model for surface water routing. The variable infiltration curve method is used for the water budget calculations (Zhao et al., 1995), which

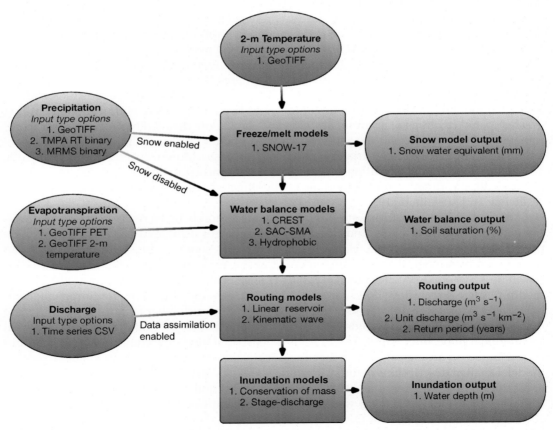

FIG. 14.1 The Ensemble Framework for Flash Flood Forecasting (EF5) with different hydrologic modeling components.

includes the calculation of excess rainfall, which is separated into its surface and subsurface elements through a theoretical method based on hydraulic conductivity (Wang et al., 2011b). Recently, the surface excess rainfall component was routed as overland flow with an implementation of the kinematic wave model for a wide, shallow flow (Vergara et al., 2016).

Another recent flood forecasting system at the conterminous United States (CONUS) scale is the NOAA National Weather Service (NWS)-implemented National Water Model (NWM). The NWM is being developed through close collaboration between NOAA NWS's Office of Water Prediction and National Center for Atmospheric Research (NCAR) and Federal Integrated Water Resources Science and Services partners. This hydrologic forecasting system provides some unique features in comparison to all the other national-level flood forecasting systems. Some of the advancements of this system include simulation of streamflow produced for 2.7 million river reaches and other hydrologic information on 1 km and 250 m grids. In addition to reach-based routing, the model executes short-range forecasts on an hourly basis while medium-range forecasts out to 10 days are produced once per day. A daily ensemble long-range forecast out to 30 days is also produced.

14.3 HYDROLOGICAL MODELING FRAMEWORK

14.3.1 Hydrometeorological and Geospatial Data Sets

Floods are caused by slow-moving or large-areal-extent convective storms producing extreme precipitation over a specific area for a long duration. Therefore improved accuracy such as higher spatial and temporal resolution of quantitative precipitation estimates (QPEs) and quantitative precipitation forecasts (QPFs) can provide the greatest potential benefit to the hydrologic community, since rainfall is the primary factor affecting the timing and stage of river discharge, which is critical for flood prediction. The impacts of improved QPE and QPF can be realized only through close coupling between the atmospheric and land-surface model components, data assimilation, and the development of combined ensemble forecasting systems to provide uncertainty information for early warning and decision making. The near real-time availability of these data over vast regions, many of which are entirely ungauged, has spawned the capability for systematic rainfall monitoring and subsequent flood and drought monitoring.

In this study the key forcing data sets used for EF5/CREST are the satellite precipitation product from the TRMM Multisatellite Precipitation Analysis (TMPA) (Huffman et al., 2007) and the evapotranspiration from the Famine Early Warning Systems Network (FEWS NET) (http://earlywarning.usgs.gov/fews/global/index.php). Digital Elevation processed from the Shuttle Radar Topography Mission (SRTM) (Rabus et al., 2003) is used to generate flow direction, flow accumulation, and contributing basin area. The model is implemented using DEM derived from SRTM data (Smith and Sandwell, 2003). SRTM data have been hydrologically corrected and conditioned as part of the Hydrological Data and Maps Based on Shuttle Elevation Derivatives at Multiple Scales (Hydro-SHEDS) project (Lehner et al., 2008). Using Geographic Information System (GIS) tools, the original Hydro-SHEDS DEMs, available at 30-arc-s (approximately 1 km at the equator) horizontal resolution, are scaled to the desired 0.125° latitude by 0.125° longitude resolution. Then the resulting DEM is processed to remove any pits or sinks following the procedure outlined in Jenson and Domingue (1988). A drainage direction map (DDM) is generated using the eight-direction flow-direction algorithm; this DDM is used to produce a flow accumulation map.

14.3.1.1 Observed Daily Streamflow

The spatial domain of the Indus River Basin (IRB) for this study is between latitudes 22°N and 38°N and longitude 66°E and 83°E (Fig. 14.2). The observed streamflow data was obtained from the Federal Flood Commission (FFC) that archives daily gauged river discharge and water levels in the reservoirs in coordination with the Pakistan Meteorological Department (PMD). The river discharge data is selected based on the length of the record and the completeness and reliability of the data. The available discharge gauges with average streamflow values that are maintained by PMD are shown in Table 14.1 and Fig. 14.2. The time period for this analysis was selected from 2008 to 2011, which includes the devastating 2010 flood.

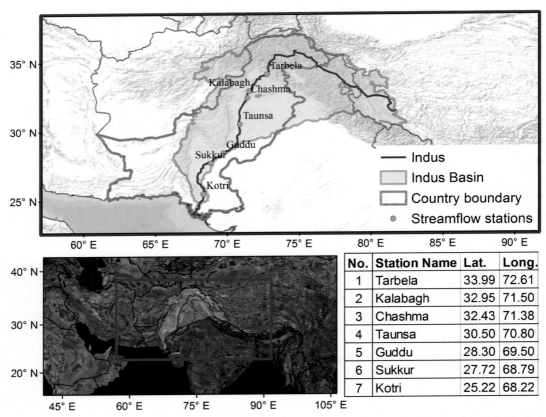

No.	Station Name	Lat.	Long.
1	Tarbela	33.99	72.61
2	Kalabagh	32.95	71.50
3	Chashma	32.43	71.38
4	Taunsa	30.50	70.80
5	Guddu	28.30	69.50
6	Sukkur	27.72	68.79
7	Kotri	25.22	68.22

FIG. 14.2   Map of the Indus River Basin showing river discharge stations used in the distributed hydrologic modeling.

TABLE 14.1 River Discharge From the Selected Gauge Stations Over Indus River Basin for Monsoon and Pre-Monsoon Seasons Based on Daily River Flows

Gauge Station	Lat.	Long.	Elevation (m)	2008–2009 Discharge (m³/s)		2010 Discharge (m³/s)	
				Monsoon	Pre-Monsoon	Monsoon	Pre-Monsoon
Tarbela	34.12	72.75	430	5297	2410	7773	1595
Kalabagh	32.92	71.50	200	5909	3331	8989	2506
Chashma	32.43	71.39	180	6329	3354	10,020	2377
Taunsa	30.50	70.86	120	5664	2861	9677	1995
Guddu	28.41	69.71	75	4838	1987	11,476	1294
Sukkur	27.69	69.71	60	3222	1506	10,488	1017
Kotri	25.47	68.31	22	1644	328	7606	156

In addition to the observed daily streamflow and precipitation data sets, the other hydrometeorological and geophysical data sets are used to set up the hydrological model. The data sets required to set up the model included the digital elevation model, soil types, land use, and evapotranspiration and are all derived from remote sensing data and archival data in public domains. Some of these geospatial data used to implement the hydrological model are listed in Table 14.2 and are briefly discussed below.

14.3.1.2 Quantitative Precipitation Estimates (QPE)

The recent development of the remote sensing technique (e.g., ground- and space-based satellite precipitation estimation) and physically based numerical models demonstrates the potential for flood estimation at the macro scale and enables potential for flood detection system at the global scale (Westerhoff et al., 2013; Wu et al., 2012; Yilmaz et al., 2010; Proud et al., 2011; Brakenridge et al., 2007; Hong et al., 2007). Precipitation estimates at varying spatial and temporal scales are vital for climatic and hydrologic studies. With global coverage, high resolution, and near real-time availability, satellite-based precipitation data are critical for a wide range of sectoral applications studies such as terrestrial hydrology, climate change, agriculture, and natural hazards monitoring.

TABLE 14.2 Summary of the Geospatial Data Products Used to Set Up the Hydrologic Model

Type	Period	Resolution	Description	Source
TRMM	1997–	3 h/0.25°	Gridded rainfall	NASA-GSFC
GPM Day-1	2017–	3 h/0.1°	Global precipitation	NASA-GSFC
MODIS (MOD16)	2000–	1 d/1 km	Global ET	NASA-GSFC
ASTER DEM	2002–	30 m–1 km	Topography/watershed	USGS
MODIS imagery	2000–	1 d/30 m–1 km	Land cover/inundation maps	NASA-GSFC

The underlying principle in precipitation estimation from remote sensing is to observe the backscatter from different hydrometeor types (rain, hail, snow, ice crystals) in the atmosphere. Satellite-based precipitation estimation started with the use of visible (VIS) and infrared (IR) instruments, by looking at the cloud top temperatures (Krajewski and Smith, 2002; Ba and Gruber, 2001; Petty, 1995; Hong et al., 2004). Since the late 1970s infrared satellite remote sensing techniques were used for precipitation estimation (Arkin et al., 1994; Meisner and Arkin, 1987). The majority of algorithms attempt to correlate the surface rain rate with IR cloud-top brightness temperatures (Tb) using the information obtained from IR imagery. The algorithms developed to date may be classified into three groups depending on the level of information extracted from the IR cloud images: cloud-pixel-based, cloud-window-based, and cloud-patch-based (Yang et al., 2011). Several examples of these algorithms may clarify this classification further.

In the past decade, a number of quasi-global-scale estimates were developed, including the National Aeronautics and Space Administration (NASA)'s Tropical Rainfall Measuring Mission TRMM–based Multi-Satellite Precipitation Analysis (TMPA) (Huffman et al., 2007), the Naval Research Laboratory Global Blended-Statistical Precipitation Analysis (NRL-blended) (Turk and Miller, 2005), the Climate Prediction Center morphing algorithm (CMORPH) (Joyce et al., 2004), the Precipitation Estimation from Remotely Sensed Information using Artificial Neural Networks (PERSIANN) (Hsu et al., 1997), and the Global Satellite Mapping of Precipitation (Kubota et al., 2007).

The satellite precipitation products from the TMPA have been widely applied to hydrological predictions at both regional and global scales (Hong et al., 2007; Stisen and Sandholt, 2010; Su et al., 2008; Wu et al., 2012; Yilmaz et al., 2010; Yong et al., 2010; Zhang et al., 2013). These and other quantitative precipitation estimations (QPEs) from Earth-orbiting satellites have shown potential for water resource and hazard monitoring in remote regions and ungauged watersheds. The advancement of satellite sensors has provided better measurement capability to estimate precipitation at a high spatial and temporal resolution at a global scale. Contrary to space-based platforms, estimates from both rain gauge station networks and weather radars have higher accuracy but lack the complete coverage, particularly over watersheds in remote regions and complex terrains. In the past three decades numerous methods have been developed to estimate precipitation through satellite and in situ remote sensing techniques, which include both spaceborne and ground-based radar systems. Despite the great achievement in the TRMM era, TRMM data have some inherent limitations associated with spatio-temporal coverage and uncertainty of solid or light precipitation estimations over higher latitudes and altitudes (Yong et al., 2012; Hou et al., 2014; Huffman et al., 2007; Yong et al., 2014). The TRMM (3B42V7) requires validation before it is used to study the rainfall diurnal variation over the study domain; therefore the spatial error structure of surface precipitation is systematically studied by comparing it with rain gauge observations. The authors studied spatial and temporal characteristics of the Asian monsoon over the IRB using TRMM Version 7 product (3B42V7) (Khan et al., 2014).

The Global Precipitation Measurement (GPM) mission will further stimulate and sustain this demand. Designed as the successor to the Tropical Rainfall Measurement Mission (TRMM), the GPM Core satellite will enable the formation of a constellation, with a variety of precipitation-measuring satellites from NASA and its partners, and with GPM Core as the common calibration standard. The precipitation retrievals from these satellites (Level-2 data) are merged to create user-level products (Level-3), such as NASA's Integrated Multi-satellitE Retrievals for GPM (IMERG) (Huffman et al., 2014) (Fig. 14.3) and NOAA's CPC Morphing technique (CMORPH).

14.3.1.3 Quantitative Precipitation Forecast (QPF)

The QPF product used in this study is the Global Forecasting System (GFS) produced by the National Centers for Environmental Prediction (NCEP) and the National Oceanic and Atmospheric Administration (NOAA). The GFS precipitation is available from a new model run every 6 h starting about 3 h after the analysis time of each run; the model runs with a 1-h time step. GFS precipitation rate data are available at a horizontal resolution of 0.25° latitude by 0.25° longitude. EF5 uses the first 5 days of forecast precipitation from each model run. The precipitation rate data are a sum of the convective and stratiform rain components in the GFS model. The GFS GRIdded Binary version 2 (GRIB2) files are transformed into GeoTIFF files for use in the hydrologic model by helper scripts which download the new data when it is available.

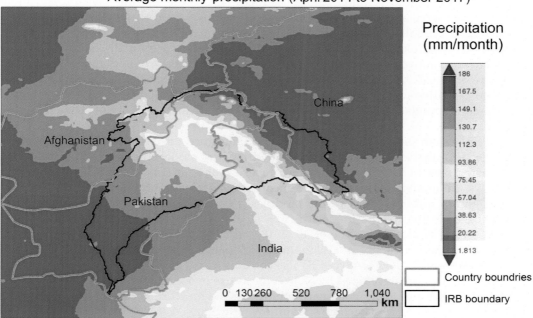

FIG. 14.3 Merged satellite-gauge precipitation estimates (GPM 3IMERGM v05) showing average monthly precipitation from April 2014 to November 2017.

14.3.2 Distributed Hydrological Modeling

Numerical models have been used for operational flood forecasting since the development of the first watershed hydrological model, the Stanford Watershed Model in 1966 (Crawford and Linsley, 1966). Hydrological models have evolved from lumped-process models (Williams and Hann, 1978) to semi-distributed models (Beven and Kirkby, 1979; Zhao et al., 1980) and to fully distributed models (Abbott et al., 1986a,b; Wang et al., 2011a; Wigmosta et al., 1994). However, studies on dynamically coupling hydrological processes predicted by distributed hydrological models with soil physics and mechanics determining slope stability are still in a very early stage (Bogaard and Greco, 2014) due to lack of knowledge on interactions between these processes and differences in the spatio-temporal scales of the flood events.

Geospatial data assimilation within a hydrologic modeling has provided new insights into hydrologic processes such as precipitation and evapotranspiration patterns and the impact on runoff response. The near real-time availability of in situ and remote sensing data over vast regions, many of which are ungauged, has spawned the capability for systematic rainfall monitoring and subsequent flood and drought monitoring. However, assimilation of remote sensing data into distributed hydrologic models is marred with uncertainties related to data estimates and numerical models. In this regard two recent studies on flood monitoring have argued about the issue of hydrologic model recalibration by using satellite rainfall data (Vergara et al., 2013, Nikolopoulos et al., 2013). These studies showed that the coarse rainfall resolution associated with some of the main satellite products (e.g., 3-hourly and 25 km) reduces the skill of the hydrologic model particularly at decreasing drainage areas and high values of streamflow, calibrating model parameters at rainfall resolutions representative to satellite products could significantly improve the hydrologic prediction skill of satellite-based rainfall estimates. Others examined the impact of satellite pixel resolution on hydrologic model calibration (Maggioni et al., 2013, Gourley et al., 2011). They also found that hydrologic model parameters were sensitive to pixel resolution, and it was recommended to use a geospatial data set that corresponds to the resolution of the numerical model to be used for flood forecasting.

The EF5 system is a numerical modeling architecture developed at the University of Oklahoma (OU) that encompasses multiple hydrological model cores and additional related software modules. EF5's modular architecture also enables frequent updates to the software and allows the modeling framework to incorporate new scientific advances as necessary. Fig. 14.1 shows the flow chart of the modules' input and output data contained in EF5. One of the main hydrologic models in EF5 is the Coupled Routing and Excess Storage (CREST) (Wang et al., 2011a; Xue et al., 2013), which is a distributed hydrological model developed by the University of Oklahoma in cooperation with the National Aeronautics and Space Administration (NASA). EF5 also contains the Sacramento Soil Moisture Accounting (SAC-SMA) model developed by NOAA for its use in operational forecasting. SAC-SMA contains an upper and lower zone partitioning scheme for storing water in the ground and subsequently releasing it to become streamflow. The Snow Water Equivalent is modeled in EF5 using the coupled Snow-17 model (Anderson, 1976), which accounts for freezing and melting using the air temperature as an additional data input into the modeling system.

Snow-17 models the entire snow accumulation and melting process including accounting for factors such as wind and rain on snow through parameterization. EF5 couples the water balance processes to two different routing schemes, either linear reservoir routing or kinematic wave routing. These routing schemes are approximations of the full dynamic wave equations suitable when the hillslope is large.

We discuss the quality of the results from EF5/CREST in the Indus Basin in Section 14.4.2. The model simulates runoff generation, evapotranspiration, infiltration, and surface and subsurface routing at each grid cell in the model domain. In EF5/CREST the infiltration and runoff are partitioned via the variable infiltration curve concept (Liang et al., 1994; Zhao et al. 1980), while surface and subsurface water are routed using the kinematic routing approximation. The EF5/CREST model has been used to monitor streamflow, soil moisture, Snow Water Equivalent, and evapotranspiration using input from gridded meteorological forcing fields. A detailed description of the EF5/CREST model structure and training material can be found at http://hydro.ou.edu/research/crest. This model has been used to predict streamflow, soil moisture, Snow Water Equivalent, and evapotranspiration using inputs from gridded meteorological forcing fields.

14.4 INDUS RIVER BASIN FLOOD MONITORING SYSTEM

This chapter covers the history of the joint University of Oklahoma (OU) and National Aeronautics and Space Administration (NASA) capacity-building efforts in the Indus Basin region. Then we describe the data sets needed to fulfill the modeling goals and the various methods used to produce output for the Indus Basin Flood Dashboard (http://flash.ou.edu/pakistan). The dedicated online web page shows the flood forecasting system, which is how the results of the project reach a broad audience.

The Indus River is one the largest rivers in the world with a drainage area of about 1 million km2 and a mean annual discharge of ($7900\,\mathrm{m^3/s}$). It is shared by Afghanistan, China, India, and Pakistan; however, the largest portion of the Indus River watershed is in Pakistan (52%) and India (33%). The Upper Indus River rumbles through the mountainous terrain of the Hindu Kush, Karakorum, and Himalayan mountain ranges; while the Lower Indus River meanders through the southern plains. The Indus Basin receives much of its precipitation during two seasons: (1) heavy summer rains in the northeastern part from monsoon currents and (2) late winter and early spring rains due to disturbances in the mid-latitude westerlies (Khan, 1993;Wang et al., 2011c).

14.4.1 Hydrologic Model Evaluation Metrics

To quantify the performance of the satellite-based QPE for streamflow simulations, the widely used statistical metrics were used and are explained next.

The Pearson correlation coefficient (CC) shown in Eq. (14.2) is used to assess the agreement between the simulated and observed data value of simulated data (streamflow as "simulated data"; "SIM" is used in the formula) and the observed data (such as observed streamflow in this study; "OBS" is used in the formula).

$$CC = \frac{\sum\limits_{i=1}^{n}(OBS_i - \overline{OBS})(SIM_i - \overline{SIM})}{\sqrt{\sum\limits_{i=1}^{n}(OBS_i - \overline{OBS})^2 \sum\limits_{i=1}^{n}(SIM_i - \overline{SIM})^2}} \qquad (14.1)$$

Here i is the ith time step observed $\overline{OBS}$ and simulated $\overline{SIM}$ data, respectively.

The CC is used to assess the degree of linear association between simulated and observed streamflow values.

The relative Bias (%) defined in Eq. (14.2) is used to measure the agreement between the averaged value of simulated data (streamflow as "simulated data"; "SIM" is used in the formula) and observed data (such as observed streamflow in this study; "OBS" is used in the formula).

$$Bias = \left[\frac{\sum\limits_{i=1}^{n}SIM_i - \sum\limits_{i=1}^{n}OBS_i}{\sum\limits_{i=1}^{n}OBS_i}\right] \times 100 \qquad (14.2)$$

The root mean square error (RMSE) shown in Eq. (14.3) was selected to evaluate the average error magnitude between simulated and observed data.

$$RMSE = \sqrt{\frac{1}{n}\sum\limits_{i=1}^{n}(OBS_i - SIM_i)^2} \qquad (14.3)$$

Here n is the total number of pairs of simulated and observed data; i is the ith values of the simulated and observed data; $\overline{SIM}$ and $\overline{OBS}$ are the mean values of simulated and observed data, respectively. The mean absolute error (MAE) measures the average magnitude of the error as shown in Eq. (14.4).

$$MAE = \frac{1}{n}\sum\limits_{i=1}^{n}|OBS_i - SIM_i| \qquad (14.4)$$

The Nash-Sutcliffe Coefficient of Efficiency (NSCE) as defined by Nash and Sutcliffe in 1970 is used to assess the performance of model simulation and observation, as shown in Eq. (14.5). The NSCE determines the relative magnitude of residual variance in the simulations compared with observed variance, and to assess the predictive power of a hydrological model.

$$NSCE = 1 - \frac{\sum\limits_{i=1}^{n}(OBS_i - SIM_i)^2}{\sum\limits_{i=1}^{n}(OBS_i - \overline{OBS})^2} \qquad (14.5)$$

The NSCE can range from $-\infty$ to 1. A value of 1 indicates perfect agreement between simulated and observed streamflow. Positive NSCE values are generally viewed as acceptable levels of performance, whereas negative values indicate that the model is a worse predictor than the observed mean, which is often deemed unacceptable.

14.4.2 Distributed Model Simulations

In this section we introduce some of the efforts to achieve this objective by implementing a distributed hydrologic modeling framework discussed in Section 14.3. To generate streamflow simulation the parameters of the EF5/CREST model need to be calibrated. In this study the calibration procedure for EF5/CREST calibration was performed using an auto-calibration method based on the Differential Evolution Adaptive Metropolis (DREAM) method (Pronzato et al., 1984; Brooks, 1958), by maximizing the NSCE value between the simulated and observed daily streamflow. The DREAM and other related shuffled complex evolution (SCE) (Duan et al., 1992) techniques have been applied by many in hydrologic model calibrations (Ma et al., 2006; Bekele and Nicklow, 2007; Lin et al., 2006) due to their effectiveness and robustness.

Here we discuss the assessment of the quality of hydrological model simulation skill in the IRB employing precipitation products, namely the TRMM 3B42v7. The EF5/CREST is calibrated using daily observed discharge data for the period between 2005 and 2012. A one-year period (2004) is used for warming up the model states. To quantify the accuracy of the hydrologic model, we used several statistical indexes, including CC, Bias, RMSE, and MAE. In addition NSCE was used to assess the hydrologic model fit between simulated and observed streamflow. Detailed descriptions on these statistical indexes are shown in Eqs. (15.1)–(15.5) and are explained in Section 14.4.1.

The hydrologic model was tested for seven Indus River Basin stations, namely Tarbela, Kalabagh, Chashma, Taunsa, Guddu, Sukkur, and Kotri, as shown in Fig. 14.4A–G, respectively. The model evaluation statistics for all the streamflow gauge stations mentioned are listed in Table 14.3. Overall the hydrologic model calibration result showed good performance for the station at Tarbela with CC 0.87, 4.4% bias, and NSCE of 0.74 (Fig. 14.4A). The second-best calibration performance was at the Chashma station with CC 0.84, −7% bias, and NSCE of 0.70 (Fig. 14.4C). Similarly, the model performance for the Kalabagh and Taunsa stations were acceptable with CC 0.84, and NSCE of 0.67 (Fig. 14.4C). The optimized parameter combination resulted in CC of 0.85 with 15% bias and NSCE of 0.67 for the station at Taunsa (Fig. 14.4D). The EF5/CREST model performance for the other stations (Guddu, Sukkur, and Kotri) was relatively low, as shown in Table 14.3 and in Fig. 14.4E, Fig. 14.4F, and Fig. 14.4G, respectively. The model results clearly show an overestimation of simulated discharge at these streamflow gauge locations. This can be attributed to the limitation of the hydrologic model to account for water abstraction upstream of the watershed above these locations. The hydrologic model-based simulation showed overestimation during the summer months, particularly at the Lower Indus Basin. One of the main reasons for the noise in model simulations can be linked to the increase in human activities in the catchment area during recent years.

Representation of the dynamic operations of hydraulic conditions of the water channels, reservoirs, canals, and other means of water abstraction for agriculture and irrigation is crucial for flood forecasting systems. In this study these processes are not fully represented due

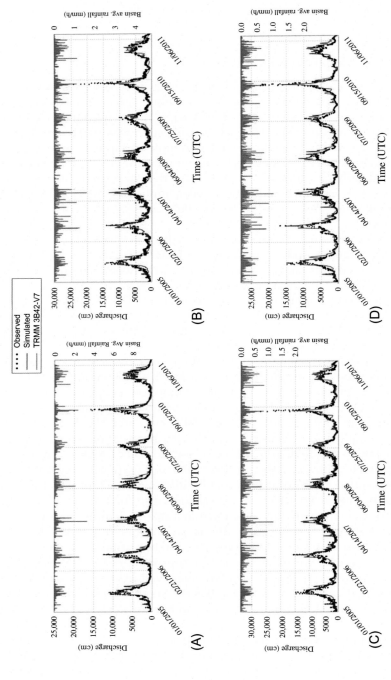

FIG. 14.4 Hydrologic model EF5/CREST comparisons with gauge-observed discharge data at the following gauge stations: (A) Tarbela, (B) Kalabagh, (C) Chashma, (D) Taunsa,

Continued

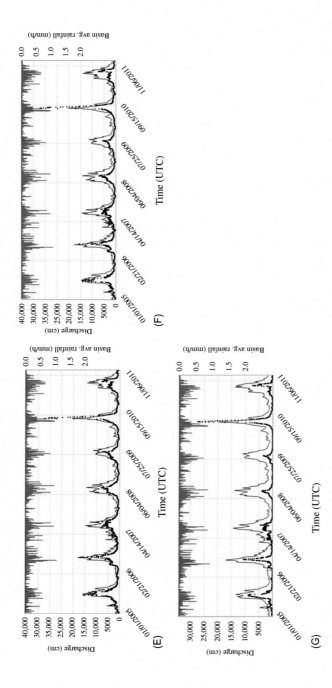

FIG. 14.4, Cont'd (E) Guddu, (F) Sukkur, and (G) Kotri. Model evaluation statistics for all these sub-basins are listed in Table 14.2. Model calibration was done for the period from January 1, 2005 to January 1, 2012.

TABLE 14.3   Model Evaluation Statistics for Monthly Discharge for Indus River Sub-Basins

Figure Number	Station Name	CC	Bias	NSCE	MAE	RMSE	Peak ERR	Peak Time ERR
Fig. 14.4A	Tarbela	0.87	−4.4	0.74	791	1267	−6615	−8.0
Fig. 14.4B	Kalabagh	0.84	−7.0	0.6.7	1013	1504	−9150	−7.0
Fig. 14.4C	Chashma	0.84	−1.1	0.70	1039	1604	−11,106	−7.0
Fig. 14.4D	Taunsa	0.85	15	0.67	1078	1629	−6067	−9.0
Fig. 14.4E	Guddu	0.78	61	0.25	1998	2862	11,904	−2.0
Fig. 14.4F	Sukkur	0.67	95	0.13	2413	3359	−11,682	−2.0
Fig. 14.4G	Kotri	0.67	286	−1.83	3451	4314	−7607	3.0

to limited availability of data and limited understanding of the IRB's complex water management system. The other hydrologic processes that need an in-depth understanding and physical representation in the hydrologic model are glacier melt and snowmelt. The main source of discharge in the Upper Indus Basin (UIB) is winter precipitation from snow that melts the following summer. The other source of runoff volume is glacier melt, and the main driver is the summer temperature. Some of the aspect of representing these process in numerical terms are discussed by Daniele Bocchiola and Andrea Soncini in Part I and Chapter 1 of this book.

14.5 DISCUSSION AND CONCLUSION

Hydrometeorological disasters occur due a combination of multidimensional processes, which are not only atmospheric phenomena but are also influenced by static and dynamic hydrologic conditions (e.g., soil types and their volumetric water contents), the intersection of natural hazards with infrastructure, modifications by city planning, and societal values and behaviors. Over the past two decades powerful computing resources, distributed computing such as cloud infrastructure, and an increase in spaceborne instruments that provide Earth-observation data have benefitted the hydrologic research community. It is now feasible, even routine, to access estimates of precipitation, potential evapotranspiration, land use and land cover, topography, and other variables collected from space in a timely fashion. Moreover, these data sets are often available to the public regardless of geopolitical boundaries. The high-resolution precipitation estimates from spaceborne sensors have provided understanding of rainfall patterns and runoff response over vast regions such as the IRB (Fig. 14.5). This unprecedented near real-time data availability for ungauged regions has spawned systematic hydrometeorologic monitoring and subsequent flood modeling capabilities. These advances are critical for issuing early warnings for flooding events, especially in undeveloped regions with insufficient hydrometeorological data.

To achieve reliable hydrologic forecast, instantaneous and seamless high-spatio-temporal data and models are employed to estimate precipitation, soil moisture, evapotranspiration,

FIG. 14.5 Hydrologic model EF5/ CREST simulation of distributed streamflow over the Indus River Basin.

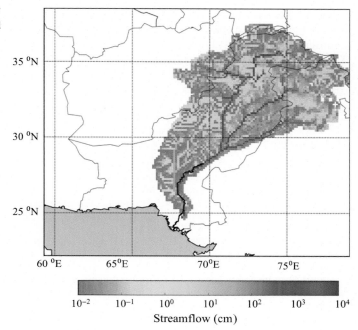

groundwater, snowmelt, and runoff. Advances in sensor technology for data capture, high-performance computing, and distributed hydrologic modeling frameworks are critical for coupling of dynamic Earth system processes, visualization, and atmospheric and hydrologic models where the terrestrial movement of water impacts the heat fluxes and mass balances. In this context, applied research activities were conducted to couple a global forecast system model with a distributed hydrologic model for high-impact hazard prediction, such as large-scale floods over the Indus River Basin. However, challenges still exist in large river-basin-scale hydrological models in terms of model parameterization, the accuracy and resolution of hydrometeorological data estimates, quantification and representation of water management processes, and the computational efficiency of high-resolution hydrological modeling. In addition, integration of the Hydrologic Ensemble Forecasting System (HEFS) to minimize uncertainties in the hydrologic and atmospheric models can play a significant role in hydrologic forecasting and water management.

Improved hydrologic modeling, real-time data access, and engagement of key stakeholders are the common denominators for flood forecasting and water resources management. A multidisciplinary approach based on physical science and engineering as well as social intelligence solutions can inform us about the geophysical processes and help provide better security to people who live in areas prone to flooding. Nonstructural measures such as vulnerability and risk assessments, floodplain zoning, and land-use planning and enforcement generally are not included in management practices in the Indus Basin. Therefore, in addition to providing flood monitoring and forecasting services to communities at risk, and having greater flood preparedness on the part of the communities, there is a need for integrated flood management planning. Collaboration between flood management agencies, academia, and civil society is essential for developing coordinated and integrated flood management programs and activities. Moreover, existing systems can be improved considerably. Together these measures will help water and flood management authorities by improving and supporting decision-making processes required for the management of floods and droughts in the Indus Basin.

References

Abbott, M.B., Bathurst, J.C., Cunge, J.A., O'Connell, P.E., Rasmussen, J., 1986a. An introduction to the European hydrological system—Systeme Hydrologique Europeen, "SHE", 1: history and philosophy of a physically-based, distributed modelling system. J. Hydrol. 87, 45–59.

Abbott, M.B., Bathurst, J.C., Cunge, J.A., O'Connell, P.E., Rasmussen, J., 1986b. An introduction to the European hydrological system—Systeme Hydrologique Europeen, "SHE", 2: structure of a physically-based, distributed modelling system. J. Hydrol. 87, 61–77.

Adams III, T.E., Pagano, T.C., 2016. Flood forecasting: a global perspective. In: Flood Forecasting. Academic Press, Boston.

Adhikari, P., Hong, Y., Douglas, K., Kirschbaum, D., Gourley, J., Adler, R., Robert Brakenridge, G., 2010. A digitized global flood inventory (1998–2008): compilation and preliminary results. Nat. Hazards 55, 405–422.

Anderson, E.A., 1976. A point energy and mass balance model of a snow cover. NOAA Tech. Rep. NWS 19. Natl. Oceanic and Atmos. Admin., Silver Spring, MD, 150 pp.

Arkin, P.A., Joyce, R., Janowiak, J.E., 1994. The estimation of global monthly mean rainfall using infrared satellite data: the GOES precipitation index (GPI). Remote Sens. Rev. 11, 107–124.

Ba, M.B., Gruber, A., 2001. GOES multispectral rainfall algorithm (GMSRA). J. Appl. Meteorol. 40, 1500–1514.

Bartholmes, J.C., Thielen, J., Ramos, M.H., Gentilini, S., 2009. The European flood alert system EFAS—Part 2: statistical skill assessment of probabilistic and deterministic operational forecasts. Hydrol. Earth Syst. Sci. 13, 141–153.

Bekele, E.G., Nicklow, J.W., 2007. Multi-objective automatic calibration of SWAT using NSGA-II. J. Hydrol. 341, 165–176.

Beven, K.J., Kirkby, M.J., 1979. A physically based, variable contributing area model of basin hydrology/Un modèle à base physique de zone d'appel variable de l'hydrologie du bassin versant. Hydrol. Sci. Bull. 24, 43–69.

Bogaard, T., Greco, R., 2014. Preface "Hillslope hydrological modelling for landslides prediction". Hydrol. Earth Syst. Sci. 18, 4185–4188.

Brakenridge, G.R., Nghiem, S.V., Anderson, E., Mic, R., 2007. Orbital microwave measurement of river discharge and ice status. Water Resour. Res. 43, W04405.

Brooks, S.H., 1958. A discussion of random methods for seeking maxima. Oper. Res. 6, 244–251.

Changnon, S.A., Pielke Jr., R.A., Changnon, D., Sylves, R.T., Pulwarty, R., 2000. Human factors explain the increased losses from weather and climate extremes*. Bull. Am. Meteorol. Soc. 81, 437–442.

Clark III, R.A., Flamig, Z.L., Vergara, H., Hong, Y., Gourley, J.J., Mandl, D.J., Frye, S., Handy, M., Patterson, M., 2017. Hydrological modeling and capacity building in the Republic of Namibia. Bull. Am. Meteorol. Soc. 98, 1697–1715.

Crawford, N.H., Linsley, R.S., 1966. Digital simulation in hydrology: The Stanford Watershed Model IV. Technical Report No. 39, Department of Civil Engineering, Stanford University, Palo Alto, CA.

Duan, Q., Sorooshian, S., Gupta, V., 1992. Effective and efficient global optimization for conceptual rainfall-runoff models. Water Resour. Res. 28, 1015–1031.

Flamig, Z., Vergara, H., Clark, R., Gourley, J., Kirstetter, P., Hong, Y., 2015. The ensemble framework for flash flood forecasting: global and CONUS applications. In: AGU Fall Meeting Abstracts.

Gourley, J.J., Hong, Y., Flamig, Z.L., Wang, J., Vergara, H., Anagnostou, E.N., 2011. Hydrologic evaluation of rainfall estimates from radar, satellite, gauge, and combinations on Ft. Cobb Basin, Oklahoma. J. Hydrometeorol. 12, 973–988.

Gourley, J.J., Flamig, Z.L., Hong, Y., Howard, K.W., 2014. Evaluation of past, present and future tools for radar-based flash-flood prediction in the USA. Hydrol. Sci. J. 59, 1377–1389.

Gourley, J.J., Flamig, Z.L., Vergara, H., Kirstetter, P.-E., Clark III, R.A., Argyle, E., Arthur, A., Martinaitis, S., Terti, G., Erlingis, J.M., Hong, Y., Howard, K.W., 2017. The FLASH project: improving the tools for flash flood monitoring and prediction across the United States. Bull. Am. Meteorol. Soc. 98, 361–372.

Hong, Y., Hsu, K.-L., Sorooshian, S., Gao, X., 2004. Precipitation estimation from remotely sensed imagery using an artificial neural network cloud classification system. J. Appl. Meteorol. 43, 1834–1853.

Hong, Y., Adler, R.F., Hossain, F., Curtis, S., Huffman, G.J., 2007. A first approach to global runoff simulation using satellite rainfall estimation. Water Resour. Res. 43, W08502.

Hou, A.Y., Kakar, R.K., Neeck, S., Azarbarzin, A.A., Kummerow, C.D., Kojima, M., Oki, R., Nakamura, K., Iguchi, T., 2014. The global precipitation measurement mission. Bull. Am. Meteorol. Soc. 95 (5), 701–722.

Hsu, K.-l., Gao, X., Sorooshian, S., Gupta, H.V., 1997. Precipitation estimation from remotely sensed information using artificial neural networks. J. Appl. Meteorol. 36, 1176–1190.

Huffman, G., Bolvin, D., Nelkin, E., Wolff, D., Adler, R., Gu, G., Hong, Y., Bowman, K., Stocker, E., 2007. The TRMM multisatellite precipitation analysis (TMPA): quasi-global, multiyear, combined-sensor precipitation estimates at fine scales. J. Hydrometeorol. 8 (Online).

Huffman, G., Bolvin, D., Braithwaite, D., Hsu, K., Joyce, R., Kidd, C., Sorooshian, S., Xie, P., 2014. Early examples from the integrated multi-satellite retrievals for GPM (IMERG). In: EGU General Assembly Conference Abstracts, p. 11232.

Hunter, N.M., Bates, P.D., Neelz, S., Pender, G., Villanueva, I., Wright, N.G., Liang, D., Falconer, R.A., Lin, B., Waller, S., Crossley, A.J., Mason, D.C., 2008. Benchmarking 2D hydraulic models for urban flooding. Proc. ICE: Water Manag. 161, 13.

Jenson, S., Domingue, J., 1988. Extracting topographic structure from digital elevation data for geographic information system analysis. Photogramm. Eng. Remote. Sens. 54, 1593–1600.

Joyce, R.J., Janowiak, J.E., Arkin, P.A., Xie, P., 2004. CMORPH: a method that produces global precipitation estimates from passive microwave and infrared data at high spatial and temporal resolution. J. Hydrometeorol. 5, 487–503.

Khan, J.A., 1993. The Climate of Pakistan. Rehbar Publishers.

Khan, S.I., Hong, Y., Gourley, J.J., Khattak, M.U.K., Yong, B., Vergara, H.J., 2014. Evaluation of three high-resolution satellite precipitation estimates: potential for monsoon monitoring over Pakistan. Adv. Space Res. 54, 670–684.

Krajewski, W.F., Smith, J.A., 2002. Radar hydrology: rainfall estimation. Adv. Water Resour. 25, 1387–1394.

Kubota, T., Shige, S., Hashizume, H., et al., 2007. Global precipitation map using satellite-borne microwave radiometers by the GSMaP project: production and validation. IEEE Trans. Geosci. Remote Sens. 45, 2259–2275.

Kunkel, K.E., Pielke Jr., R.A., Changnon, S.A., 1999. Temporal fluctuations in weather and climate extremes that cause economic and human health impacts: a review. Bull. Am. Meteorol. Soc. 80, 1077–1098.

Lehner, B., Verdin, K., Jarvis, A., 2008. New global hydrography derived from spaceborne elevation data. Eos Trans. AGU 89 (10), 93–94.

Liang, X., et al., 1994. A simple hydrologically based model of land surface water and energy fluxes for general circulation models. J. Geophys. Res. Atmos. 99 (D7), 14415–14428.

Lin, J.-Y., Cheng, C.-T., Chau, K.-W., 2006. Using support vector machines for long-term discharge prediction. Hydrol. Sci. J. 51, 599–612.

Ma, H.-b., Dong, Z.-c., Zhang, W.-m., Liang, Z.-m., 2006. Application of SCE-UA algorithm to optimization of TOPMODEL parameters [J]. J. Hohai Univ. (Nat. Sci.) 4, 001.

Maggioni, V., Vergara, H.J., Anagnostou, E.N., Gourley, J.J., Hong, Y., Stampoulis, D., 2013. Investigating the applicability of error correction ensembles of satellite rainfall products in river flow simulations. J. Hydrometeorol. 14, 1194–1211.

Meisner, B.N., Arkin, P.A., 1987. Spatial and annual variations in the diurnal cycle of large-scale tropical convective cloudiness and precipitation. Mon. Weather Rev. 115, 2009–2032.

Nikolopoulos, E.I., Anagnostou, E.N., Borga, M., 2013. Using high-resolution satellite rainfall products to simulate a major flash flood event in northern Italy. J. Hydrometeorol. 14, 171–185.

Pappenberger, F., Bartholmes, J., Thielen, J., Cloke, H.L., Buizza, R., de Roo, A., 2008. New dimensions in early flood warning across the globe using grand-ensemble weather predictions. Geophys. Res. Lett. 35.

Petty, G.W., 1995. The status of satellite-based rainfall estimation over land. Remote Sens. Environ. 51, 125–137.

Pronzato, L., Walter, E., Venot, A., Lebruchec, J., 1984. A general-purpose global optimizer: implementation and applications. Math. Comput. Simul. 26, 412–422.

Proud, S.R., Fensholt, R., Rasmussen, L.V., Sandholt, I., 2011. Rapid response flood detection using the MSG geostationary satellite. Int. J. Appl. Earth Obs. Geoinf. 13, 536–544.

Rabus, B., Eineder, M., Roth, A., Bamler, R., 2003. The shuttle radar topography mission—a new class of digital elevation models acquired by spaceborne radar. ISPRS J. Photogramm. Remote Sens. 57 (4), 241–262.

Ramos, M.-H., Bartholmes, J., Thielen-del Pozo, J., 2007. Development of decision support products based on ensemble forecasts in the European flood alert system. Atmos. Sci. Lett. 8, 113–119.

Sene, K., 2012. Flash Floods: Forecasting and Warning. Springer Science & Business Media.

Smith, B., Sandwell, D., 2003. Accuracy and resolution of shuttle radar topography mission data. Geophys. Res. Lett. 30, 1467. https://dx.doi.org/10.1029/2002GL016643.

Soares-Frazão, S., Lhomme, J., Guinot, V., Zech, Y., 2008. Two-dimensional shallow-water model with porosity for urban flood modelling. J. Hydraul. Res. 46, 45–64.

Stisen, S., Sandholt, I., 2010. Evaluation of remote sensing based rainfall products through predictive capability in hydrological runoff modelling. Hydrol. Process. 24, 879–891.

Su, F., Hong, Y., Lettenmaier, D.P., 2008. Evaluation of TRMM multisatellite precipitation analysis (TMPA) and its utility in hydrologic prediction in the La Plata Basin. J. Hydrometeorol. 9, 622–640.

Thielen, J., Bartholmes, J., Ramos, M.H., de Roo, A., 2009. The European flood alert system—Part 1: concept and development. Hydrol. Earth Syst. Sci. 13, 125–140.

Turk, F.J., Miller, S.D., 2005. Toward improved characterization of remotely sensed precipitation regimes with MODIS/AMSR-E blended data techniques. IEEE Trans. Geosci. Remote Sens. 43, 1059–1069.

Van Der Knijff, J.M., Younis, J., De Roo, A.P.J., 2010. LISFLOOD: a GIS-based distributed model for river basin scale water balance and flood simulation. Int. J. Geogr. Inf. Sci. 24, 189–212.

Vergara, H., Hong, Y., Gourley, J.J., Anagnostou, E.N., Maggioni, V., Stampoulis, D., Kirstetter, P.-E., 2013. Effects of resolution of satellite-based rainfall estimates on hydrologic modeling skill at different scales. J. Hydrometeorol. 15, 593–613.

Vergara, H., Kirstetter, P., Gourley, J., Flamig, Z., Hong, Y., Arthur, A., Kolar, R., 2016. Estimating a-priori kinematic wave model parameters based on regionalization for flash flood forecasting in the conterminous United States. J. Hydrol. 541, 421–433.

Vieux, B.E., Bedient, P.B., 2004. Assessing urban hydrologic prediction accuracy through event reconstruction. J. Hydrol. 299, 217–236.

Wang, J., Hong, Y., Li, L., Gourley, J.J., Khan, S.I., Yilmaz, K.K., Adler, R.F., Policelli, F.S., Habib, S., Irwn, D., Limaye, A.S., Korme, T., Okello, L., 2011a. The coupled routing and excess storage (CREST) distributed hydrological model. Hydrol. Sci. J. 56, 84–98.

Wang, J.H., Yang, H., Li, L., Gourley, J.J., Sadiq, I.K., Yilmaz, K.K., Adler, R.F., Policelli, F.S., Habib, S., Irwn, D., Limaye, A.S., Korme, T., Okello, L., 2011b. The coupled routing and excess storage (CREST) distributed hydrological model. Hydrol. Sci. J. 56, 84–98.

Wang, S.Y., Davies, R.E., Huang, W.R., Gillies, R.R., 2011c. Pakistan's two-stage monsoon and links with the recent climate change. J. Geophys. Res.-Atmos. 116.

Westerhoff, R.S., Kleuskens, M.P.H., Winsemius, H.C., Huizinga, H.J., Brakenridge, G.R., Bishop, C., 2013. Automated global water mapping based on wide-swath orbital synthetic-aperture radar. Hydrol. Earth Syst. Sci. 17, 651–663.

Wigmosta, M.S., Vail, L.W., Lettenmaier, D.P., 1994. A distributed hydrology-vegetation model for complex terrain. Water Resour. Res. 30, 1665–1679.

Williams, J.R., Hann Jr., R.W., 1978. Optimal Operation of Large Agricultural Watersheds with Water Quality Constraints. Technical Report No. 96. In: Texas Water Resources Institute, Texas A&M University, College Station, TX.

WMO (World Meteorological Organization), 2011. Weather Extremes in a Changing Climate: Hindsight on Foresight. WMO-No. 1075. WMO, Switzerland.

Wu, H., Adler, R.F., Hong, Y., Tian, Y., Policelli, F., 2012. Evaluation of global flood detection using satellite-based rainfall and a hydrologic model. J. Hydrometeorol. 13, 1268–1284.

Xue, X., Hong, Y., Limaye, A.S., Gourley, J.J., Huffman, G.J., Khan, S.I., Dorji, C., Chen, S., 2013. Statistical and hydrological evaluation of TRMM-based multi-satellite precipitation analysis over the Wangchu Basin of Bhutan: are the latest satellite precipitation products 3B42V7 ready for use in ungauged basins? J. Hydrol. 499, 91–99.

Yang, H., Sheng, C., Xianwu, X., Gina, H., 2011. Global precipitation estimation and applications. In: Multiscale Hydrologic Remote Sensing. CRC Press.

Yilmaz, K.K., Adler, R.F., Tian, Y., Hong, Y., Pierce, H.F., 2010. Evaluation of a satellite-based global flood monitoring system. Int. J. Remote Sens. 31, 3763–3782.

Yong, B., Ren, L.L., Hong, Y., Wang, J.H., Gourley, J.J., Jiang, S.H., Chen, X., Wang, W., 2010. Hydrologic evaluation of multisatellite precipitation analysis standard precipitation products in basins beyond its inclined latitude band: a case study in Laohahe basin, China. Water Resour. Res. 46.

Yong, B., Hong, Y., Ren, L.-L., Gourley, J.J., Huffman, G.J., Chen, X., Wang, W., Khan, S.I., 2012. Assessment of evolving TRMM-based multisatellite real-time precipitation estimation methods and their impacts on hydrologic prediction in a high latitude basin. J. Geophys. Res. 117(D9).

Yong, B., Chen, B., Gourley, J.J., Ren, L., Hong, Y., Chen, X., Wang, W., Chen, S., Gong, L., 2014. Intercomparison of the Version-6 and Version-7 TMPA precipitation products over high and low latitudes basins with independent gauge networks: Is the newer version better in both real-time and post-real-time analysis for water resources and hydrologic extremes? J. Hydrol. 508, 77–87.

Zhang, Y., Hong, Y., Wang, X., Gourley, J.J., Gao, J., Vergara, H.J., Yong, B., 2013. Assimilation of passive microwave streamflow signals for improving flood forecasting: a first study in Cubango River Basin, Africa. IEEE J. Sel. Top. Appl. Earth Observ. Remote Sens., 6(6), 2375–2390.

Zhao, R.J., Zhuang, Y.G., Fang, L.R., Liu, X.R., Zhang, Q.S., 1980. The Xinanjiang model. In: Hydrological Forecasting Proceedings Oxford Symposium. Oxford University, IAHS Publication, pp. 351–356.

Zhao, R.J., Liu, X.R., 1995. The Xinanjiang model. In: Singh, V.P. (Ed.), Computer Models of Watershed Hydrology. Water Resources Publications, Littleton, Colorado, USA, pp. 215–232.

Annual Flood Monitoring Using Synchronized Floodwater Index in 2010 Indus River Flood

Young-Joo Kwak, Jonggeol Park

International Centre for Water Hazard and Risk Management (ICHARM-UNESCO), Public Works Research Institute, Tsukuba, Japan, Department of Environmental Information, Tokyo University of Information Sciences, Chiba, Japan

15.1 INTRODUCTION

15.1.1 Background

Reducing risk to floods is an essential component in integrated water resources management (IWRM) and involves assessment and implementation of national and international development projects, as described in the United Nations Sustainable Development Goals (SDGs) (United Nations, 2014, 2015). Among the 17 goals, the following three are strongly interconnected to mitigate disaster risks: Goal 6 ("Ensure availability and sustainable management of water and sanitation for all"), Goal 11 ("Make cities and human settlements inclusive, safe, resilient, and sustainable"), and Goal 13 ("Take urgent action to combat climate change and its impacts by regulating emissions and promoting developments in renewable energy") (United Nations, 2014, 2017). Although holistic in nature, future IWRM strategies need to focus on flood risk resilience and the coping capacity at local, regional, national, and global levels. Doing so is important because the number and intensity of extreme disaster events may increase in a changing climate, as underscored by the Global Water Partnership (GWP, 2000).

Asia is the world's most disaster-prone region, where the greatest number of people were affected by flood disasters during the period 1980 through 2010. In particular, human exposure to floods has been increasing in South Asia, East Asia, and the Pacific region (UNISDR, 2015). Kwak et al. (2012) demonstrated that the Asia-Pacific region will experience increasing

---

flood risks throughout the 21st century, as more extreme rainfall will cause higher water depths leading to more floods in many areas. Moreover, historically, the Indus River Basin endured 21 major flood events between 1950 and 2010. However, the 2010 flood caused the greatest damage due to abnormally intensive monsoon rainfall, although the flooded area was smaller than the areas involved in the 1950, 1956, 1973, 1976, 1992, and 1998 floods. Thus in response to flood risk reduction and sustainable development, transboundary river basins in Asia should be taken as the highest priority in IWRM.

Among applications of remote sensing data related to disaster risk management, multiple satellite-based flood mapping and monitoring are the traditional models used in spaceborne remote sensing in hydrometeorological disaster studies. For instance, optical and synthetic aperture radar (SAR) data are used to detect surface water in applied research and big-data analysis domains using water detection algorithms based on a statistical classifier and the best combination of spectral bands, such as the Normalized Difference Water Index (NDWI) (Gao, 1996; McFeeters, 1996).

International communities working on satellite-based disaster mapping and crisis response have already been initiated and are actively providing operational-level emergency support and response services with international collaboration (i.e., International Chart, Sentinel Asia, and the United Nations Institute for Training and Research). In current efforts toward space-based rapid risk mapping at the global level, the two main contributors to near real-time flood mapping are operational flood monitoring services and research-level flood monitoring development.

15.1.2 Operational Flood Monitoring: Rapid Mapping System

To begin, the National Aeronautics and Space Administration (NASA)-based Goddard Space Flight Centers' Hydrological Sciences Laboratory and the Dartmouth Flood Observatory (NASA, 2010; DFO, 2010) provide examples of efforts to operationalize near real-time global flood mapping as the most notable application of optical imagery for flood detection on a daily or bi-daily basis (NASA, 2010). The Advanced Rapid Imaging and Analysis (ARIA) project, sponsored by the California Institute of Technology (Caltech) and NASA through the Jet Propulsion Laboratory (JPL), aims to bring geodetic imaging capabilities to an operational level to support local, national, and international hazard response communities (NASA-JPL, 2015). Operational since 2003, UNOSAT (the UNITAR Operational Satellite Applications Programme) rapid mapping provides satellite image analysis, such as derived maps, GIS-ready data, statistics, and reports, during humanitarian emergencies, for both natural disasters and conflict situations (United Nations, 2003).

Since the start of its operations in February 2015, the Copernicus Emergency Management Service (EMS) has provided rapid mapping and risk and recovery mapping in a wide range of emergency situations resulting from natural or manmade disasters. Copernicus EMS-Mapping consists of on-demand and fast provision of geospatial information to support and provide a set of mapping services with worldwide coverage funded by the European Commission. Also, the European Space Agency (ESA) has continually provided the operational service of Grid Processing on Demand (G-POD) for Earth Observation applications. The ESA G-POD is a generic environment where specific data handling applications can be seamlessly plugged into a system which is currently operated in testing mode including SAR-based mapping tools to retrieve an automatically generated flood map (European Space Agency (ESA), 2017).

Finally, the Center for Satellite-based Crisis Information (ZKI) at the German Aerospace Center (DLR) has been monitoring flood disasters by activating its TerraSAR-X satellite, as part of the International Charter (DLR, 2013).

In addition to the preceding types of operational-level flood monitoring, many spatio-temporal image-processing algorithms, including ones for flood change detection, have been developed and improved to achieve potential flood forecasting in the future as well as rapid flood mapping as an effective emergency response in the early stage of a flood disaster. Observation-based inundation maps and their final products after validation are widely used to assess the magnitude and extent of a flood risk, not only to support relief services but also to calibrate hydrological model-based flood simulations.

This case study introduced a new type of flood monitoring based on the combination of multiple-index and in situ data associated with floodwaters, using annual time-series multiple data. A remote sensing-based flood detection algorithm was designed to identify surface water and floodwater in a conceptually simple way, relying mainly on a hybrid conditional process that was coupled with the Synchronized Multiple-Floodwater Index (SfWI^2). The main purpose of this study was to accurately monitor changes in river water, including the maximum flood extent in a transboundary river basin using multiple satellite data to optimize evaluation of floodwater in annual flood frequency maps and to minimize the known data limitations of optical and multispectral images.

15.2 STUDY AREA

The Indus River Basin is one of the largest river basins in Asia, being about 3180 km long with an area of 960,000 km^2. It has an average annual river discharge of 7900 m^3/s with an annual sediment load of 291 million tons/year at the river mouth. There are many water infrastructures comprising 7 dams and 3 main reservoirs (Mangla, Tarbela, and Chashma), 19 barrages, 12 interriver link canals, and 45 canals extending about 60,000 km with 110,000 km of irrigation, drainage, and flood-protection facilities in the Indus River Basin in Pakistan. Fig. 15.1 shows the mega river network (blue lines) with water infrastructures, for example, dams, reservoirs, and barrages (red points), whereas the yellow pixels indicate vulnerable riverine floodplains located in lower lands as flood risk zones. Kwak et al. (2012) simulated the risk zone using hydrological data and elevation models based on the SHuttle Elevation Derivatives at multiple Scales (HydroSHEDS), which originated from a combination of the Shuttle Radar Topography Mission (SRTM) Digital Elevation Model (DEM) (Lehner et al., 2006; Farr et al., 2007). More than 138 million people in the Indus River Basin in Pakistan depend on irrigated agriculture for their livelihood, with the cultivated areas encompassing about 14 million hectares in the floodplains of the Indus River and its 5 main tributaries. A rising population, climate change, and degradation of riparian ecosystem services have accelerated flood risks, which are further exacerbated by inadequate flood planning and management. The selected study area suffered from a devastating flood caused by

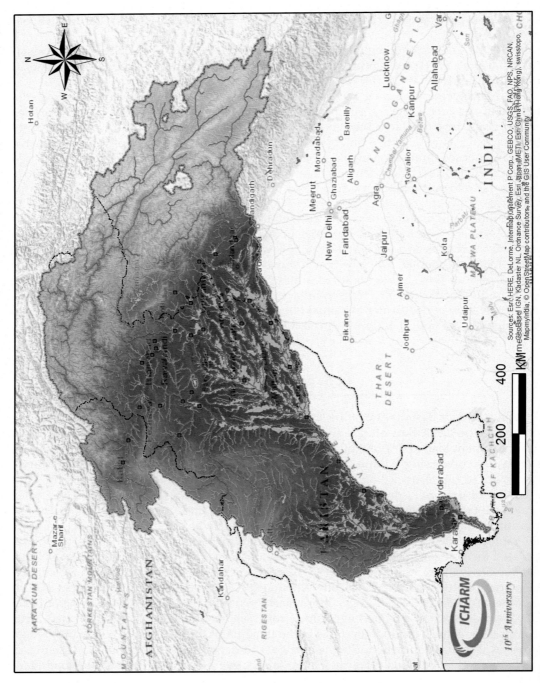

FIG. 15.1 Indus River network (*blue lines*) with water infrastructures, for example, dams, reservoirs, and barrages (*red points*), and vulnerable riverine floodplains in lowland areas (*yellow pixels*).

an unprecedented heavy rainfall from late July to early August in 2010. According to the National Disaster Management Authority (NDMA) of the Pakistani government, over 21 million people were directly affected, reportedly with over 1900 dead, 2966 injured, and damaged property worth more than $9.5 million; that is, approximately 2 million homes were damaged or destroyed over an area of 3.6 million hectares (36,000 km^2) (Ali, 2013).

The overall precipitation pattern and its intensity during the 2010 monsoon season was captured by the latest satellite precipitation estimates, that is, daily Tropical Rainfall Measuring Mission (TRMM) Level 3B42 Version 7 products, the Advanced Microwave Scanning Radiometer (AMSR-E) from NASA (Huffman et al., 2007), and ground rain-gauge stations from the Pakistan Meteorological Department (PMD, 2011). Based on the comparison of precipitation intensity from NASA's TRMM data between August 1, 2010, and August 9, 2010 (Khan et al., 2014), Fig. 15.2 shows the comprehensive intensity of abnormal rainfall, in particular around the Punjab province, Hyderabad in the Lower Indus Valley, and the Upper Chenab River in the northeastern mountain area. The dark blue pixels in Fig. 15.2 indicate as much as 24 mm of rainfall per day above normal (average) daily rainfall. For example, abnormal rainfall in the Upper Chenab River broke the record with more than four times the normal monthly rainfall from 1971 to 2000 (PMD).

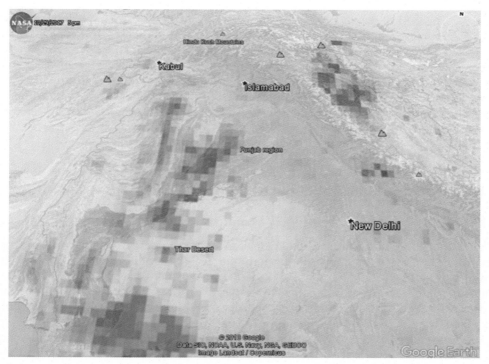

FIG. 15.2 The spatial distribution of abnormal rainfall from NASA's TRMM data between August 1, 2010, and August 9, 2010, during Indus River flood, based on Google Earth (NASA, 2017).

15.3 DATA USED

Two primary annual time-series data models were used for long-term flood monitoring on a large scale in the transboundary Indus River Basin. Both land surface reflectance (LSR) and temperature (LST) data were acquired from the Moderate Resolution Imaging Spectrometer (MODIS) and the Land Processes Archive Center (LPDAAC) within the NASA Earth Observing System Data and Information System (EOSDIS). The LPDAAC is located at the USGS Earth Resources Observation and Science (EROS) Center, Sioux Falls, South Dakota (NASA, 2014).

First, in order to detect floodwaters for dynamic flood mapping, we used time-series MODIS data from 2010, which is a type of the MYD09A1/Aqua level-3, 8-day composite surface reflectance with 500 m spatial resolution (swath: 2000 km), selecting the best possible value of the surface reflectance considering cloudiness, cloud shadows, a low solar zenith angle for each grid during an 8-day period (NASA, 2014). The surface reflectance bands of the MYD09A1 are a 7-band product computed from level-1B land band, band 1 (red: 620–670 nm), band 2 (NIR: 841–876 nm), band 3 (blue: 459–479 nm), band 4 (green: 545–565 nm), band 5 (NIR: 1230–1250 nm), and bands 6 and 7 (SWIR: 1628–1652 nm and 2105–2155 nm, respectively). Second, the level-3 MODIS land surface temperature (LST) and emissivity 8-day composite products (MYD11A2/Aqua) were employed to improve the accuracy of floodwater detection. The LST 8-day composite products of MYD11A2/Aqua were simple average values of all corresponding daily LST products with 1000 m spatial resolution collected within the selected 8-day period. The LST 8-day composite products were improved from validation and correction by using a split-window algorithm with comprehensive regression analysis with both the daytime and night-time surface temperature bands and their quality indicator layers. Because the LST products were contaminated by clouds and heavy aerosols, the average values per pixel were newly generated from the LST data of clear sky collected within the 8-day period.

15.4 RESEARCH-BASED FLOOD MONITORING AND MAPPING

15.4.1 Multisatellite-Derived Flood Mapping

To illustrate rapid and accurate flood detection and mapping, Fig. 15.3 shows the proposed conceptual framework of national and international flood monitoring processes based on a multisatellite-derived global water index between two main-streaming, operational flood monitoring services and research-level flood monitoring development. After selecting cloud-free images from multitemporal spaceborne sensors, the Synchronized Multiple-Floodwater Index (SfWI^2) was developed to improve more accurate detection of spatio-temporal floodwater changes. We considered annual water changes and floodwater extents, correlated with the sensitivity of surface water. For rapid global flood mapping, automatic multisatellite data processing is essential to provide risk information in an operational flood monitoring system.

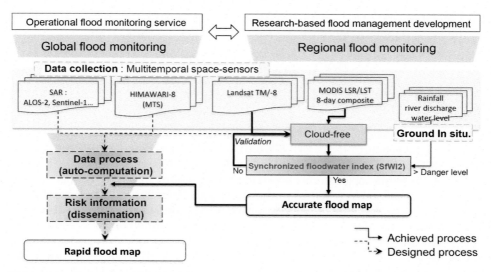

FIG. 15.3 Conceptual framework of national and international flood monitoring using Earth Observation multisatellite data and synchronized multiwater index.

15.4.2 The Synchronized Multiple-Floodwater Index

For maximum flood mapping and monitoring of an international river basin, the Synchronized Multiple-Floodwater Index (SfWI^2) was applied to the 2010 flood in the Indus River Basin. The multiple satellite-based SfWI^2 is a novel flood detection algorithm coupled with a dynamic floodwater index to generate more accurate flood maps than can a single use of a MODIS-derived index, that is, NDWI and MLSWI (Modified Land Surface Water Index). The SfWI^2 was already proposed and evaluated for Bangladesh floods as the best pilot case study during the flood monsoon season (Kwak, 2017). However, the conceptual SfWI needs improving by applying the dynamic threshold optimization based on a convergent global-threshold method considering different flood conditions and surface complexity at the local and regional flood scales, such as the characteristics of significant changes from MLSWI values and land surface temperature (LST) within the same floodwater pixels in floodplains. The SfWI^2 was formulated based on the conditional presence of flood pixels, as in Eq. (15.1)

$$\mathrm{SfWI}^2(t) = \begin{cases} \mathrm{WoI}(t), \text{if } x < Th_{WoI} \\ \mathrm{NDVI}(t), \text{if } y < Th_{VI} \\ \mathrm{LST}(i), \text{if } z \geq Th_{WST} \\ \mathrm{MLSWI}(t), \text{if } wi \geq Th_{DL} \\ 0, \text{if}, x \geq Th_{WOI} \cup y > Th_{VI} \cup z < Th_{wst} \cup wi < Th_{DL} \end{cases} \quad (15.1)$$

where WoI is the white index to remove clouds (0 < WoI < 1), NDVI is the normalized difference vegetation index (−1 < NDVI < 1), LST is the land surface temperature, Th_{WoI} is the optimized threshold value of cloud masking, Th_{VI} is the threshold value of vegetation index for flooded vegetation area, Th_{WST} is the threshold value of mean water surface temperature, and Th_{DL} is the flood danger level of river water when a flood occurs. Thus,

$$\text{WoI} = \frac{\rho_R + \rho_B + \rho_G - 1.5 \times \rho_{SWIR}}{210} \times 100 \tag{15.2}$$

$$MLSWI_{2\&7} = \frac{1 - \rho_{NIR2} - \rho_{SWIR7}}{1 - \rho_{NIR2} + \rho_{SWIR7}} \tag{15.3}$$

where ρ is the atmospherically corrected surface reflectance of MODIS, ρ_{R1} is band 1 (Red: 620–670 nm), ρ_{NIR2} is band 2 (NIR: 841–876 nm), ρ_{B3} is band 3 (Blue: 459–479 nm), ρ_{G4} is band 4 (Green: 545–565 nm), and ρ_{SWIR7} is band 7 (SWIR: 2105–2155 nm).

Eq. (15.2) is the White-object Index (WoI) to remove cloud cover. After collecting 46 images (8-day composite: MOD09A1, 500 m) of the study area from 2010, we first used the WoI because cloud-free images are essential to improve flood inundation mapping during the monsoon season. The optimal threshold value of WoI was determined on 90% clearance of clouds by a supervised binary (two-class) classification to remove cloud cover using the band combination of MODIS. Eq. (15.3) is the band combination of MLSWI between bands 2 and 7 for floodwater detection (Kwak, 2017; Kwak et al., 2015). The MLSWI can be developed and applied to temporal processing to extract floodwaters using annual time-series data such as the cloud-free 46 images from 2010. Also, LST data collected from the 46 images (8-day composite: MOD11A1, 1 km) were used for improvement of water detection to reduce the range ambiguity of a threshold value from MLSWI ($0 < MLSWI \leqq 1$) in a mixed pixel using the mean value of water surface temperature. To fill the missing and error values of time-series LST data, FFT (fast Fourier transform) filtering, a conventional algorithm in trigonometric series simplified by dealing with complex exponential signals, was employed to interpolate the missing data with periodic oscillation.

15.4.3 Flood Map Validation

To map the maximum flood extent with annual flood monitoring, time-series floodwater detection by SfWI^2 was simulated to identify the annual frequency of occurrence of floodwater pixels after the Indus River flood peak between late July and early August in 2010. To confirm the pixel-based floodwater extent, the MODIS-derived SfWI^2 performance was validated with two high-resolution spaceborne sensors, for example, the Phased Array type L-band polarimetric Synthetic Aperture Radar (PALSAR) aboard the Advanced Land Observing Satellite (ALOS) launched by the Japan Aerospace Exploration Agency (JAXA) in 2005 (HH polarization and 12.5 m spatial resolution) and the Landsat 5 TM (30 m spatial resolution) launched by NASA in 1984. The validation process was applied to the selected training sites based on a supervised land classification with visual analysis to identify floodwater pixels. The MODIS-derived floodwater pixels were compared with PALSAR-detected floodwater pixels (processing level 1.5), which were acquired on August 5, 2010, and Landsat 5 composite image, which was acquired on August 12, 2010, respectively.

15.5 RESULTS

15.5.1 Long-Term Annual Flood Mapping

Transboundary river basin-scale flood maps showing annual flood extents and propagation patterns were estimated and demonstrated with the $SfWI^2$ before and after the flood peak between late July and early August in 2010. As a result of pixel-based flood detection by the $SfWI^2$, although the spatial distribution of flood extent varied depending on its location with a surface mixture complexity, the flood extent distinctly represented the topographic characteristics of the lowland area. In particular, it clearly identified the boundary of floodwater pixels over the lowland area corresponding to the ground surface undulation based on SRTM-DEM. The resultant annual flood frequency map showed how many times the floodwater pixel appeared at the same pixel during 2010. In other words, the annual flood frequency map presented the superior capability of flood duration in spatio-temporal flood dynamics during the monsoon period. Therefore, among the important flood risk impact factors, flood duration is a key factor for measuring flood severity defined by magnitude, frequency, and intensity. Fig. 15.4A shows large-scale flood detection over the entire Indus River Basin and represents an effective snapshot of the maximum flood extent using the single water index (e.g., MLSWI) during August 5, 2010 through August 12, 2010; whereas Fig. 15.4B indicates a dynamic flood situation with different periods of flood duration at different locations in the Lower Indus River after worsening of the flood peak in 2010. In Fig. 15.4B the blue pixels indicate pixels of water, which are counted once from the 8-day composite data from 2010, and the red pixels indicate pixels counted more than five times (at least 40-day flood duration including permanent water) from the 8-day composite data after the flood peak.

15.5.2 Flood Map Comparison and Validation

Transboundary river basin-scale flood maps with annual flood extent and propagation were validated with high-resolution spaceborne sensors of both optical (Landsat TM-5) and SAR (ALOS-PALSAR) data. Taking advantage of two satellite images that were taken at different times by different types of sensors during flooding and using different band combinations and SAR-backscattering intensity-derived change detection, it was possible to better understand an actual flood situation.

After validation we found that the MODIS-derived flood map was in good agreement with the two different types of sensors, as shown in Fig. 15.5. Two different cases of confluence points were selected as the representative validation between the mainstream of the Indus River and the tributary rivers in 2010 after the peak floods on August 5 (located at the Chashma barrage, Punjab) and on August 22 (located at the Guddu barrage, Rajanpur), respectively. Fig. 15.5A–E shows that the blue pixels, derived from the MLSWI, indicate the locations with one water-pixel appearance based on the 8-day composite images, which estimate the maximum flood duration despite the partial underestimation and overestimation. As a result of the comparison with the high spatial-resolution ALOS-PALSAR (12.5m spatial resolution) acquired on August 5, 2010, we found that the MODIS-derived flood map partially overestimated because of the data limitations of the

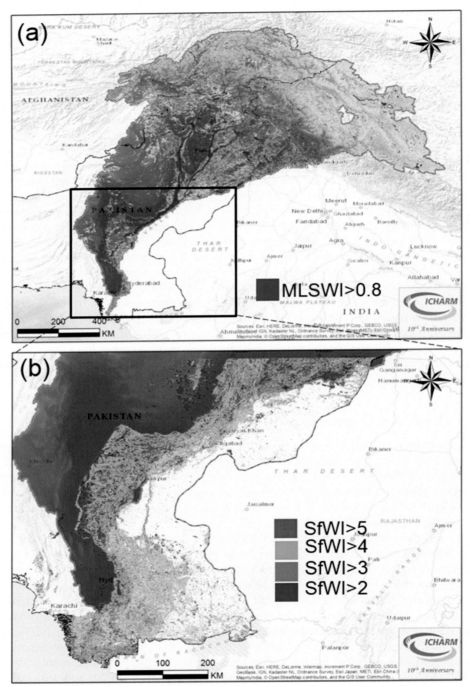

FIG. 15.4   MODIS-derived flood maps; (A) maximum flood extent map *(blue pixels)* during August 5–12, 2010, over the entire Indus River Basin using a single use of floodwater index and (B) annual flood frequency map (SfWI^2 2–5 in four color pixels) in the Lower Indus Valley using the synchronized multiwater index (SfWI^2).

IV. WATER EXTREMES IN INDUS RIVER BASIN

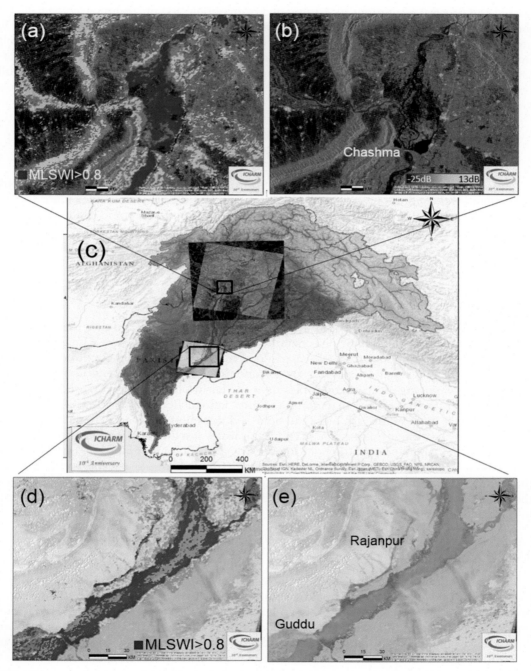

FIG. 15.5 A comparison between MODIS-derived flood map and high spatial-resolution images. *Blue pixels* indicate floodwater pixels in A, C, D; (A, D), enlarged MLSWI-derived flood maps on the ALOS-PALSAR (12.5 m) data acquired on August 5, 2010 (B) and on the Landsat-5 (30 m) RGB composite image acquired on August 28, 2010 (E).

spatial and spectral resolutions that originated from undetected micro-topographical characteristics, such as small tributary rivers and braid or point bars in a river, shown in Fig. 15.5A and B. Fig. 15.5D and E, however, shows that the MODIS-derived flood map partially underestimated in the case of the Lower Indus Valley in comparison with the Landsat-5 (30 m spatial resolution) RGB composite image acquired on August 28, 2010. These results confirm that the floodplain near the Indus River network is the area most vulnerable to flood hazard and risk in the Indus River Basin.

15.6 DISCUSSION

The advantages of rapid annual flood monitoring are discussed mainly in terms of the improvement of the simple processing and application of multiple-water index algorithms in the mega-delta floodplain. Although there are many important issues and challenges related to accuracy and efficiency in the flood mapping process, SfWI^2 can contribute to large-scale flood detection over a transboundary river basin with a high data processing speed and scalability.

15.6.1 Improvements

It is possible to achieve more accurate and rapid flood mapping and monitoring, as indicated in Fig. 15.3, and automatic near real-time forecasting of flood risks. The use of operational flood monitoring and research-based flood management should continue to improve accuracy in detecting spatio-temporal flood inundations by focusing on annual flood monitoring and maximizing the use of multiple advanced Earth Observation (EO) products. Under the new conceptual framework of national and international flood monitoring, SfWI^2 needs to be improved so as to reduce the ambiguity of flood detection between various EO products and multiple algorithms, taking into account the water fraction and the changes in annual, long-term water spatial distribution. Moreover, dynamic flood mapping at a transboundary river basin scale needs to employ the best DEM data from advanced spaceborne sensors with three-dimensional detecting capability, because the best DEM data effectively help discern the extent of floodwater events, even in mixed vegetation areas where it is typically difficult to differentiate water from land.

15.6.2 Challenges

End users of flood risk management services call on the world's scientific and administrative sectors to develop and coordinate emergency disaster services to deal with large flood events at the transboundary river basin scale. The consequences of large flood events, both positive and negative, affect the ecosystem and human lives, and these consequences vary depending on the floods' location, duration, depth, and velocity, as well as the exposed vulnerability associated with human activity. Large floods have various advantages and disadvantages in socioeconomic and environmental processes, with end users focusing first on making intuitive and simple solutions to reduce disaster risk and its damages. Thus the flood mapping and monitoring process could be operated as a one-button service using a fully automated calculation of a flood detection algorithm as a possible improvement with validation of flood extents. Moreover, other relevant floodwater indicators could be created, modified, and merged as alternative measures from various flood-related index-based approaches to characterize a flood using spaceborne optical, multispectral, and SAR sensors.

Although there are many challenges to overcome in various EO products and services, decision makers responsible for flood monitoring and operational emergency response need to collaborate more closely with local, national, and international stakeholders, with all players using the same framework of disaster risk reduction and resilience building, not only for flood monitoring but also for disseminating reliable risk information using advanced remote sensing technologies.

15.7 CONCLUSIONS

This chapter introduced an accurate and dynamic flood mapping technique using the synchronized multiple-water index (SfWI^2). The SfWI^2 is a key algorithm for producing an annual flood frequency map for multiple satellite-based dynamic flood monitoring with spatial and temporal changes in flood events. The following are the primary conclusions resulting from the study presented in this chapter.

(1) Seek more comprehensive understanding of the complexity of the relationship between large flood events and humans and ecosystems, as well as a rapid and dynamic long-term flood monitoring system appropriate for international river basin management, taking into account surface-water routing processes. In mega deltas and floodplains like those in the Indus River Basin, the dynamic flood process is complicated by the interactions between large floods and their environmental effects, because floods are influenced by many factors, such as micro-topography (e.g., levees, trees, roads, and railways), river networks (e.g., canals, reservoirs, and drainages), water infrastructures (e.g., dams, barrages, and artificial canals), and heterogeneous land cover with mixed areas.

(2) Seek to maximize the optimal operation of annual flood monitoring using the limited spaceborne-multispectral sensor data as snapshots in time-series data. It is also important to interlink dynamic flood mapping and flood hydraulic characteristics as much as possible, for example, river discharge, water level, and floodwater velocity. The annual flood frequency maps derived from the SfWI^2 play an important role in disseminating risk information on the dynamics of spatio-temporal flood inundations at large transboundary river basins, on both temporal and spatial dimensions. Therefore improved flood maps can be useful for validating hydrological simulations as well as flood risk management in the early stage of a flood disaster. At the same time, even if they are indirect products of EO data, flood hydraulic characteristics should be considered as flood risk-induced parameters (i.e., flood volume with river discharge, floodwater depth, flow velocity, and water level rise-rate).

(3) For rapid global flood mapping, improvement of the algorithm of flood monitoring is inevitable. The more improvements are made to fill the gap between multiple satellite-based data and simple or complex index-based algorithms to achieve information sharing

of emergency response at disaster operations, the more accurate and efficient will be the detection of mega-flood dynamic changes within the framework of national and international flood monitoring.

(4) To reduce the risk of flood disasters and for sustainable development in both urban and rural areas, the dynamic annual flood monitoring and rapid mapping system can be used. This system can help provide more reliable risk information to stakeholders via web-GIS technology and mobile applications with near real-time big data disaster management capabilities, not only to support policy and decision makers at local, regional, national, and international levels but also to strengthen community-driven efforts and flood resilience in high-risk flood areas.

Acknowledgments

This study was supported by the Japan Society for the Promotion of Science (JSPS) KAKENHI Grant-in-Aid for Scientific Research B: 15H05136. The authors express our thanks to NASA EOSDIS Land Processes DAAC, USGS Earth Resources Observation and Science (EROS) Center, Sioux Falls, South Dakota (NASA, 2014) for providing data set and final products.

References

Ali, A., 2013. Indus Basin Floods: Mechanisms, Impacts, and Management. Asian Development Bank, Mandaluyong City. 55p. Available from: https://www.adb.org/sites/default/files/publication/30431/indus-basin-floods.pdf.

Dartmouth Flood Observatory (DFO), 2010. Dartmouth Atlas of Global Flood Hazard. http://floodobservatory.colorado.edu/. (Accessed 10 October 2010).

European Space Agency (ESA), 2017. Grid Processing on Demand (G-POD) for Earth Observation (EO), Emergency Management Service (EMS)–Mapping of floods. http://emergency.copernicus.eu/mapping/ems/emergency-management-service-mapping. (Accessed 17 July 2014).

Farr, T., Rosen, P., Caro, E., Crippen, R., Duren, R., Hensley, S., Kobrick, M., Paller, M., Rodriguez, E., Roth, L., Seal, D., Shaffer, S., Shimada, J., Umland, J., Werner, M., Oskin, M., Burbank, D., Alsdorf, D., 2007. The shuttle radar topography mission. Rev. Geophys. 45, 1–33. RG2004.

Gao, B., 1996. NDWI—a normalized difference water index for remote sensing of vegetation liquid water from space. Remote Sens. Environ. 58, 257–266.

German Aerospace Center (DLR), 2013. The Center for Satellite based Crisis Information (ZKI). http://www.dlr.de/eoc/en/desktopdefault.aspx/tabid-12797/. (Accessed 17 October 2005).

Global Water Partnership, 2000. Integrated Water Resource Management (IWRM) at a Glance: Technical Advisory Committee. Global Water Partnership Secretariat, Stockholm. Available from: https://www.gwp.org/globalassets/global/toolbox/publications/background-papers/04-integrated-water-resources-management-2000-english.pdf.

Huffman, G.J., Bolvin, D.T., Nelkin, E.J., Wolff, D.B., Adler, R.F., Gu, G., Hong, Y., Bowman, K.P., Stocker, E.F., 2007. The TRMM multisatellite precipitation analysis (TMPA): quasi-global, multiyear, combined-sensor precipitation estimates at fine scales. J. Hydrometeorol. 8, 38–55. https://pmm.nasa.gov/data-access/downloads/trmm. (Accessed 17 November 2011).

Khan, S.I., Hong, Y., Gourley, J.J., Khattak, M.U., De Groeve, T., 2014. Multi-sensor imaging and space-ground cross-validation for 2010 flood along Indus River, Pakistan. Remote Sens. 6, 2393–2407.

Kwak, Y., 2017. Nationwide flood monitoring for disaster risk reduction using multiple satellite data. ISPRS Int. J. Geoinf. 6, 203–215.

Kwak, Y., Takeuchi, K., Fukami, J., Magome, J., 2012. A new approach to flood risk assessment in Asia-Pacific region based on MRI-AGCM outputs. Hydrol. Res. Lett. 6, 55–60.

Kwak, Y., et al., 2015. Prompt proxy mapping of flood damaged rice fields using MODIS-derived indices. Remote Sens. 7, 15969–15988.

Lehner, B., Verdin, K., Jarvis, A., 2006. HydroSHEDS Technical Documentation. Version 1.0 World Wildlife Fund US, Washington, DC, pp. 1–27.

McFeeters, S.K., 1996. The use of the normalized difference water index (NDWI) in the delineation of open water features. Int. J. Remote Sens. 17, 1425–1432.

NASA, 2014. EOSDIS Land Processes DAAC. USGS Earth Resources Observation and Science (EROS) Center, Sioux Falls, SD.https://lpdaac.usgs.gov/data_access/data_pool. (Accessed 10 October 2005).

NASA Goddard's Hydrology Laboratory, 2010. Near real-time global flood mapping. https://floodmap.modaps.eosdis.nasa.gov/. (Accessed 10 September 2015).

NASA-JPL, 2015. The Advanced Rapid Imaging and Analysis (ARIA) Project. https://aria.jpl.nasa.gov/. (Accessed 15 February 2006).

NASA, 2017. NASA Earth Observatory 2010, near-real time rainfall by TRMM. Available from: https://earthobservatory.nasa.gov/images/45177/unusually-intense-monsoon-rains (Accessed 3 May 2015).

Pakistan Meteorological Department (PMD), 2011. Flood Forecasting Division. http://ffd.pmd.gov.pk/cp/floodpage.htm. (Accessed 11 October 2015).

United Nations, 2003. The UNITAR Operational Satellite Applications Programme (UNOSAT). http://www.unitar.org/unosat/. (Accessed 16 May 2003).

United Nations, 2014. Open Working Group proposal for Sustainable Development Goals. p. 24.

United Nations, 2015. Transforming Our World: The Agenda for Sustainable Development. p. 41.

United Nations, 2017. The Sustainable Development Goals (SDGs) report. A. Guterres, New York. 60p. Available from:https://www.un.org/sustainabledevelopment/sustainable-development/. (Accessed 17 October 1930).

United Nations Office for Disaster Risk Reduction (UNISDR), 2015. The Human Cost of Weather Related Disasters 1995–2015. p. 30.

Further Reading

Eric Vermote, 2015. NASA GSFC and MODAPS SIPS, NASA MYD09A1 MODIS/Aqua Surface Reflectance 8-Day L3 Global 500m SIN Grid. NASA LP DAAC. doi:https://doi.org/10.5067/MODIS/MYD09A1.006 (Accessed 15.06.24)

PART V

WATER MANAGEMENT IN INDUS RIVER BASIN

Water Management in the Indus Basin in Pakistan: Challenges and Opportunities

Muhammad Basharat

International Waterlogging and Salinity Research Institute (IWASRI), Pakistan Water and Power Development Authority (WAPDA), Lahore, Pakistan

16.1 SURFACE WATER AVAILABILITY

All the major rivers that flow into the Indus River system enter Pakistan from its neighboring countries, India, China, and Afghanistan. Flow forecasting of such a large transboundary river basin is necessary in order to optimize the resources in its downstream areas. The Indus Basin Irrigation System (IBIS) serves a combined irrigated area of 26.02 mha (million hectares) in India and Pakistan. Issues that impede implementation of Integrated Water Resources Management (IWRM) strategies include the transboundary nature of the Indus River, limited availability of data, and inadequate data sharing. These problems result in a gap between water demands and supply in various water-use sectors.

The main source of water in Pakistan is the Indus River system, as shown in Fig. 16.1. The system resembles a funnel, with a number of water sources at the top that converge into a single river that flows into the Arabian Sea, east of Karachi. The average annual inflow of the western and eastern rivers and their tributaries at the rim stations is 180 bcm (billion cubic meters) (146.01 maf—million acre feet)—of which about 128 bcm is diverted to the distribution system. The Indus River and its five major tributaries form one of the world's largest contiguous irrigation systems. The Indus Basin irrigation network in Pakistan stretches over an area of 22.14 mha.

Based on the Indus Water Treaty (IWT) of 1960 with India, Pakistan was allocated the flow of three western rivers (Indus, Jhelum, and Chenab), with occasional spills from the Sutlej and Ravi rivers, and India was allocated the flow of three eastern rivers (Ravi, Beas, and Sutlej).

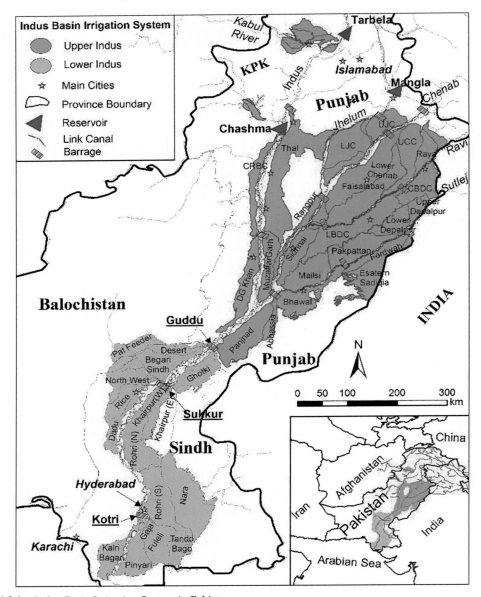

FIG. 16.1 Indus Basin Irrigation System in Pakistan.

A system consisting of two storage dams, eight interriver link canals, and six barrages was constructed as replacement works under the treaty to transfer water from western rivers to canal systems, which were dependent on the eastern rivers. The network has three major reservoirs: the Tarbela, Mangla, and Chashma. It also includes 19 barrages or headworks, 12 link canals, 43 canal commands, and over 107,000 watercourses. Irrigation demands over

TABLE 16.1 Annual Average Flows (bcm) to Indus River and Its Tributaries

River	Annual Average Flows (bcm)		
	1922–61	1985–95	2000–09
Kabul	32.1	28.9	23.3
Indus	114.7	77.3	100.3
Jhelum	28.4	32.8	22.8
Chenab	32.1	33.9	27.8
Ravi	8.6	6.2	1.4
Sutlej	17.3	4.4	0.5
Annual total	233.1	183.5	176.0

the past 150 years have resulted in very large diversions of water. At the same time, the three reservoirs are losing storage capacity due to sedimentation.

The flows of the Indus and its tributaries vary widely from year to year and within a year, as shown in Table 16.1, for different periods.

The second source of water in Pakistan is rainfall, with an annual quantum of about 50 bcm; the third source is groundwater with about 50–60 bcm annual usage. An additional 20 bcm of groundwater can be further developed if both the surface and groundwater are combined and there is proper planning at basin and local levels. In the past underuse of groundwater resulted in waterlogging and salinity issues in many areas of the Indus Basin. In contrast, overuse of groundwater is now triggering groundwater mining, saltwater intrusion, and increased surface salinity. Fortunately, because surface irrigation water is sufficient to avoid pumping of ground water, no saltwater intrusion has been detected in the aquifer adjoining the Arabian Sea.

16.2 INSTITUTIONAL SETUP

Irrigation systems in Pakistan started in the middle of the 19th century, integrating the physical infrastructure with system design and management principles. The first weir controlled diversion to the Upper Bari Doab Canal was constructed under British rule. Until the independence of Pakistan in 1947, barrages and canals were built by the Irrigation Works Department, Chief Engineer. With the division of the Punjab irrigated area between India and Pakistan after independence, five additional barrages and eight interriver link canals were constructed under the Indus Basin Replacement Works Project. This was done so that the irrigated areas deprived from river flows of Ravi, Beas, and Sutlej rivers could be compensated by flow from the Chenab, Jhelum, and Indus rivers. These works were executed by the Pakistan Water and Power Development Authority (WAPDA) with the help of international consultants and contractors. WAPDA is the federal agency responsible for monitoring, assessment, and distribution of expected flows in the river system to the provinces. WAPDA also executed almost all the construction related to irrigation during the last century, that is, installation of SCARP tubewells and construction of surface drains, which were transferred to Provincial Irrigation Departments (PIDs) after about 1 year of operation by WAPDA.

Today several federal and provincial agencies and departments are managing the water distribution and irrigated agriculture in the country. After the Water Apportionment Accord (WWA) of 1991, the Indus River System Authority (IRSA) was established to monitor the distribution patterns among the provinces. WAPDA still has the major role (and the technical capacity) of monitoring, data management, and assessment of water availability in the Indus River System. However, decisions regarding surface water management and distribution among the provinces now rests with IRSA. While WAPDA still manages the storage dams, it carries out operations in consultation with IRSA and the PIDs.

Prior to the 18th Amendment, groundwater was primarily a federal subject and all the development was being undertaken by the WAPDA, except local authorities handled for domestic and industrial sectors. The 18th Amendment specifically elaborates that "Subject to the existing commitments and obligations, mineral oil and natural gas within the province or the territorial waters adjacent thereto shall vest jointly and equally in that province and the Federal Government." Currently, the surface water is apportioned and distributed by the Federal Government; since the groundwater is a by-product of surface water, groundwater can be considered the provincial resource.

16.3 RIVER WATER DISTRIBUTION AT BASIN SCALE

Constitutionally, Pakistan has a federal system of government, where both federal and provincial governments share power. The Federal Ministry of Water and Power is responsible for water sector policy formulation. This ministry originally executed most of the projects as an autonomous agency, WAPDA, for developing water resources, including main dams, barrages, and link canals. All the tubewell and drainage projects were executed by WAPDA until 2000. Subsequently, provincial governments began rehabilitating irrigation and drainage systems. However, WAPDA still manages the multipurpose reservoirs on the Indus and its tributaries and operates them in consultation with the Indus River System Authority (IRSA) and Provincial Irrigation Departments (PIDs) according to the water rights and seasonal allocations to the provinces. An account of the provincial-level canal flows for all the canals is maintained on a 10-day basis by the respective PIDs, IRSA, and H&WM of WAPDA.

According to the accord, the three online reservoirs at Tarbela, Mangla, and Chashma and the interriver link canals are the key structural facilities for Indus Basin water management. The allocation of reservoir water shared by provinces was centralized, using "suggested operation criteria" established on a 10-day basis. According to the formula for distributing water from the IRS, the total water available in the system was estimated to be 114.35 maf below rim stations. It was allocated as 55.95 maf for Punjab, 48.76 maf for Sindh, 5.78 maf for Khyber Pakhtunkhwa, and 3.87 maf for Balochistan. The accord provided for the distribution of any surpluses and the shortages as well. The agreement left water discharge to the sea unresolved, subject to a study; however, it allocated 10 maf in the interim for discharge to the sea (Ministry of Water and Power, 1991). Later, in 2005, a study by consultants for the Ministry of Water and Power recommended that 8.6 maf of water annually be released below Kotri Barrage (Federal Flood Comission, 2005).

16.4 HISTORICAL PERSPECTIVE

The riparian conflict over Indus waters originated in the earlier part of the last century, when the Sutlej Valley Project and the Sukkur Canal Project were being planned by the colonial government of Her Majesty. During the division of the subcontinent, the international border between India and Pakistan did not take into consideration the irrigation system boundaries. Rivers were divided but borders were not divided. All the head works on the boundary between India and Pakistan were given to India, so the state could divert river water for the areas on the Indian side. Areas falling downstream and on the Pakistan side lost control of diversion of the Sutlej and Ravi rivers. Consequently, India stopped all the water from entering into canals irrigating Pakistani parts. The two governments could not resolve the dispute bilaterally. The dispute was finally resolved after proactive negotiations led by the World Bank with signing of the Indus Water Treaty in 1960.

16.4.1 Water Apportionment Accord

Table 16.2 shows the share each province receives from the available river flows in the IBIS as agreed upon in the Water Apportionment Accord of 1991. Areas irrigated from canals and wells are also given in Table 16.2. MacDonald et al. (1990) estimated that 79% of the area in Punjab and 29% of that in Sindh have groundwater that is suitable for irrigation. Therefore keeping in view increasing water demands, varying groundwater depth, and quality in irrigated areas, conjunctive use of surface and subsurface reservoirs needs to be pursued much more systematically than in the past. A study, Mara and Duloy (1984) suggested that large gains in agricultural production and employment are possible, given better policies as well as more efficient allocation and management of surface and ground water resources.

TABLE 16.2 Surface Water Allocation in WAA of 1991 and Irrigated Area on Provincial Basis

Province	Water Allocation (maf)			Irrigated Area (mha)
	Kharif	Rabi	Total	
Punjab	37.07	18.87	55.94	15.09
Sindh	33.94	14.82	48.76	5.18
KP	5.28	3.50	8.78	0.85
Baluchistan	2.85	1.02	3.87	0.399
Total	79.14	38.21	117.35	21.52

16.5 DISTRIBUTION AT PROVINCIAL SCALE

At the provincial level, a special regulation program is prepared by the respective Regulation Directorate for each crop season (two seasons in each year) and for all canals. These programs are prepared on a 10-day basis and are conveyed to the headworks where the canals off-take. The main canal flows are monitored at the headworks and conveyed to the Regulation Directorate. At the federal level, the Indus River System Authority (IRSA) distributes the water according to the apportionment accord between the provinces.

16.6 DISTRIBUTION AT CANAL COMMAND

The quantity of water to be released to a canal is based on the allocations made by the Regulation Directorate of the Department, or the indent placed by the canal management, whichever is less. This quantity is then released by the engineer in charge of the headworks who monitors the canal flow and informs the Regulation Directorate about variations from the schedule. A subdivision is the basic administrative unit of the network and a subdivisional officer (SDO) is the "Regulation Officer" of the subdivision. Each SDO works out the requirement (indent) of the area under his or her supervision. The water requirement from the tail end to the head end conveyed by the respective SDOs is expected to include the effects of events such as rainfall, canal breaches, and so on. When the supplies are less than the demands, an 8-day rotation is usually adopted. The tail end of the distributary would run under this system for 7 days.

16.7 DISTRIBUTION AT WATERCOURSE COMMAND

"Warabandi" is a rotational method for equitable distribution of available water in an irrigation system by turns fixed according to a predetermined schedule that specifies the day, time, and duration of supply to each farmer in proportion to his landholding under the command of the outlet (Singh, 1981; Malhotra, 1982). The term *warabandi* means "turn" (wari) that is already fixed. Warabandi used to be a 10-day rotation. But days (time) were counted only during the channel flow period, so it was called "kacha" warabandi (variable warabandi). Due to inherent conflicts arising in varying timings and channel discharges, "pacca" warabandi (fixed waribandi) using a 7-day rotation was adopted gradually in all the command areas three or four decades ago. This warabandi has official backing, with fixed timing during the 7-day rotation, which is changed by 12 h at the end of each year to coordinate night-times with day times.

The main objectives of canal operations are to achieve as much equity as possible and to ensure supplies to the tail-end farmers. The state management ends at the turnout (outlet). Waribandi is formulated for all the farmers along a watercourse. Within the limitations set by surface water availability, the farmers have the authority to decide most matters related to crop production and cropping intensity. Farmers also manage groundwater usage, which they are free to share. The shareholders are expected to operate and maintain the watercourse and implement the warabandi system. In the case of a dispute, the IPD management intervenes and fixes the start and duration of the turn for each cultivator according to his land-holding size. The underlying assumption, which does not hold true in an unlined system, is that there is no loss in conveyance of water from mogha (outlet) to the farm gate.

Equity in water distribution in the Indus Basin system is essential to increase and sustain a high level of production in agriculture. The stagnation in agricultural productivity is mainly due to inequitable distribution of water and the related problems of waterlogging and increased salinity, which are caused by an inefficient irrigation system. The irrigation system was designed to allocate water at the canal head, distributary head, and at the mogha. Unfortunately, inequity exists at all levels. Furthermore, the concept of "pacca" warabandi needs revision, as equity in quantity of water has to be ensured instead of duration of water availability (time). A water loss function needs to be added to the warabandi formula. There is sufficient knowledge to initiate the warametric system in Pakistan, which is based on volume equity. The only requirement is the development of a system of water measurement in the Provincial Irrigation Departments. The volume equity will also help to use the scarce water resources judiciously.

16.8 EFFORTS TO IMPROVE DISTRIBUTION EQUITY

Equity in water distribution at the distributary head is extremely poor. Some of the distributaries are getting 50% less than their allocated water, and others are getting 50% more water than allocated. The delivery performance efficiency measured by the International Water Management Institute (IWMI) in the Punjab and Sindh provinces indicate that the ratio of actual versus design discharge varies widely, from 0.43 to 1.57. Inequity may be associated with factors such as lack of maintenance of the irrigation system, lack of interest of the operating agency, theft of water by the water users, lack of objective-oriented operations, and rent seeking. The inequity is neither site-specific nor time-specific and can be observed any time at any point along the system. Basin-wide studies conducted by the WAPDA during the late 1980s indicated average watercourse conveyance losses of 40%. Recently published results by the IWMI in southeastern Punjab on the Sirajwah distributary indicate losses that are even higher than 40% in certain watercourses.

So that the designed hydraulic regime could be brought back to its original levels, urgent desilting and remolding of the old, neglected irrigation system were required. This is being accomplished mainly in the Sindh and Punjab provinces with the help of the World Bank and the Asian Development Bank. With these efforts, tail ends of the minor and distributary channels now have more equity as the gauge data, particularly of the tails, is being monitored and transmitted to irrigation secretaries. However, equity among individual outlets along the channels has not improved much because the irrigation officials are continuing the practice of rent seeking by adjusting the outlet discharges to under- or over-design values. The inequity is linked to the overall corruption norms in the society.

16.8.1 Program Monitoring and Implementation Unit—Punjab

In 2006 the Government of Punjab launched a new program to maintain a computerized database for irrigation system releases to improve irrigation management, reduce rent seeking, and increase transparency in water use (http:/irrigation.punjab.gov.pk). Under this program flows in the main canal and its off-takes are monitored all along the system to the tails of secondary level channels (distributaries, minors, and subminors). The records are relayed to the relevant officers in charge. Discharge records are maintained for heads and tails of almost all distributaries by the specifically created Program Monitoring and Implementation Unit (PMIU) at the PID Secretariat in Lahore. Though recording discharges is meticulous, the quality of the data remains suboptimal due to the absence of automatic measuring devices and poor repair and maintenance of the secondary level channels. However, with the launching of these monitoring activities, the situation is improving except where there is lack of financial investment in operation and management. In Sindh, such a program is still needed.

16.9 CHALLENGES AND OPPORTUNITIES

The provinces unanimously agreed to the Water Apportionment Accord of 1991, wherein the shortage of surface storages was recognized and each province was given the right to develop water resources falling in its own territory. But due to a lack of commitment and national interest being overcome by provincial interests and regional politics, additional reservoirs could not be built. Although the Diamir Bhasha Dam project was initiated in 2011, to date the only substantial progress is in land acquisition for the site of dam because no donor agency has committed funds. Since the Water Apportionment Accord of 1991, no significant progress has been made for developing additional storages. However, many old irrigation channels were rehabilitated, and field-scale water conservation assistance has been provided to the farmers in the form of subsidies and technical assistance.

16.9.1 Surface Storages

Pakistan has a storage capacity of 14.46 maf (after the Mangla Raising), which is only 10% of annual inflows (144 maf) at rim stations. With a total population of 180 million and a storage capacity of 17.84 bcm, the per person storage capacity is only $100 \, m^3$. In comparison the United States and Australia have per person storage capacity of 5000 and $2200 \, m^3$, respectively. The dams of Colorado and Murray Darling can store 900 days of river runoff. Pakistan's neighboring country India can store 120–220 days of water in peninsular rivers. In contrast, Pakistan can only store river flows of about 30 days in the Mangla, Tarbela, and Chashma reservoirs (Brisco and Qamar, 2005).

Storages are not aligned in series, which if they were, would provide an opportunity for extra storage utilization, that is, by depleting the upper and storing in the lower one. Storage capacities of 30 days or less is insufficient for a country as densely populated as Pakistan, as the quantity can neither meet water supply demands during droughts nor hold enough water during the summer monsoon season to mitigate flooding.

16.9.2 Improving Water Allocation

The IBIS is more than a century old. Water allowances and canal water distributions responded to increasing crop water requirements in a southward direction, for example, higher water allowance in Sindh as compared to Punjab. But within a province the canal water supplies do not address differences in irrigation demand. The consequence is unprecedented groundwater depletion in certain areas, for example, Bari Doab, and consistent waterlogged conditions in certain other canal commands, such as Rice Canal.

16.9.2.1 The IBIS Scale

The IBIS is the largest contiguous irrigation system in the world. In Pakistan it starts from Khyber Pakhtunkhwa (KP) and runs through Punjab to Sindh, and there is drastic variation in climatic parameters, for example, temperature and rainfall. The canal system that exists today was started in the middle of the 19th century under the British colonial administration. The IBIS was designed and executed in stages. The first major modern canal to be constructed was the Upper Bari Doab Canal (UBDC), the survey for which was started in 1850 (Pakistan National Committee of ICID, 1991). By virtue of its design, the irrigation system, which is more than a century old, has allowed variable allocation of surface supplies. These canal water allocations have been reviewed many times on a provincial basis in the context of establishing water rights on Indus River flows among the four provinces. With these reviews, the provinces agreed on the Water Apportionment Accord (WAA), which is based on historical uses (1977–82) of each canal command (Indus River System Authority, 1991).

16.9.2.2 Provincial Scale

Within each province canal commands have more or less established water rights and withdrawals based on their designed water allowance and established historical uses. However, the designed water allowance and canal supplies have never been reviewed to determine whether they are logical relative to the demand for crop water in these canal commands. For example, the Punjab Private Sector Groundwater Development Project PPSGDP (2000) reported that in Punjab the areas with deeper groundwater levels are generally located in the tail ends of the canal system. Later Basharat and Tariq (2013) studied the impact of spatial climate variability and reported on its impact on irrigated hydrology in Punjab, Pakistan. All the variabilities are due to the remarkable variation in reference crop evapotranspiration (ETo) and rainfall across the canal commands in Punjab.

16.9.2.3 Canal Command Scale

Even within the same canal command, ETo requirements and annual normal rainfall vary so much that water reallocation is needed. Research conducted by the International Irrigation Management Institute (IIMI) (presently IWMI) in Pakistan indicates that unreliable tail supplies are related more to the type of outlet than to the distance from the distributary head (Basharat and Tariq, 2013). Because groundwater is generally far more saline and sodic than canal water, the increased reliance on groundwater in certain areas is adding more salts to the soils, causing deterioration of both soil and groundwater quality. Vander Velde and Kijne (1992) established this scenario for the tail ends of two distributaries in the Punjab, but they also reported that in other distributaries fresh groundwater recharge was better at the tail ends. In their work, they looked at various possibilities of redistributing surface water supplies along one distributary channel so as to improve groundwater quality in the middle and tail reaches, and they concluded that there was insufficient surface water for the canal to achieve such an objective. Reallocation between and within canal distributaries is moreover politically unacceptable at present. Yet given the need for integrating surface and groundwater management, there needs to be more flexibility in redirecting surface supplies.

Ahmad et al. (2009) investigated remote sensing analysis of actual evapotranspiration (ETa) in Rechna Doab Irrigation System in Punjab, Pakistan, and concluded that the adequacy and reliability of combined surface water and groundwater deliveries decline toward the tails of the canal and toward the central and downstream parts of the Doab. The average ETa over Rechna Doab was estimated as 850 mm/year, which is almost double the average rainfall and almost 60% of the reference crop evapotranspiration.

The depth of the water table continuously increases from head to tail in most of the canal commands in Punjab, Pakistan, in spite of equitable canal water supplies. In the wake of increased groundwater use over time, the tail end farming community and irrigation management institutions must confront this emerging issue following detailed analysis and thoughtful consideration. To better comprehend this dilemma, canal water availability, crop water requirements, and groundwater recharge across the Lower Bari Doab Canal command were analyzed. It was found that annual rainfall decreases toward the tail end (212 mm) as compared with the head end (472 mm). In contrast, the annual gross and net crop water requirements at the tail end are 10.2% and 32.5% higher, respectively, as compared with the head end (Basharat and Tariq, 2013). Groundwater recharge rates are considerably lower at the tail end than at the head end. As a result, groundwater mining is taking place at the tail end by 0.34 m/year, whereas in head end areas, the depth of the water table is stable. Spatial climate variability across the command is the main cause for these inequities. Reallocating canal water and enhancing recharge to groundwater in these relatively more water-stressed areas during wet years is needed; otherwise, no groundwater management activity in this regard will have technical and social viability. Ignoring climatic variability within the canal command is one of the serious issues in irrigation system design that prevents optimal conjunctive water use; thus it prevents achieving the highest potential agricultural output as well.

16.9.3 Inequitable Distribution

Analysis of the distribution of irrigation water has shown that inequity of distribution exists at various levels in the irrigation system (Bhutta and Van Der Velde, 1992; Van Der Velde, 1992; Kuper and Kijne, 1992; Latif and Sarwar, 1994; Habib and Kuper, 1996; Khepar et al., 2000; Halcrow Pakistan (Pvt) Ltd, 2006). Although the situation is improving due to monitoring efforts by the irrigation departments, overall degradation in governance and authority of the government is again having a negative impact. As a result of this inequity, resources are used inefficiently, which is contributing to an observed decline in the discipline of users. Causes for inequity of supplies are (1) the bad state of repair of the irrigation channels and regulating structures, (2) the bad state of repair of many of the watercourses resulting in high losses, and (3) the inadequacy of flow measuring devices and the breakdown of discipline that leads to widespread tampering with regulating structures and stealing of water. Without proper discipline, groundwater governance cannot be supported and its management implemented only with great difficulty. For example, in canal commanding in the Adhi Kot area, water use efficiency could be increased substantially through improved water reallocation.

16.9.3.1 Watercourse Command Scale

The major disadvantage of the warabandi system presently in use is that watercourse losses are not taken into account in water balance calculations. In practice, the watercourse diversion time for a land holding is based on the number of acres and ignores the location of the land, that is, the distance between the outlet on the watercourse and the head of the watercourse at its parent channel. As a result, tail-enders receive a reduced supply or none at all. Therefore modifications are required at least at the watercourse level so that all farmers receive their due share by virtue of their basic equitable rights. At the same time, the system of revenue collection creates another inequity. Revenue (abiana) collection is based on the canal flat rate rather than on actual volume of water delivered to the farmers, which varies depending on the location of the farmer on the irrigation system. Canal water is almost free of cost compared to pumped groundwater. Farmers at the tail end receive less canal water and must pump more groundwater and thus pay much more for raising the same amount of crop acreage than do farmers at the head of the watercourse.

16.9.4 Transboundary Water Issues

Partitioning large contiguous areas of the IBIS between India and Pakistan, especially partitioning of the Punjab, led to a division of rivers. That is, the water rights of three eastern rivers (Sutlej, Beas, and Ravi) were fully given to India, and the water rights of the remaining three western rivers (Chenab, Jhelum, and Indus) were fully reserved for Pakistan. The World Bank led the negotiation and building of a network of barrages, link canals, and three storage reservoirs with the purpose of transferring surplus water from the western rivers to the eastern rivers, allowing diversions from the Ravi and Sutlej rivers.

All this planning and execution was carried out before there was need for environmental flows or any pressure on aquifers for groundwater pumping. Since that time the population has increased manyfold while the availability of water has decreased from about 5000 to 1000 m^3/person/year in both neighboring countries. This has caused a decrease in water inflows from about 22 maf annually to about 2 maf in Ravi and Sutlej.

The data on the Ravi and Sutlej flows for three different periods are shown in Table 16.3, as follows: (1) 40 years before Indus Water Treaty (1960), that is, 1922–61; (2) 22 years after completion of the IWT works, that is, 1977–99; and (3) the latest 11 years, that is, 2000–11. According to flow changes in the Ravi and Sutlej, the impact of the IWT emerged with vigor in the last decade, giving India the capacity to divert almost all the flow from both the Ravi River and the Sutlej River.

TABLE 16.3 Annual Average Flows (maf) in Ravi and Sutlej Rivers for Different Periods (Basharat and Tariq, 2014)

Ravi			Sutlej		
1922–61	1976–99	2000–11	1922–61	1976–99	2000–11
7.0	5.51	1.20	14.0	3.51	0.78

The reduction in eastern river inflows is causing a significant reduction in recharge to the aquifer, and consequently groundwater levels are falling by 16–55 cm/year in the central and lower parts of Bari Doab (Basharat and Tariq, 2014).

16.9.5 Waterlogging and Salinity Issues

Because water allocation continues to occur without any change in the basic criteria for the irrigation requirements for canal water, waterlogging and salinity issues are continuing unabated where drainage and groundwater are concerned.

16.9.6 Improving Water Use Efficiency at Field Scale

16.9.6.1 Promoting High Efficiency Irrigation Systems

Many irrigation systems are in the promotion stage. An example is drip irrigation, which has been subsidized by the Pakistani government through the World Bank. Drip irrigation for orchids is an excellent example. No progress has been made in the on-farm water storage sector, although talks about these activities are in progress in the print media. Universities all over Pakistan have started schemes allocating land to landless agriculture graduates so that they can put their skills to practical use. An example is the renowned University of Agriculture, Faisalabad, which started projects to promote improved agricultural and irrigation practices.

16.9.7 Water Pollution and Treatment

The Government of Pakistan is not inclined toward addressing water management and treatment. In fact, a large amount of water is pumped out of the aquifer and discharged into rivers without any treatment.

16.9.7.1 Domestic Sector

More than 50% of the population in Pakistan does not receive treated water. Only people in the upper class who live in the more upscale areas receive treated water; for example, only a select population receives arsenic-treated water for drinking purposes.

16.9.7.2 Agriculture and Industrial Sector

Neither the government nor the farming community give attention to the treatment of water for agricultural purposes. Moreover, many factories whose owners have no regard for proper waste management and harmful chemicals like lead are situated near farmlands; thus such chemicals are making their way into agricultural products. Similarly, while industries bound by foreign laws and regulations treat water accordingly, the government does not have treated-water requirements for indigenous industries.

16.9.7.3 Water Treatment at Source

The Rawal Lake Water Treatment Plant (RLWTP) uses the Rawal Lake in Islamabad as its source and provides treated water to the city of Rawalpindi, and the surrounding communities benefit greatly from it. Some cities get treated water from stored individual dams, such as storage from the Rawal Dam.

16.9.8 Water Supply Sustainability in Mega Cities

In large cities, entities are free to pump and deliver their desired amount of groundwater without treatment or binding to do so by any governmental department. There is no central authority for evaluating or planning the sustainability of water supplies in cities. Instead water is given freely at a flat rate to those who can afford it, which leads to a waste of water in cities. Meanwhile, the poor do not have access to clean water, even for drinking purposes.

16.10 CONCLUSIONS

Following are conclusions and recommendations drawn from the current study:

* Unlike pumped groundwater, canal water is supplied almost free of cost.
* Reallocation of canal water should flow from the head of the canal toward the tail of the canal.
* Similarly, necessary steps should be taken to reallocate canal water distribution from head to tail.
* Water conservation measures should be implemented to control warabandi by taking into account seepage losses in the channel.
* Canal water in areas with high water tables needs to be reallocated to areas with low water tables.
* Canal water reallocation is recommended from north to south and along the watercourse.
* Steps should be taken under the 1960 treaty to control groundwater on both boundaries across the Indus River Basin.
* An additional 20 bcm of groundwater can be further developed if surface and groundwater are used in conjunction and if proper planning at both basin and local scales is put in place.

References

Ahmad, M., Turral, H., Nazeer, A., 2009. Diagnosing irrigation performance and water productivity through satellite remote sensing and secondary data in a large irrigation system of Pakistan. Agric. Water Manag. 96 (4), 551–564.

Basharat, M., Tariq, A., 2013. Spatial climatic variablility and its impact on irrigates hydrology in a canal command. Arab. J. Sci. Eng. 38 (3), 507–522.

Basharat, M., Tariq, A., 2014. Command scale integrated water management in response to spatial climate variablity in LBDC irrigation system. Water Policy J. 16 (2), 374–396.

Bhutta, M.N., Van Der Velde, E.J., 1992. Equity of water distribution along secondary canals in Punjab, Pakistan. Irrig. Drain. Syst. 6, 161–177.

Brisco, J., Qamar, U., 2005. Pakistan Water and Economy: Running Dry. The World Bank, Oxford University Press, Oxford.

Federal Flood Comission, 2005. Kotri Study 1: Releases below Kotri Barrage to Check Sea Water Intrusion Through River Indus and Groundwater. Ministry of Water and Power.

Habib, Z., Kuper, M., 1996. Performance assessment of the water distribution in the Chishtian sub-division at the main canal level. IIMI Pakistan Report, Volume R-059.

Halcrow Pakistan (Pvt) Ltd, 2006. Punjab irrigated agriculture development sector project, draft final report Annex 3—Water balance studies. *Asian Development Bank*. TA-4642-PAK.

Indus River System Authority, 1991. Apportionment of Waters of Indus River System Between the Provinces of Pakistan, Agreement 1991 (A Chronological Expose). Indus River System Authority, Government of Pakistan, Islamabad.

Indus Water Treaty, 1960. Site Resources. World Bank, pp. 1–24.

Khepar, S., Gulati, H., Yadav, A., Brar, T., 2000. A model for equitable distribution of canal water. J. Irrig. Sci. 19, 191–197.

Kuper, M., Kijne, J.W., 1992. Irrigation Management in the Fordwah Branch Command Area South East Punjab, Pakistan. International Irrigation Management Institute (IIMI), Colombo, pp. 1–24.

Latif, M., Sarwar, S., 1994. Proposal for equitable water allocation for rotational irrigation in Pakistan. Irrig. Drain. Syst. 8 (1), 35–48

MacDonald, M., et al., 1990. Water Sector Investment Planning Study. vol. 1 Government of Pakistan, Harza Engineering International and National Engineering Services of Pakistan.

Malhotra, S., 1982. The Warabandi and its Infrastructure. Central Board of Irrigation and Power, India.

Mara, G., Duloy, J., 1984. Modelling efficient water allocation in a conjunctive use regime: the Indus Basin of Pakistan. Water Resour. Res. 20 (11), 1489–1498.

Ministry of Water and Power, 1991. Water Apportionment Accord. Ministry of Water and Power.

Pakistan National Committee of ICID, 1991. Irrigation and Drainage Development in Pakistan. Islamabad.

PPSGDP, 2000. Consultants, draft technical report No. 45, groundwater management and regulation in Punjab. Punjab Private Sector Groundwater Development Project, Groundwater Regulatory Framework Team, Project Management Unit, Irrigation and Power Department, Government of Punjab.

Singh, K., 1981. Warabandi for Irrigates Agriculture in India. Publication No. 146, Central Board of Power and Irrigation, India.

Van Der Velde, E.J., 1992. Performance Assessment in a Large Irrigation System in Pakistan: Opportunities for Improvement at the Distributary Level. International Irrigation Management Institute (IIMI), Colombo, pp. 203–237.

Vander Velde, E.J., Kijne, J.W., 1992. Salinity and Irrigation Operations in Punjab, Pakistan: Are There Management Options? International Irrigation Management Institute (IIMI), Colombo, pp. 203–237.

17

Integrated Irrigation and Agriculture Planning in Punjab: Toward a Multiscale, Multisector Framework

Ayesha Shahid*, Afreen Siddiqi†, James L. Wescoat, Jr.†

*Masters of City Planning, Massachusetts Institute of Technology, Cambridge, MA, United States
†Massachusetts Institute of Technology, Cambridge, MA, United States

17.1 THE PROBLEM OF RESOURCE MANAGEMENT IN THE INDUS RIVER BASIN

Drawing from the rich history of the "garden" metaphor in the Indus River Basin under the Mughals and the evolving modes of planning since the creation of Pakistan, James Wescoat proposed a conceptualization of the Indus River Basin "…not as a single basin garden – but as a living mosaic of gardens that flourish on a subcontinental scale in the 21st century" (Wescoat Jr., 2012, p. 40). This garden of gardens encapsulates within it many different visions, planners, actors, and levels of expertise, with each element playing an important role in shaping the Indus River Basin. Together these actors can address the issues related to the unsustainability of resources needed for irrigated agriculture in Pakistan. The garden of gardens metaphor is useful because it implies that for the garden to achieve its full potential, its parts must be coordinated and managed by enabling all its "gardeners"—or for the purposes of this present study, the water planners—to work with each other in various capacities. However, given the scale and complex spatial structure of irrigated agriculture in the Punjab province, planning and managing one of the largest contiguous irrigated systems in the world is a daunting challenge.

Agriculture accounts for more than 40% of Pakistan's employment and roughly 21% of its GDP, based on 2010 estimates (Bennmessaoud et al., 2013, p. 10). Because most of the agricultural land in Pakistan depends on irrigation for water, the irrigation system plays an integral role in maintaining the agricultural sector. Due to expansions in irrigation infrastructure the cultivable area of land increased from 25 million acres in 1947, at the time of independence, to roughly 43 million acres in 2010 (Agriculture Census Organization, 2010, p. 4; Briscoe and Qamar, 2005, p. 40). However, the level of Pakistan's crop yields remains well below what has been achieved globally and regionally. Due to a shortage in storage for surface water used for food production, farmers rely on groundwater to supplement canal water supplies, which has led to a decrease in water tables (Briscoe and Qamar, 2005, p. 41). Groundwater is also the main source of energy generation in Pakistan's agriculture sector, with up to 20% of the Punjab's energy usage being consumed in agriculture by way of tubewells (Siddiqi and Wescoat Jr., 2013, p. 580). Furthermore, it has been found that the use of tubewells and the inefficiencies in surface canal water infrastructures are related to water logging and salinity in the soil, to the point that poor water quality has caused a loss of cultivable land.

Canal water use and to some extent groundwater use are managed by the Punjab Irrigation Department (PID), while other agricultural inputs and food production are supported and managed by the Punjab Agriculture Department (PAD). To prevent overlapping of responsibilities, these two departments have always been clearly separated jurisdictions, which has made integration across sectors, alignment of goals, and coordination of developments quite difficult. Indeed, this is one of the core unresolved irrigation planning problems in the Indus River Basin. It is what renders Punjab a mosaic of gardens rather than a garden of gardens.

Several attempts have been made to at least partially address this situation. *Pakistan Vision 2025* was published by the national government in 2014 as a blueprint for the country's development. The document identified seven key pillars for achieving this goal, of which Pillar IV was Security, described as "sufficient, reliable, clean and cost-effective availability of energy, water and food security – for now and the future"(Planning Commission, 2014, p. 58). However, the report goes on to recognize that "these sectors have suffered historically from severe failings of integrated policy and execution" (Planning Commission, 2014). A focus on integration of different water uses as well as the food-water-energy (WEF) nexus has been identified by the province of Punjab in various policy and planning documents, for example, *Pakistan Vision 2025,* the National Water Policy (draft), and the Punjab Water Policy (draft). Integrated water resources management (IWRM) and the water-energy-food nexus have been explicitly called out as paradigms guiding the way forward for provincial policy and planning in the draft provincial water policy (Government of the Punjab, 2016, pp. 7 and 33). The Government of Punjab has begun giving an institutional shape to this idea with the creation of a water resources management wing.

This chapter investigates the idea of integrated planning of food, water, and energy in irrigated agriculture to explore how equitable, efficient, and environmentally sustainable resource management might be achieved in Punjab, Pakistan. The first section explores current concepts of integration in academic literature and shows that integrated management and nexus-thinking paradigms are ambiguous on governance and implementation mechanisms. Specifically, processes of governance are complex and multidimensional and are not easily modified for integrated or multisectoral management. The second section uses a historical lens to make the case that efforts at integration appear like a tussle between departments and hierarchical tiers, as centers of planning and management have shifted within and between departments over time. This historical analysis leads to the third section, which studies current mechanisms for integrated planning through an institutional analysis and finds that departmental officials make a distinction between planning, management, and operational functions, with each taking a different meaning on integration. This section specifically studies planning as the process that sets up the framework within which management and operational functions are carried out. Sectoral priorities, future infrastructure investments, and policy implementation mechanisms are all delineated in long-term, mid-term, and annual plans. Focusing on the planning dimension, the final section proposes a way forward to realize integrated planning such that departments are able to plan irrigated agriculture in Punjab in a multisectoral, multidimensional, and gardenlike manner.

17.2 IWRM AND NEXUS THINKING

Interactions among food, water, and energy in irrigated agriculture most clearly manifest at the farm level since agricultural output (or food) is a clear outcome of water (in the form of surface irrigation) and energy (used in pumping groundwater). In economic terms, the farmer is assumed to constantly weigh the trade-offs involved in prioritizing one set of inputs over others to optimize production (e.g., using energy for groundwater extraction costs substantially more than the nominally priced canal water). However, the relevant government departments for these inputs and outputs function in sectoral silos. To address this problem Dr. Muhammad Abid Bodla, board member of the Punjab Planning and Development Department (PND), with responsibility for infrastructure development, summarized that the current provincial focus is on achieving integration by rationalizing water use between sectors and reducing inequities in water access: "On the macro scale, we want to understand how to allocate water between different uses, and on the meso and micro scale, we want to look at the food-water-energy nexus and understand how to reduce head-tail inequities to address rural poverty." At the macro or provincial scale, the major planning problem the PND foresees is in the competing demands for water between agricultural, domestic, industrial, and environmental uses. The meso and micro scales refer to the subprovincial district, tehsil, village, and farm tiers where head-tail inequities in water conveyance are commonly understood to be major causes of the yield gap and poor agricultural output in Punjab (Briscoe and Qamar, 2005, pp. 31–32). After more than a decade of consultations and reiterations, a provincial water policy seeking to address these problems is near finalization and is expected to be approved nationally and provincially. Government of the Punjab (2016) draft Water Policy has put forward a multipronged approach to water resources management that seeks to combine ideas of integrated water resources management, food-water-energy nexus thinking, cutting edge water informatics tools, and deliberate thinking around water governance and institutional reform to tackle the challenges confronting Punjab today.

It is important to note that while the Punjab Water Policy explicitly calls out recognition of the food-water-energy nexus as a goal, the policy prioritizes integrated water resources management, with nexus development as a subgoal to pursue demand management. The "IWRM approach," explains the policy document, "moves away from single sector water planning to multi-objective planning and integrated planning of land and water resources, recognizing the wider social, economic and development goals and entailing cross sectoral coordination" (Government of the Punjab, 2016, p. 33). This is important because while both IWRM and nexus thinking aspire to cross-sectoral integration, they can mean very different things for how they translate into institutional arrangements. The nexus approach "treats different sectors—water, energy, food and climate security—as equally important (i.e., multicentric)."

Giordano points out in his analysis of IWRM's limitations: "IWRM is flawed because it puts water at the center though it is only one aspect of holistic problem management" (Giordano and Shah, 2014, p. 365). Benson et al. (2015) highlight that while nexus thinking encourages multitiered institutional arrangements and mechanisms for joint decision making, it is silent on conception of how governance should occur (p. 760). In contrast, IWRM proposes principles of "good governance" such as "transparency, collaborative decision making and the use of specific policy instruments" (Benson et al., 2015, p. 760). Therefore no clear prescription for operationalizing ideals of multiobjective, cross-sectoral, joint decision making exists.

Studying examples of successful management of common pool resources like canal and groundwater, Elinor Ostrom notes that farmer behavior is determined by the "rules-in-use" set up by the particular institutional mix in an area (Ostrom, 2010, p. 418). This requires close scrutiny of mechanisms of cross-departmental coordination, vertical integration within departments, and participation of individual farmers and irrigators in planning and management of resources. Scott et al. (2015, p. 32) summarize: "for successful environmental outcomes at the level of watershed to be replicated at the basin scale would require robust feedback loops that support both vertical and horizontal institutional linkages that can respond to vagaries of both socio-economic heterogeneity and also bio-physical change and variability." In Punjab even while policy and planning decisions are made predominantly at the provincial level, much of the interface between government institutions and users of the irrigated agriculture system occurs at the meso scales of districts and canal distributaries. This is also where political forces, problems of elite capture, and farmer rivalry manifest most explicitly and is therefore an important part of setting up an integrated system of governance. Benson et al. (2015) explain that the idea of integration raises "institutional challenges" with both "opportunities and impediments to joint decision-making" and that "because 'resource coupling' is played out at different institutional levels, 'multi-tiered institutional arrangements' are required to govern it" (Benson et al., 2015, p. 760).

Perhaps the most fundamental hurdle to cross-departmental integration is the misalignment of administrative boundaries of the Punjab Agriculture and Irrigation Department (Fig. 17.1). The Punjab Irrigation Department (PID) follows a canal command hierarchical structure that is area-based, with the entire irrigated system divided among five to six chief engineers (CE) overseeing zones. Each zone is split into two to three canal circles headed by a superintendent engineer (SE). Each circle is a unit with a complete canal system managed by the superintendent engineer. The circle is broken into divisions along major distributaries managed by the executive engineer (XEN). The division is further divided along minor distributaries into subdivisions, headed by the subdivisional officer (SDO). In contrast, agricultural management is done along administrative divisions with the province of Punjab divided into 9 divisions and 36 districts. Each district is further subdivided into tehsils and union councils, based on population. Note that while the provincial Agriculture and Irrigation Departments' institutional tiers are clearly demarcated, groundwater management and regulation, especially after the devolution of authority under the 18th Amendment to the Constitution, is considered absent by the departments. While the Provincial Irrigation and Drainage Authority (PIDA) Act, 1999, drafted rules for groundwater management, implementation has been minimal (Qureshi et al., 2010, p. 1560). No one clear entity manages groundwater as different functions related to groundwater management have been assigned to PID, PAD, and PIDA at different times. However, a new groundwater policy is under development that may create more formal mechanisms for groundwater governance.

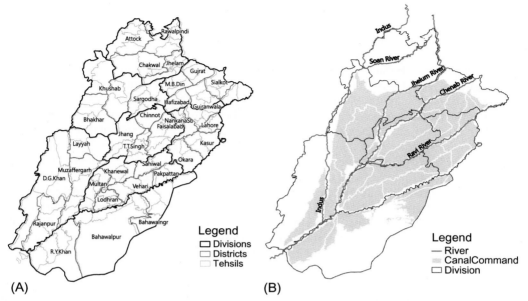

FIG. 17.1 (A) The Punjab Agriculture Department follows an administrative structure based on division and district boundaries as conventionally followed by different departments in the Punjab government. (B) The Punjab Irrigation Department has an administrative structure based on the irrigation network with each canal command subdivided along major and minor distributaries.

The Punjab water policy draft highlights improved irrigation efficiency, water productivity, and implementation of cropping patterns as means by which the effectiveness of water use in agriculture can be improved. It recommends measures such as agro-ecological zoning, extensification over intensification, and head-tail equity as ways to manage demand for water (Government of the Punjab, 2016, pp. 29–31). However, the separation between the PID and the PAD, as well as assumptions of incompatibility behind the different tools and levers available to the two departments, means that the cross-sectoral integration required to implement such policies does not exist at the moment, encouraging the adoption of one-size-fits-all solutions across the province. Focusing on surface and groundwater (with energy used in its abstraction) as inputs to agricultural production, this chapter investigates integrated planning of irrigation and agriculture in Punjab in order to understand how planning has evolved over time to reach its current form. The chapter then proposes a way forward to reimagine irrigation and agriculture planning for holistic and effective problem solving in the Indus River Basin of Punjab.

17.3 HISTORY OF INTEGRATION IN THE INDUS RIVER BASIN

The development of the Indus River Basin under British rule was rooted firmly in the tradition of the "science of empire." Science of empire refers to the British practice of linking the colonial state to the local community through settling land within "a framework of local relationships" and "by the legal and administrative language of the colonial state" (Gilmartin, 1994, p. 1133). A basic canal infrastructure was developed by "[mobilizing] local elites and their followers in canal digging" so as to produce "communities" of sharers and maintainers of waters, and as a way of linking the state with local elites (Gilmartin, 1994, p. 1132). A larger philosophical reason behind situating the Indian population within an agricultural property rights regime was to locate them within a local social order of community and custom associated with the colonial state. David Gilmartin explains: "On one level, of course, this was because of the close connection in this arid region between the local control of water and settlement on the land. But on another level, it reflected the fact that control of irrigation was a hinge between the power of the local "community" and that of the state" (Gilmartin, 1994, p. 1134). This had two possible consequences. Some argued that it "tended to legitimize the exercise of local 'privilege' in the management of irrigation water, even though this often undercut effective water management…" which led to "oppression on the part of the headmen of villages" (Gilmartin, 1994, p. 1136), because inequalities were hard-coded into the irrigation infrastructure as it was developed by certain local elites to their own advantage. However, a conflicting argument was made by some British officials for recognizing the importance of local knowledge and customs in order to encourage self-governance. Gilmartin points out that most of this discussion was based around the use of *cherr* labor provided by irrigators during the winter season for silt clearance. This struggle represented the start of an ongoing push and pull between a multilevel management of the irrigation infrastructure that included farmers engaged in self-governance versus an engineer-based view of management that prioritized efficiency by use of scientific and technical knowledge (Gilmartin, 1994, p. 1136). In the early 19th century the cherr system of labor was abolished, marking a win for imperial science in favor of the engineer-led paradigm of governance. Gilmartin (1994, p. 1137) identifies the Canal and Drainage Act of 1873 as the ultimate codification of the bureaucratization and centralization of canal network management, and explains that "the achievement of British Indian engineers in the late nineteenth and early twentieth centuries was…to define a new, 'scientific' way of looking at the problem that would justify excluding local 'communities' from a role in the administration of major canal channels."

Around the same time that the Canal and Drainage Act of 1873 institutionalized the bureaucratization of irrigation management, the British Indian government also set up the Department of Agriculture (established in 1871). This was followed by the establishment of a number of different crop research institutes meant to study and solve problems in Indian agriculture (Gill and Mushtaq, 1998, pp. 3–4). The Irrigation Department was officially created in 1905 after separating the responsibilities of civil works and canal development from the Planning Works Department and handing them over to the Irrigation Department. Its primary responsibility continued to be the development of irrigation infrastructure until 1958, when the Pakistan Water and Power Development Authority (WAPDA) was created for the implementation of the Indus Basin Project (directed by technical and financial assistance from foreign donors) (Gill and Mushtaq, 1998, pp. 10–11). The creation of the WAPDA at the national level to implement the Indus River Basin Project in the aftermath of the 1960 Indus Water Treaty solidified a science-based paradigm of water management that was large-scale, top-down, and infrastructure heavy.

In the 1950s when waterlogging and salinity emerged as a challenge to agriculture, groundwater became another important variable requiring agricultural water management.

To address these problems the Salinity Control and Reclamation (SCARP) project was launched in the 1960s to install thousands of state-managed deep tubewells, often located at the head of a tertiary canal, to supplement the surface water supply and simultaneously address problems of waterlogging. With the SCARP project the irrigation engineer was forced to step outside of irrigation's "technical system of flows" (Gilmartin, 2015, p. 239) and develop a technical vision for groundwater management at the village and community level. What neither the national government nor the donor organizations foresaw was the explosion of private tubewells that followed, enabling rapid crop intensification by year-round crop cultivation. For the first time the engineers were confronted with a tubewell-based demand-driven system "rooted in the independent actions of the individual irrigators themselves" (Gilmartin, 2015, p. 240) as opposed to a supply-driven system based on large-scale infrastructure.

The issue of integration across the Irrigation Department and the Agriculture Department was thus recognized as early as the 1960s, as highlighted in a 1960 report by the Food and Agriculture Commission (FAC). The report succinctly captured the tone of the split between the Irrigation Department and the Agriculture Department that can be found echoed through their corridors even today. It explained:

> Irrigation, of course, increased agricultural production, but one of the first points to note about the Irrigation Department is that it is essentially an engineer's affair, supplying water at field outlet point and leaving the farmer to distribute and use it as he likes. The best agricultural usage of water is not the department's business. Its staff are purely there to supply the water in the canals, see what crops are grown and collect the water charges. *(Gill and Mushtaq, 1998, pp. 16–17)*

The report went on to note: "Unfortunately, none of this contributed to the increase of production," and the "…engineers were not agriculturalists and the agriculturalists had not applied themselves sufficiently to the problems of irrigated agriculture" (Gill and Mushtaq, 1998, p. 17). The report stressed that the existing irrigation and agriculture jurisdictional boundaries should be reconsidered to decide "where to draw the line between Irrigation and Agriculture" (Gill and Mushtaq, 1998).

The general consensus was that either the Irrigation Department staff that inspects the farmer's field to collect revenue-related information can be given the additional function of improving water usage at the farm level or this function should be added to the existing agricultural extension services because on-farm water distribution is "more a matter of agriculture rather than engineering" (Gill and Mushtaq, 1998, p. 17). The FAC suggested that because the biggest issues confronting farmers at the time (salinity and drainage) were best addressed at the farm level, these functions should be handed to the Department of Agriculture so the irrigation staff could focus on decreasing canal head-tail discrepancies for impartial distribution of water. This theory was put into practice and culminated in the formation of the On-Farm Water Management Wing within the Department of Agriculture. However, in spite of the envisioned role of the On-Farm Water Management wing as solving problems of irrigation at the farm level and a general appreciation of the big strides it had made in improving watercourses and encouraging water conservation technology use among farmers, the 1998 case study of the Department of Agriculture argued that the sharp separation between the Irrigation Department and the Agriculture Department at the watercourse level remained a great divide, "which runs through the framework from field level operators to provincial level departments, and to federal level ministries" (Gill and Mushtaq, 1998, p. 86).

The Irrigation Department case study of 1998 similarly identified the lack of interagency coordination as a major hurdle to improving the performance of the irrigation system. The report included coordination with the Department of Agriculture as well as the WAPDA, the Punjab Revenue Department, and the Finance and Planning and Development departments in the list of relevant departments. It further emphasized the need for farmer participation in irrigation management for "efficient operation of canal systems, control over water theft, and transparency in revenue assessment and collection" (Ul-Haq, 1998, pp. 97–98). The report highlighted two efforts to realize interagency cooperation: the Revised Action Plan of 1979 and the Water Sector Investment Planning Study (WSIPS) of 1990 conducted by the WAPDA and the World Bank, and went on to conclude that neither was very successful at realizing the goal of better coordination among the concerned departments. The WAPDA and World Bank WSIPS study updated the original Indus Basin Model that became the basis for future modeling efforts. However, the preface for the report essentially declared it useless at the time, explaining that the model was not effectively transferred to the Government of Pakistan due to lack of technical capacity and expertise within the department to run it (Ahmad et al., 1990, Preface). Even while departmental analyses highlighted the need for coordination within and between the concerned departments, the nature of the basin-wide models used and their complexity made coordination with other agencies and farmers quite impossible. Later efforts to streamline the model have addressed some of these challenges, but it remains a tool available only to the provincial departments.

The government responded to the challenge of irrigation inequity and poor revenue collection by investing in the development of local community organizations at the village level as a "critical strategic resource" (Gilmartin, 2015, p. 243). It was as a result of this thinking that the Water User Association Act was promulgated in 1981 based on initial pilots with on-farm water management under the Punjab Agriculture Department that sought to involve farmers in delivering watercourse infrastructure improvements. Water User Associations (WUAs) were thus formed at the watercourse level to involve farmers more actively in management of the improvements. As explained by experts in the 1990s, "It is not enough to try to create a *sense* of local ownership in WUAs...The organizations must *belong* to the water users in fact" (Gilmartin, 2015, p. 245). However, the WUAs' experiment was considered of limited success, since the "...World Bank's post-project evaluations later confirmed that the projects achieved their physical components (water losses in watercourses were reduced from about 40% to 25–30%), but failed in most of their institutional objectives (World Bank, 1996)" (Bandaragoda, 2006, p. 57). The Irrigation Department was reluctant to adopt a similar approach for distributary-level canal management. However, creation of the Provincial Irrigation and Drainage Authorities (PIDAs) at the behest of donor organizations in 1999 created an obligation for the Irrigation Department to also create farmer organizations (FO) for participatory irrigation management (PIM). This program sought to transfer the management and operation of irrigation infrastructure to farmer organizations (FOs) and area water boards (AWBs) formed in canal command areas at distributary and circle levels, respectively. While the FO and AWB experiment by the PID is relatively recent in comparison to the WUAs (as formed by PAD), the initial experience has not achieved the desired objectives for similar reasons as with WUAs. Gilmartin (2015) summarizes: "And this reflected, at least in part, the reluctance of engineers to formally involve them [i.e. the farmers] in playing such roles" (p. 243).

With the 18th Constitutional Amendment in 2010, most federal responsibilities were devolved to the provinces, making them much more powerful than at any point in Pakistan's history. However, even as different tiers of government have been empowered through devolution at different times, some scales of irrigation and governance remain conspicuously absent from the discussion. This is the scale at which "lower employees," or the mid- to low-level bureaucracy of the provincial departments, act as the implementing arms of PID and PAD. For both interdepartmental and intradepartmental coordination, this scale has been identified as an important point of linkage. For example, the 1998 case study of the Department of Agriculture highlighted the need for this tier to be strengthened in order to improve irrigated agriculture in Pakistan through the use of the One Window Operation (OWO) and canal councils. OWOs at the tehsil or union council level were proposed to serve as facilities where "…the Agriculture Department, Irrigation and Power Department and the Farming Community participate equally to ensure sustainable agricultural production in their respective area" (Gill and Mushtaq, 1998, p. 100). Such a multidisciplinary approach to irrigated agricultural management was considered urgently needed "to increase per capita yield" (Gill and Mushtaq, 1998, p. 100). Focusing further on improvements in irrigation, the agriculture case study further suggested:

> "…institutions, such as Canal Councils, ought to be assigned an active role to the farmers [sic] in assisting the Irrigation Department for operation and maintenance of the canal system. Therefore, it is proposed to set up a committee comprising of [sic] a Public Representative, [Sub-Division Officer] of Irrigation Department, Water Management Specialist, [Extra Assistant Director Agriculture] of Agriculture Department and Assistant Commissioner, to strictly monitor the operation and maintenance activities of the canal network. The same will help to ensure equity and reliability of irrigation water to every water user. *(Gill and Mushtaq, 1998, p. 102)*

However, the role of these organizations continues to be envisioned as being in operations and maintenance, and planning decisions are considered to be a provincial responsibility. The next section explains how planning is currently carried out in the province of Punjab and why integrated planning is lacking in its processes and outcomes vis-à-vis irrigated agriculture today.

17.4 IRRIGATED AGRICULTURE PLANNING IN PUNJAB TODAY

In thinking about integrated water resources management, it is important to unpack the term "management," identify what it entails, and how it relates to "planning." Conversations with members of PAD and PID reveal three different types of decision-making categories in the day-to-day functions of the departments: planning, management, and operation. A simple distinction is that management involves ongoing decision making while planning is forward looking, generally on a timescale of at least one to 5 years and longer. Elaborate manuals for how the annual and mid-term programs are to be produced by the national Planning Commission and the provincial Planning and Development departments exist. They provide clear deadlines, departmental obligations, and annual guidelines based on priorities identified by the government of the day. The outputs of this planning process are mid-term and annual development plans with lists of projects to be implemented by concerned departments.

The line agency for planning functions is the Planning and Development Department (PND) of the Government of Punjab. PND serves as the central node and coordinating agency for compiling all departmental plans and the final decision-making entity through which projects are approved based on financial constraints (Aberman et al., 2013, p. 16).

Given its central position, the PND serves as a key link between Punjab Irrigation and Agriculture Departments' plans as a way of integrating their priorities and goals. Further, as the hub for coordinating government and donor programs, identifying provincial goals and visions, and making financial trade-offs regarding which projects will be funded, the PND is also responsible for ensuring that all provincial projects fit within the larger development agenda envisioned by the provincial government. In this role as the natural coordinator among different departments and the institutional platform where senior management of all concerned departments can provide input on future plans, the PND is ideally placed to "bridge interorganizational networks" for integrated planning (Siddiqi and Wescoat Jr., 2013, p. 46). Especially since devolution, the provincial planning departments have become the key agencies for coordinating donor activities (Aberman et al., 2013, p. 17), as well as for aligning Punjab's developmental goals with national strategy priorities as highlighted in the Pakistan Vision 2025 or provincially in documents like the Punjab Growth Strategy 2018. The PND has in recent years also asked each department to develop sectoral plans that can guide the way forward for individual departments. Sectoral plans exist for both PAD and PID; however, only the agriculture sectoral plan is available publically.

While clear guidelines and manuals exist for how annual and mid-term development plans are to be developed, planning at subprovincial levels gets murkier. When asked at the district level what planning documents exist, district departmental officials reference the district budgets and their contribution to the provincial annual development plan. Districts compile information on current and development project expenses from tehsils and union councils, which is forwarded up the chain of command and compiled by departments at the provincial level to inform the provincial development plan.

17.4.1 Multiscale Planning

An analysis of planning decisions made at the different institutional tiers in the Punjab Irrigation Department (PID) and the Punjab Agriculture Department (PAD) reveals that planning remains concentrated at the provincial level with poor mechanisms for integrating meso- and micro-level stakeholders within and outside the departments in the planning process. Since PID and PAD have different organizational structures, the terms meso and micro are used to refer to similar institutional levels. For the PAD the meso scale can be taken to refer to the division and district levels, with micro scale referring to the farm and village level. Tehsil and union councils are connectors between the micro and meso scales for the PAD. For the PID the zone and circle tier may be taken as the meso scale, and the farmer or watercourse can be taken as the micro scale. Currently, the meso scale, or the subprovincial tiers of the government, are taken largely as information-gathering and execution arms for departments to make and implement provincial plans. Understanding integration of planning vertically within the PID and PAD requires understanding (1) how information for planning decisions is collected from the subprovincial tiers for making provincial plans, (2) what planning, management, and operational decisions are made at the different institutional tiers within the provincial Irrigation and Agriculture departments, and (3) how information and decisions flow up and down the vertical tiers and across horizontal linkages.

17.4.1.1 Irrigation

The flow of influence within irrigation is upward with the lowest tier officials identifying gaps and the need for rehabilitation within the irrigation infrastructure, thus acting as arbiters for water conflict resolution and working to ensure that allocated canal water requirements are met by influencing the PND, donors, and other reform agendas. PID officials remain insistent that problem identification, proposals for new projects, and other information all flows from the bottom up and contest that planning in the PID is a top-down affair. As explained by Syed Mahmood Ul Hassan, exchief engineer for the Punjab Irrigation Department and current project management advisor for an Asian Development Bank flood resilience project, "Everything comes from down to up, rarely is it going top-down." For the Annual Development Plan of the Irrigation Department, the chief engineers and superintending engineers identify problematic areas and projects that are required and address them based on input from their field formation, that is, the executive engineers (XENs) and subdivisional officers (SDOs). The projects identified by the PID are then pitched to the Planning and Development Board, and once approved these projects are incorporated into the Mid-Term Development Framework and the Annual Development Plans by the PND. Hassan admitted that the department does not have proactive planning: "Generally it's reactive planning and mostly focused on maintenance of existing irrigation infrastructure." Efforts at developing a proactive planning mechanism within the department have been ongoing since 2006 when the Strategic Planning and Reform Unit (SPRU) was set up. The SPRU conducted a detailed system requirements analysis for irrigation infrastructures and identified the replacement cost of irrigation assets as more than $20 billion and estimated the rehabilitation and upgrading requirements as more than $2 billion worth of investment. The donor organizations in this narrative act as the asset financiers necessary to raise funds to implement these mega barrage and canal system rehabilitation projects. However, this infrastructure-heavy approach meant that the micro tiers were left out of the conversation. As Hassan explained, "We were able to plan big projects but were not able to go down to root level and that's what needs to be done."

This follows the pattern identified by Bandaragoda in a 2006 analysis of water sector reforms in Pakistan: "Sporadic changes were introduced to the irrigation management organization, based on ad-hoc project-based requirements, making the management structure rather ineffective in a fast changing socio-economic context" (p. 56). Combined with the absence of a coherent planning vision, with the impetus for the reforms coming from donors instead of the farmers and primary system users, these reforms have lacked the necessary institutional support to achieve their objectives. Bandaragoda (2006) goes on to summarize this discrepancy: "…efforts to achieve stability through enhanced physical infrastructure and technological inputs were mostly subverted by poor institutional support, resulting in low agricultural yields, widespread irrigation misconduct, severe tail-end deprivation, low productivity of manpower and financial resources" (p. 56).

The establishment of AWBs at the canal level and farmer organizations (FOs) at the distributary levels at the behest of donor encouragement was meant to solve institutional failures of efficiently operating and maintaining the irrigation infrastructure by transferring management to the farmers' hands. It is worth noting that the AWBs and FOs were primarily envisioned as being involved in the operation of the canal system to ensure fairness in water distribution, resolve disputes, and collect revenue, but never as having a planning function. Not unexpectedly, officials of the PID showed skepticism in the idea of participatory irrigation management, since it was set up as a parallel system excluding the PID rather than as a way for developing strong linkages between the PID and the farmers. An official of the Strategic Planning and Reforms Unit of the PID, Dr. Muhammad Javed (director of SPRU), explained, "PIDA is managing five Area Water Boards where Irrigation Department has no responsibility and Farmer Organizations are managing the system. Being non-technical organizations, they failed to manage the system."

Between the macro provincial tier and the micro individual irrigator/farmer tier, the zone and circle levels of the Punjab Irrigation Department play a key operational and management role in the irrigation system. They manage the existing canal system, identify needed projects, and operate the canal system on a day-to-day basis to maintain the system as it exists at optimal efficiency. While planning documents cannot be developed without their contribution, planning outputs in the form of annual or mid-term development plans remain provincial in nature rather than being zone- or circle-specific. As Arif Nadeem, former Secretary to the Government of Punjab in the departments of Agriculture and Irrigation, explained:

> The executive engineer's (XEN) role is to operate the system as it exists at optimal efficiency. The Superintendent Engineer (SE) is the supervising authority of the XEN. He makes development decisions on canal structure changes, water course development, silting etc. The Chief Engineer (CE) is the morale booster or policy formulator. It is at the Chief or Superintendent Engineer level that planning should happen.

The CE or SE level roughly corresponds to the meso scale that fulfills the important function of providing a bridge between individual irrigators and the provincial government, managing problems, identifying needed developmental works, and supervising the operation of the canal system. This also makes this level of actors ideally placed to produce zone- and circle-specific plans.

17.4.1.2 Agriculture

The Punjab Agriculture Department (PAD) is distinct from the PID in that its subprovincial tiers are structured along administrative boundaries of district, tehsils, and union councils. Focusing on input services related to provisioning, such as extension services or on-farm water management (excluding price setting and services related to agricultural markets), the PAD begins its planning cycle every year once the Budget Call letter is issued by the finance department. All directorates then ask their subordinates for concept notes and proposed projects on water management, agriculture extension services, and so on. The office of the departmental head then "compiles, deliberates and prioritizes based on estimated availability of resources and developmental budget," as explained by a provincial official. Once the list of ongoing and proposed projects are compiled, the secretary of agriculture, in coordination with the PND and donor input and provincial leadership, determines which of the proposed projects will receive final approval. The PAD has published a detailed sectoral plan that sets departmental strategy for the future. For the PAD, according to departmental officials, two considerations help determine future priorities for the department: (1) import substitution and export enhancement and (2) productivity enhancement and food security. For example, both provincial and district officials reported that when a rapid increase in the import bill for pulses was noticed, the department decided to encourage their cultivation. The departmental sectoral policy has also prioritized increasing the export of fruit and vegetables, which will perform the dual function of increasing farmers' incomes and generating foreign currency through exports (Agriculture Department, 2015, pp. 30 and 36).

There is an effort by the PAD to involve stakeholders at all administrative levels, such as chambers of commerce, farmer groups, and so on, through consultations; however, no formal mechanism for incorporating their feedback into planning documents exists. Lacking direct levers to guide farmers' choices, the PAD has little control over what crops farmers choose to plant, except for tools like supporting prices to incentivize the growth of certain crops. Wheat is grown extensively in the province, and as the major grain in the area, it is used to generate income and for subsistence. Because it is difficult for small farmers to invest in more land and lucrative crops that require heavier investment, they are not able to respond to the PAD's efforts to steer them toward diversifying and growing alternate crops, such as horticultural crops. The PAD relies on market-based mechanisms or its Extension Directorate's extensive field services to incentivize farmers to change their behavior or adopt better technologies and farming practices for improved productivity. Biophysical and socioeconomic elements determine the crop options available to farmers, and the PAD sees its role as enabling farmers to make the most of the inputs and resources available through technology, sharing knowledge, and providing credit. Among other things, the department provides agricultural credit measures and disseminates technology such as laser leveling on a self-application basis. Farmers who are aware of these practices can apply for them in their local district and tehsil offices and then utilize the services. Once a minimal level of take-up is achieved, the PAD expects the private sector to step in and take the lead in providing such technologies. One example that achieved success involved spreading the use of agricultural machinery, and now the uptake of laser levelers is progressing similarly well.

Given the market-based levers available to the PAD for influencing farmers' choices and behavior, the strategic vision for agriculture in Punjab is determined at the provincial level and transferred downward. At the district level, where the agriculture officials historically reported to the district coordinating officer (DCO) as well as the provincial agriculture department, the Executive District Officer for agriculture's responsibilities are limited to supervising and approving field plans for extension staff and supporting the DCO in districtwide activities. The subprovincial tiers also play the very important role of collecting data on agriculture in the province. Data on annually cultivated areas, crop yields, and output are collected by revenue *patwaris* (crop reporting services) and communicated to the PAD and the provincial Bureau of Statistics (Bureau of Statistics, 2015, p. 29). As this information flows upward in the PAD, crop targets and implementation details of provincial projects, subsidies, and schemes flow downward. Crop targets are communicated to the district and division offices, with each target developed with a slight factor of growth on the previous year's output. Even though the PAD's field formations work closely with farmers, given the current institutional set up, the PAD limits its role to one of information and technology dissemination or troubleshooting problems as they are confronted (e.g., the cotton pest outbreak of 2015 was managed on an emergency basis by all tiers of the Agriculture Extension Directorate). As one divisional official for PAD explained, "In Pakistan, the PC-1 (project concept document) demand comes from the top. The secretary orders, experts are called overnight and figures are derived and planned. Political influence is over bearing [sic]. What is needed is for farmers to be included in any planning process (small, medium, or big)." Planning decisions therefore are almost exclusively taken at the provincial level.

With the latest move on the part of the provincial government to enact the local government system, the division tier situated between the district and province has been strengthened, such that district agriculture officials will report only to their respective provincial counterparts. Touted as a good thing by the PAD district officials, this change allows district officials and subprovincial tiers of the agriculture department to be better connected with their provincial department both in terms of departmental structure and flow of funding. However, this does not change the fact that district-level officials are one degree removed from the planning process and are thought of as responsible for implementing line functions only. District officials frequently cite the impetus for departmental plans to be ad hoc, which comes from demands of donor organizations or provincial tiers. It is not surprising then that one of the largest ongoing PAD projects is the World Bank-assisted Punjab Irrigated-Agriculture Productivity Improvement Project (PIPIP), which seeks to introduce water conservation technologies in irrigated agriculture across Punjab. The closest thing to a plan at the district tier remains the monthly extension plan that sets a work plan for the Agriculture Extension's field staff. Nothing resembling a formal resource management or cropping plan at the district or division level exists.

17.4.2 Multiscale and Cross-Sectoral Planning

The organizational analysis of the PID and PAD shows that one hurdle to multiscale planning is that planning, as manifested in documents like sectoral or annual development plans, is thought of as a provincial function by officials at all governmental tiers. However, even at the provincial level, coordination between the PID and PAD for integrated planning is poor. A detailed network mapping and analysis exercise conducted on irrigated agriculture in Pakistan found the Punjab Irrigation Department (PID) to be "the powerful owner of the water in Punjab" (Aberman et al., 2013, p. 16), even while in terms of centrality and influence, the PAD shows one more degree of influence than the PID (Aberman et al., 2013, p. 12). The network map for Punjab showed that all entities influencing the PID and PAD are the same except for one difference: while the PAD influences farmers, the PID influences the Indus River System Authority (IRSA—the national authority responsible for implementing the Water Apportionment Accord among the provinces). As explained by Chaudhry Ashraff, former director general of the On-Farm Water Management Program: "The separation between irrigation and agriculture is very clear. Irrigation has all canal infrastructure up till the water course. Agriculture has the water-course and the farm. There is no need for more coordination than that."

This shows a critical difference between the PID and the PAD: the PID is an upward-pointing entity, while the PAD is a downward-pointing one. The field formations of the

PID, without interfering at the farm or watercourse level of irrigation infrastructure, identify infrastructural gaps and problem areas, which are conveyed upward through the vertical chain of command and then consolidated at the provincial level and submitted to the PND. Based on available finances and with feedback from the PAD, the national government, donor organizations, and the IRSA, new projects are planned and approved and appropriately implemented by the PID. Armed with the latest information on irrigation works and policy decisions, as well as the seasonal water allocations provided by the IRSA, the PAD devises its plans and policies, which are communicated down the vertical chain of command to its field formations, which then use the tools available to the PAD to influence farmer behavior. Thus hierarchically, due to the limitation of the PID managing a supply-based infrastructure, the PAD becomes subservient to the limitations of the PID. Even while agriculture officials accept that the largest hurdle to improving yield in Punjab is reliable availability of water, they find themselves helpless in influencing this variable other than by encouraging the use of water conservation techniques at the farm level. In spite of the various reform efforts, due to the water-centric nature of irrigated agriculture in Punjab, agriculture will continue to find itself at a hierarchically lower plain in planning and management decisions. Mahmood Ul Hassan, former chief engineer at the PID explained this as the key difference between the PID and the PAD:

> We have a supply driven system designed for equitable distribution of available supply among cultivators. That is a constraint in optimal development of agriculture. In other countries, irrigation is demand based…farmers provide their demand on 10-daily basis for the amount of water they will need. In Pakistan, cultivators manage whatever water is being supplied.

Interviews with departmental officials thus reveal that many feel the PAD's responsibilities are nested within and hierarchically below those of the PID and that agricultural performance is fundamentally constrained by irrigation limitations. This is also a constraint in achieving integration between the PID and the PAD (Fig. 17.2).

The geographical misalignment between the departmental structures makes identifying relevant counterparts between the PID and the PAD at the subprovincial tiers difficult. Hassan explained that irrigation was not devolved in 2002 because of the physical discrepancies between irrigation and agriculture: "Agriculture does not have its own infrastructure. Compared to that irrigation has a huge infrastructure that cannot be integrated at district level." Commenting on possible integration with the PAD, he continued: "Relevance of agriculture department is in operations only as information disseminators. What more can be done with agriculture at this point? Unless if PAD decides to implement some sort of cropping pattern according to climate, that's when coordination with PID will have to happen." Interviews with the district staff of the PAD similarly revealed that planning is taken to be the responsibility of the province, and so is integration of planning goals between the two departments. At the subprovincial levels, the only mechanisms for operational coordination between the two departments exist in the form of advisor committee meetings that are primarily held to resolve issues of water theft and disputes in the presence of all stakeholders (farmers and representatives of the PID and the PAD) (Fig. 17.2). These meetings are meant to provide a forum for farmers to highlight problems and conflicts. However, as one

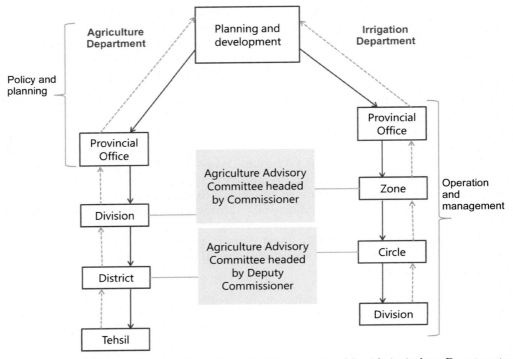

FIG. 17.2 Roles and linkages between Punjab Irrigation Department and Punjab Agriculture Department with *dotted arrows* showing flow of information and *solid arrows* showing flow of decisions. Note that PAD tiers are shown as hierarchically below those of PID to capture the greater power and influence held by PID.

senior PID official pointed out with regard to the advisory committee meetings, "Sure those meetings happen, but whether the irrigation representatives attend or not is another matter."

At the provincial level, there is an attempt to rethink institutional arrangements to develop more formal mechanisms of coordination between the different water users in the province. Dr. Javed (director of SPRU) explained:

> Presently no one department is managing the water resource. Even the Irrigation Department is doing water use management, not resource management. The Irrigation Department is going to be revamped to act as a water resources management department. A new legal entity will be designated where all the stakeholders will put up their water requirements. This new entity will do resource management and not only use management as is the business as usual.

With the establishment of a water resources management entity, a policy for groundwater regulation can also be developed. Dr. Javed elaborated further that the idea of a water resources management entity is crucial to bringing conjunctive water management within the jurisdiction of the government: "Currently no one owns ground water so it cannot be managed. The new policy under development gives government the right to manage groundwater but ownership will continue to belong to the land owners." A future groundwater management framework might then use tools like reallocation of surface water (instead of the current fixed time-based allocations) or regulation of groundwater in critical areas of Punjab for conjunctive water management.

One of the most recent and innovative development projects supported by the Asian Development Bank focuses on a multipronged overhaul of the irrigation system (Asian Development Bank, 2006, p. 5). This substantive reform effort has targeted the Lower Bari Doab Canal (LBDC) system in its first phase for physical rehabilitation as well as implementation of participatory irrigation management. This project seeks to encourage conjunctive water management, with an emphasis on improving agricultural productivity, from the farm level all the way up to the canal system level. Building on earlier participatory management models, it envisions farmer participation to be through a series of organizations based on the irrigation infrastructure, with the Khal Panchayat or Water User Association managing the watercourse level, the farmer organizations (FOs) managing the distributaries, and the irrigation management units (IMUs—overseeing several FOs) to serve as the technical interface between the FOs and the AWB (responsible for a canal command with the PID's Superintendent Engineer acting as the Board's Chief Executive). Unlike previous participatory management efforts that excluded the PID, this model seeks to involve the PID, the PAD, and the farmers in a multiscale, cross-sectoral effort (Fig. 17.3).

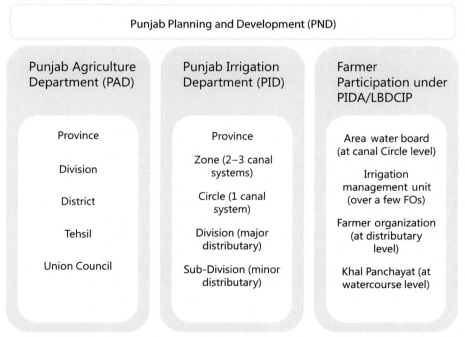

FIG. 17.3 A diagram depicting the revised participatory irrigation management institutional structure under the LBDC Improvement Project; which has introduced the irrigation management units to lessen the divide between PID, PAD, and farmers.

Introduction of the irrigation management unit particularly seems to be resonant of the canal councils proposed over two decades ago as a platform to bring all stakeholders to the same table. As the ADB project document explains:

> The IMUs will respond to needs of the *khal panchayats* and farmer organizations, and provide the following technical services: (i) O&M [Operations and Maintenance] and engineering guidance for minor and distributary canals; (ii) groundwater, conjunctive use, and OFWM [On Farm Water Management] strategies; and monitoring of these activities; (iii) agricultural services; and (iv) administration and financial management including billing and record keeping for *abiana* [charges]. *(Asian Development Bank, 2006, pp. 12–13)*

Perhaps what is more promising is that the project will ensure that farmer organizations are involved in the identification and proposal of civil works in consultation with the Project Management Unit, consult with the farmers in survey and design, and collect feedback from the farmers on their concerns with regard to the project (Asian Development Bank, 2006, p. 10). One component of the project design is dedicated to on-farm water management and agriculture through the creation of demonstration farms for each FO, capacity building on the latest technologies, conjunctive water management, and soil maintenance (Asian Development Bank, 2006, p. 11).

With the LBDC Improvement Project, the Punjab Irrigated Agriculture Investment Program presents an innovative strategy for bridging the gap between the PID, the PAD, and the farmers to empower all stakeholders to achieve integrated management. It provides a solid foundation and perhaps a pilot effort of sorts to start reimagining planning of irrigated agriculture in Punjab. The conversion of the program's components into permanent institutional mechanisms for the PID and the PAD and successful capacity building of all stakeholders will be a test of the project's success, as well as an opportunity and an early blueprint for developing an integrated system for planning of irrigated agriculture in Punjab.

17.5 REIMAGINING PLANNING OF IRRIGATED AGRICULTURE IN PUNJAB

17.5.1 Empower the Meso Scale to Create Local Plans

The institutional analysis reveals that not only is there a sharp geographical and jurisdictional divide between Punjab's Agriculture and Irrigation Departments, there is also a separation between the types of functions performed by the different institutional tiers. Planning is clearly identified and accepted by all institutional tiers to be a provincial function. Key structural issues like the fixed supply, the time-based water allocation system in irrigation, the lack of tools available to the Punjab Agriculture Department to notify cropping zones or influence farmer choices, and an absent groundwater policy hinder efforts at integrated planning. However, it must be noted that subprovincial tiers of government carry out important planning functions like data gathering and implementing planning decisions. The PID clearly considers the zone and canal level and chief engineers and superintendent engineers to be ideally located for identifying planning projects. In the PAD, even while planning remains largely a top-down affair, the provincial department inevitably relies on district-level administration and field formations to identify priority areas. For example, Chaudhry Abdul

Hameed, director of the Agriculture Extension for the Division of Faisalabad explained, "We realized that pulses in Faisalabad are very expensive and were being grown on little area. We told the (provincial) department that farmers did not get seeds so a new project was started to subsidize seeds and seed drills and teach farmers the method to use them." As a result, the amount of area where pulses are grown in Faisalabad grew substantially.

Another divisional official of the PAD explained that as a policy they ask farmers to use as much groundwater as they need for their crops. The decision by the department to not regulate groundwater abstraction for now is in itself a policy decision that has been taken keeping farmer welfare in mind, regardless of the variable levels of groundwater availability across the province. Such one-size-fits-all province-wide policies and decisions highlight the absence of a formal planning output at the meso scale. This is not to say the meso scale is absent from planning, rather that plans for the meso scale do not exist. In spite of the function of the meso scale in the Irrigation Department and the Agriculture Department as the connecting link for flow of information from the farmers to the provincial tiers and flow of policy and planning decisions from the provincial tiers down to the farmers, all *plans* remain provincial in nature.

Sectoral and annual development plans are key examples of planning outputs that address problems at a provincial level, unable to take local and contextual variabilities into account at required levels of granularity such as at the district or tehsil level. Decisions to guide departmental actions, for example, technology dissemination or infrastructural civil works, are communicated downstream for implementation, but since no explicitly laid out plans for the meso scale exist, strategies and tools available for achieving these targets remain unclear. For example, even while cropping targets are disseminated downstream from the provincial tier to the district and tehsil levels within the Agriculture Extension Directorate, the only tool available to extension workers to achieve these goals is through sharing Extension knowledge with farmers. Equivalent functionaries at the PID remain unaware of these goals, focusing instead on civil work improvements or operation and maintenance of the irrigation infrastructure. This is particularly inadequate given the variation in yield and water used by farmers within and across different canal commands and districts (Fig. 17.4).

Without formal district, division, zonal, or canal plans, coordination between different departments and farmers to solve important problems becomes infeasible. However, as envisioned by the LBDC Improvement Project, the meso scale can be empowered to adopt a cross-sectoral approach to managing irrigated agriculture for canal command scales. Once the meso scale is imagined to have a planning function specific to areas within each administrative tier's jurisdiction, similar to the one carried out by the PND at the provincial level, the number of tools and strategies available to the PID and the PAD at the meso tier can expand. The planning outputs in this case can then be zone/circle or division/district plans, generated and informed by active participation of the PID and PAD staff as well as the farmers.

17.5.2 Boundary Spanning for Integrated Planning

Given the limited levers and tools available to the PID and PAD in improving agricultural productivity and solving problems of irrigation inequity and groundwater abstraction, departmental plans at the meso scale will be insufficient until they are multisectoral and

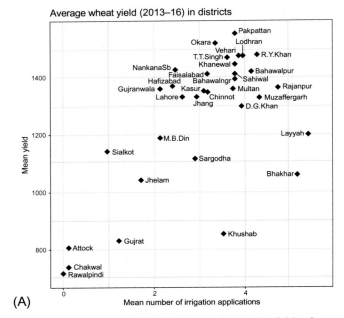

(A)

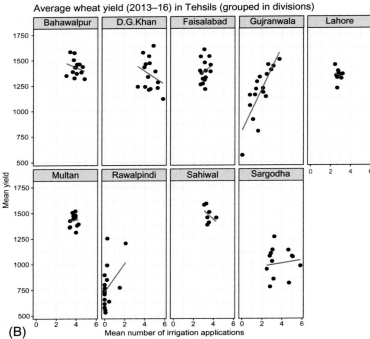

(B)

FIG. 17.4 Plots of average yield (kg/acre) versus average number of irrigation application in Rabi season for wheat (2013–16) by district (A) and tehsil (B). Both plots show that substantial variation in yields and the number of irrigation applications within and across districts and divisions exists. *Data from Crop Reporting Services, Punjab Agriculture Department.*

integrated in nature. Planning mechanisms at the meso scale therefore need to be boundary-spanning in nature, much like the role the PND plays at the provincial level in bridging gaps between disparate departments. Boundary roles are key for creating links between different stakeholders by information processing and external representation: "Information from external sources come into an organization through boundary roles, and boundary roles link organizational structure to environmental elements, whether by buffering, moderating or influencing the environment. Any given role can serve either or both functions" (Aldrich and Herker, 1977, p. 218). Boundary roles and organizations act as "mediators" between outside organizations (e.g., different departments) or groups (e.g., farmer groups) (Aldrich and Herker, 1977, p. 220).

According to the Director General of Agriculture, Extension & Adaptive Research, Punjab, Dr. Muhammad Anjum Ali, one of the most important needs for boundary spanning problems in integration between the PID and PAD is bridging the technical capacity gap:

> Water is a vital input for agriculture but PID is being managed by civil engineers who manage water supplies and take care of infrastructure, not irrigation. Whereas, PAD has a fleet of agriculture engineers and agronomists working on correct use of water at farms and physiological needs of water at critical stages of the crop. Strengthening of linkages through joint functions and programs will be beneficial to economize the use of water in all its forms, *i.e* surface, sub-surface and rains from the source to the point of consumption.

Canal councils and irrigation management units can address only part of the problem since they are focused on operations and management, and are not going so far as to proactively develop plans. Unsurprisingly then, the problem of an adequate water supply, as identified by agriculture officials to be the biggest hurdle to improving yield, cannot be easily solved by simply improving the operation or management of existing institutions and infrastructures. A district official for PAD explained: "Without new reservoirs, canal lining cannot achieve much. There is no solution to theft. Tail end is not getting any water and advisory councils are unwilling to take action." He pointed to challenges in coordination both vertically and horizontally in current planning processes: "There is a lack of ownership of irrigation infrastructure by stakeholders because no feedback is asked from them. I am unaware of all projects running in the irrigation department so what will a common farmer know of what projects are going on?" The PAD provides subsidies on inputs, technical knowledge, and micro-credit to farmers to support agriculture, however, he explained: "There is a top to bottom sort of implementation. Allocation of subsidies is on the basis of crop acreage, we are not given or asked for any feedback." The role of the meso scale tiers in the PID and PAD therefore remains limited to implementing provincial schemes and projects, leaving cross-sectoral coordination to the province.

District and division officers as well as the circle and zonal officials are uniquely positioned as the eyes and ears of the provincial departments while simultaneously acting as government-farmer interfaces. In this unique dual role, the meso scale can be reimagined as a boundary spanning scale that can (1) play a crucial role in meso-scale planning catering to the particular socioecological variations within that particular division or zone, while (2) simultaneously including farmer voices at the design and planning stage for bottom-up ownership, and (3) connecting them to the macro or provincial tier for information dissemination. Considering this tier as the "boundary spanning" scale thus will allow planning to be reconfigured in the institutional setup as a multilevel process happening simultaneously at the provincial and district/division or circle/zone levels. Integrated planning bodies, with stakeholders from the PID, the PAD, and local farmer groups working at the meso scale will enable thinking of planning as a dynamic cross-sectoral process, empowering all stakeholders to use multiple tools and levers to improve water productivity, achieve higher yields, and practice sustainable use of resources. Plans for irrigated agriculture in Punjab can then be developed and tailored to specific needs and priorities of each division, district, or zone and circle, using multipronged strategies that combine infrastructural improvements with strategies like smart crop selection, technology use, subsidies, and micro-credit schemes. Once sufficiently developed, such boundary spanning organizations might even take the shape of targeted structural reform programs such as land consolidation and collective organization of farmers to encourage a shift toward higher profit agricultural products like fruits and vegetables. Unlike the provincial scale, that by nature of the scale of its operation is limited to providing broad policies and strategies, the boundary spanning meso-scale planning organization will provide opportunities to operationalize multidimensional plans that use cross-sectoral levers for holistic problem solving (Fig. 17.5).

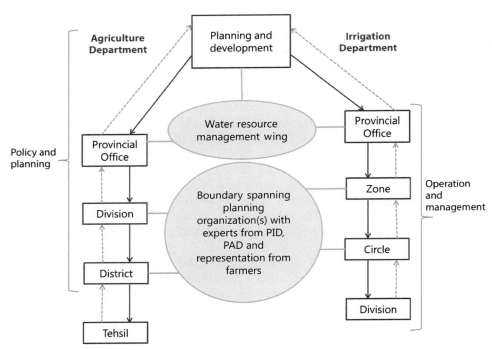

FIG. 17.5   This figure depicts proposed boundary spanning organization(s) for integrated planning through cross-sectoral boundary spanning structures and tools.

17.5.3 Boundary Spanning Objects for Integrated Planning

As boundary functions can be institutionalized in boundary organizations, "boundary objects" based on integrated analyses can be developed as unifiers for irrigation and agriculture experts to sit at the same table and align their strategies to achieve departmental goals. Boundary objects are defined as "...collaborative effort outputs that 'are both adaptable to different viewpoints and robust enough to maintain identity across them'" (Cash et al., 2003, p. 8089). One such object in enabling boundary spanning functions can be integrated analysis developed by the PAD and the PID in collaboration bringing "multiple types of expertise to the table" and "[enhancing] legitimacy by providing multiple stakeholders with more, and more transparent, access to the information production process" (Cash et al., 2003, p. 8089). This analysis can provide the foundation for development of cross-sectoral plans (also boundary objects) at the meso scale, with the involvement of farmer organizations for unified understanding and ownership of the agreed upon strategies, goals, and priorities among all stakeholders.

Efforts in recent years to improve data collection by the Programme Monitoring and Implementation Unit (PMIU) in the PID and the Crop Reporting Services in the PAD have empowered the departments to carry out more accurate and granular analysis of departmental performance in irrigation and agriculture. The biggest challenge in integrated irrigation and agriculture analysis at the meso scale is the spatial and geographical misalignment between the boundaries of the two departments. However, a number of methodologies have been devised to map agriculture output measured on district boundaries to canal command areas so that water discharges can be included in agricultural output and yield analyses (Tahir and Habib, 2001 and Kirby and Ahmad, 2016). Since 2012 the Crop Reporting Services (CRS) department of the Provincial Agriculture Department has also formalized crop yield and production estimation methods with detailed data on inputs from a district-level representative sample of over 1200 villages from all over Punjab. Since this data set includes information on the amount of water used by farmers and expenditure on tubewell usage, it allows for estimating the relationship between agricultural output, water availability, and energy usage at each of the departmental tiers of management. Useful measures such as agricultural productivity in different canal commands or districts for different crops, water productivity of surface water versus groundwater, availability of groundwater versus surface water (while controlling for other measures like soil type, fertilizer and pesticide use, technology use, etc.) can be analyzed using regularly collected departmental data (see Fig. 17.6 for an example).

Measures like water productivity per unit expenditure or land are particularly useful in allowing both the Irrigation Department and the Agriculture Department to identify common leaders and laggards. These can be defined in terms of agricultural productivity or water and energy productivity across the province, but also within districts, divisions, and canal command areas, allowing meso-scale plans to prioritize irrigation infrastructure development, groundwater replenishment, and agricultural input support as per the needs of different areas. More developed versions of these plans could include larger reform efforts like development of new water reservoirs, notification of cropping zones, land consolidation with farmer input, and conjunctive water management with ownership of farmers as well as the PAD and PID. Expanding planning power downward is especially useful to allow for more responsive management and planning for the variable socioeconomic and biophysical

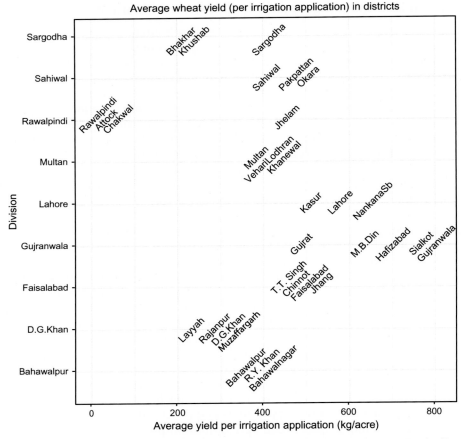

FIG. 17.6 Metric of water productivity, such as wheat yield per irrigation application (2013–16), allows leaders and laggards within a division to be identified for development of district-level integrated plans. Similar analyses can also be done at the tehsil level. *Data from Crop Reporting Services, Punjab Agriculture Department.*

conditions. In a recent paper on modeling the agricultural food-water-energy nexus in the Indus River Basin in Pakistan, Yang et al. (2016) use the revised Multi-Year Indus Model Basin (developed by the World Bank) to evaluate the impacts of alternate water allocation mechanisms to the current system. They find that a more flexible interprovincial and intraprovincial surface water allocation policy will increase surface water use in the basin, while groundwater and energy use will be lower (Yang et al., 2016).

The challenge of planning to manage the food-water-energy nexus can be partly addressed by combining disparate indicators in one analysis, allowing the trade-offs between using different agricultural or irrigation strategies to be considered simultaneously. Plans developed using boundary objects (such as jointly developed models, analyses, and reports) by boundary spanning organizations will empower planners and policymakers in the PID and the PAD as well as mid-level officials to focus on solving farmers' problems specific to their areas. Most importantly, a boundary spanning approach to planning that enables plans to be created for multiple tiers of the province simultaneously, will (1) allow all stakeholders (PID, PAD, and farmers) to have an equal voice in setting priorities and strategies for problem solving and thus achieve a greater sense of ownership and (2) allow all stakeholders to participate in processes that establish clear feedback loops and enable two-way information sharing and reiterative problem solving.

17.6 CONCLUSIONS

In an iconic lecture summarizing the key lessons of how complex economic systems may be governed using diverse institutional arrangements, Elinor Ostrom highlighted that no standard principles of good institutional design exist. In systems such as that of irrigated agriculture in Punjab, there are many different centers of decision making that function with varying degrees of independence and interdependence. Ostrom highlights: "Building trust in one another and developing institutional rules that are well matched to the ecological systems being used are of central importance for solving social dilemmas" (Ostrom, 2010, p. 435). Both within the Agriculture Department and the Irrigation Department and in the halls of provincial and federal planning and economic development departments, a need for reforming current planning and decision-making processes is recognized. With the first drafts of the national and provincial water policies nearing official approval, the Government of Punjab is on the path to reform irrigated agriculture in the province by prioritizing IWRM and nexus thinking. The establishment of the Water Resources Management wing, enactment of a groundwater policy, and setting up of a Decision Support System (while all three are in initial stages) will equip the PID and the PAD with detailed information systems to map on-the-ground conditions, monitor the operations of the irrigation system, and support decision making. Province-wide initiatives such as the soil survey and mapping of groundwater levels and quality as well as revamping of participatory management efforts (as in the LBDC Improvement Project) are also underway.

Given the ongoing institutional reform effort, there is an opportunity for the Government of Punjab to be even more ambitious in its reform agenda. The government should elevate the meso scale and reimagine the role of the agriculture department vis-à-vis irrigation and the farmers to truly achieve integrated planning in the province. The historical and institutional analysis in this chapter shows that emphasizing the subprovincial, and especially the meso scale of governance, is key to achieving vertical and horizontal integration. As evident from conversations with department officials, farmers' capacities, the PID's financial constraints and the PAD's institutional limitations are all considered hurdles to improving irrigated agriculture. A boundary spanning approach that is able to bring all stakeholders to the same table with different capabilities and constraints will allow technical and capacity gaps to be filled and potentially innovative strategies and solutions to emerge. With a push toward geographic information systems (GIS) and data-driven tools for decision support, there is increasing capacity within the PID and the PAD to develop technologies and processes that incentivize integrated planning. ICT tools are especially useful for making information sharing easier among disconnected stakeholders and GIS and Remote Sensing can be used to fill data and information gaps where they exist. Once boundary spanning tools and organizations develop, greater reforms may be achieved by providing these organizations with resources for flexible water allocations, credit and subsidy distributions, and development of farmer cooperatives. If these steps are successfully implemented, each garden in the Indus River Basin of Punjab may achieve its highest potential and develop into a bountiful, beautiful, and sustainable garden of gardens.

References

Aberman, N.L., Wielgosz, B., Zaidi, F., Ringler, C., Akram, A.A., Bell, A., Issermann, M., 2013. The Policy Landscape of Agricultural Water Management in Pakistan. (IFPRI—Discussion Papers).

Agriculture Census Organization, 2010. Agricultural Census 2010: Pakistan Report.

Agriculture Department, 2015. Punjab Agriculture Sectoral Plan 2015.

Ahmad, M., Brooke, A., Kutcher, G., 1990. Water Sector Investment Planning Study.

Aldrich, H., Herker, D., 1977. Boundary spanning roles and organization structure. Acad. Manag. Rev. 2, 217. https://dx.doi.org/10.2307/257905.

Asian Development Bank, 2006. Punjab Irrigated Agriculture Investment Program.

Bandaragoda, D.J., 2006. Limits to donor-driven water sector reforms: insight and evidence from Pakistan and Sri Lanka. Water Policy 8, 51–67.

Bennmessaoud, R., Basim, U., Cholst, A., Lopez-Calix, J.R., 2013. Pakistan—The Transformative Path (English). World Bank, Washington, DC. http://documents.worldbank.org/curated/en/579401468286506572/Pakistan-The-transformative-path.

Benson, D., Gain, A., Rouillard, J., 2015. Water governance in a comparative perspective: from IWRM to a 'nexus' approach? Water Alternat. 8 (1), 756–773.

Briscoe, J., Qamar, U., 2005. Pakistan's Water Economy: Running Dry. World Bank, Washington, DC. http://documents.worldbank.org/curated/en/989891468059352743/Pakistans-water-economy-running-dry.

Bureau of Statistics, 2015. Punjab Development Statistics 2015.

Cash, D.W., Clark, W.C., Alcock, F., Dickson, N.M., Eckley, N., Guston, D.H., Jäger, J., Mitchell, R.B., 2003. Knowledge systems for sustainable development. Proc. Natl. Acad. Sci. 100, 8086–8091.

Gill, M.A., Mushtaq, K., 1998. Case Study of Agriculture Department.

Gilmartin, D., 1994. Scientific empire and imperial science: colonialism and irrigation technology in the Indus Basin. J. Asian Stud. 53, 1127. https://dx.doi.org/10.2307/2059236.

Gilmartin, D., 2015. Blood and Water: The Indus River Basin in Modern History. University of California Press, Oakland, CA.

Giordano, M., Shah, T., 2014. From IWRM back to integrated water resources management. Int. J. Water Resour. Dev. 30, 364–376. https://dx.doi.org/10.1080/07900627.2013.851521.

Government of the Punjab, 2016. Punjab Water Policy (Draft).

Kirby, M., Ahmad, M., 2016. Time series (1980–2012) crop areas in the districts and canal commands of Pakistan. (No. csiro:EP163815)CSIRO.

Ostrom, E., 2010. Beyond markets and states: polycentric governance of complex economic systems. Transnatl. Corp. Rev. 2, 1–12.

Planning Commission, 2014. Pakistan 2025: One Nation—One Vision, Planning Commission.

Qureshi, A.S., McCornick, P.G., Sarwar, A., Sharma, B.R., 2010. Challenges and prospects of sustainable groundwater management in the Indus Basin, Pakistan. Water Resour. Manag. 24, 1551–1569. https://dx.doi.org/10.1007/s11269-009-9513-3.

Scott, C., Kurian, M., Wescoat Jr., J., 2015. The water-energy-food nexus: enhancing adaptive capacity to complex global challenges. In: Kurian, M., Ardakanian, R. (Eds.), Governing the Nexus: Water, Soil and Waste Resources Considering Global Change. Springer International Publishing, Dresden, pp. 15–38.

Siddiqi, A., Wescoat Jr., J.L., 2013. Energy use in large-scale irrigated agriculture in the Punjab province of Pakistan. Water Int. 38, 571–586. https://dx.doi.org/10.1080/02508060.2013.828671.

Tahir, Z., Habib, Z., 2001. Land and Water Productivity: Trends Across Punjab canal Commands, Working Paper. International Water Management Institute, Lahore.

Ul-Haq, I., 1998. Case Study of the Punjab Irrigation Department.

Wescoat Jr., J., 2012. The Indus River Basin as a garden. Gartenkunst 24, 33–41.

World Bank, 1996. Pakistan - On-Farm and Command Water Management and Irrigation Systems Rehabilitation Projects. World Development Sources, WDS 1996. World Bank, Washington, DC.

Yang, Y.C.E., Ringler, C., Brown, C., Mondal, M.A.H., 2016. Modeling the agricultural water–energy–food nexus in the Indus River Basin, Pakistan. J. Water Resour. Plan. Manag. 1424016062. https://dx.doi.org/10.1061/(ASCE)WR.1943-5452.0000710.

Further Reading

Agriculture Department, 2011. Punjab Irrigated-Agriculture Productivity Improvement Project. Directorate General Agriculture (Water Management), Lahore.

Developing Groundwater Hotspots: An Emerging Challenge for Integrated Water Resources Management in the Indus Basin

Muhammad Basharat

International Waterlogging and Salinity Research Institute (IWASRI), Pakistan Water and Power Development Authority (WAPDA), Lahore, Pakistan

18.1 INTEGRATED WATER RESOURCES MANAGEMENT

Challenges most countries face in their struggle for economic and social development are increasingly related to water. Groundwater availability, water shortages, quality deterioration, and flood impacts are among the problems that require greater attention and action. Integrated Water Resources Management (IWRM) is a process that can help countries deal with water issues in a cost-effective and sustainable way. The concept of IWRM—in contrast to traditional, fragmented water resources management, is fundamentally concerned with the management of water demand as well as with its supply. Thus for real IWRM to occur in the Indus River System, first and foremost, water availability and water demand in both the spatial and temporal context must be linked up. However, achieving these objectives requires contemporary technology combined with capacity building and human resources development.

As in many other developing communities in South Asia, agriculture in the Indus River Basin (IRB) depends on groundwater irrigation to sustain the current level of crop production. Because canal irrigation systems do not provide farmers in Pakistan with adequate water or enough control over irrigation deliveries, the majority of the farmers have turned to groundwater as a sole or supplemental source of irrigation. However, groundwater availability and use are not considered in designing the canal distribution network or water allocation needs throughout the canal distribution system. The sale and purchase of groundwater through informal water markets offer other farmers, particularly nonowners of private tubewells, the opportunity to use groundwater. "The factors affecting private tubewell development and the emergence of groundwater markets are complex and interlinked" (PIES, 2001) and include physical, economic, and social factors. The increased use of private tubewells has increased the total water availability for crop production and also has provided on-demand control over irrigation supplies at the farm level. Pakistan, an agrarian country where irrigation is used on 75% of the agricultural land, mainly in the Indus Basin, uses all of these management practices. Unfortunately, this increasingly important supplement to canal water is threatened by overdevelopment and quality deterioration in many of the irrigated areas of the Indus Basin. The Punjab province in particular is facing unprecedented groundwater depletion rates (NESPAK/SGI, 1991; PPSGDP, 2000; Basharat, 2012; Basharat and Tariq, 2013). Shah (2006) argues that "sustaining the massive welfare gains groundwater development has created without ruining the resource is a key water challenge facing the world today."

18.2 GROUNDWATER DEPLETION AND EMERGING HOTSPOT AREAS

The cropping intensity ratio was 102.8%, 110.5%, and 121.7% in 1960, 1972, and 1980, respectively (Ahmad, 1995). Now it is operating at about 172% (Mirza and Latif, 2012) and is even higher in certain areas. Thus due to higher abstraction rates than corresponding recharge rates, groundwater mining is reported with increasing frequency in the literature (NESPAK/SGI, 1991; Steenbergen and Olienmans, 1997; Basharat and Tariq, 2013; Cheema et al., 2014). With a dramatic increase in the intensity of groundwater exploitation in the past three decades, the policy landscape for Pakistan has changed. "The main policy issues now relate to environmental sustainability and welfare" (Steenbergen and Olienmans, 1997). These issues include declining groundwater tables, deteriorating groundwater quality in fresh groundwater areas, and inequitable access to this increasingly important natural resource.

18.2.1 Urban Areas

Urbanization, a shift of the population from a rural to an urban society, is an outcome of social, economic, and political developments that lead to immense pressure on the natural and built environment. Pakistan is already an arid and water-scare country; therefore protecting and sustaining aquifers is particularly critical in its increasingly populated urban areas.

The shift to urban growth in mega cities such as Lahore continues. According to the 1998 population census, there were 6,318,745 persons in the Lahore district, and the population density was 3565.9 persons per km^2. Within the district the urban population was 82.44%, and the rural population was 17.56%. In 1998 the average literacy rate was 64.7%. It was 69.1% in the urban population and 41.7% in the rural population. During 2007–08 physical access to drinking water within a dwelling was available to 98% of residents, and improved sources for drinking water sources were available to 99% of the residents (Government of Punjab, 2011). The estimated population of the Lahore district (including rural population) was 8.83 million as of December 31, 2011 (Government of Punjab, 2011), with 84% of the population residing in the metropolitan city area (Government of Pakistan, 2011). Thus population expansion, particularly in metropolitan areas, seems to be the biggest driver of increasing groundwater use. A recent hydrograph showing groundwater behavior in Lahore is shown in Fig. 18.1 (Basharat et al., 2015).

Similarly, Quetta is the biggest population center in the Baluchistan province. The city's population depends entirely on groundwater storage in the aquifer. The city's population increased from 0.26 million in 1975 to 1.452 million in 2014. The increasing population of the city and the unplanned use of groundwater has contributed to a depleted water table in recent decades. The Water and Power Development Authority (WADPA) noticed a 0.25 m/year overdraft in 1989 and as a result increased observation of the network of wells.

Lahore and Quetta are among the emerging hotspots in small cities. Basharat et al. (2015) predicted that in 2050 the groundwater requirement for the total population (81,249,000) in the 25 major cities of Pakistan will be 8.672 bcm (billion cubic meters).

18.2.2 Agricultural Areas

The irrigation system was initially designed to bring as much land under canal command as possible and to provide settlement opportunities. The designed annual cropping intensities were generally kept low, at 60%–80% (Jurriens and Mollinga, 1996). According to the latest agro-economic farm survey carried out in 2010–11, encompassing 200 watercourses spread all over the Indus Basin Irrigation System (IBIS), the cropping intensities on average increased from 129% in 1988 to 172% in 2011 (Mirza and Latif, 2012). Groundwater use has continuously accelerated since the 1960s due to increased intensity of cropping. For example, the Lower Bari Doab Canal (LBDC) irrigation system was designed originally for a relatively low cropping intensity of about 67%. The peak crop water demand is about 8 mm/day at the farm head, which the Pakistan Meteorological Department for Multan estimated by using 30-year average reference evapotranspiration (ETo) (Pakistan Meteorological Department (PMD), 2006), indicating a net peak flow requirement of about 1.0 lps/ha. The canal flow of 0.23 lps/ha (3.3 cusecs/1000 acres) at watercourse head, for the current cropping intensity of about 160%, cannot meet peak demand, thus leading to the use of tubewell irrigation.

As is the case all over Punjab, there has been an exponential growth of tubewells within the past four decades in the LBDC command. The reported number of tubewells in the canal command from 1994 to 1995 was about 20,000. It rose in 2005 to 48,102 (NESPAK, 2005). This phenomenal increase in the number of tubewells (installed through farmers' investments) has also been the prime driver of increasing cropping intensity. To supplement canal water supplies, over the past 40 years farmers in the LBDC have transitioned to a heavy reliance on groundwater. An underground reservoir that was recharged by a newly built irrigation system with low cropping intensities is now being overexploited due to the increased intensity of cropping, and as a result groundwater levels are getting deeper. All this change can be attributed to overpopulation.

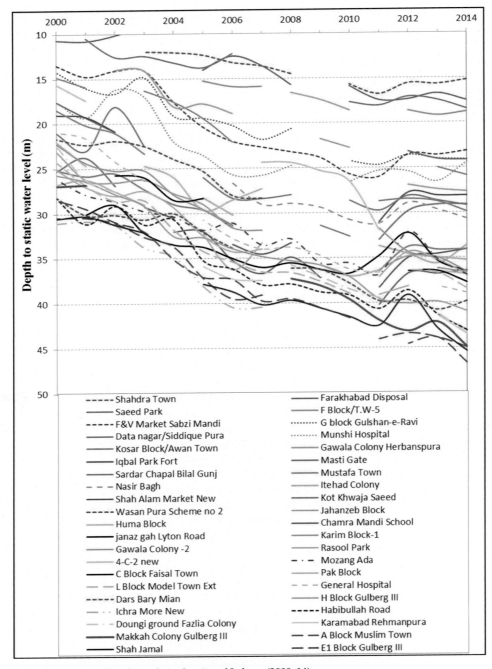

FIG. 18.1   Groundwater level trends in the city of Lahore (2000–14).

18.3 WATERLOGGING AND SALINITY—STILL A CONSISTENT THREAT

Currently, almost 43% of the area in the IBIS is classified as waterlogged with a depth to water table of <3m. The province of Sindh has the largest percentage of the IBIS's area (81%) that is classified as waterlogged. In the last few decades, while the waterlogged area has increased in the province of Sindh, it has decreased considerably in the province of Punjab, mainly due to the extraction of large amounts of groundwater from both public and private tubewells (NESPAK, 2005).

Three basin-wide soil salinity surveys were conducted during the periods 1953–54, 1977–79, and 2001–03. According to the first survey, conducted by the Colombo Plan Assistance in 1953–54, 44% of the area surveyed was affected by excess surface salinity. The second survey, conducted by the WAPDA in 1977–79, indicated that only 28% of the total surveyed area was affected by surface salinity. This reduction in the area of surface salinity was due to the installation of the public Salinity Control and Reclamation Project (SCARP) and private tubewells in the fresh and marginal groundwater zones. The latest salinity surveys, conducted during 2001–03 by the Soil and Reclamation Directorate of the WAPDA, showed that 27% of the area is affected by surface salinity, which is almost the same as in the survey of 1977–79 (WAPDA, 2006).

18.4 INTEGRATION NEEDS: SURFACE WATER-GROUNDWATER NEXUS IN PAKISTAN

Pakistan has an established system of water sharing and water rights for agricultural use in the Indus Basin Irrigation System (IBIS) that have remained essentially unchanged since the country's birth in 1947. This established system has some positive and some negative aspects. For example, over the past few decades, groundwater has been the single most important resource enabling increased agricultural production, particularly in the Upper Indus Plain. However, over time the sustainability of agriculture, especially in the Punjab province, has become linked to the sustainability of groundwater resources. Unfortunately, at some locations unprecedented mining of groundwater has caused waterlogging, surface salinity, and groundwater depletion. Because the provisions for increasing the water supplied to the canals are limited, continuous improvement in the performance of irrigated agriculture must be approached by revisiting policies on management and allocation of both the surface water and the groundwater.

Knowledge of the requirements for crop water use is necessary for planning and operating an irrigation system as large as the IBIS. In the agriculture sector, with only meager options for expansion in land and water resources in the country, there is a need to focus on increasing the efficiency of the existing land and water resources based on a rational allocation of scarce resources. Thus within the context of the changing water availability, groundwater regimes, and river flow conditions, this chapter examines how following the "business as usual" model for sharing and using irrigation is creating a huge strain on the groundwater conditions in some areas in Punjab.

18.4.1 Spatial Variation in Water Demand and Supply

The Indus Basin Irrigation System is more than a century old. Water allowances and canal water distributions responded to increasing crop water requirements in a southward direction, for example, higher water allowance in Sindh as compared to Punjab. But within a province the canal water supplies do not address the issue of differences in irrigation demands. The consequence is unprecedented groundwater depletion in Bari Doab and waterlogging in certain other canal commands. After the Indus Waters Treaty of 1960, gradually reduced flows and the ultimate desiccation of the eastern rivers have also contributed to falling groundwater levels of adjoining aquifers. In an earlier study Basharat et al. (2014) compared water allocations in the Water Apportionment Accord of 1991, the annual average canal water diversions, and the irrigation demands for canal commands in Punjab. Rainfall, the ultimate source of water, has a beneficial impact on both canal water and groundwater. The study concluded that the efficiency of existing irrigation systems can be improved by adopting the concept of IWRM. For example, to avoid waterlogging and groundwater depletion, reallocation of canal water supplies among the irrigation units in Punjab, in proportion to the relative irrigation water demand and cropping intensities, is recommended (Basharat et al., 2014). A similar study is recommended for Sindh canal commands.

18.4.2 Spatial Variation in Rainfall

The climate in Pakistan is more arid in the northeast to southwest direction, the orientation of the irrigation system in the IBIS. Although temperature, solar radiation, and wind all affect evapotranspiration demand, rainfall is also an important determinant of irrigation demand. According to the rainfall contours developed by the Pakistan Meteorological Department, the mean annual precipitation ranges from about 100 mm in parts of the Lower Indus Plain to over 1000 mm near the foothills in the Upper Indus Plain. On the other hand, lake evaporation increases in the north-south direction, from 1270 mm at Peshawar to 2800 mm at Thatta (Ahmad, 1982). Computations by Ullah (2001) indicate that at Sialkot (32.52°N, 74.53°E), the extreme northeast of the Indus Plain has the lowest value of reference evapotranspiration (1210 mm/year). In contrast, Jacobabad (28.28°N, 68.46°E) has the highest value of reference evapotranspiration (2112 mm/year).

The relative contribution of rainfall in most of the canal commands is low compared with the other two sources of irrigation water, that is, canal water and groundwater. However, the canal commands lying in the upper parts of Punjab—for example, the upper parts of Chaj, Rechna, and Bari Doabs (land between two rivers)—do receive enough rainfall, mostly in July to September, to specifically contribute to crop water requirements. The excess goes to groundwater recharge in varying extents, depending on the intensity of the rainfall and its distribution in time and space.

18.5 REALLOCATION—A CHALLENGE IN ITSELF, AND THE ONLY OPTION

Despite a combination of low overall production, a sustainability threat, and a variety of inequalities in irrigated agriculture, water reallocation and management in the Indus Basin is possible without large physical interventions (Habib, 2004). Reallocation of surface water supplies from fresh groundwater areas with relatively low surface irrigation duties to fresh groundwater areas with high irrigation duties and saline groundwater areas would help cut groundwater mining in the latter areas and waterlogging in the former areas (Ahmad and Kutcher, 1992). Each province could optimize authorized water allocations of various canal commands and reallocate water allowance based on evapotranspiration, cropping pattern, and cropping intensity with the goal of long-term sustainability. Optimizing the allocation of water will dramatically increase the economic productivity of water in both deficit and excess canal commands. Implementation of defined entitlements from the national level to the farm level and eventually implementation of defined entitlements to cover both surface water and groundwater need to be considered. Such a management system will help in reallocation of high value use and in the emergence of voluntary and consensual approaches, making reallocation both politically attractive and practical (World Bank, 2005). The present pattern of unsustainable water allocation, extracting water from nature for agricultural uses, must be reversed (Molle et al., 2010), and definition and application of environmental flow provide a good starting point. This calls for the redesign and effective implementation of the water accord (1991) allocations within the framework of the present Indus River System Authority, as noted in the article "Reversing the Indus Basin Closure," by Saifal (2011).

18.5.1 System-Scale Smart Surface Water Reallocation

In >70% of the irrigated areas of Pakistan, groundwater is providing an on-demand irrigation water source and is the single most important factor in sustaining the increased cropping intensities of the past few decades. But this supply of irrigation water is at stake due to an inequitable and insufficient supply of canal water and the resulting overuse of groundwater, mostly in tail end areas of the canal commands. To address this problem many national and international experts have advocated integrated and conjunctive management of surface and groundwater. In the past groundwater use was monitored in both cities and agricultural areas, but unfortunately this is no longer the case, particularly in Sindh and Khyber Pakhtunkhwa (KP).

18.6 RECOMMENDED INNOVATIONS

Surface and groundwater monitoring activities in cities and agricultural areas should be continued with vigor. Groundwater monitoring activities should be continued in at least four irrigated provinces of Pakistan.

18.6.1 Groundwater Governance

18.6.1.1 Urban Sector: Volume-Based Water Charging

Delivery of urban water should be monitored everywhere. Instead of selling it at a flat rate, water should be charged based on volume.

18.6.1.2 Industrial Sector

Withdrawal of water by industries should be monitored. The industrial sector is polluting urban wastewater with salts. To avoid excess salinity, water should be treated before being discharged into urban surface drains.

18.6.1.3 Agricultural Sector

The level of groundwater used for agricultural water is declining in many areas, and therefore farmers will have to pump deeper for water.

18.6.1.4 Managing Cropping Intensities and Patterns

New cropping zones need to be developed, and crops and cropping intensity standards need to be developed for these new zones. Government departments should create criteria for growing crops and should strictly monitor these criteria for compliance.

18.7 CONCLUSIONS AND RECOMMENDATIONS

- The urban sector needs to use volume-based billing rather than flat rate billing.
- The industrial sector should either self-treat water or be charged for the cost of minimum treatment.
- New cropping zones need to be developed, and crops and cropping intensities should be developed for these new zones.
- Reallocation of water from the agricultural sector to the urban sector should be done from high water table areas.
- Reallocation of water in the agricultural sector should be done from high water table areas to low water table areas, for example, Bari Doab.

References

Ahmad, N., 1982. An Estimate of Water Loss by Evaporation in Pakistan. Irrigation Drainage and Flood Control Research Council.

Ahmad, N., 1995. Groundwater Resources of Pakistan (Revised). Lahore.

Ahmad, M., Kutcher, G., 1992. Irrigation Planning with Environmental Considerations: A Case Study of Pakistan's Indus Basin. World Bank Technical Paper 166World Bank, Washington, DC.

Basharat, M., 2012. Intergration of Canal and Groundwater to Improve Cost and Quality Equity of Irrigation Water in a Canal Command (Ph.D. thesis). Centre of Excellence in Water Resources Engineering, University of Engineering and Technology, Lahore.

Basharat, M., Tariq, A., 2013. Spatial climatic variability and its impact on irrigated hydrology in a canal command. Arab. J. Sci. Eng. 38 (3), 507–522.

Basharat, M., Umair Ali, S., Azhar, A.H., 2014. Spatial variation in irrigation demand and supply across canal commands in Punjab: a areal integrated water resources management challenge. Water Policy J. 16 (2), 397–421.

Basharat, M., Sultan, S.J., Malik, A., 2015. Groundwater Management in Indus Plain and Integrated Water Resources Management Approach. IWASRI Publication No. 303.

Cheema, M., Immerzeel, W., Bastiaanssen, W., 2014. Spatial quantification of groundwater abstraction in the Irrigated Indus Basin. Ground Water 52 (1), 25–36.

Government of Pakistan, 2011. Statistic of Pakistan. Government of Pakistan, Islamabad.

Government of Punjab, 2011. Statistic of Punjab. Government of Punjab, Lahore.

Habib, Z., 2004. Scope for Reallocation of River Water for Agriculture in the Indus Basin *(Ph.D. thesis)*. Ecole Nationale du Génie Rural, des Eaux et Forets Centre de Paris, Paris.

Jurriens, R., Mollinga, P., 1996. Scarcity by Design: Protective Irrigation in India and Pakistan. International Institute for Land Reclamation and Improvement, Cornell University, New York.

Mirza, G., Latif, M., 2012. Assessment of current agro-economic conditions in Indus Basin of Pakistan. In: *In proceedings of International Conference on Water, Energy, Environment and Food Nexus: Solutions and Adaptations under Changing Climate.*

Molle, F., Wester, P., Hirsch, P., 2010. River basin closure: processes, implications and responses. Agric. Water Manag. 97, 569–577.

NESPAK/SGI, 1991. Contribution of Private Tubewells in the Development of Water Potential. Planning and Development Division, Ministry of Planning and Development, Islamabad.

NESPAK, 2005. Punjab Irrigated Agriculture Development Sector Project. Water and Agricultural Sector Project. Water and Agricultural Studies. Lower Bari Doab Canal Command.

Pakistan Meteorological Department (PMD), 2006. Pakistan Meteorological Department Bulletins. PMD, Islamabad.

PIES, 2001. Punjab private sector groundwater development project—final project impact evaluation report. MM Pakistan (Pvt) Ltd. In association with Mott MacDonald Limited.

PPSGDP, 2000. Consultants, Draft Technical Report No. 45, Groundwater management and regulation in Punjab, Punjab Private Sector Groundwater Development Project, Groundwater Regulatory Framework Team, Project Management Unit. Irrigation and Power Department, Government of the Punjab.

Saifal, A., 2011. Reversing the Indus Basin Closure. ARPN J. Agric. Biol. Sci 6, 36–46.

Shah, T., 2006. Groundwater and Human Development: Challenges and Opportunities in Livelihoods and Environments. Tushaar, India.

Steenbergen, V.F., Olienmans, W., 1997. Groundwater Resources Management in Pakistan.

Ullah, 2001. Spatial Distribution of Reference and Potential Evapotranspiration Across the Indus Basin Irrigation Systems.

WAPDA, 2006. Survey by Soil and Reclamation Directorate of WAPDA.

World Bank, 2005. Pakistan's Water Economy: Running Dry. The World Bank in Pakistan. IWMI.

Further Reading

Bureau of Statistics, 2012. Punjab Development Statistics. Government of the Punjab, Lahore.

NESPAK, 1991. Punjab Irrigated Agriculture Development Sector Project. Water and Agricultural Sector Project. Water and Agricultural Studies. Lower Bari Doab Canal Command.

Quantitative Estimation of Resource Linkages in Water Infrastructure Planning

Akhtar Ali*, Muhammad Akram†

*RAS Knowledge Hub, Lahore, Pakistan †Water and Power Development Authority, Lahore, Pakistan

19.1 INTRODUCTION

The Indus River Basin (IRB) is approximately 3180 km long and is shared by four riparian countries—Afghanistan, China, India, and Pakistan. It serves about 300 million people who live in the basin area. Like other large basins in the world, the IRB contributes much to the ecology and people there, including water for food, clean energy, and household uses; for economic development and social well-being, as well as for ecological functions. The IRB's performance of these services depends on how its resources are used and on those who use and manage those resources. Currently, the basin faces several challenges. Overall, it is a water-stressed basin, now suffering from a water shortage in its middle and downstream areas in Pakistan. The per capita water availability has already dropped below the threshold level. The basin's spatial and temporal water variability heightens water shortages and causes conflicts over use of the water. Climate change, declining water quality, and projected glacial retreat are eminent threats to the water security and sustainability of the IRB and its services. Water-related disasters put millions of people at risk, in terms of both their lives and their livelihood; and increased stress has already caused visible migration. Moreover, the 2010 floods demonstrated the destructive and disruptive power of the Indus River, causing a direct loss of over $10 billion and 1600 deaths and affecting an area of 38,600 km^2 (Ali, 2013).

Development of the IRB's water infrastructure took place gradually over a 150-year period, from 1850 to 2000. This development was often driven by specific water needs, and thus the IRB did not benefit from holistic water infrastructure planning (WIP). Over the past two decades, ever-increasing water demands and water shortages, water-related disasters, and awareness of the eminent threat of climate change have forced water users and managers to shift their focus from water resource development to river basin management. Thus water infrastructure development is now based more on WIP. WIP, however, requires careful assessment of the linkages between the resources and the services the basin provides, which is particularly important since development at one location or in one sector may be done at the cost of other locations or sectors.

This chapter identifies the IRB's resources and their linkages in the context of upstream and downstream and cross-sectoral interactions. The chapter's primary focus is to highlight the role of resource linkages in water infrastructure planning at the basin scale. It also deals with the regional sociopolitical context as it relates to water. Hydro-politics, that is, politics affected by the availability of water and water resources, is considered as an imminent threat to the optimal and sustainable use of the IRB's resources and services. The book also warns that the deepening water conflicts may lead to water wars and regional instability and that reliability and transparency in information gathering and sharing are the means for resolving water conflicts.

19.2 CONCEPT OF RIVER BASIN RESOURCE LINKAGES

River basins, with their river flows, floodplains, and fertile soil, have been a primary source of people's livelihood and economic development for centuries. The four main river valley civilizations—the Mesopotamian civilization in the Tigris-Euphrates valley (3500–2000 BC), the Egyptian civilization in the Nile River valley (3200–1000 BC), the Harappan civilization in the Indus River valley (3200–1300 BC), and the Ancient Chinese civilization in the Yellow River valley (Huang He) (4000–1500 BC)—were mostly agrarian, and they developed along floodplains, where the rivers sustained them (Wright, 2009). In early societies people used rivers for communication, transportation, and trade. On the other hand, rivers and the rich land around them have also been the source of conflict among nations, and several wars have been fought to acquire them and the diverse resources they offer.

All streams and rivers originate from some type of precipitation, with the main sources being rainfall, glaciers, and snowpacks. Because these sources have an annual hydrologic cycle, they are renewable and sustainable. Glaciers and snowpacks act as water reservoirs by storing and releasing water under varying temperature regimes. When melting occurs, the extra surface water constitutes runoff that flows downstream into the basin. The downstream flows serve to the groundwater recharge, irrigation and maintaining the important ecosystems. Naturally flowing water is then managed by governmental agencies and the people who inhabit the land.

A typical river basin has linkages between (1) the resources (water, land, and vegetation, all of which are affected by climate), (2) utilization of the resources (water—for both household uses and irrigation—energy, food production and livestock, transportation, development, recreation, and the environment), and (3) users of the resources (i.e., consumptive and environmental users). In the context of upstream and downstream linkages, changes in how water is used and managed upstream may result in changes downstream.

Changes over time, including lessening amounts of glacial melt and snowmelt due to climatic changes, can affect the availability of water resources in a given location and also downstream. Also, land use changes in catchments and upstream water diversions can significantly alter the downstream flow of water and sediments, so it is important to know the accumulative effects of such changes so that they do not significantly alter the quality and renewable aspects of these resources. For example, degraded land can result in low infiltration rates and low capacities for storing rainwater, which can rapidly deplete the resource, causing water shortages, soil erosion, and heavy sediment loads that lead to water quality problems. The disposal of marginal-quality water into freshwater streams is also strongly linked with the quality of water downstream and the resulting health hazards. In addition, overexploitation can challenge the sustainability of natural resource systems (NRSs). Therefore it is important that water resources be planned and managed, including particularly water in river basins, which generations of people will continue to depend on for sustenance and well-being.

Because management can result in inadequate services or inequitable gains for individuals or groups, a holistic plan is needed. Holistic river basin planning can be defined as a process of assessing basin resources and resource linkages and a basin's capacity to be further developed, protected, or augmented for the continuity of the services it provides in an equitable and sustainable manner.

19.3 INDUS BASIN RESOURCES

19.3.1 The People and the Land

The Indus River Basin (IRB) is a source of livelihood for a population of about 300 million (at 2018 level) in four neighboring countries: Afghanistan, China, India, and Pakistan. Pakistan shares 61% of the population (183 million) followed 35% by India (105 million) (Laghari et al., 2012; Syaukat, 2012).[1] Syaukat (2012) indicates that the IRB's total area of 1.12 million km^2 is distributed between Pakistan, 47% (520,000 km^2); India, 39% (440,000 km^2); China, 8% (88,000 km^2); and Afghanistan, 6% (72,000 km^2).[2] The IRB comprises a delta area at about mean sea level and high mountains with an elevation over 5000 m. Twelve large cities are located in the basin—Kabul in Afghanistan; Amritsar in India; and Peshawar, Islamabad, Rawalpindi, Faisalabad, Gujranwala, Lahore, Multan, Sukkur, Hyderabad, and Karachi in Pakistan. The Indus Basin supports a unique ecosystem and aquatic life in cold mountains and hot and humid plains.

[1] The population is projected from 237 million in 2005 to 300 million in 2018 following reference by Laghari et al. (2012).

[2] Sinha (2010) reported in Water and Energy: A Flashpoint in Pakistan-India Relations? The IRB area of 1,170,838 km^2 in four neighboring countries including Pakistan (632,954 km^2; 54.06%), India (374,887 km^2; 32.02%), China (86,432 km^2; 7.38%) and Afghanistan (76,542 km^2; 6.54%).

19.3.2 The Climate and the Water Resources

The IRB's climate varies from subtropical arid and semiarid to temperate subhumid on the plains of the Sindh and Punjab provinces to alpine in the mountainous highlands of the north. Annual precipitation over the IRB ranges between 100 and 500 mm in the lowlands to a maximum of 2000 mm on mountain slopes. Snowfall at higher altitudes (above 2500 m) accounts for most of the river runoff (Syaukat, 2012). Summer monsoonal rains in the Upper Indus Basin (UIB) contribute on average 20%–30% of the total precipitation (ADB, 2010).

The designed live storage capacity (at time of construction in 1960s) of the three large hydropower reservoirs in Pakistan's Indus Basin was 22.98 km^3 (Tarbela 11.96 km^3; Mangla 10.15 km^3, which includes a recent increase of 3.58 km^3; and Chashma 0.87 km^3). The current live storage capacity (reflects 2014–2015) of these three large reservoirs is 17.89 km^3, representing an overall storage loss of 22%. Pakistan can barely store 30 days of water (Briscoe et al., 2006). The total Indus Basin's inflow from China to India was estimated at 181.62 km^3 a year. The mean annual inflow into Pakistan from India through the western tributaries (Jhelum and Chenab rivers), under the Indus Water Treaty (1960) between India and Pakistan, amounts to 170.27 km^3 (World Bank Group, 1960). The mean annual natural inflow into Pakistan through the eastern rivers (Ravi, Beas and Sutlej rivers) was estimated at 11.1 km^3. However, the treaty reserved this flow for India (Syaukat, 2012).

The total inflow from Afghanistan to Pakistan was estimated at 21.5 km^3. The Kabul River Basin (KRB) contributes 15.5 km^3 (of which 10 km^3 comes from the Kunar River, which first enters Afghanistan from Pakistan and then flows back to Pakistan after joining the Kabul River), and 6 km^3 is contributed by other tributaries (Panjshir, Gomal, Margo, Shamal, Kurram) (Syaukat, 2012). The mountain headwaters of the Indus Basin contribute approximately 60% of the mean annual flow. Approximately 80% of this volume enters the river system during the summer from June to September (Syaukat, 2012).

The total water withdrawal in the IRB is estimated at 299 km^3, of which Pakistan accounts for approximately 63%, India 36%, Afghanistan 1%, and China barely 0.04%. Irrigation withdrawal accounts for 278 km^3, which constitutes 93% of the total available freshwater. Surface water and groundwater account for 52% and 48% of total withdrawals in the Indus Basin, respectively (FAO, 2010).

19.3.3 Energy

The IRB has a total hydropower potential of 55,000 MW (megawatts) (ADB, 2010). The Indus Basin in Pakistan has a hydropower potential of approximately 41,722 MW. In 2018 the total installed capacity was about 9500 MW, which included the recently commissioned Neelum Jhelum project (969 MW), the Tarbela Four project (1410 MW), and the Golen Gol project (108 MW). Together they represent only 17% of the total hydropower potential, so the geothermal potential of the IRB may be another resource component. Nevertheless, hydropower is helping to make the IRB a rapidly developing region.

Afghanistan is constructing 12 dams on the Kabul River with a total storage capacity of 5.2 km^3 and a total hydropower potential of 1890 MW. The dams will irrigate 16,400 ha and will increase the existing storage capacity for the annual surface water from 3% to 24% (Vincent, 2014). The riparian countries' cooperative efforts for these projects can benefit both Afghanistan and Pakistan in terms of energy, trade, food production, and flood mitigation (LEAD, 2017).

In the Indian IRB the hydropower potential has been estimated as being 19,988 MW at 60% load. India has also developed hydropower infrastructures in the basin, including eight dams with 3556.8 MW projects on the Sutlej River, 11 dams with 2015.5 MW projects on the Beas River, six dams with 1738.4 MW projects on the Ravi River, four dams with 1565.1 MW projects on the Chenab River, and five dams with 736.6 MW projects on the Jhelum River (Modi et al., 2012). With on-going projects on Chenab and Jhelum rivers, India will have developed about 9612 MW hydropower.

19.3.4 Irrigation and Food Production

Syaukat (2012) shows a total area equipped for irrigation in the entire IRB as being approximately 26.3 million ha, of which Pakistan accounts for approximately 19.08 million ha (72.7%), India for 6.71 million ha (25.6%), Afghanistan for 0.44 million ha (1.7%), and China for 0.03 million ha (0.1%). The area irrigated by surface water accounts for 53%, while groundwater accounts for 47%. The total water withdrawal in the Indus River Basin is estimated at 299 km^3, of which Pakistan accounts for approximately 63%, India for 36%, Afghanistan for 1%, and China for barely 0.04%. Irrigation withdrawal accounts for 278 km^3, or 93% of the total. Surface water and groundwater account for 52% and 48% of total withdrawals in the Indus River Basin, respectively (Syaukat, 2012).

In Pakistan the Indus Basin's irrigated agriculture produces about 80% of the country's food and uses >90% of its renewable water resources. The other water uses remain within 10% of total water availability.[3] Groundwater uses in agriculture are about 50% in Punjab and 20% in Sindh. The associated water infrastructure comprises the Mangla reservoir with 9.12 km^3 capacity and the Tarbela reservoir with 11.9 km^3 capacity, 19 barrages, 14 interriver link canals, 45 irrigation canals about 60,800 km in length, and 1.6 million km–long watercourses. About 36% of the area is irrigated by water from the Mangla reservoir and 64% from the Tarbela reservoir (World Commission on Dams, 2000). Water released from a barrage in the downstream river channels is meant for downstream uses, including environmental flows. Pakistan's agriculture accounts for 25% of its national gross domestic product (GDP), 47% of its total employment, and >60% of its annual earnings from national foreign exchanges.

India has developed about 6.7 million ha of irrigated land. The associated infrastructure includes the Bhakra Dam on the Sutlej River (capacity of 9340 million cubic meters), the Pong Dam on the Beas River (capacity of 8570 million cubic meters), and the Thein Dam on the Ravi River (capacity of 3280 million cubic meters). The other infrastructures include interlinking the Ravi, Beas, and Sutlej rivers through canal networking, diversions, and tunnels; modification of the Upper Bari Doab Canal system and the Ropar and Madhopur barrages. India also constructed the Bist Doab canal system, the Ferozepur feeder, the head regulator of the Rajasthan feeder, and the Indira Gandhi Canal project (Central Water Commission, India, 2010).

In Afghanistan agriculture contributes 25% of the country's GDP and provides 60% of the workforce (around 4.5 million workers). Agro-processing accounts for 90% of the total

---

[3]This estimate includes both surface and groundwater. In the Indus Basin, groundwater mainly depends on the recharge through the Indus Basin Irrigation System.

manufacturing (Syaukat, 2012). The KRB drains 12% of Afghanistan's total area of 652,000 km^2, contributes 26% of the country's river flows, and serves 35% of the country's population (Favre and Kamal, 2004; World Bank, 2010).[4] The Kabul River in Afghanistan has three main tributaries: the Logar, Panjsher, and Kunar rivers. The Kunar River originates in Pakistan, enters into Afghanistan downstream of the town Arandu, and then joins the Kabul River in Afghanistan near Jalalabad. At its outlet at Dacca near the Pakistan border, the KRB's flow has been estimated at 19.2 Mm^3 per year with contribution from the Kunar River subbasin and 5224 Mm^3 without contribution from Kunar River subbasin. An estimated inflow of 10 km^3 per year enters from Pakistan via the Kunar River (Akhtar, 2017). The KRB has an intensively irrigated area of 196,606 ha and an intermittently irrigated area of 96,074 ha. A further potential for land development of about 59,330 ha has also been reported (World Bank, 2010).

19.4 RESOURCE LINKAGES

19.4.1 Water Links the Basin's Other Resources

World Water Report 4 (WWR4) recognizes the centrality of water and its global dimensions and delivers the three main messages (UNESCO, 2012):

- Water underpins all aspects of development. It is the only medium that links sectors and through which major crises can be jointly addressed.
- A coordinated approach to managing and allocating water across competing sectors to meet multiple goals also helps ensure that progress made in one sector is not offset by decline in other sectors.
- Global interdependencies will increasingly be woven through water. If no immediate action is taken, regions and sectors without enough water to meet their own needs will need to rely more heavily on the resources in other regions and sectors.

WWR4 highlights the importance of water as the key factor linking the other resources, specifically within river basins. Further, several internal (within the basin) and external (outside the basin) factors can influence water services (water, energy, and food) and can cause water-related disasters (water quality, health hazards, floods, and drought). Assessing these links in the IRB can help us better understand the governing processes that are shaping the basin's status and its capacity to deliver services to communities.

In the IRB, agriculture, hydro-energy, households and industry, and aquatic ecosystems are the main users of the water. An ever-increasing population, changes in lifestyle, the effect of climate change, and the growing importance of water for the ecosystem have created competition among users of water and how they use it. Still, water has the capacity to integrate the people and the land with an aim toward achieving sustainable services. Water-related disasters (floods, droughts, water quality, and health risks) affect the performance of water infrastructures, and their management has been a matter of high concern.

---

[4]Kabul River is a tributary of the Indus River Basin, which generates in Afghanistan.

The resource linkages between climate, water, land, and population through a water-food-energy-ecosystem nexus are many and complex. Interactions among the resources and their uses are mainly governed by biophysical, hydrological, and anthropogenic processes. Over time, these processes change proximate and distant areas. Take the case of upstream areas. The effects of upstream changes (changes in water quality and quantity, pollution, sediment, floods) are transferred downstream through river channels and are subjected to further modification along the way. The cumulative impact of these changes can significantly alter the downstream environment. The effects of these changes in the Indus Basin are already significant, and further aggravation is projected in the "business as usual" scenario practiced in management of the basin.

19.4.2 Indus Basin Has Strong Upstream-Downstream Linkages

The Upper Indus Basin (UIB) comprises a glacier and snowpack accumulation zone at high elevations and a high rainfall and steep topographic zone at relatively lower elevations. Because the former zone lacks the ability to support human settlements, the major changes there are not directly manmade but result from changes in temperature that lead to glacial melt and snowmelt, although changes are occasionally due to natural disasters, such as earthquakes, landslides, and the generation of or changes in young streams. In the high rainfall and steep topographic zone, the people are settled and raise livestock, fruit trees, and other agricultural products in valleys where such opportunities exist. Cutting of forests for energy use in harsh winters and for economic reasons, overgrazing, and cultivation of crops on slopes reduce the land's capacity to absorb water, which causes high peaks in runoff and quick dissipation of water and soil erosion. Because of its suitable topography and water availability, this zone offers opportunities for water storage for subsequent uses and hydropower generation.

Most of the UIB lies in China, India, and Afghanistan. The Chinese part of the IRB has a few on-site opportunities for development and is less likely to impact downstream resources, unless a massive water transfer is undertaken across the basin. Land use changes, water storage, and other upstream developments in India and Afghanistan have strong linkages with the level of water shortage and sediments and floods downstream in Pakistan. The hydro-political environment between upstream India and Afghanistan and downstream Pakistan makes this an area for potential conflict. Climate change, the quest for clean and cheap energy, and environmental flows are further complexities.

The Middle Indus Basin (MIB) is an area of economic development, disturbances, and pollution. The MIB can be divided further into an area dominated by high rainfall and steep topography followed by the Upper Indus Plain (UIP). The MIB zone has regular rivers, moderately steep slopes, fertile soil, and the floodplains of the Indus River tributaries. This zone has a settled population and a significant amount of economic development. Flooding and water pollution start in this zone. The UIP is a zone of high population and economic development and is prone to frequent and costly flood damage. This is a well-settled, irrigated agricultural area. Numerous Interriver link canals and irrigation canals often interrupt drainage. Also, a poorly planned road network has changed the hydraulics mechanisms of the floodplain and further constrained drainage of flood water. Irrigated agriculture dominates most of the arable land. Pakistan's largest cities and towns are located in the MIB.

The Lower Indus Basin (LIB) comprises the Lower Indus Plain (LIP) and the delta area. The LIP is densely populated and mainly depends on irrigated agriculture along the Indus River; it is protected from floods through the extensive use of dikes. The delta is divided into two main areas: the canal irrigation area and the coastal area. The canal irrigation area is well settled, and its agricultural productivity is high. However, the drainage is poor, the water quality is declining, and damages from flooding are high. The coastal area is also mostly irrigated by canals, but faces problems related to poor drainage, intrusion of saline water, and low productivity.

Moellenkamp (2007) reports success in addressing upstream-downstream issues in the Rhine and Elbe river basins. This provides an example of the two rivers basins (Rhine and Elbe), where EU's Water Framework Directive (WFD) helped managing the basins in the upstream and downstream context. The WFD mitigated water quality problems in the classical upstream-downstream setting, applying new technical measures for water treatment. Upstream-downstream problems have already been diminished by cooperation in an integrated environment. Moreover, enhanced cooperation in international river basins has led to solidarity in managing river basins. Downstream countries are now placed at a level equal to that of other countries in the basin and have the same rights and duties as upstream countries. The European Commission has taken over an additional role as an instance of control and inspection in river basin management.

Nepal et al. (2014) indicate that a complex relationship exists between upstream activities and processes and their influence on downstream areas. The upstream-downstream relationships have multiple facets, which also reflects the complex relationships between the natural environment and the people who live within it. The level of human-environment complexity is especially high in the Himalayan region as a result of its inaccessibility, fragility, marginality, and diversity (Jodha et al., 2004). Thus hydrology and sedimentation, the impact of climate change, regional heterogeneity, the need for an integrated systems analysis approach, and the need to address upstream-downstream linkages are receiving more emphasis. A process-based explanation and linked diagram is shown in Fig. 19.1.

19.4.3 Water and Energy Links Are Strong

Water and energy are strongly linked in the IRB. Water is used to generate hydro-energy and to keep the thermal and nuclear energy plants cool. The energy is also used to convey, distribute, and transport water. Groundwater pumping, industrial processes, and virtual water trade involve energy. Desalinization is also anticipated and will require energy. The general perception is that hydro-energy does not compete with other water demands, which is partly true, but prioritizing it for water releases from reservoirs creates competition.

19.4.4 Human Migration Is Linked With Water Stress

Wrathall et al. (2018) link human migration with climate change, water stress, and agricultural problems. They suggest that migration is usually a delayed response to water stress. As illustrated in the case of Mexico, the migration rate, however, stabilizes if water shocks are not repeated for an extended period (Nawrotzki et al., 2015).

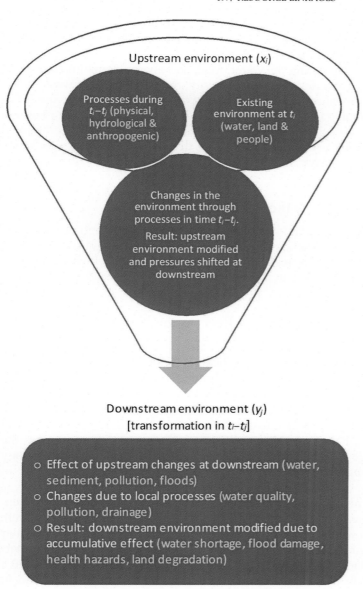

FIG. 19.1 Upstream and downstream linkages with associated governing processes.

19.4.5 Irrigation and Urban Water Uses Are in Competition

Ensink et al. (2002) estimate that >40 million people depend on irrigated water for their domestic water supply in the Indus Basin, Pakistan. This number is projected to double by 2030. The use of irrigated water to meet domestic water needs has also been reported in Morocco (Laamrani et al., 2000); India (Yoder, 1983); Sri Lanka (Van Der Hoek et al., 1999);

and Nepal, Jordan, and Mexico (Ault, 1981), and it seems to be common in semiarid and arid countries. Vector-borne diseases like malaria and schistosomiasis and their negative impacts are often linked with irrigation systems (Oomen et al., 1990).

Coelho et al. (2012) indicate that meeting the water, food, and energy demands needed by 2050 will depend on increasing water and land productivity; reducing loss and waste in the food chain, from production to consumption; and reducing the amount of energy used. Ensuring water for all is fundamental to trade-offs in water use. However, appropriate planning will require knowledge of science, ecology, and economics, as well as ethics and international cooperation. The trade-offs in water use involve numerous, diversified activities, which may conflict with hydropower, thermoelectric, and bioenergy production.

19.4.6 Change in Land Use Impacts the Water Availability

Change in land use, especially cutting in forested areas and cropping in high rainfall areas with steep slopes or along streams, influences the terrestrial phase of the hydrological cycle and precipitation pattern and also contributes a significant sediment load, which changes the pattern of streams and deposits sediments in reservoirs.

19.4.7 Climate Change Has Overwhelming Effect on the Natural Resource System

Chaudhry et al. (2009), assessing climate change indicators for Pakistan, found a mean annual temperature increase of 0.47°C and an annual maximum temperature increase of 0.87°C during 1960 and 2007 and an increased heat wave duration of 31 days. They also showed an increase of 61 mm in the average annual precipitation. The temperature rise in the Indus Basin, Pakistan, is projected to rise from 4.29°C (for A1B scenario) and 4.81°C (for A2 scenario) from 2071 to 2100 (ADB, 2010). Pakistan's National Climate Change policy recognizes the diverse impacts climate change has on water resources, agriculture, energy, and ecology and recommends a two-pronged adaptation and mitigation policy (Government of Pakistan, 2012).

Climate changes in the IRB are directly or indirectly linked with water, food, floods, energy, and the environment. The temperature rise will increase glacier melt and snowmelt, thus reducing the stock of fresh water, which will in the short-term increase the flow of water and sediments downstream, but in the long term will lead to a water shortage. The temperature rise may also increase glacial lake outburst floods (GLOFs) and landslides. In 2010 a landslide in the Hunza valley in the Upper Indus Basin created the Attabad Lake and in the process killed 20 people, displaced 6000 more, and affected the lives of 25,000. It also rerouted the Karakoram Highway 19 km, which is the only road between Pakistan and China. The effects of climate change will require prioritization, strategic planning and investment, and improved management.

The effects of climate change are linked to physical and economic systems. In physical systems, changes in temperature and precipitation affect hydrological systems. Similarly, evaporation can directly affect the amount of water stored in reservoirs. In the Indus delta and coastal areas, a rise in sea level may cause saltwater intrusion, which could result in drainage problems. Climate change links to physical systems can be assessed mathematically, whereas climate change links to economic systems are largely indirect. For example, a temperature rise and change in precipitation can affect cropping pattern, crop yield, and crop calendar. Similarly, a change in demand for water and energy can be attributed to climate variability. However, all of these factors indirectly affect the economy in multiple ways.

19.4.8 Transboundary Issues and Hydro-Politics Affect the Basin's Services

The Indus Basin's water is largely shared between upstream Afghanistan and India and downstream Pakistan. Upstream storages and diversions greatly influence water availability downstream. Regional geopolitical situations may drive the riparian countries to attempt to control the shared waters for political gain. The level of upstream storages and regulations are likely to be used as a political instrument. India and Pakistan are already facing tension related to upstream water storages. Regional political stability and synergy for the use of the IRB's resources can offer amicable solutions to the transboundary issues and reduce tension about the shared waters.

In Pakistan's part of the IRB the reservoir operation strategy (ROS) prioritized water releases for irrigation.[5] Hydro-energy is generated as a by-product of water releases from the two reservoirs. Household and industrial water uses are thought to be small and are largely met via the use of groundwater. Social and environmental water uses are largely ignored. Cross-sectoral water competition, the energy crisis, and water shortages over the past two to three decades prompted a movement from considering an irrigation-based water policy to considering diversified water services.

19.5 WATER INFRASTRUCTURE PLANNING

19.5.1 Context of WIP

In the basin context WIP is carried out either to enhance benefits from the water or to reduce harm, or both. For an underdeveloped basin where water resources exceed demand and where there is no significant environment pressure, WIP may proceed by analyzing the natural resource system (NRS) through a single objective, such as technical viability or cost-benefit analysis of a water supply project. Nevertheless, for a basin facing water stress and environmental pressure, a careful analysis of the NRS is needed to determine net gains and cumulative negative impacts. In the latter case resource linkages and competition among sectors and multiple users for use of the water become important in establishing a trade-off between social, economic, and ecological objectives.

---

[5]Tarbela and Mangla water storage reservoirs in Pakistan.

The IRB now faces several challenges: (1) water competition among sectors (households, food, energy, environment) and users, (2) water competition among riparian countries and states within the countries, (3) water-related disasters (flood, drought, health hazards), and (4) climate change. The resource linkages within the basin are complex, and the NRSs experience high water variability and uncertainty. Changes due to climatic, hydrological, and anthropogenic processes and hydro-politics further complicate the analysis and planning of NRSs. Future WIPs in the IRB will be based on strategic planning, including social, economic, and environmental scenarios and trade-offs. Most important, availability of scientific information will guide informed decision making on critical issues such as water reallocation and priority setting.

19.5.2 Infrastructural Development in the History

The Indus Basin's water infrastructure emerged from a flood-based agricultural society to a regulated irrigation-based agricultural society. In the early days, people settled along the rivers and depended on the water and land for their livelihood. Monsoonal floods and inflowing fertile soil provided the water and nutrients required for raising summer crops. Carryover moisture from the summer often supported low-water-demanding winter crops, mainly barley and wheat. Until 1850 the people used water from canals inundated by floods to irrigate land along the river.

Population growth and changing lifestyles increased pressure for adequate and safe water and more food, which drove the expansion of agriculture, diversification, and a stabilized crop yield. The second era of water infrastructure development started during the second half of the 18th century through the construction of water diversion structures on the rivers. In 1859 a regulating structure for the Upper Bari Doab Canal (UBDC) was constructed at Madhopur Headworks on the Ravi River. Following construction of the UBDC head regulator, the following projects were completed: the Sirhind Canal at Rupar Headworks on the Sutlej River in 1872, the Sidhnai Canal at the Sidhnai barrage on the Ravi River in 1886, the Lower Chenab Canal at Khanki Headworks on the Chenab River in 1892, and the Lower Jhelum Canal at the Rasul barrage on the Jhelum River in 1901. The Lower and Upper Swat canals on the Swat River and the Kabul River Canal system on the Kabul River were completed from 1885 to 1914. These projects helped transform the inundated canals into perennial irrigation canals and occasional cropping into a conventional agriculture regime.

Water shortages in the Ravi River resulted in the construction of the first interriver water transfer project from 1907 to 1915, which transferred surplus water from the Jhelum River to the Chenab River and then to the water-deficient Ravi River. The Sutlej Valley Project constructed four barrages and two canals on the Sutlej River in 1933. After separation of the two countries in 1947, upstream India stopped the flow of water downstream to Pakistan, which caused a water dispute between the two countries. The World Bank-sponsored, Indus Water Treaty (IWT) between India and Pakistan in 1960 resulted in the Indus Basin Project and facilitated construction of the Mangla and Tarbela dams, five barrages, eight interriver link canals, and one syphon in Pakistan. As a result, the irrigated area almost doubled in both countries during the second half of the 19th century (ADB, 2010). The drivers for water sector infrastructural development with context and timeframe are shown in Table 19.1. A detailed

TABLE 19.1 Stages of Indus Basin Water Infrastructure Development

Period	Driver of Change	Main Infrastructure Development Features
Before 1850 (an era of flood dependency)	Low population and limited needs; opportunist farmers raised summer crops on flood-based inundation in summer and winter crop on residual soil-moisture.	Inundation canals were constructed to carry floodwater for farming in the farther areas.
1850–1900 (an era of flow regulation and regular farming)	A need for yield stabilization and crop diversification drove construction of water diversion, and regulation structures on canal heads ensured regular supply of irrigation water.	The structures completed were: • UBDC regulating structure at Madhopur Headworks on Ravi River in 1859; • Sirhind Canal from Rupar Headworks on Sutlej River in 1872; • Sidhnai Canal at Sidhnai barrage on the Ravi River in 1886; • LCC at Khanki Headworks on Chenab River in 1892; • LJC at Rasul barrage on Jhelum River in 1901; • Lower and Upper Swat canals on Swat River from 1885 to 1914; and • Kabul River Canal System on Kabul River from 1885 to 1914.
1900–50 (an era of expansion and advancement)	• Water shortage in Ravi River drove interriver water transfer from Jhelum River to Ravi River; and • Need for expansion of irrigated agriculture.	• Interriver link canals from Jhelum to Chenab and then Chenab to Ravi rivers were constructed from 1907 to 1915. • The Sutlej Valley Project, constructed four barrages and two canals on Sutlej River in 1933.
1950–2000 (an era of tackling water divide problems and agricultural expansion)	• Water cut by India from UBDC head regulators deprived downstream riparian Pakistan from water. • World Bank sponsored Indus Water Treaty (1960), facilitating massive interriver water transfer. • Extensive irrigation system without drainage caused widespread water logging and salinity.	• Mangla Dam was constructed in 1968 on Jhelum River. • Tarbela Dam was constructed on Indus River in 1976. • Water transfer from Indus, Jhelum, and Chenab rivers to Ravi, Beas, and Sutlej rivers under Indus Basin replacement works (1960–80). • Construction of five barrages, eight interriver Link canals and one syphon. • Salinity Control and Reclamation Project was implemented from 1964 to 2000 to control water logging and salinity.

Continued

V. WATER MANAGEMENT IN INDUS RIVER BASIN

TABLE 19.1   Stages of Indus Basin Water Infrastructure Development—cont'd

Period	Driver of Change	Main Infrastructure Development Features
2000–18 (an era of solving complex river basin problems and cross-sectoral water issues)	Pace of horizontal expansion significantly reduced and focus shifted from hard-core infrastructure development to management of complex water system, cross-sectoral water allocation, and basin governance-related issues.	• Drainage problem. • Groundwater governance issue. • National drainage program started and failed to meet the envisaged objectives. • Water strategy and water policy prepared. • Devolution of powers in 2010 assigned water management responsibility to the provinces.

chronology of the IBWI is shown in Appendix 20.A. A schematic map of the Indus Basin and major water infrastructures is shown in Fig. 19.2.

The rapid growth of the water infrastructure, however, witnessed multidimensional and diversified problems of water shortages in some areas, waterlogging and salinity, groundwater that first rose and then declined, water quality degradation and health hazards, and water-related management, social, and political problems. Nevertheless, water releases for irrigation remained a priority following the use of hydropower as by-product of the irrigation releases.

19.5.3 Guiding Principles for the Water Infrastructural Planning

The issue for the 2012 Mekong 2 Rio Conference in Thailand was "how transboundary rivers can best meet the water, energy and food needs of riparian populations while minimizing negative impacts." This issue remains valid for WIP in the IRB. Increasing population, changes in lifestyle, climate change, transboundary noncooperation, and poor water governance challenge the basin's water security and sustainability. Given the uses for the water—agriculture and food (>90%), energy, households and municipalities (domestic water uses and sanitation), the environment (ecological and aquatic systems, pollution dilution, salt leaching), and social (recreational and religious)—the IRB suffers from water competition among users, lack of appropriate water resource planning, lack of proper cross-sectoral synergies and water allocations, and a decline in water quality. An obvious example of competition is between use of irrigation and hydro-energy, where irrigation needs have priority in operations at the Tarbela and Mangla reservoirs. Water variability (too much or too little, seasonal fluctuations), and water uncertainties (precipitation, glacier melt, shifting rainfall) and related risks (floods and landslides, health hazards) further challenge the area's water security.

Traditionally, the priority for WIP has been placed on the irrigation water requirements for food production, followed by the requirements for hydro-energy. Irrigation uses about 90% of the total water resources. Due to their small share (<10% of the total water uses), household and industrial water uses do not significantly affect flow regulations. Environmental water flows have been a low priority, mainly due to lack of awareness of the cost of the damage they do.

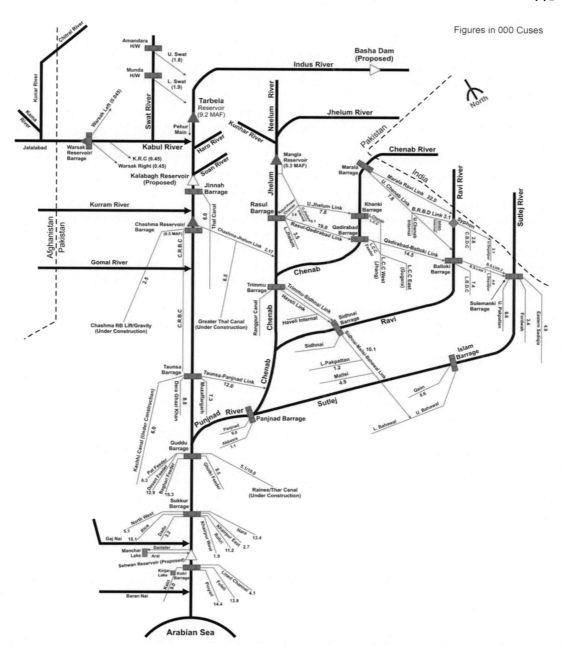

FIG. 19.2 Schematic diagram of Indus River Basin's water storage and conveyance system.

Water in the Indus Basin is linked to a socioeconomic trajectory. The Indus Basin's WIP is defined by its objective and is limited by boundary conditions. The objective of the IRB's WIP may be defined as "adequate, good quality water to meet the intended demand throughout the year and for future generations without undermining the resource-base." The water infrastructure should provide regulatory flexibility for optimized benefits and safeguard against extreme conditions (flood and drought). Multiobjective criteria—that is, water for food, energy, households, the environment, and economic development—can help to better assess proper water reallocation. However, the increasing scarcity of water and uncertainty about its availability will shape future planning.

In theory, water use for hydro-energy does not compete with other types of water use (households, food production, industrial uses, and environmental flows). In Pakistan, for example, the water release pattern for the two major reservoirs, that is, Tarbela and Mangla, is decided on the basis of available water storage and downstream irrigation water requirements. Water used for the generation of hydro-energy results from reservoirs' outflows as a by-product, which does not match with the peak energy demand. The reservoirs' winter releases (November to March) remain low due to dropping reservoir levels and low irrigation demands. This is the lean period for hydro-energy generation, which creates competition. Further, the cooling water requirements of fuel-based energy generation also compete with other uses.

Similarly, water competition for other uses, such as the competing demand between households and irrigation, or irrigation versus environmental flows, and so on, are expected to intensify in the future. Rapid groundwater decline in some of the urban aquifers requires reallocation of surface water from irrigation to the household and industrial sectors. The obvious example is the city of Lahore, where the state government is considering such an allocation. Revival of the Ravi and Sutlej rivers is also being considered by the state government, which will culminate in water reallocation from one use to another use. Appropriate WIP based on an informed decision will be needed.

The IRB's services for water, food, energy, and the environment are strongly linked through water availability, uncertainty and operational flexibility, and the land and the people who depend on it for their livelihood. Resource linkage-based WIP may drive future thinking (Fig. 19.3). However, climate change, transboundary issues, and regional sociopolitical conflicts need to be included in the process. This process may require science-based information and its transparent use, regional integrity, and political settlements and resolutions of disputes.

19.6 QUANTITATIVE ESTIMATION OF THE RESOURCE LINKAGES

19.6.1 Assessment Tools

As with the world's other major transboundary basins—the Nile, Mekong, and Ganges basins—several tools were developed to assess resources and their linkages in Pakistan's IRB. The main focus of the IRB tools include (1) combined crop and hydrology models to optimize the design and operation of irrigation infrastructures in the IBIS, (2) groundwater models to assess water availability and irrigation-related water logging and salinity, (3) flood

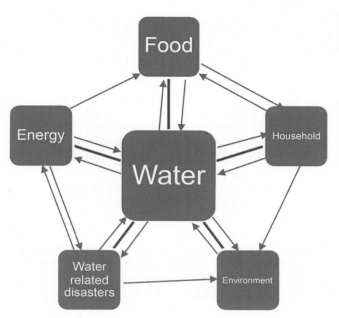

FIG. 19.3 Resource linkages for water infrastructural planning in the Indus River Basin.

forecasting, and (4) rainfall runoff models to estimate inflows into the river system as inputs to all water resources studies (Johnston and Smakhtin, 2014).

Since the late 1970s the World Bank has supported the Indus Basin Model (IBM) and its revisions (IBMR) (Bisschop et al., 1982). Johnston and Smakhtin (2014) provide a brief description of the types of models, their changes, and their capacities. The IBMR-III, an agro-economical and irrigation optimization model, was transferred to the WAPDA in 1989. The model is capable of distributing UIB inflows into 12 agro-climatic zones and 45 main canal commands vis-à-vis linking a hydrologic model and aquifer system with an economic model of agricultural production. The IBMR-III was used for several studies, including ranking of new irrigation projects after the Water Apportionment Accord in 1991, designing the Kalabagh Dam (Ahmad et al., 1990) and alternative salinity management projects (Rehman et al., 1997), optimizing cropping patterns for the Rechna Doab subbasin (Jehangir and Ali, 1997), and assessing future requirements for water in the Lower Chenab Canal command area (Jehangir et al., 2003). The IBMR was also updated to model raising the height of the Mangla Dam (Alam and Olsthoorn, 2011). Also, the model was modified as IBMR-2012, which was used to evaluate the impact of climate change on water allocation and food security (Yang et al., 2013).

The Global Change Impact Studies Centre, Islamabad, has been developing databases for the IRB for the application of DHSVM.[6] The Pakistan Meteorological Department (PMD), the Water Resources Management Directorate, and the WAPDA's Snow and Ice Hydrology

---

[6]http://www.gcisc.org.pk/Water.aspx.

Project provide seasonal and 10 daily flow forecasts, based primarily on meteorological data. A list of major modeling studies between 2000 and 2012 as compiled by Johnston and Smakhtin (2014) is shown in Appendix 20.B.

An evaluation of the Indus Basin's assessment tools showed that the Indus Basin Planning Model was not operational and that none of the models have the capability to provide high-quality information on critical planning or operation-related issues (Planning Commission, Government of Pakistan, 2011). The WAPDA does not have the required expertise to manage the model, and its operational use is limited (UNEP-DHI and UNEP, 2016). Improvements in the PCRWR's SWAT model and the Delft flood forecasting and early warning system (FEWS) model are being explored. Johnston and Smakhtin (2014) provide a detailed discussion on the main reasons for the models' failure to produce accurate results, including scarcity as well as uncertainties in the hydrometeorological data, which are directly linked to unreliable model simulations. Climate and river flow data for the UIB fall far short of the WMO recommendation for density in rugged mountainous areas.

UNEP-DHI and UNEP (2016) presented the conceptual framework for assessment of transboundary river basins. This framework addresses (1) water quantity for agricultural, human, and environmental water stresses, (2) water quality related to nutrient and wastewater pollution, (3) ecosystems related to wetlands' lack of connectivity, dams, and extinction of species, (4) governance of legal framework for hydro-political tensions and enabling the environment, and (5) socioeconomics related to economic dependence on water resources, societal well-being, and exposure to flood and drought. In view of these thematic areas and their indicators, in many aspects, the Indus Basin model lacks the capability to be a basin-scale model. Expanding needs and their associated complexities and reliable information are two main challenges for a comprehensive Indus Basin model. The Water Apportionment Accord (Government of Pakistan, 1991) among the four provinces allocates water shares to the provinces, which adds another dimension and needs to be addressed in the assessment model. A comprehensive Indus Basin model should be able to guide management planning beyond the sectors, which poses several challenges in the absence of a dedicated institution and qualified experts. A comprehensive Indus Basin planning model that is backed by adequate and accurate data from all the riparian countries and that is capable of addressing the emerging issues related to global changes, hydro-politics, and the impact of climate change will be required.

19.6.2 The Quantification Challenges

Quantification and accurate assessment of the resources and resource linkages face many challenges in the IRB, including the following:

- Uncertainty and complexity are primary challenges constraining accurate analysis of the NRS.
- Authenticated information is not available for a transparent and fair analysis.
- Comprehensive assessment tools and associated skills to use the tools are not available.
- A common platform is not available for knowledge sharing and interactions among the riparian countries.

- International and regional synergies and cooperation are missing, and countries do not share data.
- Data gaps are big, and effective monitoring needs substantial funding.
- A constructive and positive dialogue for conducive approaches, effective institutional mechanisms, and improved water governance has not yet started.
- The stakeholders' lack confidence; therefore information sharing is at a minimum.

19.6.3 The Quantification Guidelines

Given the dynamic nature of the resources and their complex interactions and quantification challenges, straightforward guiding principles for the IRB are advised. The overall accuracy of the quantification will depend on the accuracy of information and conceptual models to link the resources at different regional and temporal scales. Depending on the similarity analysis and with some adjustments, the guiding principles listed below may also be useful for other river basins in estimating the linkage of their resources.

1. Divide the river basin into logical hydrological, ecological, and management units.
2. Put the process together and screen the relevant information for adequacy, accuracy, and reliability.
3. Identify and assess the resources within those units. The main resources can be precipitation, glacier, snow, stream flows, wetlands, land, soil water, groundwater, graywater, vegetation, and so on.
4. Develop a conceptual model that includes all the resources and their interactions and complexities. Identify sensitive parameters and divide the conceptual model into a workable thematic and regional or smaller boundaries.
5. Assess the resource uses and past trends at the unit or regional scales.
6. Assess the basin governance, geopolitics, and strong interest groups at the units.
7. Assess the balances and imbalances within the units.
8. Model the river basin on the basis of outcomes from the assessment of the units. Make sure that water has a central role.
9. Develop likely scenarios on the basis of past trends and future predictions.
10. Develop a baseline planning framework for optimum resource uses through simulating the possible changes.
11. Negotiate with the stakeholders on the plan and modify it as needed.
12. Establish a mechanism for accurate and transparent information gathering, analysis, and sharing.
13. Develop a common communication platform for scientific information management and informed decision making.
14. Continue periodic monitoring, evaluating, and revising the plan within an overall and agreed upon framework.

In addition to these quantification guidelines, a matrix for quantitative assessment of resource linkages is suggested in Table 19.2 for researchers as a starting point on hydrologic research in the Indus River Basin.

TABLE 19.2 Guiding Matrix for Quantitative Assessment of Resource Linkages

Indus Basin Zones	Resources	Linking Factors	Linkage and Impact Process	Main Determinant	Quantification Methodology	Main Requirements
Upper Indus Basin						
i. Glaciers and snowpack zone	• Glaciers and snowpack (virtually, no population in this area)	Glacier retreat attributes to water shortage downstream.	• Climate change and temperature rise can increase glacier retreat and melt snowpack faster than it can recover causing water shortage downstream in longer term.	○ Temporal variation of the river water flows	○ Remote sensing and GIS-based assessment of changes ○ Climate change modeling ○ Monitoring of glaciers and river flow data	○ Data measurement, monitoring and management/sharing mechanism ○ Increased financial support for research and the glacier change assessment efforts ○ Glacier and snowpack monitoring ○ Water monitoring in streams and lakes
ii. High rainfall and mountainous zone	• Population • Climate and precipitation • Land and vegetation • Water and soil • Topography and geology	The actions and changes in this zone are linked with change in water quality and quantity downstream.	• Poorly planned socioeconomic development in this area can cause pollution and water shortage downstream. • Deforestation for energy in harsh winter and economic reasons can cause high soil erosion and sediment yield impacting rivers and reservoirs. Capacities downstream • Change in climate and land use in this area can affect the runoff, soil loss, pollution load, and livelihood downstream. • Landslide can affect downstream water and sediment load. • Suitably developed water storages and regulation can increase hydropower and reliability of irrigation water.	○ Vegetation cover ○ Water quantity ○ Water quality ○ Soil loss and sediment transport ○ Pollutant and pollution load ○ Groundwater quantity and quality ○ Flood ○ Food ○ Hydropower	○ Remote sensing and GIS-based spatial assessment and changes ○ Rainfall-runoff modeling ○ River flow modeling ○ Groundwater modeling ○ Water quality modeling ○ Real-time data measurement, monitoring and management	○ Indus Basin resources information gathering and sharing mechanism ○ Cross-sectoral and interregional vis-à-vis transboundary communication and coordination ○ Water monitoring in streams and lakes ○ Climatic data ○ Land use data (satellite imageries) ○ Water uses

Middle Indus Basin

Hilly and rolling topography zone	• Population • Climate and precipitation • Land and vegetation • Water and soil • Topography	• The actions and changes are linked with change in water quality and quantity downstream.	• Population pressure for more water and food and changing lifestyle cause resource depletion on-site and declining water quality and availability downstream. • Cross-sectoral water competition and inequitable distribution cause water shortages or make access to clean water difficult. • Poorly planned development and misuse of resources can cause pollution and water shortage downstream. • Change in climate and land use can affect runoff, soil loss, pollution load, and livelihood downstream. Landslide can affect downstream water and sediment load. Suitably developed water storages and regulation can increase hydropower and reliability of irrigation water.	○ Vegetation cover ○ Water quantity ○ Water quality ○ Soil loss and sediment transport ○ Pollutant and pollution load ○ Groundwater quantity and quality ○ Flood ○ Food ○ Hydropower ○ Tributaries' inflows	○ Remote sensing and GIS-based spatial assessment and changes ○ Rainfall-runoff modeling ○ River flow modeling ○ Groundwater modeling ○ Water quality modeling ○ Real-time data measurement, monitoring and management	○ Indus Basin resources information gathering and sharing mechanism ○ Cross-sectoral and interregional vis-à-vis transboundary communication and coordination ○ Water monitoring in streams and lakes ○ Climatic data ○ Land use data (satellite imageries) ○ Water uses ○ Groundwater monitoring
Upper Indus Plain	• Population • Climate • Land • Water and soil • Crops	• Urban, industrial and agricultural pollution • Water allocation and withdrawals • Sediment and pollution transport • Drainage effluent	• Population pressure for more water and food and changing lifestyle cause resource depletion on-site and declining water quality and availability downstream. • Cross-sectoral water competition and inequitable distribution cause water shortages or make access to clean water difficult. • Poorly planned development and misuse of resources can cause pollution and water shortage downstream. • Change in climate, cropping pattern, and land use can affect the resources and livelihood	○ Drainage surplus ○ Water quantity ○ Water quality ○ Sediment transport ○ Pollutant and pollution load ○ Groundwater quantity and quality ○ Flood ○ Food ○ Hydropower ○ Water for energy	○ Remote sensing and GIS-based spatial assessment of productivity and changes ○ River flow modeling ○ Groundwater modeling ○ Water quality modeling ○ Real-time data measurement,	○ Indus Basin resources information gathering and sharing mechanism ○ Cross-sectoral and interregional vis-à-vis transboundary communication and coordination ○ Transparency and accuracy of information collection and sharing

Continued

TABLE 19.2 Guiding Matrix for Quantitative Assessment of Resource Linkages—cont'd

Indus Basin Zones	Resources	Linking Factors	Linkage and Impact Process	Main Determinant	Quantification Methodology	Main Requirements
			opportunities on-site and downstream. • Groundwater overexploitation and poor drainage can cause land degradation and decline in agricultural productivity. • Water conservation and safe use of marginal quality water can help save precious resources.	○ Environmental flows ○ Tributaries inflows	monitoring and management	○ Water monitoring in streams and lakes ○ Climatic data ○ Land use data (satellite imageries) ○ Water uses ○ Groundwater monitoring
Lower Indus Basin						
Lower Indus Plain	• Population • Climate • Land • Water and soil • Crops	• Urban, industrial and agricultural pollution • Water allocation and withdrawals • Sediment and pollution transport • Drainage effluent	• Population pressure for more water and food and changing lifestyle cause resource depletion on-site and declining water quality and availability downstream. • Cross-sectoral water competition and inequitable distribution cause water shortages or make access to clean water difficult. • Poorly planned development and misuse of resources can cause pollution and water shortage downstream. • Change in climate, cropping pattern, and land use can affect the resources and livelihood opportunities on-site and downstream. • Groundwater overexploitation and poor drainage can cause land degradation and decline in agricultural productivity. • Water conservation and safe use of marginal quality water can help save precious resources.	○ Drainage surplus ○ Water quantity ○ Water quality ○ Sediment transport ○ Pollutant and pollution load ○ Groundwater quantity and quality ○ Flood ○ Food ○ Hydropower ○ Water for energy ○ Environmental flows	○ Remote sensing and GIS-based spatial assessment of productivity and changes ○ River flow modeling ○ Groundwater modeling ○ Water quality modeling ○ Real-time data measurement, monitoring and management	○ Indus Basin resources information gathering and sharing mechanism ○ Cross-sectoral and interregional vis-à-vis transboundary communication and coordination ○ Transparency and accuracy of information collection and sharing ○ Water monitoring in streams and lakes ○ Climatic data ○ Land use data (satellite imageries) ○ Water uses ○ Groundwater monitoring

Delta and coastal zone	• Population • Climate • Land • Water and soil • Crops • Aquatic life • Wind energy	• Urban, industrial and agricultural pollution • Water allocation and withdrawals • Sediment and pollution transport • Drainage effluent • Saline water intrusion	• Population pressure for more water and food and changing lifestyle cause resource depletion on-site and declining water quality and availability downstream. • Cross-sectoral water competition and inequitable distribution cause water shortages or make access to clean water difficult. • Poorly planned development and misuse of resources can cause pollution and water shortage. • Change in climate, cropping pattern, and land use can affect the resources and livelihood opportunities. • Groundwater overexploitation and poor drainage can cause land degradation and decline in agricultural productivity. • Water conservation and safe use of marginal quality water can help save precious resources.	o Drainage surplus o Water quantity o Water quality o Sediment o Pollutant and pollution load o Groundwater quantity and quality o Flood o Food o Water for energy o Environmental flows	o Remote sensing and GIS-based spatial assessment of productivity and changes o River flow modeling o Groundwater modeling o Water quality modeling o Saline water intrusion modeling o Real-time data measurement, monitoring and management	o Indus Basin resources information gathering and sharing mechanism o Cross-sectoral and interregional vis-à-vis transboundary communication and coordination o Transparency and accuracy of information collection and sharing o Glacier and snowpack monitoring o Water monitoring in streams and lakes o Climatic data o Land use data (satellite imageries) o Water uses o Groundwater monitoring

Notes

(i) Upper Indus Basin (UIB) covers area upstream of Tarbela Dam and includes (1) glaciers and snowpack area up to Bhasha and (2) high rainfall and steep topography between Bhasha and Tarbela.

(ii) Middle Indus Basin (MIB) covers area between Tarbela and Guddu, including (1) rolling and hilly topography from Tarbela to Chashma and (2) Upper Indus Plain from Chashma to Guddu.

(iii) The Lower Indus Basin (LIB) covers area between Guddu Barrage and sea, including (1) Lower Indus Plain from Guddu to Kotri Barrage and (b) delta and coastal area from Kotri Barrage to sea.

19.7 CONCLUSIONS

The IRB is a water-stressed basin and faces the challenges of ever-increasing demands for water, food and energy; and downstream impacts from upstream development that lead to degraded ecosystems, declining livelihoods, and water-stress-based migration. Climate change and transboundary politics and hydro-politics further heighten the declining natural resource system.

The basin's resources (water, food, energy, and environment) are strongly linked through on-site and upstream-downstream processes. Upstream water resource development now causes downstream water shortages. Water variability and uncertainty further intensify water shortages, at places causing conflict and migration. Dealing with and estimating uncertainty and complexity are important components of WIP.

The IRB lacks authenticated information, information sharing, and cooperation among the riparian countries, which is one of the major constraints for the basin's water security, for its sustainability, and for the services it provides. It also lacks an appropriate tool for assessing resource linkages, as well as the institutional capacity to plan and implement an informed decision-making support system.

The IRB will need a robust, fair, and transparent information gathering and sharing system among the riparian countries and within the region. The information provided here provides a basis for quantification of the resources through their linkages and services. A shift in focus from water resource development to integrated river basin management may be required to address water security and sustainability of the basin's services.

Future WIP in the IRB will depend greatly on cross-sectoral synergies as well as cooperation over water issues and good water governance. In order to make the right decisions and take appropriate actions, science-based evidence must be used to overcome the weaknesses inherent in current assessments and the related uncertainties. A newly formed "Indus Water Institute" could help streamline information sharing and provide a common platform for making informed decisions.

APPENDIX 20.A

Historical Development of the Indus Basin Water Infrastructure

Year	Event	River	Country	Purpose	Main Function
1859	Completion of the Upper Bari Doab Canal	Ravi	India	Irrigation	First canal with controlled water intake at Madhopur Headworks
1872	Completion of the Sirhind Canal	Sutlej	India	Irrigation	Water intake at Rupar Headworks
1886	Completion of the Sidhnai Canal	Ravi	Pakistan	Irrigation	Water intake at Sidhnai Headworks
1892	Completion of the Lower Chenab Canal	Chenab	Pakistan	Irrigation	Water intake at Khanki Headworks

Historical Development of the Indus Basin Water Infrastructure—cont'd

Year	Event	River	Country	Purpose	Main Function
1901	Lower Jhelum Canal	Jhelum	Pakistan	Irrigation	Water intake at Rasul Barrage
1885–1914	Lower and Upper Swat, Kabul River and Paharpur Canal	Swat and Kabul	Pakistan	Irrigation	Offtakes: Lower Swat Canal from Munda at Swat, Upper Swat Canal from Amandara at Swat and Kabul River Canal from downstream of Warsak Dam at Kabul River
1893	Kabul River Canal	Kabul	Pakistan	Irrigation	–
1907–15	Triple Canal Project	Jhelum, Chenab, and Ravi	Pakistan	Interrivers water transfer	Transfer surplus water from Jhelum River to Chenab and then water-deficient Ravi River
1918	Amandara Headworks	Swat	Pakistan	Irrigation	Upper Swat Canal offtakes
1920	Indus Discharged Committee	–	–	Monitoring discharge data	Committee formed to record discharge data
1930s	Sukkur barrage and its system completed	Indus	Pakistan	Irrigation	Water diversion from Indus River to Barrage's command area
1933	Completion of Sutlej Valley Project	Sutlej		Irrigation	Four barrages and two canals and Bhakra reservoir plan.
1939	Completion of Haveli and Rangpur Canals	Chenab	Pakistan	Irrigation	Water diversion at Trimmu Barrage
1947	Completion of the Thal Canal	Indus	Pakistan	Irrigation	Water diversion at Kalabagh Headworks
1948	Water disputes between India and Pakistan		India and Pakistan		India unilaterally cut off supplies to Pakistan on river headworks in India
1954	Completion of the Nangal Dam	Sutlej	India	Irrigation and Hydro-energy	
1955	Completion of the Kotri barrage	Indus	Pakistan	Irrigation	Controlled flow regulation
1958	Completion of the Taunsa barrage	Indus	Pakistan	Irrigation	Water diversion along both river banks
1960	Indus Water Treaty (IWT)		World Bank, India, and Pakistan	Irrigation and hydro-energy	Resolved water disputes
1960	Warsak Dam	Kabul	Pakistan	Irrigation and hydro-energy	243 MW (1100 GWh) Original capacity was 160 MW. Upgraded in 1980

Continued

V. WATER MANAGEMENT IN INDUS RIVER BASIN

Historical Development of the Indus Basin Water Infrastructure—cont'd

Year	Event	River	Country	Purpose	Main Function
1962	Completion of the Guddu barrage	Indus	Pakistan	Irrigation	Water diversion for irrigation
1963	Completion of the Bhakra Dam	Sutlej	India	Irrigation and hydro-energy	Water storage and regulation
1960–68	Naghlu Dam	Kabul	Afghanistan	Hydro-energy	Design capacity 100 MW
1960–76	Indus Basin Project	Multirivers	Pakistan	Irrigation and hydro-energy	IWT projects, including Mangla Dam, five barrages including Chashma reservoir, one syphon and eight interriver link canals completed in 1971; Tarbela Dam completed from 1975 to 1976
1964	Darunta Dam	Kabul	Afghanistan/ Jalalabad	Hydro-energy	Originally 40 MW reduced to 11 MW by silting of dam
1965	Mailsi syphon	Sutlej	Pakistan	Irrigation	Under Sutlej River
1965	Sidhnai barrage	Ravi	Pakistan	Irrigation	Water diversion for irrigation
1967	Qadirabad	Chenab	Pakistan	Irrigation	Water diversion for irrigation
1967	Rasul	Jhelum	Pakistan	Irrigation	Water diversion for irrigation
1968	Mangla Dam completed	Jhelum	Pakistan	Irrigation and hydro-energy	Water storage and regulation. Part of Indus Basin Project
1968	Marala barrage	Chenab	Pakistan	Irrigation	Water diversion for irrigation
1971	Chashma Barrage	Indus	Pakistan	Irrigation	Intermediate storage and water diversion
1974	Completion of Pong Dam	Sutlej	India	Irrigation and hydro-energy	Water storage and regulation
1976	Tarbela Dam completed	Indus	Pakistan	Irrigation and hydro-energy	Water storage and regulation, part of Indus Basin Project
1977	Completion of Pandoh Dam	Sutlej	India	Irrigation and hydro-energy	Water storage and regulation
1986	Completion of Salal Dam	Chenab	India	Hydro-energy	Water storage and regulation
2008	Completion of Baglihar Dam	Chenab	India, Pakistan, World Bank	Hydro-energy	Water storage and regulation; disputes since the construction began in 1999; World Bank finally authorizes construction
2018	Golen Gol Dam	Mastuj/ Chitral	Pakistan	Hydro-energy	106 MW (436 GWh annually)

Updated from FAO Aquastat, 2010. Information system on water and agriculture. http://www.fao.org/nr/water/aquastat/main/index.stm (Accessed 13 May 2018).

APPENDIX 20.B

List of Hydrological and Water Resources Models for Indus Basin Since 2000

Model	Type	Area	Source	Indus Application
IBMR-III (Indus Basin Model Revised)	Hydrologic network model with agro-economic and irrigation optimization	Indus subbasins (Rechna Doab, Lower Chenab Canal) Annual/monthly/10 days	Rehman et al. (1997)	Scenarios for the management of salinity, including economic feasibility and reallocation of irrigation supplies
Updated IBMR		Indus basin (surface water only), 14 agro-climatic zones Annual/monthly/10 days	Alam and Olsthoorn (2011)	Assess level of increase of crest for Mangla Dam; based on conjunctive surface and groundwater use
WEAP	Water accounting and allocation	14 subbasins, monthly	de Condappa et al. (2009)	Assessment and planning of water resources, contribution of glacier melt, impact of climate change
SWAT AVSWAT	Rainfall-runoff model using 11 years of daily data from 22 climate stations	Indus Basin, 346 HRUs daily	Rehman et al. (1997)	Pakistan Council for Research in Water Resources (PCRWR), ongoing: potential of Indus for extension of irrigation and generation of hydropower to meet the increasing demand for food and energy in the country
SWAT	Rainfall-runoff model	Indus Basin Daily	Mccartney et al. (2012)	Basin scale assessment of agricultural water use
SWAT		Upper Indus and Mangla Basin, Daily (1994–2003)	UNESCO-IHE (2012) http://bit.ly/1lMuPjR	Climate change impact on the hydrological cycle and on water resources
UBC model	Semidistributed hydrological model	Indus Basin, Daily	University of British Colombia Quick and Pipes (1977)	Short-term forecasting of river inflows by WAPDA (generate inputs for MODSIM and others for WAPDA)
HBV linked to PRECIS RCM	Semidistributed, hydrological model linked to global climate model	Hunza, Gilget, and Astore subbasins	Akhtar et al. (2008)	Impact of CC on flows
SWAM	Stream water availability model based on topological	Upper Indus Basin	Mukhopadhyay and Dutta (2010)	Stream water availability in ungauged basin (including snowmelt)

Continued

List of Hydrological and Water Resources Models for Indus Basin Since 2000—cont'd

Model	Type	Area	Source	Indus Application
	model and global climate data			
DHSVM	Distributed-hydrology-soil-vegetation model	Pakistan	Global Change Impact Studies Centre, Pakistan http://www.gcisc.org.pk/Water.aspx	Real-time hydrological forecasting by integrating DHSVM with MM5 (meso-scale atmospheric model)
PWRI	Distributed surface–groundwater, river flow model	Upper and middle Indus Basin 2 hourly (2008–10)	Aziz and Tanaka (2011)	Flood forecasting
Delft-FEWS	Hydrological	Indus Basin	http://www.itc.nl/flooding-and-pakistan http://www.deltares.nl/en/software/479962/delft-fews Werner et al. (2004)	Flood forecasting/early warning
Indus 21 WCAP	Flow forecasting model	Mangla Basin, Upper Indus	http://www.aht-group.com/index.php?id=622	Improved snowmelt and flow forecast for Mangla Basin
FEGWST	3D finite element groundwater and salinity model	Lower Indus	Chandio et al. (2012)	Groundwater flow and saline intrusion in response to groundwater pumping
FEFLOW	Finite element, 2D and 3D	Irrigated area of the Rechna Doab	Sarwar and Eggers (2006)	Conjunctive use model to evaluate alternative management options for surface and groundwater resources.
FEFLOW 3D		Upper Chaj Doab	Ashraf and Ahmad (2008)	Regional groundwater decline since 1999; scenarios for impact of extreme climatic conditions (drought/flood) and variable groundwater abstraction on the regional groundwater system
MODFLOW, MT3D	Groundwater model		Ali et al. (2004)	Salinity; performance of scavenger wells

Johnston, R., Smakhtin, V., 2014. Hydrological modeling of large river basins: how much is enough? Water Resour. Manag. 28 (10), 2695–2730. Published online: 9 June 2014 # Springer Science+Business Media Dordrecht 2014. Web. https://link.springer.com/article/10.1007/s11269-014-0637-8 (Accessed 24 May 2018.

References

ADB, 2010. Islamic Republic of Pakistan: glacial melt and downstream impacts on indus dependent water resources and energy., 2010. ADB RETA No. 6420-PAK promoting climate change impact and adaptation in Asia and the Pacific. Report prepared by International Center for Integrated Mountain Development Kathmandu, Nepal.

Ahmad, M., Brooke, A., Kutcher, G.P., 1990. Guide to the Indus Basin Model Revised. World Bank, Washington, DC.

Akhtar, F., 2017. Water Availability and Demand Analysis in the Kabul River Basin, Afghanistan. (doctoral dissertation). University of Bonn, Germany.

Akhtar, M., Ahmad, N., Booij, M.J, 2008. Use of regional climate model simulations as input for hydrological models for the Hindukush? Karakorum? Himalaya region. Hydrol. Earth Syst. Sci. Discuss. 5 (2), 865–902. https://hal.archives-ouvertes.fr/hal-00298939/.

Alam, N., Olsthoorn, T.N., 2011. Sustainable conjunctive use of surface and ground water: modelling on the basin scale. Ecopersia 1, 1–12.

Ali, A., 2013. Indus Basin Floods: Mechanisms, Impacts, and Management. Asian Development Bank, Mandaluyong City.

Ali, G., Asghar, M.N., Latif, M., Hussain, Z., 2004. Optimizing operational strategies of the scavenger wells in lower Indus Basin of Pakistan. Agric. Water Manag. 66 (3), 239–249.

Ashraf, A., Ahmad, Z., 2008. Regional groundwater flow modelling of Upper Chaj Doab of Indus Basin, Pakistan using finite element model (Feflow) and geoinformatics. Geophys. J. Int. 173 (1), 17–24. https://ieeexplore.ieee.org/abstract/document/8143292.

Ault, S.K., 1981. Expanding Non-agricultural Uses of Irrigation for the Disadvantaged: Health Aspects. The Agricultural Development Council Inc., New York, NY. 86p.

Aziz, A., Tanaka, S., 2011. Regional parameterization and applicability of Integrated Flood Analysis System (IFAS) for flood forecasting of upper-middle Indus River. Pak. J. Meteorol. 8, 21–38. http://www.pmd.gov.pk/rnd/rnd_files/vol8_issue15/3_Regional%20Parameterization%20and%20Applicability%20of%20Integrated%20Flood%20Analysis%20System.pdf.

Bisschop, J., Candler, W., Duloy, J.H., O'Mara, G.T., 1982. The Indus basin model: a special application of two-level linear programming. In: Applications. Springer, Berlin, Heidelberg, pp. 30–38.

Briscoe, J., Qamar, U., Contijoch, M., Amir, P., Blackmore, D., 2006. Pakistan's Water Economy: Running Dry. Oxford University Press, Karachi.

Central Water Commission, India, 2010. Annual report 2010. http://cwc.gov.in/main/downloads/17_10_2011Final%20Draft%20AR%202010-11_rev.pdf. (Accessed 15 November 2017).

Chandio, A.S., Lee, T.S., Mirjat, M.S., 2012. . The extent of waterlogging in the lower Indus Basin (Pakistan)—a modeling study of groundwater levels. J. Hydrol. 426, 103–111. https://www.sciencedirect.com/science/article/pii/S0022169412000601.

Chaudhry, Q.-u.-Z., Mahmood, A., Rasul, G., Afzaal, M., 2009. Climate Change Indicators of Pakistan. Pakistan Meteorological Department.

Coelho, S., Agbenyega, O., Agostini, A., Erb, K.H., Haberl, H., Hoogwijk, M., Lal, R., Lucon, O., Masera, O., Moreira, J.R., 2012. Land and water: linkages to bioenergy. In: Global Energy Assessment-Toward a Sustainable Future. Cambridge University Press/The International Institute for Applied Systems Analysis, Cambridge; New York, NY/Laxenburg, pp. 1459–1525.

de Condappa, D., Chaponnière, A., Lemoalle, J., 2009. A decision-support tool for water allocation in the Volta Basin. Water Int. 34 (1), 71–87.

Ensink, J.H., Aslam, M.R., Konradsen, F., Jensen, P.K., van der Hoek, W., Ensink, J.H., 2002. Linkages Between Irrigation and Drinking Water in Pakistan. vol. 46. IWMI.

FAO Aquastat, 2010. Information system on water and agriculture. http://www.fao.org/nr/water/aquastat/main/index.stm. (Accessed 13 May 2018).

Favre, R., Kamal, G.M., 2004. Watershed Atlas of Afghanistan. In: Water Resources and Environment. Government of Afghanistan, Ministry of Irrigation, Kabul.

Government of Pakistan, 2012. National Climate Change Policy of Pakistan. Ministry of Environment, Government of Pakistan, Islamabad.

Government of Pakistan, 1991. Water Apportionment Accord Between Government of Pakistan and the Four Provincial Governments on Allocation of Provinces Share. Islamabad, Pakistan.

Jehangir, W.A., Ali, N., 1997. Salinity Management Alternatives for the Rechna Doab, Punjab, Pakistan: Volume Six-Resource Use and Productivity Potential in the Irrigated Agriculture.

Jehangir, W.A., Ashfaq, M., Rehman, A., 2003. Modeling for efficient use of canal water at command level. Pak. J. Water Resour. 7 (1), 43.

Jodha, N.S., Bhadra, B., Khanal, N.R., Richter, J., 2004. Poverty alleviation in mountain areas of China. In: Proceedings of the International Conference Held From ll-l5 November, QOO2, in Chengdu, China. ICIMOD/lnWENT—Capacity Building International Germany, Kathmandu/Feldafing.

Johnston, R., Smakhtin, V., 2014. Hydrological modeling of large river basins: how much is enough? Water Resour. Manag. 28 (10), 2695–2730.

Laamrani, H., Khallaayoune, K., Laghroubi, M., Abdelilah, T., Boelee, E., Watts, S.J., Gryseels, B., 2000. The Metfia in Central Morocco. Water Int. 25 (3), 410–417.

Laghari, A.N., Vanham, D., Rauch, W., 2012. The Indus basin in the framework of current and future water resources management. Hydrol. Earth Syst. Sci. 16 (4), 1063–1083.

LEAD, 2017. Prospects for Benefit Sharing in the Transboundary Kabul River Basin: Investigating the Social, Economic and Political Opportunities and Constraints. LEAD Pakistan Discussion Paper Series, www.lead.org.pk. (Accessed 13 May 2018).

Mccartney, M., Forkuor, G., Sood, A., Amisigo, B., Hattermann, F., Muthuwatta, L., 2012. The Water Resource Implications of Changing Climate in the Volta River Basin. International Water Management Institute (IWMI), Colombo, Sri Lanka, pp. 1–33. IWMI Research Report 146. https://doi.org/10.5337/2012.219.

Modi, S.N., Javadekar, S.P., Rajagopalan, S.V., Garbyal, S.S., 2012. South Asia Network on Dams, Rivers and People. https://sandrp.in/themes/hydropower-performance/. (Accessed 13 May 2018).

Moellenkamp, S., 2007. The "WFD-effect" on upstream-downstream relations in international river basins? insights from the Rhine and the Elbe basins. Hydrol. Earth Syst. Sci. Discuss. 4 (3), 1407–1428.

Mukhopadhyay, B., Dutta, A., 2010. A stream water availability model of Upper Indus Basin based on a topologic model and global climatic datasets. Water Resour. Manag. 24 (15), 4403–4443. https://www.researchgate.net/publication/225389753_A_Stream_Water_Availability_Model_of_Upper_Indus_Basin_Based_on_a_Topologic_Model_and_Global_Climatic_Datasets.

Nawrotzki, R.J., Hunter, L.M., Runfola, D.M., Riosmena, F., 2015. Climate change as a migration driver from rural and urban Mexico. Environ. Res. Lett. 10 (11), 114023.

Nepal, S., Flügel, W.A., Shrestha, A.B., 2014. Upstream-downstream linkages of hydrological processes in the Himalayan region. Ecol. Process. 3 (1), 19.

Oomen, J.M.V., de Wolf, J., de Jobin, W.R., 1990. Health and Irrigation. Publication 45, International Institute for Land Reclamation and Improvement, Wageningen.

Planning Commission, Government of Pakistan, 2011. Modelling, Simulation and Data Assimilation for Indus River Basin Management. SimIndus Group of Pakistani Researchers. Planning Commission, Government of Pakistan, Islamabad.

Quick M.C., Pipes A. 1977. University of British Colombia, Watershed Model. Hydrological Sciences-Bullettin-des Sciences Hydrolgiques, XXII, 1 3/1977. Department of Civil Engineering. http://hydrologie.org/hsj/220/hysj_22_01_0153.pdf.

Rehman, G., Jehangir, W.A., Rehman, A., Aslam, M., Skogerboe, G.V., 1997. Principal Findings and Implications for Sustainable Irrigated Agriculture: Salinity Management Alternatives for the Rechna Doab, Punjab. Pakistan, International Water Management Institute, Colombo, Sri Lanka.

Sarwar, A., Eggers, H., 2006. International FEFLOW User Conference, Berlin, Germany. http://www.feflow.info/uploads/media/feflow_conference_2006_03.pdf.

Sinha, U.K., 2010. Water and Energy: A Flashpoint in Pakistan-India Relation. US Congressional Research Service Report, August 2010.

Syaukat, Y., 2012. Irrigation in Southern and Eastern Asia in Figures. FAO the United Nation.

UNEP-DHI and UNEP, 2016. Transboundary River Basins: Status and Trends. United Nations Environment Programme (UNEP), Nairobi. twap-rivers.org. (Accessed 25 May 2018).

UNESCO, 2012. World Water Development Report 4 (WWDR 4). 1 rue Miollis75015 Paris, France. www.unesco.org/new/en/natural-sciences/environment/water/wwap/.../wwdr4-2012/.

UNESCO-IHE, 2012. National Report on IHE Activities. Hydrological modelling of the Upper Indus River Basin, Pakistan. http://www.unesco.org/new/fileadmin/MULTIMEDIA/HQ/SC/temp/IHP-IGC-XX/Pakistan_NatRep_2012.pdf.

Van Der Hoek, W., Konradsen, F., Jehangir, W.A., 1999. Domestic use of irrigation water: health hazard or opportunity? Int. J. Water Resour. Dev. 15 (1–2), 107–119.

Vincent, T., 2014. Afghanistan and Pakistan: A Decade of Unproductive Interactions Over the Kabul-Indus Basin. https://www.thethirdpole.net/en/2014/07/07/afghanistan-and-pakistan-a-decade-of-unproductive-interactions-over-the-kabul-indus-basin/. (Accessed 12 May 2018).

Werner, M., van Dijk, M., Schellekens, J., 2004. DELFT-FEWS: an open shell flood forecasting system. In: Hydroinformatics (in 2 volumes, with CD-ROM) (pp. 1205–1212). https://www.worldscientific.com/doi/abs/10.1142/9789812702838_014.

World Bank, 2010. Afghanistan—Strategic Scoping Options for Development of the Kabul River Basin: A Multisector Decision Support System Approach. Washington, DC.

World Bank Group, 1960. Indus Waters Treaty (1960) Between the Government of India and Government of Pakistan. https://siteresources.worldbank.org/INTSOUTHASIA/.../IndusWatersTreaty1960.pdf.

World Commission on Dams, 2000. Dams and Development: A New Framework for Decision-Making: The Report of the World Commission on Dams. Earthscan.

Wrathall, D.J., Van Den Hoek, J., Walters, A., Devenish, A., 2018. Water Stress and Human Migration: A Global, Georeferenced Review of Empirical Research. Land and Water Discussion Paper 11, Food and Agriculture Organization of the United Nations, Rome.

Wright, R.P., 2009. The ancient Indus: urbanism, economy, and society. vol. 10. Cambridge University Press. Reported in Wikipedia. River Valley Civilization, https://en.wikipedia.org/wiki/Indus_Valley_Civilisation. (Accessed 22 October 2017).

Yang, X., Gao, W., Shi, Q., Chen, F., Chu, Q., 2013. Impact of climate change on the water requirement of summer maize in the Huang-Huai-Hai farming region. Agric. Water Manag. 124, 20–27.

Yoder, R., 1983. Non-agricultural Uses of Irrigation Systems: Past Experience and Implications for Planning and Design. Irrigation Management Network Paper-Overseas Development Institute.

Further Reading

UN Water, 2012. Managing water under uncertainty and risk. The United Nations world water development report 4, UN Water Reports, World Water Assessment Programme.

Index

Note: Page numbers followed by "*f*" indicate figures, "*t*" indicate tables, and "*b*" indicate boxes.

Printed in the United States
By Bookmasters